Universitext

Universitext is a series of textbooks that presents material from a wide variety of mathematical disciplines at master's level and beyond. The books, often well class-tested by their author, may have an informal, personal even experimental approach to their subject matter. Some of the most successful and established books in the series have evolved through several editions, always following the evolution of teaching curricula, into very polished texts.

Thus as research topics trickle down into graduate-level teaching, first textbooks written for new, cutting-edge courses may make their way into *Universitext*.

More information about this series at https://link.springer.com/bookseries/223

James McKee · Chris Smyth

Around the Unit Circle

Mahler Measure, Integer Matrices and Roots of Unity

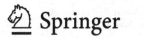 Springer

James McKee (iD)
Egham, UK

Chris Smyth (iD)
Edinburgh, UK

ISSN 0172-5939   ISSN 2191-6675   (electronic)
Universitext
ISBN 978-3-030-80030-7   ISBN 978-3-030-80031-4   (eBook)
https://doi.org/10.1007/978-3-030-80031-4

Mathematics Subject Classification: 11C08, 11C20, 11R06, 05A05, 05C20, 05C22, 11R18, 11S05, 15B36

This Springer imprint is published by the registered company Springer Nature Switzerland AG
The registered company address is: Gewerbestrasse 11, 6330 Cham, Switzerland

You shall see them on a beautiful quarto page, where a neat rivulet of text shall meander through a meadow of margin.

Sheridan, School for Scandal.

Preface

Consider an algebraic number α. By definition, it is the zero in some field of an irreducible polynomial P_α with integer coefficients. The complete set of zeros of such a polynomial constitutes a *conjugate set of algebraic numbers*. The study of such sets is the subject of this book. We embed them in $\mathbb{C}$ and study them there. We study them embedded in $\overline{\mathbb{Q}}_p$, the algebraic closure of the field $\mathbb{Q}_p$ of p-adic numbers.

Very often, we restrict to *algebraic integers*, where P_α, the minimal polynomial of α, is *monic* (has leading coefficient 1). We study conjugate sets of algebraic integers as eigenvalues of matrices with integer entries. More generally, we study algebraic integers by attaching combinatorial objects, such as graphs or directed graphs (digraphs), to them. We study conjugate sets when they are all real. We study them when they are all sums of roots of unity.

We are also interested in conjugate sets that are small, in some sense. For this purpose, the Mahler measure (a height function) is the main tool. For conjugate sets of totally real algebraic integers, their span (diameter) is also important. For totally positive algebraic numbers, the *trace* is studied, too, both as the sum of the elements of a conjugate set and as the sum of the eigenvalues of an integer symmetric matrix. The *house* (maximum modulus of the elements of a conjugate set) is another important height function. It has attracted renewed attention recently with Dimitrov's surprising proof of the Schinzel–Zassenhaus Conjecture. We present his proof.

The range of values assumed by a height function is an interesting topic in its own right. We study the set of these values for the Mahler measure of polynomials with integer coefficients, in any number of variables. This includes 'Lehmer's Conjecture', concerning the question of whether for noncyclotomic polynomials such Mahler measures are bounded away from 1. For algebraic integers in a cyclotomic field, we do the same for their *Cassels height* (the mean square modulus of their conjugates).

Certain sets of square integer matrices are a particularly interesting source of algebraic integers, through their eigenvalues. While every algebraic integer is an eigenvalue of the companion matrix of its minimal polynomial, when we restrict to symmetric or other sets of such matrices, we get some challenging questions. Some of these are discussed in the book.

We now outline the contents of the chapters. Chapter 1 contains basic facts about the Mahler measure of polynomials in one variable and some applications. We also discuss small values of the Mahler measure of integer polynomials. Tables of these, and also their presumed small limit points, are given in Appendix D.

Chapter 2 contains much material about the Mahler measure of integer polynomials in several variables. This includes a discussion of Boyd's Conjecture that the set of such Mahler measures is closed. It also includes specific closed form evaluations of particular Mahler measures.

Chapter 3 contains a version of Dobrowolski's proof of the strongest unconditional result in the direction of the proof of Lehmer's Conjecture. We essentially follow the proof of Cantor and Straus, which uses the confluent Vandermonde determinant.

Chapter 4 gives an account of Dimitrov's recent proof of the Schinzel–Zassenhaus Conjecture.

Chapter 5 concerns roots of unity. One topic is finding *cyclotomic points* (points whose coordinates are roots of unity) on plane curves. Another topic is a discussion of Robinson's two problems and five conjectures concerning cyclotomic integers.

Many of our chapters concern integer square matrices. They have a visual interpretation as signed multidigraphs. So for $A = (a_{ij})$, the entry $a_{ij} \neq 0$ for $i \neq j$ can be regarded as a directed edge from vertex i to vertex j of weight a_{ij}. For $i = j$ the entry $a_{ii} \neq 0$ represents a charge a_{ii} on the ith vertex. When A is symmetric with all entries either 0 or 1 and all diagonal entries 0, we simply have a graph (A being its adjacency matrix). If A is symmetric with all entries -1, 0 or 1 and all diagonal entries 0, we have a signed graph. If we allow nonzero diagonal entries we have a charged graph or charged signed graph. Throughout the book, we move freely between the matrix and graphical interpretations of these objects.

Chapters 6 and 7 concern *cyclotomic matrices*. These are integer symmetric matrices whose eigenvalues are all of the form $\omega + \omega^{-1}$ for some root of unity ω. It turns out that essentially all of them are (adjacency matrices of) graphs, signed graphs or charged signed graphs. After describing how cyclotomic matrices can be 'grown', the important tool of Gram vectors is introduced. The Classification Theorem for cyclotomic matrices is then stated. The classification includes signed graphs that tessellate a torus and charged signed graphs that tessellate a cylinder. Its proof occupies Chap. 7.

Chapter 8 is devoted to a description of the structure of the set of all Cassels heights of cyclotomic integers as what we call a Thue set. Further, a complete description of the values of the Cassels heights of integers in a cyclotomic field $\mathbb{Q}(\omega_p)$, where ω_p is a pth root of unity for a prime p, is given.

In Chap. 9, some structure theory is developed to show which (charged, signed) graphs can be embedded in the toroidal or cylindrical tessellations of Chap. 6. This will be useful in Chap. 13 for describing minimal noncyclotomic matrices.

Chapter 10 gives basic facts about the transfinite diameter of subsets of the complex plane, as well as some applications to describing which such subsets can contain infinitely many conjugate sets of algebraic integers. It also summarises some results concerning the integer transfinite diameter and the monic integer transfinite diameter. Some of these results are used in Dimitrov's proof in Chap. 4 of the Schinzel–Zassenhaus Conjecture.

In Chap. 11, we prove or survey various lower bounds for Mahler measures that apply when the integer polynomials have their zeros restricted in some way. One such restriction is to bound the number of monomials of the polynomial. Another, the *totally p-adic* case, is to restrict to polynomials that split completely over the field of p-adic numbers. A third is to assume that the polynomials define abelian extensions of the rationals. A fourth is to restrict the zeros to a compact set not containing all of the unit circle (Langevin's Theorem). In these situations, stronger results can be obtained than in the unrestricted general case. In Chap. 12, another restricted case, that of nonreciprocal polynomials, is studied.

In Chap. 13, all minimal noncyclotomic matrices are found. These are noncyclotomic integer square matrices for which the removal of any row and corresponding column gives a cyclotomic matrix.

The explicit auxiliary function method, described in Chap. 14, is used to find the smallest mean values of a given function f on the values of various restricted sets of conjugate algebraic integers or numbers. Knowing that certain resultants are non-zero expresses the problem as a linear programming problem that, when dualised, gives an auxiliary function to be minimised. This then often gives a discrete spectrum of the smallest mean values of f on these restricted sets. Two important examples of this method are $f(x) = x$ and $f(x) = \max(1, \log x)$ for $x > 0$. The first gives a spectrum of small values of the mean trace of totally positive algebraic integers (the Schur–Siegel–Smyth trace problem), while the second does the same for the absolute Mahler measure of such algebraic integers.

In Chap. 15, the analogous trace problem for positive definite integer symmetric matrices is studied. In contrast to the trace problem of the previous chapter, this problem can be completely solved: the trace of such an $n \times n$ matrix is at least $2n - 1$, and this bound is attained.

The small-span problem for conjugate sets of totally real algebraic integers is a classical one: which such sets lie in an interval of length less than 4? In Chap. 16, the analogous problem is solved for integer symmetric matrices.

The next three chapters concern symmetrizable matrices. Chapter 17 describes the structure of symmetrizable matrices, as well as presenting other relevant material needed for the following chapters. In Chap. 18, all (nonsymmetric) maximal symmetrizable cyclotomic matrices are found, and Chap. 19 concerns the trace problem for nonsymmetric symmetrizable integer matrices.

Salem graphs are a special kind of graph that can be associated with a Salem number or quadratic Pisot number in a specific way. In Chap. 20, families of these are constructed using certain rational functions called rational interlacing quotients and circular interlacing quotients. Special cases of these constructions have an application to the following chapter on minimal polynomials of integer symmetric matrices.

Exercises in the book are distributed throughout the text, including the appendices. They are of very variable difficulty, ranging from the almost trivial to the challenging. An example of an easy one is the following.

Exercise 0.1 (Serre [Ser19, (A.7.5)]) Show that when the complex numbers w and z are of modulus 1 then

$$|(w - z)(\overline{w} - z)| = |(w + \overline{w}) - (z + \overline{z})|.$$

If we feel that an exercise is even trickier, and perhaps that we do not know the answer, it is marked as a Problem. Some exercises require light use of a computer, and some require more substantial programming: these are flagged as computational exercises. Unsolved problems are presented either as Research Problems or Open Problems, the former generally being more open-ended and less notorious.

Most chapters end with a section of Notes, followed by a Glossary of terms used in that chapter. The Notes include references to the sources of some of the results and proofs that have been presented, as well as references to related work. For results that have been stated in the text, but not proved, references are given beside the result.

We end this preface with a list of some books and survey articles containing material related to that in our book, which could be consulted. First of all, we highly recommend the recent text on Mahler measure by Brunault and Zudilin [BZ20]. It contains some advanced material that we have referred to only briefly: for instance, a detailed account of Deninger's work, and treatment of Mahler measure of elliptic curves in much greater depth than we have described. Other texts containing a significant to amount of Mahler measure material are those of Schinzel [Sch00] and Bombieri and Gubler [BG06]. For results on real and complex polynomials in general, the scholarly book of Rahman and Schmeisser [RS02] is a very valuable source of results and references.

While we treat some aspects of the spectra of graphs, the topic is well covered in the books by Godsil and Royle [GR01] and Brouwer and Haemers [BH12]. We refer to Pisot and Salem numbers only briefly—they are equal to their Mahler measures! For a more extensive treatment, see the book [BDGGH+92] by Bertin *et al*. The classic text [Sal63] by Salem is also well worth reading.

In our treatment of roots of unity and cyclotomic integers in Chap. 5, the considerable literature on vanishing sums of roots of unity is not covered. See for instance Zannier's survey paper [Zan95] for an account of this area.

The idea of producing a book of this sort has been with us for several years. Right from the start, Rémi Lodh at Springer gave us encouragement to pursue the dream, and was patiently supportive of our slow progress. We owe him a huge debt for nudging us at the right moments and continuing to believe in the project. We are also appreciative of the substantial help and encouragement provided by the reviewers of the manuscript. We thank Igor Pritsker for detailed feedback on the chapter on the transfinite diameter. To all these and to all our friends and colleagues who have helped us through this work, thank you: we are glad it is now finished!

Royal Holloway, UK   James McKee
Edinburgh, UK   Chris Smyth
May 2021

Contents

Chapter 1
Mahler Measures of Polynomials in One Variable

1.1 Introduction

In this chapter, we introduce some basic results about the Mahler measure of polynomials in one variable and some of its applications. In later chapters, there are further results for one-variable Mahler measures: Dobrowolski's Theorem (Chap. 3) and Mahler measures for restricted sets of polynomials (Chaps. 11 and 14) and for nonreciprocal polynomials (Chap. 12).

1.1.1 Polynomials over the Field $\mathbb{C}$ of Complex Numbers

Polynomials form a class of fundamental mathematical objects. By introducing an unknown variable, say z, they represent the first step in extending the basic numerical operations of addition and multiplication to a larger domain than fields of numbers like the complex numbers $\mathbb{C}$.

Taking a typical monic polynomial of degree d,

$$P(z) := z^d + a_1 z^{d-1} + \cdots + a_d, \tag{1.1}$$

say, with coefficients $a_1, , \ldots, a_d \in \mathbb{C}$, the Fundamental Theorem of Algebra (see [DF04, 14.6]) tells us that $P(z)$ factorises over $\mathbb{C}$ as

$$(z - \alpha_1)(z - \alpha_2) \cdots (z - \alpha_d), \tag{1.2}$$

say, where $\alpha_1, \alpha_2, \ldots, \alpha_d$ are the *zeros* of $P(z)$. This factorisation is unique. By comparing coefficients of powers of z in (1.1) and (1.2), we can express the a_i ($i = 1, \ldots, d$) in terms of the α_j ($j = 1, \ldots, d$). Thus, we have a bijection between $\mathbb{C}^d$ and the d-element multisets of $\mathbb{C}$, defined by the mapping

© Springer Nature Switzerland AG 2021
J. McKee and C. Smyth, *Around the Unit Circle*, Universitext,
https://doi.org/10.1007/978-3-030-80031-4_1

$$(a_1, \ldots, a_d) \mapsto \{\alpha_1, \ldots, \alpha_d\}. \tag{1.3}$$

1.1.2 Polynomials over the Field $\mathbb{Q}$ of Rational Numbers

When we restrict this mapping (1.3) to $\mathbb{Q}^d \subset \mathbb{C}^d$, its image is a countable set of d-element multisets of $\mathbb{C}$ whose elements are called **algebraic numbers**. When $P(z)$ as in (1.1) is irreducible over the rational field $\mathbb{Q}$, then $\{\alpha_1, \ldots, \alpha_d\}$ is a **conjugate set of algebraic numbers**. A fundamental problem is to describe these sets. For instance, how are they distributed in the complex plane? Because the a_i are all real, we have from (1.1) and (1.2) that

$$\prod_{j=1}^d (z - \alpha_j) = P(z) = \overline{P}(z) = \prod_{j=1}^d (z - \overline{\alpha_j}),$$

so that the multiset $\{\alpha_1, \ldots, \alpha_d\}$ is stable under complex conjugation.

By dropping the requirement that $P(z)$ be monic, we can multiply it by the lcm of the denominators of the a_i, and we can instead assume that $P(z) = a_0 z^d + \cdots + a_d$ has integer coefficients and **content** (the gcd of its coefficients a_i) equal to 1. When $a_0 = 1$, the zeros α_j are **algebraic integers**. Then we know that there is a monic polynomial with integer coefficients, call it $P_\alpha(z)$, which has α as a zero, and is irreducible over $\mathbb{Q}$, but, over $\mathbb{C}$, splits into linear factors

$$P_\alpha(z) = \prod_{j=1}^d (z - \alpha_j),$$

say, where $\alpha = \alpha_1$. This polynomial is the **minimal polynomial** of α. Denote by C_α the set $\{\alpha_1, \ldots, \alpha_d\}$. The α_j are **conjugates** of α, and sets of the type C_α are conjugate sets, as defined above. By Exercise 1.2, these are indeed sets—they have no repeated elements.

What distinguishes sets C_α from arbitrary sets of d complex numbers? Firstly, because $P_\alpha(\overline{\alpha_j}) = \overline{P_\alpha(\alpha_j)} = 0$, we know that C_α is stable under complex conjugation. In particular, it has an even number of nonreal elements. Geometrically, it is symmetric about the real axis. Also, we know that the integers $(-1)^i a_i$ $(i = 1, \ldots, d)$ are the elementary symmetric functions of the α_j. Specifically, $(-1)^i a_i$ is the sum of all $\binom{d}{i}$ possible products of i distinct conjugates α_j. But the sequence $(a_1, \ldots, a_d)$ is not arbitrary, since it must provide the coefficients of the irreducible polynomial $P_\alpha(z)$. For example, a_d cannot be 0 for $d > 1$. More generally, $P_\alpha(z)$ can have no irreducible factor of degree $d' \le \lfloor d/2 \rfloor$.

Exercise 1.1 Prove that the minimal polynomial of an algebraic number is unique and has no multiple zeros.

Exercise 1.2 Prove that two conjugate sets either coincide or are disjoint and that the field of algebraic numbers $\overline{\mathbb{Q}}$ is a countable disjoint union of conjugate sets.

Next, note that either $P_\alpha(z) = z$ or a_d is a nonzero integer, implying that

$$|\alpha_1 \cdots \alpha_d| = |a_d| \geq 1.$$

This shows that z is the only such polynomial having all its zeros in the open disc $|z| < 1$. Furthermore, if $|a_d| = 1$ and all α_j are in the closed unit disc $|z| \leq 1$, then they are actually all on the unit circle $|z| = 1$. For the closed disc $|z| \leq 1$, moreover, we have the first of Kronecker's two theorems, to which we now turn.

1.2 Kronecker's Two Theorems

Theorem 1.3 (Kronecker's First Theorem [Kro57]) *Suppose that α is a nonzero algebraic integer that lies, with its conjugates, in the unit disc $|z| \leq 1$. Then α is a root of unity.*

Proof Suppose that α has degree d, with minimal polynomial

$$P_\alpha(z) = z^d + a_1 z^{d-1} + \cdots + a_d,$$

say. Then, because $(-1)^k a_k$ is the sum of all possible distinct k-tuples of the zeros of P, all of which have modulus at most 1, we have $|a_k| \leq \binom{d}{k}$ $(k = 0, \ldots, d)$. The same bounds apply to the coefficients of the polynomial P_r whose zeros are the rth powers of those of P. Because by Proposition A.22 α^r is also an algebraic integer, the coefficients of P_r are integers, so there are only finitely many possibilities for the coefficients of these polynomials. Hence, there are finitely many possibilities for all the zeros of all the P_r, and so in particular, there are only finitely many possibilities for the α^r. Therefore, two of them must be equal, say $\alpha^r = \alpha^s$ with $r < s$. Then $\alpha^{s-r} = 1$, and we see that α is a root of unity. $\qquad\square$

Theorem 1.4 (Kronecker's Second Theorem [Kro57]) *Suppose that β is a real algebraic integer that lies, with its conjugates, in the interval $[-2, 2]$. Then β is conjugate to $2\cos(2\pi/\ell)$ for some positive integer ℓ.*

Proof Let $Q(x)$ be the minimal polynomial of β, of degree d' say. Then the polynomial $z^{d'} Q(z + 1/z)$ has all its zeros on $|z| = 1$, so, by the First Theorem, these are all roots of unity. Hence, $\beta = \omega_\ell + \omega_\ell^{-1}$ for some primitive ℓth root of unity ω_ℓ. By Proposition 5.2, ω_ℓ is conjugate to $e^{2\pi i/\ell}$, showing that β is conjugate to $2\cos(2\pi/\ell)$. $\qquad\square$

Kronecker's theorems suggest that we should try to describe all 'small' (suitably defined) algebraic integers that are not covered by his theorems.

One way to do this is using the Mahler measure.

The **Mahler measure** $M(P)$ of a polynomial $P(z) = a_0 z^d + \cdots + a_0 \in \mathbb{Z}[z]$ with $a_0 \neq 0$, and zeros $\alpha_1, \ldots, \alpha_d \in \mathbb{C}$, is defined as

$$M(P) := |a_0| \prod_{i=1}^{d} \max(1, |\alpha_i|).$$

So $M(P)$ measures, in a single value, how far outside the unit circle the zeros of P are. Thus, when $a_0 = 1$, the Mahler measure $M(P)$ is the product of the moduli of those zeros of P that lie in the region $|z| > 1$ (empty products being 1). Clearly $M(P) \geq |a_0| \geq 1$. The definition extends immediately to any **Laurent polynomial** (negative powers of z allowed). Such polynomials can be written as $z^k P(z)$, where P is a regular polynomial as in (1.1), and k is any (possibly negative) integer. Then $M(z^k P(z))$ is defined simply as $M(P)$.

The *n*th **cyclotomic polynomial** $\Phi_n(z)$ is defined as the minimal polynomial of all primitive nth roots of unity. (From Proposition 5.2(b), we know that they all have the same minimal polynomial.) These polynomials are discussed in Chap. 5. Throughout the book, we use the term **cyclotomic polynomial** to refer to any monic integer polynomial whose zeros are all roots of unity. Thus, such polynomials are products of polynomials $\Phi_n(z)$ for a multiset of values of n.

The following is almost immediate.

Exercise 1.5 Show that, for a nonzero algebraic integer α, the mean modulus of its conjugates is at least 1, with equality if and only if α is a root of unity.

The *n*th Chebyshev polynomial (of the first kind) $T_n(x)$, though usually defined so that all its zeros are in $[-1, 1]$, is more naturally defined so that they lie in $[-2, 2]$. So for $n = 1, 2, 3, \ldots$ we define $T_n(x)$ by $T_n(z + z^{-1}) := z^n + z^{-n}$, and then $T_n(x)$ is monic with integer coefficients. Its zeros are clearly at $2\cos((2j - 1)\pi/n)$ ($j = 1, \ldots, n$). The following formula gives a convenient way of computing $T_n(x)$—for a discussion of the resultant, see Appendix A.

Exercise 1.6 Show that the resultant $\mathrm{res}_z(z^2 - xz + 1, z^{2n} - yz^n + 1)$ is equal to $(y - T_n(x))^2$. Thus, $T_n(x)$ can be obtained from the coefficient of y in this expression.

Exercise 1.7 The *n*th Chebyshev polynomial of the second kind $U_n(x)$ is defined by $U_n(z + z^{-1}) := (z^n - z^{-n})/(z - z^{-1})$. Find the zeros of $U_n(x)$ and show that

$$\mathrm{res}_z(z^2 - xz + 1, z^{2n} - y(z^{n+1} - z^{n-1}) - 1) = (y - U_n(x))^2(4 - x^2).$$

In number theory, Mahler measure arose first in a paper of Lehmer [Leh33] in 1933, as a way of estimating the growth rate of integer sequences defined by a linear recurrence. In this paper, he noted that the smallest Mahler measure $M(P) > 1$ for an integer polynomial P that he could find was for $P(z) = L(z)$, where

$$L(z) := z^{10} + z^9 - z^7 - z^6 - z^5 - z^4 - z^3 + z + 1,$$

for which $M(L) = 1.17628082\cdots$, known as **Lehmer's number**. He asked whether such $M(P)$ were always at least c, for some absolute constant $c > 1$. To this day, $M(L)$ is the smallest $M(P) > 1$ that has been found, but his question has not been answered. That the answer to this question is 'yes' is now often called 'Lehmer's Conjecture'. That the answer is 'yes' and with $c = M(L)$ is the Strong Lehmer Conjecture. If true, the Strong Lehmer Conjecture would be an extension of Kronecker's First Theorem, as it would show that any irreducible monic integer polynomial $P(z)$ with $M(P) < M(L)$ must be either z or a cyclotomic polynomial.

The best unconditional result in the direction of a proof of Lehmer's Conjecture is due to Dobrowolski [Dob79], who proved that if $M(P) > 1$, then for any $\varepsilon > 0$ there is a $d_0(\varepsilon)$ so that

$$M(P) > 1 + (1 - \varepsilon) \left(\frac{\log \log d}{\log d} \right)^3 \qquad (1.4)$$

for all P of degree $d \geq d_0(\varepsilon)$. A little later, Cantor and Straus [CS83] proved a version of this inequality with $(1 - \varepsilon)$ replaced by $1/1200$, but valid for all $d \geq 2$. Later, Voutier [Vou96], improved the constant $1/1200$ to $1/4$, again valid for $d \geq 2$. A version of Dobrowolski's result is proved in Chap. 3, based on the method of Cantor and Straus.

In any study of Lehmer's question, we of course need to make use of the fact that P is not cyclotomic, so that the zeros α_i of P are not roots of unity. Thus, we know that, for all positive integers n, the product $\prod_{i=1}^{d} (\alpha_i^n - 1)$, being nonzero and a symmetric function of the α_i, is a nonzero integer. Using only this arithmetical information, Blanksby and Montgomery [BM71] and later Stewart [Ste78b] used this fact (and very different methods!) to prove for some positive constant c and $d \geq 2$ that

$$M(P) > 1 + \frac{c}{d \log d}.$$

In their proof, Blanksby and Montgomery used the Fourier analysis, while Stewart, borrowing an idea from transcendence proofs, constructed an auxiliary function. Dobrowolski, in the proof of (1.4), made use of the slightly deeper fact that, for all primes p, we have that $\prod_{i=1}^{d} \prod_{j=1}^{d} (\alpha_i^p - \alpha_j)$ is a nonzero integer multiple of p^d. In Voutier's proof, he uses some additional arithmetic information concerning the modulus of integer polynomial discriminants.

If P is not self-reciprocal (see Sect. A.1), then it is known from Theorem 12.1 that $M(P) \geq M(z^3 - z - 1) = 1.3247\ldots$, so, when addressing Lehmer's question, we can assume that $P(z)$ is self-reciprocal.

There are several ways of coming up with the polynomial $L(z)$. One way (see Exercise 1.8 below for another way) is to note that it comes from the graph E_{10}:

We take its adjacency matrix $A = (a_{ij})$ which has 1 in the (i, j)th place if vertices labelled i and j are joined by an edge, and 0 otherwise. Then

$$L(z) = z^5 \det \left((\sqrt{z} + 1/\sqrt{z})I - A \right) .$$

Exercise 1.8 Show that $P_j(z) := (z^3 - z - 1)z^n + (z^3 + z^2 - 1)$ has at most one zero, α_n say, in $|z| > 1$. Show that $M(P_n(z)) \nearrow M(z^3 - z - 1) = 1.3247\ldots$ as $n \to \infty$. Show too that $P_n(z)$ is a cyclotomic polynomial for $n = 1, \ldots, 7$, but that for $j = 8$ it equals $(z - 1)L(z)$.

It is easy to see that the Mahler measure of a polynomial with integer coefficients is an algebraic number. In fact, this algebraic number must be an algebraic integer.

Proposition 1.9 *The Mahler measure $M(P)$ of a polynomial with integer coefficients in one variable is an algebraic integer.*

Proof For each prime p, consider an algebraic closure $\overline{\mathbb{Q}}_p$ of $\mathbb{Q}_p$, armed with a valuation that extends the usual p-adic valuation on $\mathbb{Q}_p$—see A.3. Write

$$P(z) = a_0 z^d + \cdots + a_d = a_0 \prod_{j=1}^{d} (z - \alpha_i) ,$$

so that $M(P) = \pm a_0 m_1$ say, where $m_1 = \prod_{|\alpha_j| \geq 1} \alpha_j = \alpha_1 \ldots \alpha_k$ say. Now let k_p be the number of α_j that are in $|z|_p > 1$ in $\overline{\mathbb{Q}}_p$. Then, by considering $\pm a_{k_p}/a_0$ as a sum of all products of k_p of the α_j, we see that $\prod_{|\alpha_j|_p > 1} |\alpha_j|_p = |a_{k_p}/a_0|_p$, giving $|a_0|_p \prod_{|\alpha_j|_p > 1} |\alpha_j|_p = |a_{k_p}|_p$. Hence, $|a_0|_p$ times the product of any number of the $|\alpha_j|_p$ (for distinct α_j) is at most $|a_{k_p}|_p \leq 1$. In particular, $|M(P)|_p \leq 1$ and $|M(P)^*|_p \leq 1$ for all conjugates $M(P)^*$ of $M(P)$. Hence, by Exercise A.19, $M(P)$ is an algebraic integer.   □

An algebraic integer β is called a **Perron number** if it is real and positive, and all its conjugates except itself (if there are any) have modulus less than β.

Exercise 1.10 Show that the Mahler measure $M(P)$ of an integer polynomial P is a Perron number.

The Mahler measure $M(P)$ has an integral representation via Jensen's Theorem (see Exercise 1.11 below) as

$$M(P) = \exp \left\{ \int_0^1 \log |P(e^{2\pi i t})| \, dt \right\} . \tag{1.5}$$

This is the geometric mean of $|P(z)|$ for z on the unit circle.

Exercise 1.11 Prove Jensen's Theorem: that for any complex number a,

$$\int_0^1 \log |e^{2\pi i t} - a| \, dt = \begin{cases} \log |a| & \text{if } |a| > 1; \\ 0 & \text{otherwise.} \end{cases}$$

Deduce (1.5) from Jensen's Theorem.

1.3 Mahler Measure Inequalities

The inequalities in this section concern real polynomials in general—their coefficients need not be integers.

Proposition 1.12 (Gonçalves' Inequality) *Let $P(z) = a_0 z^d + \cdots + a_d$ be a polynomial with real coefficients, and with $a_0 a_d \neq 0$. Then*

$$M(P)^2 + \frac{a_0^2 a_d^2}{M(P)^2} \leq \sum_{j=0}^{d} a_j^2 .$$

Proof Factorise P over $\mathbb{C}$ as $P(z) = a_0 \prod_{i=1}^{d}(z + \alpha_i)$. Then

$$P(z) = a_0 \prod_{|\alpha_i| \geq 1}(z + \alpha_i) \cdot \prod_{|\alpha_i| < 1}(1 + \alpha_i z) \cdot f(z),$$

where

$$f(z) := \prod_{|\alpha_i| < 1} \frac{z + \overline{\alpha_i}}{1 + \alpha_i z} .$$

Also, put $b := \prod_{|\alpha_i| < 1} \alpha_i$ and $B := \prod_{|\alpha_i| \geq 1} \alpha_i$, so that $a_0 b = a_d / B$. Then $P(z) = a_0 b f(z) Q(z)$, where

$$Q(z) := \prod_{|\alpha_i| \geq 1}(z + \alpha_i) \cdot \prod_{|\alpha_i| < 1}(z + 1/\alpha_i)$$

$$= z^d + \cdots + B/b$$

$$= z^d + \cdots + \frac{a_0 B^2}{a_d} .$$

Hence,

$$\sum_{j=0}^{d} a_j^2 = \int_0^1 \left| P(e^{2\pi i t}) \right|^2 \, dt$$

$$= \int_0^1 \left| \frac{a_d}{B} f(e^{2\pi i t}) Q(e^{2\pi i t}) \right|^2 \, dt$$

$$= \frac{a_d^2}{B^2} \int_0^1 \left| Q(e^{2\pi i t}) \right|^2 \, dt \qquad (\text{as } |f(z)| = 1 \text{ on } |z| = 1)$$

$$\geq \frac{a_d^2}{B^2} \left(1^2 + \left(\frac{a_0 B^2}{a_d} \right)^2 \right)$$

$$= \frac{a_0^2 a_d^2}{M(P)^2} + M(P)^2 ,$$

since $M(P) = |a_0|B$.   □

The following weaker inequality suffices for some applications.

Corollary 1.13 (Specht's Inequality) *For $P(z) = a_0 z^d + \cdots + a_d \in \mathbb{R}[z]$,*

$$M(P) \leq \sqrt{\sum_{j=0}^{d} a_j^2}.$$

Exercise 1.14 For which polynomials P does equality in Gonçalves' Inequality occur?

Exercise 1.15 Modify Gonçalves' Inequality and its proof to cover polynomials with complex coefficients.

Lemma 1.16 (Gauss–Lucas) *For any polynomial $P(z) \in \mathbb{C}[z]$, the zeros of its derivative $P'(z)$ lie in the convex hull of the zeros of P.*

Proof Writing $P(z) = \prod_{i=1}^{d} (z - \alpha_i)$, we have the well-known partial fraction expansion

$$\frac{P'(z)}{P(z)} = \sum_{i=1}^{d} \frac{1}{z - \alpha_i}. \tag{1.6}$$

Thus, for any zero β of P', not a zero of P, it follows that

$$\sum_{i=1}^{d} \frac{1}{\beta - \alpha_i} = 0 = \sum_{i=1}^{d} \frac{1}{\overline{\beta - \alpha_i}},$$

giving

$$\sum_{i=1}^{d} \frac{\beta - \alpha_i}{|\beta - \alpha_i|^2} = 0,$$

so that $\beta = \sum_{i=1}^{d} w_i \alpha_i$ for weights

$$w_i = \frac{1}{W} \frac{1}{|\beta - \alpha_i|^2},$$

where

$$W = \sum_{j=1}^{d} \frac{1}{|\beta - \alpha_j|^2}.$$

Of course, the result is trivially true if β is also a zero of P.   □

Proposition 1.17 *Let $P(z)$ be a real monic polynomial of degree d, with Mahler measure M. Let its 'monic derivative' $\frac{1}{d}P'(z)$ have Mahler measure M'. Then $M' \leq M$.*

Proof From (1.6), we have

$$\log M' - \log M = \int_0^1 \log \left| \frac{1}{d} \sum_{i=1}^{d} \frac{1}{e^{2\pi it} - \alpha_i} \right| dt \,.$$

Now the right-hand side, considered as a function of one α_i, with the others fixed, is a subharmonic function of α_i in the open set $\mathbb{C} \setminus \{z : |z| = 1\}$ (see Appendix C). Hence, its maximum is on its boundary $|z| = 1$. Doing this for each α_i shows that $\log M' - \log M$ has its maximum when all the α_i are on $|z| = 1$. But then $M = 1$ and, by Lemma 1.16, $M' = 1$ too. So $M'/M \leq 1$ for all polynomials P. □

A related result is the following.

Proposition 1.18 *If $P(z)$ is a real polynomial of degree d and self-reciprocal with Mahler measure $M(P) > 1$, then also $M(\frac{1}{d}P')$ is greater than 1.*

This is an immediate consequence of Theorem A.8.

Exercise 1.19 For a polynomial $P(z) = a_0 z^d + a_1 z^{d-1} + \cdots + a_d \in \mathbb{C}[z]$, show that for $j = 1, \ldots, d$ we have

$$|a_j| \leq \binom{d}{j} M(P)\,.$$

Exercise 1.20 Show that if the polynomial Q, with leading coefficient q_0, is a factor of the polynomial P, with leading coefficient p_0, then $M(Q) \leq |q_0/p_0| M(P)$.

1.4 A Lower Bound for an Integer Polynomial Evaluated at an Algebraic Number

Theorem 1.21 *Given a polynomial $Q(z) \in \mathbb{Z}[z]$ of degree e and an algebraic number α of degree d, then $Q(\alpha) = 0$ or*

$$|Q(\alpha)| \geq \frac{\max(1, |\alpha|)^e}{L^{d-1} M(\alpha)^e} \,. \tag{1.7}$$

*Here, L is the **length** of Q (the sum of the moduli of its coefficients).*

Proof Let $\alpha = \alpha_1$ have minimal polynomial P with leading coefficient p_0 and conjugates α_j ($j = 1, \ldots, d$). Then from (A.2), we have

$$1 \leq |\operatorname{res}_z(P, Q)| = p_0^e |Q(\alpha)| \prod_{j>1} |Q(\alpha_j)| .$$

Also, we clearly have

$$|Q(\alpha_j)| \leq L \max(1, |\alpha_j|)^e,$$

so that

$$1 \leq p_0^e |Q(\alpha)| L^{d-1} \prod_{j>1} \max(1, |\alpha_j|)^e = \frac{|Q(\alpha)| L^{d-1} M(\alpha)^e}{\max(1, |\alpha|)^e},$$

using the definition of $M(\alpha)$. Hence, (1.7) holds.   □

When Q is linear, we have the following.

Corollary 1.22 *For any rational number p/q with $q > 0$ and algebraic number $\alpha \neq p/q$ of degree d, we have*

$$\left| \alpha - \frac{p}{q} \right| \geq \frac{\max(1, |\alpha|)}{q(|p| + q)^{d-1} M(\alpha)} .$$

In particular,

$$|\alpha - 1| \geq \frac{\max(1, |\alpha|)}{2^{d-1} M(\alpha)} .$$

1.5   Polynomials with Small Coefficients

In this section, we consider integer polynomials as factors of integer polynomials with small coefficients. The key connection with the Mahler measure is the following theorem.

Theorem 1.23 *Let $P(z) \in \mathbb{Z}[z]$ be monic. Then $P(z)$ has an integer polynomial multiple $P(z)S(z)$ say such that all coefficients of $P(z)S(z)$ are at most $\lfloor M(P) \rfloor$ in modulus.*

Very often $S(z)$ can be chosen so that $M(S) = 1$. However, this is not guaranteed by the theorem, so that maybe $M(PS) > M(P)$. Also, it is not claimed that in general PS is monic (although, of course, if $M(P) < 2$, then it must be).

Proof Let $\alpha_1, \alpha_2, \ldots, \alpha_d$ be the zeros of $P(z)$, where d is the degree of P. Note that its zeros need not be conjugate or distinct. However, we first do the case when the α_i *are* assumed to be distinct. We label the α_j so that $\alpha_1, \ldots, \alpha_{2s}$ are nonreal with $\alpha_{2j'-1} = \overline{\alpha_{2j'}}$ for $j' = 1, \ldots, s$, and $\alpha_{2s+1}, \ldots, \alpha_d$ are all real. Put $Y := 1 + \lfloor M(P) \rfloor$. We need two integer parameters N, L and further integers $y_i \in \{0, 1, \ldots, Y - 1\}$ $(i = 1, \ldots, N)$, all to be specified later.

We consider the vector $\mathbf{v} = (v_1, \ldots, v_d) \in \mathbb{R}^d$ defined as

$$\mathbf{v} := \left(\operatorname{Re} \left(\sum_{i=1}^{N} \alpha_2^i y_i \right), \operatorname{Im} \left(\sum_{i=1}^{N} \alpha_2^i y_i \right), \operatorname{Re} \left(\sum_{i=1}^{N} \alpha_4^i y_i \right), \operatorname{Im} \left(\sum_{i=1}^{N} \alpha_4^i y_i \right), \ldots, \right.$$

$$\left. \operatorname{Re} \left(\sum_{i=1}^{N} \alpha_{2s}^i y_i \right), \operatorname{Im} \left(\sum_{i=1}^{N} \alpha_{2s}^i y_i \right), \sum_{i=1}^{N} \alpha_{2s+1}^i y_i, \sum_{i=1}^{N} \alpha_{2s+2}^i y_i, \ldots, \sum_{i=1}^{N} \alpha_d^i y_i \right).$$

Now $v_j \in [-B_j, B_j]$, where

$$B_j \leq N(Y-1) M_j^N \quad (j = 1, \ldots, d), \tag{1.8}$$

with $M_j := \max(1, |\alpha_j|)$. Next, for $j = 1, \ldots, d$ divide the interval $[-B_j, B_j]$ into $\lfloor L M_j^N \rfloor$ intervals of equal length, and hence, divide the box $\prod_{j=1}^{d} [-B_j, B_j] \subseteq \mathbb{R}^d$ into smaller boxes. Since $\prod_{j=1}^{d} M_j = M(P)$, this gives us at most $L^d M(P)^N$ boxes, together containing all the possible vectors $\mathbf{v}$, with edge lengths at most

$$2N(Y-1) \frac{M_j^N}{\lfloor L M_j^N \rfloor} < \frac{2N(Y-1)}{L-1} \quad (j = 1, \ldots, d). \tag{1.9}$$

These lengths are less than $1/\sqrt{2}$ provided that

$$L \geq 1 + 2\sqrt{2} N(Y-1). \tag{1.10}$$

Now there are Y^N possible vectors $(y_1, \ldots, y_N)$, so if $Y^N > L^d M(P)^N$ then, by the Pigeonhole Principle, two vectors $\mathbf{v}_1, \mathbf{v}_2$, say, defined by distinct vectors (y_{1i}), (y_{2i}) say, must lie in the same box. Thus, it suffices to take

$$1 + 2\sqrt{2} N(Y-1) \leq L < \left(\frac{Y}{M(P)} \right)^{N/d}. \tag{1.11}$$

This can clearly be achieved by taking N sufficiently large—see Exercise 1.25 below for an estimate for such an N. Hence, the difference $\mathbf{v}' := \mathbf{v}_1 - \mathbf{v}_2$ has all components of modulus less than $1/\sqrt{2}$. It is defined by the nonzero vector $(y_{1i}) - (y_{2i}) =: (y_i')$ say, which all lie in the set $\{-(Y-1), \ldots, -1, 0, 1, \ldots, Y-1\}$. Then for

$$\mathbf{v}' = (v_1', v_2', \ldots, v_{2s-1}', v_{2s}', v_{2s+1}', \ldots, v_d'),$$

we certainly have

$$v_{2j'-1}'^2 + v_{2j'}'^2 < 1 \ (j' = 1, \ldots, s) \text{ and } |v_j'| < 1, \ (j = 2s+1, \ldots, d),$$

so that $|\sum_{i=1}^{N} \alpha_j^i y_i'| < 1 \ (j = 1, \ldots, d)$. Thus, for any α_j, we have $|\sum_{i=1}^{N} \alpha^i y_i'| < 1$ for every conjugate α of α_j. Hence, every $\sum_{i=1}^{N} \alpha_j^i y_i$ has norm less than 1, and so must be 0. Thus, all α_j are zeros of the nonzero polynomial $\sum_{i=1}^{N} y_i' z^i$, which is an integer polynomial multiple of P, given the α_j are distinct.

Finally, we outline the modifications necessary to the proof when some of the α_j are equal. If $\alpha_1 = \alpha_2 = \cdots = \alpha_k = \alpha$ say are zeros of $P(z)$, then $P(z)$ and its k'th derivatives for $k' = 1, 2, \ldots, k - 1$ are all zero at $z = \alpha$. So we need to replace the k copies of $\sum_{i=1}^{N} \alpha^i y_i$ appearing in $\mathbf{v}$ by

$$\sum_{i=1}^{N} \alpha^i y_i, \sum_{i=1}^{N} \binom{i}{1} \alpha^{i-1} y_i, \sum_{i=1}^{N} \binom{i}{2} \alpha^{i-2} y_i, \ldots, \sum_{i=1}^{N} \binom{i}{k-1} \alpha^{i-k+1} y_i.$$

Since $k \le d$, we can bound the modulus of each of these sums simply if rather crudely by $N^d (Y - 1) \max(1, |\alpha|)^N$. The effect of this is that we need only replace the 'N' appearing in (1.8), (1.9) (two places), (1.10) and the left-hand side of (1.11) by N^d. The argument is otherwise unaffected and adapts similarly if there are further sets of equal zeros. $\qquad\square$

Exercise 1.24 The theorem tells us that for every n the polynomial $(1 + z)^n$ is a factor of a polynomial with coefficients in $\{-1, 0, 1\}$. Find such a polynomial.

Exercise 1.25 Find an estimate for N in (1.11).

1.6   Separation of Conjugates

In this section, we give a lower bound for the minimum distance between conjugates of an algebraic integer. We also give a lower bound for the modulus of the derivative of the minimal polynomial of an algebraic integer, evaluated at its conjugates. Both of these bounds involve the Mahler measure.

Let α be an algebraic integer with conjugates $\alpha = \alpha_1, \alpha_2, \ldots, \alpha_d$. Define the *separation of* α, $\mathrm{Sep}(\alpha)$, by

$$\mathrm{Sep}(\alpha) = \min_{i \ne j} |\alpha_i - \alpha_j|.$$

Also, its **discriminant** $\Delta(\alpha)$ is the square of the Vandermonde determinant $\det(\alpha_j^{i-1})_{i,j=1,\ldots,d}$. Note that $\Delta(\alpha)$ is not always the same as the **field discriminant** $\Delta(\mathbb{Q}(\alpha))$ of $\mathbb{Q}(\alpha)$, which is the square of the determinant of the matrix $(\alpha_i^{(j)})$, where $\alpha^{(1)}, \alpha^{(2)}, \ldots, \alpha^{(d)}$ is a $\mathbb{Z}$-basis for the ring of integers of $\mathbb{Q}(\alpha)$, and where, with $\sigma_1, \ldots, \sigma_d$ being the embeddings of $\mathbb{Q}(\alpha)$ into $\mathbb{C}$, we have $\alpha_i^{(j)} = \sigma_i(\alpha^{(j)})$.

Exercise 1.26 Prove that $\Delta(\alpha)$ and $\Delta(\mathbb{Q}(\alpha))$ are both integers and that $\Delta(\mathbb{Q}(\alpha))$ divides $\Delta(\alpha)$.

The next lemma will immediately be applied to bound the separation of an algebraic integer from below, but it holds more generally for any d-tuple of complex numbers. The proof of this lemma appeals to Hadamard's Inequality, which we prove later as Lemma 3.2. For $\mathbf{z} = (z_1, \dots, z_d) \in \mathbb{C}^d$, we define

$$M(\mathbf{z}) := M((z - z_1) \cdots (z - z_d)).$$

Lemma 1.27 *For any* $\mathbf{z} = (z_1, \dots, z_d) \in \mathbb{C}^d$, *we have*

$$|z_1 - z_2| \geq \sqrt{3} d^{-(d+2)/2} M(\mathbf{z})^{1-d} |\det((z_j^{i-1})_{i,j=1,\dots,d})|.$$

Proof We can assume that $|z_2| \leq |z_1|$. Let $V = (z_j^{i-1})_{i,j=1,\dots,d}$. Then

$$|\det(V)|^2 = |z_1 - z_2|^2 \begin{vmatrix} 0 & 1 & 1 & \cdots & 1 \\ 1 & z_2 & z_3 & \cdots & z_d \\ z_1 + z_2 & z_2^2 & z_3^2 & \cdots & z_d^2 \\ z_1^2 + z_1 z_2 + z_2^2 & z_2^3 & z_3^3 & \cdots & z_d^3 \\ \vdots & \vdots & \vdots & \cdots & \vdots \\ z_1^{d-2} + z_1^{d-3} z_2 + \cdots + z_2^{d-2} & z_2^{d-1} & z_3^{d-1} & \cdots & z_d^{d-1} \end{vmatrix}^2$$

$$\leq |z_1 - z_2|^2 \left(\sum_{i=1}^{d-1} |i z_1^{i-1}|^2 \right) \prod_{j=2}^{d} \left(\sum_{i=0}^{d-1} |z_j|^{2i} \right) \quad \text{by Lemma 3.2}$$

$$\leq |z_1 - z_2|^2 \left(\sum_{i=1}^{d-1} i^2 \right) \max(1, |z_1|)^{2(d-2)} \prod_{j=2}^{d} \max(1, |z_j|)^{2(d-1)}$$

$$\leq |z_1 - z_2|^2 \frac{d^3}{3} d^{d-1} M(\mathbf{z})^{2(d-1)},$$

using $\sum_{i=1}^{d-1} i^2 = \frac{1}{3}(d-1)(d - \frac{1}{2})d$. This gives the required lower bound for $|z_1 - z_2|$. $\square$

We now apply this to the minimal polynomial of an algebraic integer α. For an algebraic number α, we write $M(\alpha)$ for the Mahler measure $M(P_\alpha)$, where $P_\alpha(z) \in \mathbb{Z}[z]$ is the minimal polynomial of α.

Corollary 1.28 (Mahler [Mah64]) *Let* α *be an algebraic integer of degree* d. *We have*

$$\mathrm{Sep}(\alpha) > \sqrt{3} d^{-(d+2)/2} |\Delta(\alpha)|^{1/2} M(\alpha)^{-(d-1)}.$$

Lemma 1.29 (Mahler [Mah64]) *For any* $\mathbf{z} = (z_1, \dots, z_d) \in \mathbb{C}^d$, *define the polynomial* $P(z) := (z - z_1) \cdots (z - z_d)$, *so that* $P'(z_i) = \prod_{j \neq i} (z_i - z_j)$ $(i = 1, \dots, d)$. *Then*

$$|P'(z_i)| \geq 3^{(d-1)/2} d^{-(d^2+d-1)/2} |\det(V)| \times \begin{cases} M(\mathbf{z})^{-(d-1)} & \text{if } |z_i| \leq 1; \\ \overline{|\mathbf{z}|}^{-(d-1)^2/2} & \text{if } |z_i| > 1. \end{cases}$$

Here $\overline{|\mathbf{z}|} := \max_{i=1,\ldots,d} |z_i|$.

Proof By standard column operations, we have

$$| \det(V)|^2 = |P'(z_i)|^2 \, |\det (\mathbf{c}_1 \cdots \mathbf{c}_d)|^2 \, ,$$

where $\mathbf{c}_1 = (1, z_1, \ldots, z_1^{d-1})^\mathsf{T}$ while for $j = 2, \ldots, d$,

$$\mathbf{c}_j = (0, 1, z_1 + z_j, z_1^2 + z_1 z_j + z_j^2, \ldots, z_1^{d-2} + z_1^{d-3} z_j + \cdots + z_1 z_j^{d-3} + z_j^{d-2})^\mathsf{T}.$$

Thus, by Hadamard's Inequality (Lemma 3.2),

$$|\det (\mathbf{c}_1 \cdots \mathbf{c}_d)|^2 \leq d \max(1, |z_1|)^{d-1} \left(\frac{d^{d+2}}{3} \right)^{d-1} \prod_{j=1}^d \max \left(|z_1|, |z_j| \right)^{d-2}$$

$$\leq 3^{-(d-1)} d^{(d^2+d-1)} \times \begin{cases} M(\mathbf{z})^{d-1} & \text{if } |z_i| \leq 1; \\ \overline{|\mathbf{z}|}^{(d-1)^2} & \text{if } |z_i| > 1, \end{cases}$$

which gives the required inequality. $\square$

Corollary 1.30 *For an algebraic integer α with conjugates $\alpha = \alpha_1, \alpha_2, \ldots, \alpha_d$ and minimal polynomial $P_\alpha(z)$, we have*

$$\min_{i=1}^d |P_\alpha'(\alpha_i)| \geq 3^{(d-1)/2} d^{-(d^2+d-1)/2} |\Delta(\alpha)|^{1/2} \left(\max \left(M(\alpha), \overline{|\alpha|}^{(d-1)/2} \right) \right)^{-(d-1)}.$$

1.7   The Shortness of a Polynomial

The *length* of a polynomial $P(z) \in \mathbb{Z}[z]$ is defined as usual to be the sum of the absolute values of its coefficients. We define a *short polynomial for P* to be the minimum-length product of P with a cyclotomic polynomial. We call this length the *shortness of P*, denoted by $\mathrm{sh}(P)$. Note that a polynomial and its short polynomial have the same Mahler measure. For an algebraic number α, we define the *shortness of α* to be the shortness of its minimal polynomial. A nonreciprocal polynomial clearly has shortness of at least 3, while a reciprocal noncyclotomic polynomial has shortness of at least 5, by Exercise 1.34 below.

Exercise 1.31 Show that 0 has shortness 1, and all roots of unity have shortness 2.

Exercise 1.32 Show that the shortness of $(3 + 4i)/5$ is 12.

Exercise 1.33 For an algebraic number α, show that its shortness is at least $\lceil M(\alpha) \rceil$ (the smallest integer not less than $M(\alpha)$).

Exercise 1.34 Show that all nonconstant reciprocal noncyclotomic integer polynomials have shortness at least 5.

Exercise 1.35 Is a short polynomial for an algebraic number α always unique?

Exercise 1.36 Show that the shortness of $(z - 1)^k$ is even, at most 2^k, and is 6 for $k = 3$.

Exercise 1.37 Let P be a cyclotomic polynomial and let k be the largest multiplicity of any zero of P. Show that $\text{sh}(P) \le \text{sh}((z - 1)^k)$.

Problem 1.38 (open problem)
Find the shortness of $(z - 1)^k$ for all k.

While algebraic numbers are usually specified by their minimal polynomials, there are some advantages in using short polynomials to specify them. Allowing these extra cyclotomic factors, whose Mahler measures are of course all 1, means that the length of the polynomial can be kept small, without changing the Mahler measure. For instance, the irreducible polynomial

$$
\begin{aligned}
Q(z) \\
&= z^{112} + z^{111} + z^{110} + z^{109} - z^{106} - 2z^{105} - z^{104} - z^{103} + z^{101} + z^{100} + 2z^{99} + 2z^{98} \\
&+ z^{97} - z^{95} - 2z^{94} - 2z^{93} - 2z^{92} - 2z^{91} + z^{89} + 2z^{88} + 3z^{87} + 2z^{86} + 2z^{85} + z^{84} - z^{83} \\
&- 2z^{82} - 3z^{81} - 3z^{80} - 2z^{79} - z^{78} + 2z^{76} + 3z^{75} + 3z^{74} + 3z^{73} + z^{72} - z^{70} - 3z^{69} \\
&- 3z^{68} - 3z^{67} - 2z^{66} + z^{64} + 2z^{63} + 3z^{62} + 3z^{61} + 2z^{60} - 2z^{58} - 3z^{57} - 3z^{56} - 3z^{55} \\
&- 2z^{54} + 2z^{52} + 3z^{51} + 3z^{50} + 2z^{49} + z^{48} - 2z^{46} - 3z^{45} - 3z^{44} - 3z^{43} - z^{42} + z^{40} \\
&+ 3z^{39} + 3z^{38} + 3z^{37} + 2z^{36} - z^{34} - 2z^{33} - 3z^{32} - 3z^{31} - 2z^{30} - z^{29} + z^{28} + 2z^{27} \\
&+ 2z^{26} + 3z^{25} + 2z^{24} + z^{23} - 2z^{21} - 2z^{20} - 2z^{19} - 2z^{18} - z^{17} + z^{15} + 2z^{14} + 2z^{13} \\
&+ z^{12} + z^{11} - z^9 - z^8 - 2z^7 - z^6 + z^3 + z^2 + z + 1
\end{aligned}
$$

has length 179 and Mahler measure 1.24846635. However, its shortness is only 6, with short polynomial

$$P(z) = z^{130} - z^{114} - z^{77} + z^{53} + z^{16} - 1.$$

This is #170 in Table D.1 of Appendix D. The polynomial $Q(z)$ is the noncyclotomic irreducible factor of $P(z)$. In Chap. 5, we shall discover how to identify cyclotomic factors of polynomials, and in particular from Exercise 5.7, we can see that, because $P(z)$ has no factors that are polynomials in z^4, we have $Q(z) := P(z)/\gcd(P(z), P(-z)P(z^2)P(-z^2))$. For a discussion of a connection between these short polynomials and related polynomials in 2 or 3 variables, see Sect. 2.6.1.

There are 236 known integer polynomials with Mahler measures less than 1.25, as shown in Table D.1. Computation shows that 15 of these are Mahler measures of short reciprocal pentanomials, while 214 are Mahler measures of polynomials of reciprocal hexanomials. Of the remaining seven, three are Mahler measures of heptanomials, three are Mahler measures of octanomials, while one is the Mahler measure of a decanomial. These polynomials, when not irreducible, are the product of a cyclotomic polynomial and a polynomial whose Mahler measure is the given value. All these values are in $(1, 1.25)$ and so are less than $M(Q_1) = 1.255 \cdots$, the smallest Mahler measure in Table D.2 in Appendix D. It is expected that there are infinitely many Mahler measures less than $M(Q_1)$. For more on the measures in that table, see Sect. 2.6.2.

1.7.1 Finding Short Polynomials

Currently there is no known algorithm to compute the shortness of a polynomial and its associated short polynomial. However, we can, at least in principle, compute a related polynomial that very often, indeed maybe always, turns out to be a short one.

Given a polynomial $P(z) \in \mathbb{Z}[z]$, we define a *minimum-length multiple of* P to be a nonzero polynomial of the form $P(z)T(z)$, where $T(z) \in \mathbb{Z}[z]$, and the length of PT is as small as possible among all polynomials of that form. We call the length of such a PT the *minimum length of* P. Note that if T turns out to be cyclotomic, then PT is a short polynomial for P, and the minimum length of P is actually the shortness of P. But possibly T could be noncyclotomic, so that we would have $M(PT) > M(P)$. Indeed, we cannot even rule out the possibility that T could be nonmonic.

We now describe an algorithm to find a minimum-length multiple of P in the case where P is monic, noncyclotomic and irreducible. We do not claim that the algorithm is an efficient one, merely that it is effective.

Let ℓ_P be the length of P, and $\ell_{\min}$ its minimum length. We know that $3 \le \ell_{\min} \le \ell_P$. We assume that $\ell \ge 3$ and that we have shown that $\ell_{\min} \ge \ell$. Our aim is to find out whether or not there is an integer polynomial T with PT of length ℓ and given leading coefficient v. If there is such a T, then $\ell_{\min} = \ell$ and we are finished. Otherwise we increase v by 1, if possible, or else increase ℓ by 1 and reset v to 1.

Because P is monic and noncyclotomic, we know from Theorem 1.3 that P has a zero in $|z| > 1$: call it α, say. We can choose α so that $|\alpha| = \overline{|\alpha|}$, the largest modulus of any conjugate of α (the *house* of α). Using $\overline{|\alpha|}$, we can get tighter bounds for ℓ and v.

Lemma 1.39 *If the polynomial PT has length ℓ and leading coefficient v, then*

$$\ell \ge 1 + \overline{|\alpha|} \quad and \quad v \le \frac{\ell}{1 + \overline{|\alpha|}}. \tag{1.12}$$

Proof For some positive integer v, we can write PT as

$$v z^n \pm z^{n-d_1} \pm z^{n-d_1-d_2} \cdots \pm z^{n-d_1-d_2-\cdots-d_{\ell-v-1}} \pm 1, \qquad (1.13)$$

where $d_1 > 0$ and $d_2, d_3, \ldots, d_{\ell-v-1} \geq 0$. Also, define $d_{\ell-v} \geq 0$ by

$$n = d_1 + d_2 + \cdots + d_{\ell-v-1} + d_{\ell-v}. \qquad (1.14)$$

To give a coefficient of modulus greater than 1 (other than the leading coefficient), some d_i must be 0, and the signs of $\pm z^{n-d_1-d_2-\cdots-d_{i-1}}$ and $\pm z^{n-d_1-d_2-\cdots-d_i}$ must be the same, to disallow cancellation. Then

$$|\pm \alpha^{n-d_1} \pm \alpha^{n-d_1-d_2} \cdots \pm \alpha^{n-d_1-d_2-\cdots-d_{\ell-v-1}} \pm 1| < (\ell - v)|\alpha|^{n-d_1},$$

which is less than $v|\alpha|^n$ if $d_1 > \frac{\log(\ell/v-1)}{\log|\alpha|}$. If $\frac{\log(\ell/v-1)}{\log|\alpha|} < 1$ then, since $d_1 \geq 1$, we have a contradiction. Hence, $\ell/v \geq 1 + |\alpha| = 1 + \overline{|\alpha|}$, which gives the claimed inequalities. $\square$

Now assume that PT satisfies (1.12). Then from the proof of Lemma 1.39, we have

$$d_1 \in \left\{ 1, 2, \ldots \left\lfloor \frac{\log(\ell/v - 1)}{\log|\alpha|} \right\rfloor \right\}.$$

For such a d_1, we have

$$|\pm \alpha^{n-d_1-d_2} \pm \alpha^{n-d_1-d_2-d_3} \cdots \pm \alpha^{n-d_1-d_2-\cdots-d_{\ell-v-1}} \pm 1| < (\ell - v - 1)|\alpha|^{n-d_1-d_2}$$
$$< |v\alpha^n \pm \alpha^{n-d_1}|$$

if $|\alpha|^{d_2} > \frac{\ell-v-1}{\alpha^{d_1}\pm 1}$. Hence,

$$d_2 \in \left\{ 0, 1, 2, \ldots, \left\lfloor \frac{\log(\ell - v - 1) - \log|v\alpha^{d_1} \pm 1|}{\log|\alpha|} \right\rfloor \right\},$$

where the range of values for d_2 depends on the choice of the $\pm$ sign. Continuing in this way, we have, for $i = 2, , \ldots, , \ell - v$, that

$$|\pm \alpha^{n-d_1-d_2-\cdots-d_{i+1}} \pm \alpha^{n-d_1-d_2-\cdots-d_{i+2}} \cdots \pm \alpha^{n-d_1-d_2-\cdots-d_{\ell-v-1}} \pm 1|$$
$$< (\ell - v - i)|\alpha|^{n-d_1-d_2-\cdots-d_{i+1}},$$

which is less than

$$|v\alpha^n \pm \alpha^{n-d_1} \pm \alpha^{n-d_1-d_2} - \cdots - \alpha^{n-d_1-d_2-\cdots-d_i}|$$

if

$$|\alpha|^{d_{i+1}} > \frac{\ell - v - i}{|v\alpha^{d_1+\cdots+d_i} \pm \alpha^{d_2+\cdots+d_i} \pm \cdots \pm \alpha^{d_i} \pm 1|}.$$

Hence, we obtain, for $i = 1, 2, 3, \ldots, \ell - v - 1$, that

$$d_{i+1} \in \left\{ 0, 1, 2, \ldots \left\lfloor \frac{\log(\ell - v - i) - \log |v\alpha^{d_1 + \cdots + d_i} \pm \alpha^{d_2 + \cdots + d_i} \pm \cdots \pm \alpha^{d_i} \pm 1|}{\log |\alpha|} \right\rfloor \right\}.$$

Note that the term $v\alpha^{d_1 + \cdots + d_i} \pm \alpha^{d_2 + \cdots + d_i} \pm \cdots \pm \alpha^{d_i} \pm 1$ has length $i + v < \ell$ so, by assumption, cannot be zero. Thus, for $i = 1, \ldots, \ell - v$, each d_i is bounded in terms of $d_1, \ldots, d_{i-1}$. If a polynomial of the form (1.13) is found that is 0 at $z = \alpha$ then, because P is irreducible, the polynomial will be divisible by P, and so be of the form PT. Then n is given by (1.14). Thus, the search is a finite one. It will eventually succeed in finding PT of minimum length, as it will find P itself as a last resort. This completes the algorithm description.

If T turns out to be cyclotomic, then we have found a short polynomial PT for P.

Problem 1.40 Find a monic irreducible integer polynomial P with minimum-length polynomial PT where T is *not* cyclotomic. Alternatively, prove that no such P exists.

It seems to be a challenge to extend the algorithm above to nonmonic and reducible polynomials, including those having repeated irreducible factors. Perhaps the most tractable such case could be when P is nonmonic and irreducible. By Corollary A.17, if P has no zero in $|z| > 1$ then, for some prime p, it has a zero α in $\overline{\mathbb{Q}}_p$ with $|\alpha|_p > 1$.

Problem 1.41 Modify the above algorithm so that it works when P is nonmonic and irreducible and has all its complex zeros in the disc $|z| \le 1$.

1.8   Variants of Mahler Measure

As before, for an algebraic number α, we write $M(\alpha)$ for the Mahler measure $M(P_\alpha)$, where $P_\alpha(z) \in \mathbb{Z}[z]$ is the minimal polynomial of α. Three variants of Mahler measure are used in the literature:

- The **Weil height**

$$h(\alpha) := \log \overline{M}(\alpha) = \frac{\log M(\alpha)}{\deg \alpha}. \tag{1.15}$$

- The **absolute Mahler measure** $\overline{M}(\alpha) := M(\alpha)^{1/\deg \alpha}$. Also called the **exponential Weil height**.
- The **logarithmic Mahler measure** $m(P) := \log M(P)$ is often used for polynomials P in several variables.

1.8.1   The Weil Height

The Mahler measure results for algebraic numbers α have an alternative formulation using the (absolute logarithmic) Weil height $h(\alpha)$ defined in (1.15). In particular, the following are easily proved.

Exercise 1.42 Show that for an algebraic number α, we have

- $h(\alpha) = 0$ if and only if α is 0 or a root of unity;
- $h(\alpha^{-1}) = h(\alpha)$;
- $h(\alpha + 1) \le h(\alpha) + \log 2$, with equality if and only if $\alpha = 1$;
- $h(\alpha^k) = k h(\alpha)$ for all positive integers k;
- $h(\omega\alpha) = h(\alpha)$ for any root of unity ω.

There is also a nice inequality for the Weil height, as follows.

Lemma 1.43 *For any two algebraic numbers α, β, we have*

$$h(\alpha\beta) \le h(\alpha) + h(\beta).$$

Before giving its proof, we need a preliminary lemma, with corollaries. Let us extend the p-adic valuation on $\mathbb{Q}_p$ to an algebraic closure $\overline{\mathbb{Q}}_p$ of $\mathbb{Q}_p$.

Lemma 1.44 *For a given prime p, suppose that*

$$a_0 z^d + a_1 z^{d-1} + \cdots + a_{d-1} z + a_d = a_0 \prod_{j=1}^{d} (z - \alpha_j), \qquad (1.16)$$

where $a_0 \neq 0$, all a_k are in $\mathbb{Q}_p$ and all α_j are in $\overline{\mathbb{Q}}_p$. Then

$$\sum_j \log_+ |\alpha_j|_p = \max_{k=0}^{d} \log |a_k/a_0|_p.$$

Here $\log_+ x := \max(0, \log x)$ for real positive x.

The proof follows straight from the Newton polygon of (1.16)—see Proposition A.18 of Appendix A.

Corollary 1.45 *Assume further that the coefficients a_i in (1.16) are in $\mathbb{Z}$, with $a_0 > 0$, and $\gcd(a_0, \ldots, a_d) = 1$. Then*

$$\sum_p \sum_j \log_+ |\alpha_j|_p = \log a_0,$$

where the sum is taken over all primes p dividing a_0.

Proof Suppose $p^\ell \parallel a_0$. Then $p \nmid a_k$ for some k, so that

$$\max_{k=0}^{d} \log |a_k/a_0|_p = \ell \log p \,.$$

Now do this for all such p.   □

Corollary 1.46 *For α of degree d_α over $\mathbb{Q}$, we have*

$$h(\alpha) = \frac{1}{d_\alpha} \left(\sum_j \log_+ |\alpha_j| + \sum_p \sum_j \log_+ |\alpha_j|_p \right).$$

Proof This comes straight from (1.15) and Proposition A.15.   □

Proof (of Lemma 1.43) The proof is based on the obvious inequality

$$\log_+(x_1 x_2) \le \log_+ x_1 + \log_+ x_2 \quad (x_1, x_2 > 0) \,.$$

Let α_i $(i = 1, \ldots, d_\alpha)$, β_j $(j = 1, \ldots, d_\beta)$ and $(\alpha\beta)_k$ $(k = 1, \ldots, d_{\alpha\beta})$ be the conjugates in $\mathbb{C}$ of α, β and $\alpha\beta$, respectively. Let G be the Galois group of the normal closure, N say, of $\mathbb{Q}(\alpha, \beta)$, say of degree d over $\mathbb{Q}$. Applying the d automorphisms σ of G, the numbers $\sigma\alpha$ consist of d/d_α copies of the α_i, the numbers $\sigma\beta$ consist of d/d_β copies of the β_j and the numbers $\sigma(\alpha\beta)$ consist of $d/d_{\alpha\beta}$ copies of the $\alpha_i\beta_j$. Hence, averaging the inequalities

$$\log_+ |\sigma(\alpha\beta)| \le \log_+ |\sigma\alpha| + \log_+ |\sigma\beta| \,,$$

for all $\sigma \in G$, we obtain

$$\frac{1}{d_{\alpha\beta}} \sum_k \log_+ |(\alpha\beta)_k| \le \frac{1}{d_\alpha} \sum_i \log_+ |\alpha_i| + \frac{1}{d_\beta} \sum_j \log_+ |\beta_j| \,.$$

Doing the same for each prime p, except embedding N into $\overline{\mathbb{Q}}_p$ instead of $\mathbb{C}$, we obtain

$$\frac{1}{d_{\alpha\beta}} \sum_k \log_+ |(\alpha\beta)_k|_p \le \frac{1}{d_\alpha} \sum_i \log_+ |\alpha_i|_p + \frac{1}{d_\beta} \sum_j \log_+ |\beta_j|_p \,.$$

Then adding all these inequalities and applying Corollary 1.46 gives the result.   □

1.9 Notes

The first result in the direction of Lehmer's Conjecture where, for a nonzero non-cyclotomic integer polynomial P of degree d, $M(P) - 1$ was bounded below by a function not exponentially small in d, was due to Blanksby and Montgomery [BM71]. They proved that $M(P) \geq 1 + 1/(52d \log(6d))$ for such P. Fourier analysis, in particular the nonnegativity of the Fejér kernel, played a large part in the proof. A little later Stewart [Ste78b] proved a result of similar strength, but by a transcendence-type argument, using an auxiliary function. The following result, which we state as an exercise, formed part of Blanksby and Montgomery's proof.

Exercise 1.47 ([BM71, Lemma 4]) For given real ρ satisfying $0 < \rho \leq 1$ and $z \in \mathbb{C}$ with $\rho \leq |z| \leq \rho^{-1}$, show that

$$|z - 1| \leq \rho^{-1} \left| \rho \frac{z}{|z|} - 1 \right|.$$

Proposition 1.12 (Gonçalves' Inequality) was slightly strengthened by Schinzel [Sch00, Sect. 3.2, Lemma 13]. Also, Borwein, Mossinghoff and Vaaler [BMV07] generalised the inequality from using the 2-norm to one involving the p-norm of P, where $1 \leq p \leq 2$.

For an extensive history of the Gauss–Lucas Lemma (Lemma 1.16), see Rahman and Schmeisser [RS02, pp. 91–92].

Theorem 1.21 comes from Waldschmidt [Wal00, p. 46]. For more results on the distance of α from 1, see the survey [Smy08]. Lemma 1.27 is due to Mahler [Mah64].

The 'variants' of the Mahler measure referred to in Sect. 1.8 are not true variants: they are all comparable in the sense that any given value of one variant can be used to calculate the values of the other variants. On the other hand, there are *true* Mahler measure variants. One such true variant is the ***metric Mahler measure*** $mMm(\alpha)$ of a nonzero algebraic number α defined by Dubickas and Smyth [DS01a]: it is defined to be the infimum over all $k \geq 1$ and all products $M(\beta_1)M(\beta_2) \cdots M(\beta_k)$ where $\beta_1 \beta_2 \cdots \beta_k = \alpha$. They showed that it defines a metric on the (multiplicative) group $\overline{\mathbb{Q}}^{\times}$ quotiented by the group of all roots of unity. This metric defines the discrete topology on this quotient group if and only if Lehmer's Conjecture is true. Samuels [Sam11] proved that the infimum in the definition is always attained. Thus, by Proposition 1.9, $mMm(\alpha)$ is always an algebraic integer.

1.10 Glossary

$\lfloor \; \rfloor$ The integer part, or floor function: for real x, $\lfloor x \rfloor$ denotes the greatest integer not exceeding x.

$\overline{\alpha}$.   The house of α.

$\Delta(\alpha)$, $\Delta(\mathbb{Q}(\alpha))$.   Respectively the discriminant of an algebraic integer α and the field discriminant of the field generated by α. The former is an integer multiple of the latter.

$h(\alpha)$.   The (absolute logarithmic) Weil height of an algebraic number α, namely $\log \overline{M}(\alpha)$.

$L(z)$.   Lehmer's polynomial.

$\log_+ x$.   For a positive real number x, this is the maximum of 0 and $\log x$. Thus, $\log_+ x = \log x$ if $x \geq 1$, and $\log_+ x = 0$ if $0 < x < 1$.

$M(\alpha)$.   For α an algebraic number, $M(\alpha)$ is the Mahler measure of the minimal polynomial (in $\mathbb{Z}[z]$, content 1) of α.

$\overline{M}(\alpha)$.   The absolute Mahler measure, $M(\alpha)^{1/\deg \alpha}$.

$M(P)$.   The Mahler measure of the polynomial P.

$\overline{P}(z)$, $\overline{P(z)}$.   Let $P(z)$ be a polynomial with complex coefficients. Then $\overline{P}(z)$ is the polynomial whose coefficients are the complex conjugates of those of P. Note that for a specific complex number z, the value of $\overline{P}(z)$ does not generally equal the complex conjugate of $P(z)$, which is written $\overline{P(z)}$.

$\mathrm{sh}(P)$.   The shortness of a polynomial P.

absolute Mahler measure.   If α is an algebraic number of degree d, then the absolute Mahler measure is the dth root of the Mahler measure.

absolute logarithmic Weil height.   The Weil height.

algebraic integer.   An algebraic number α is called an algebraic integer if it is a zero of a *monic* polynomial $P(z) \in \mathbb{Z}[z]$. For example, the algebraic integers inside $\mathbb{Q}$ are precisely the integers $\mathbb{Z}$, and indeed in a number field, the algebraic integers can be viewed as a generalisation of $\mathbb{Z}$ in $\mathbb{Q}$. The set of algebraic integers in fact forms a ring (Proposition A.22).

algebraic number.   A complex number α is called algebraic if it is a zero of some polynomial $P(z) \in \mathbb{Z}[z]$. There is no requirement in the definition that P be irreducible, although of course if α is algebraic, then there is an irreducible $P(z) \in \mathbb{Z}[z]$ such that $P(\alpha) = 0$.

Chebyshev polynomials.   There are two families of the Chebyshev polynomials, usually defined so as to express $\cos n\theta$ and $\sin((n+1)\theta)/\sin \theta$ as polynomials in $\cos \theta$. But we prefer the less common monic versions, and define $T_n(z)$ by $T_n(z + z^{-1}) := z^n + z^{-n}$ and define $U_n(z)$ by $U_n(z + z^{-1}) := (z^n - z^{-n})/(z - z^{-1})$. The zeros of T_n and U_n are all real and lie in the interval $[-2, 2]$.

conjugate.   If α is an algebraic number with minimal polynomial P, then the zeros of P are the conjugates of α. The set of all zeros of P is a conjugate set.

conjugate set.   If $P(z) \in \mathbb{Z}[z]$ is an irreducible polynomial, then its zeros form a conjugate set of algebraic numbers. The elements of a conjugate set are the Galois conjugates of each other.

content.   The content of an integer polynomial is the greatest common divisor of its coefficients.

cyclotomic polynomial.   Let ω_n be a primitive nth root of unity in $\mathbb{C}$. The minimal polynomial of ω_n over $\mathbb{Q}$ is called the nth cyclotomic polynomial. This polynomial is in fact the minimal polynomial of any primitive nth root of unity. When we speak simply of a cyclotomic polynomial, we mean one whose zeros are roots of unity, but we do not require irreducibility. Thus, a cyclotomic polynomial is a product of certain nth cyclotomic polynomials, allowing the possibility of more than one irreducible factor and the possibility of repeated factors.

discriminant.   The discriminant of an algebraic integer α (compare with field discriminant), written $\Delta(\alpha)$, is the square of the determinant of the Vandermonde matrix $(\alpha_j^{i-1})_{i,j=1,\ldots,d}$, where $\alpha_1, \ldots, \alpha_d$ are the conjugates of α.

field discriminant.   The field discriminant of $\mathbb{Q}(\alpha)$, written $\Delta(\mathbb{Q}(\alpha))$, is the square of the determinant of the matrix $(\alpha_i^{(j)})_{i,j=1,\ldots,d}$, where $\alpha^{(1)}, \ldots, \alpha^{(d)}$ is a $\mathbb{Z}$-basis for the ring of integers of $\mathbb{Q}(\alpha)$ and $\alpha_i^{(j)}$ is the image of $\alpha^{(j)}$ under the ith of the d embeddings of $\mathbb{Q}(\alpha)$ into $\mathbb{C}$.

house.   The house of an algebraic number α is the largest modulus of any of the conjugates of α.

Laurent polynomial.   A Laurent polynomial in z is a polynomial in z and z^{-1}. Thus, it has the shape $\sum_{i=m}^n a_i z^i$ for some $m \leq n$, where m is allowed to be negative.

Lehmer's Conjecture.   The conjecture (originally posed as a question rather than a conjecture) that there is some $c > 1$ such that any integer polynomial that has Mahler measure strictly below c has Mahler measure equal to 1.

Lehmer's number.   The Mahler measure of Lehmer's polynomial.

Lehmer's polynomial.   The polynomial $L(z) = z^{10} + z^9 - z^7 - z^6 - z^5 - z^4 - z^3 + z + 1$. Among integer polynomials, this has the smallest known Mahler measure greater than 1, namely $1.17628\cdots$.

length.   The length of a polynomial is the sum of the absolute values of its coefficients.

logarithmic Mahler measure.   The logarithm of the Mahler measure. This is particularly convenient to use for Mahler measures of polynomials in several variables, where the usual Mahler measure is defined to be the exponential of an integral.

Mahler measure.   Let $P(z) = a_0 z^d + a_1 z^{d-1} + \cdots + a_0 \in \mathbb{Z}[z]$ with $a_0 \neq 0$, and let its zeros in $\mathbb{C}$ be $\alpha_1, \ldots, \alpha_d$. If $P(z)$ has repeated zeros, then they are listed with multiplicity. The Mahler measure, $M(P)$, is defined to be the product of $|a_0|$ and all $|\alpha_i|$ for which $|\alpha_i| > 1$:

$$M(P) = |a_0| \prod_{|\alpha_i|>1} |\alpha_i|.$$

One can show that

$$\log M(P) = \int_0^1 \log \left| P(e^{2\pi i t}) \right| \, dt,$$

and this integral version of the definition will be used later to define the Mahler measure of polynomials in more than one variable.

minimum length, minimum-length multiple.   Let $P \in \mathbb{Z}[z]$. Among all integer polynomial multiples of P, say PT where $T \in \mathbb{Z}[z]$, the smallest possible length is called the minimum length of P, and an example PT that achieves that length is a minimum-length multiple of P.

Perron number.   A Perron number is a real and positive algebraic integer β such that $|\beta'| < \beta$ for any conjugate $\beta' \neq \beta$.

Pisot number.   A Pisot number is a real algebraic integer greater than 1, all of whose conjugates except itself lie in the open disc $|z| < 1$.

separation.   The separation of an algebraic integer α is the smallest modulus of the difference between two of its conjugates.

short polynomial.   Let $P(z) \in \mathbb{Z}[z]$. A short polynomial for P is a polynomial of minimum length of the shape $P(z)Q(z)$, where $Q(z)$ is a cyclotomic polynomial.

shortness.   The shortness of a polynomial $P(z) \in Z[z]$ is the length of a short polynomial for P. The shortness of an algebraic integer α is the shortness of its minimal polynomial.

Strong Lehmer Conjecture.   A more precise version of Lehmer's Conjecture, with $c = M(L(z))$.

subharmonic.   A real-valued continuous function of a complex variable, $f(z)$, is subharmonic on an open subset of the complex plane if for every closed disc in that region the function value at the centre is bounded above by the average value on the boundary of the disc: $f(z) \leq \int_0^1 f(z + re^{2\pi it})\, dt$, where z is at the centre of the disc and r is the radius. Extending $f(z)$ to its boundary by continuity (allowing $-\infty$ as a value), the maximum of $f(z)$ is attained on the boundary.

Weil height.   The Weil height of an algebraic number is the logarithm of its absolute Mahler measure.

Chapter 2
Mahler Measures of Polynomials in Several Variables

2.1 Introduction

While the set of Mahler measures of polynomials with real coefficients clearly consists of the whole of the positive real line, the situation for polynomials with integer coefficients is far more interesting. Consider the sequence $\{M(z^n - z - 1)\}_{n \geq 2}$, where

$$\log M(z^n - z - 1) = \int_0^1 \log |e^{2\pi i n t} - e^{2\pi i t} - 1| \, dt \, .$$

As $n \to \infty$, the term $e^{2\pi i n t}$ becomes increasingly uncorrelated with $e^{2\pi i t}$, so that (as we shall see)

$$\lim_{n \to \infty} \log M(z^n - z - 1) = \int_0^1 \int_0^1 \log |e^{2\pi i t_1} - e^{2\pi i t_2} - 1| \, dt_1 \, dt_2 \, .$$

Thus, if we take this integral to be the definition of $\log M(z_1 - z_2 - 1)$, then we have a convergent sequence of one-variable Mahler measures converging to a two-variable Mahler measure. This suggests a more general definition of the Mahler measure. It applies also to Laurent polynomials, so that negative exponents of the variables are allowed.

Let $k \geq 1$, $\mathbf{z}_k = (z_1, \ldots, z_k)$ and $F(\mathbf{z}_k)$ be a nonzero Laurent polynomial with integer coefficients. Then its Mahler measure $M(F)$ is defined as

$$M(F) = \exp \left\{ \int_0^1 \cdots \int_0^1 \log |F(e^{2\pi i t_1}, \ldots, e^{2\pi i t_k})| \, dt_1 \cdots dt_k \right\} . \tag{2.1}$$

© Springer Nature Switzerland AG 2021
J. McKee and C. Smyth, *Around the Unit Circle*, Universitext,
https://doi.org/10.1007/978-3-030-80031-4_2

Our example also suggests considering the set

$$\mathcal{L} := \bigcup_{k=1}^{\infty} \{M(F) \mid F \text{ is a polynomial in } k \text{ variables with integer coefficients}\}.$$

(2.2)

Then, conjecturally, the example above can be considerably extended, as follows.

Conjecture 2.1 (Boyd's Conjecture) The set $\mathcal{L}$ is a closed subset of the real line.

Note that for $M(F_1)$ and $M(F_2)$ in $\mathcal{L}$,

$$M(F_1)M(F_2) = M(F_1 F_2) \in \mathcal{L}, \qquad (2.3)$$

i.e., $\mathcal{L}$ is a multiplicative semigroup (indeed a monoid, since $1 = M(1) \in \mathcal{L}$).

Exercise 2.2 Show that for any $\rho_1, \ldots, \rho_k$ of modulus 1, we have

$$M(F(\rho_1 z_1, \ldots, \rho_k z_k)) = M(F(z_1, \ldots, z_k)).$$

(Thus, in particular, $M(F(\pm z_1, \ldots, \pm z_k)) = M(F(z_1, \ldots, z_k)))$.

We shall see that $\mathcal{L}$ has many naturally defined closed subsets.

But what, conjecturally, does $\mathcal{L}$ look like? We believe that Boyd's Conjecture 2.1 is true. This implies the truth of Lehmer's Conjecture—see Exercise 2.7.

It may be that the set $\mathcal{L}$ has far more structure. Perhaps it is even a Thue set, as defined in Sect. 8.2. The properties of $\mathcal{L}$ described in this chapter are consistent with this possibility, but the evidence so far obtained is not substantial, and we are a long way from formulating a credible conjecture. One result that would be needed is a good estimate for the difference $M\left(F_{\mathbf{r}^{(n)}}(z)\right) - M(F)$ in (2.7) of Proposition 2.17. This estimate needs to good enough to be able to show that $\mathbf{r}^{(n)}$ can be chosen so that $M\left(F_{\mathbf{r}^{(n)}}(z)\right)$ tends to $M(F)$ either from above, from below or both. This was done by Boyd [Boy81b, Appendix 2] for the particular polynomial $F = 1 + z_1 + z_2$, but it seems to be the only such example.

Let $\mathcal{L}_1 \subset \mathcal{L}$ denote the set of Mahler measures $M(P)$, for P a one-variable integer polynomial. Every $F(\mathbf{z}_k) \in \mathcal{L}$ in k variables $\mathbf{z}_k = (z_1, \ldots, z_k)$ specifies a particular subset of $\mathcal{L}_1$ by substituting monomials in the single variable z for each variable z_j. Thus, we define

$$\mathcal{M}_1(F) := \{M(F(z^{r_1}, z^{r_2}, \ldots, z^{r_k})) \mid (r_1, r_2, \ldots, r_k) \in \mathbb{Z}^k\}.$$

The closure $\overline{\mathcal{M}_1(F)}$ in $\mathbb{R}$ of this set can be described explicitly, as follows. Given $\ell \geq 0$ and an $\ell \times k$ integer matrix $A = (a_{ij})$, define the k-tuple $\mathbf{z}_\ell^A$ by

$$\mathbf{z}_\ell^A := (z_1, \ldots, z_\ell)^A := (z_1^{a_{11}} \cdots z_\ell^{a_{\ell 1}}, \ldots, z_1^{a_{1k}} \cdots z_\ell^{a_{\ell k}}) \qquad (2.4)$$

(which is $(1, 1, \ldots, 1) \in \mathbb{Z}^k$ when $\ell = 0$) and $F_A(\mathbf{z}_\ell) = F(\mathbf{z}_\ell^A)$, a polynomial in ℓ variables $z_1, \ldots, z_\ell$. Then $M(F_A)$ is defined by (2.1) with $F = F_A$ and $k = \ell$. Denote by $\mathcal{P}(F)$ the set

$$\mathcal{P}(F) := \{F_A, -F_A : A \in \mathbb{Z}^{\ell \times k}, \ell \geq 0\}, \qquad (2.5)$$

and by $\mathcal{M}(F)$ the set

$$\mathcal{M}(F) := \{M(F_A) : F_A \in \mathcal{P}(F), F_A \neq 0\}.$$

This set is the closure of $\mathcal{M}_1(F)$ in $\mathbb{R}$ (Theorem 2.5; in particular $M(F) \in \overline{\mathcal{M}_1(F)}$).

Exercise 2.3 Suppose that a real polynomial $F(z_1, \ldots, z_k)$ in k variables is self-reciprocal (see Sect. A.1 in the Appendix). Show that all polynomials in $\mathcal{P}(F)$ are self-reciprocal too.

Exercise 2.4 Show that for $F_1(x, y) = x + 1/x + 1 + y + 1/y$, $F_2(x, y) = x + 1/x - 1 + y + 1/y$ and $F_3(x, y) = x + 1/x + 1 - y - 1/y$, we have $M(F_1) = M(F_2) = M(F_3)$ but that none of the sets $\mathcal{P}(F_1)$, $\mathcal{P}(F_2)$ and $\mathcal{P}(F_3)$ are the same.

The proofs of the next two theorems will occupy Sects. 2.2–2.4 below.

Theorem 2.5 *We have*

$$\overline{\mathcal{M}_1(F)} = \mathcal{M}(F).$$

An element of a set of real numbers is said to be **isolated** if it is not a genuine limit point of the set, as defined in Sect. 2.6.2.

Theorem 2.6 *Suppose that $k \geq 1$, that the nonzero Laurent polynomial $F(\mathbf{z}_k)$ has integer coefficients and that $1 \in \mathcal{M}(F)$. Then 1 is an isolated point of $\mathcal{M}(F)$.*

Exercise 2.7 Show that if 1 were a limit point of $\mathcal{L}$, then $\mathcal{L}$ would be dense on the real interval $(1, \infty)$, $\mathcal{L}$ would not be closed and Lehmer's Conjecture would be false.

It is known that $\mathcal{L}$ is a nested union of closed sets. Specifically, define

$$F^{(n)}(\mathbf{z}_{2n}) = z_1 + z_3 + \cdots + z_{2n-1} - (z_2 + z_4 + \cdots + z_{2n}). \qquad (2.6)$$

Proposition 2.8 *The set $\mathcal{L}$ can be written as a nested union*

$$\mathcal{M}(F^{(1)}) \subseteq \mathcal{M}(F^{(2)}) \subseteq \mathcal{M}(F^{(3)}) \subseteq \cdots \subseteq \mathcal{M}(F^{(n)}) \subseteq \cdots = \mathcal{L}.$$

For the proof, see the end of Sect. 2.2. Thus, to prove that $\mathcal{L}$ itself is closed, it would be enough to show that

$$\inf_{n=2}^{\infty}\{M \mid M \in \mathcal{M}(F^{(n)}) \setminus \mathcal{M}(F^{(n-1)})\} = \infty.$$

For then any convergent sequence in $\mathcal{L}$ would lie in the closed set $\mathcal{M}(F^{(n)})$ for some n, so that its limit would again lie in that set which is, by Proposition 2.8, a subset of $\mathcal{L}$.

2.2 Preliminaries for the Proofs of Theorems 2.5 and 2.6

We now present some technical results needed for the proofs of Theorems 2.5 and 2.6. Where a result is not proved, we will either give a reference or leave it as an exercise.

Our first result shows that the Mahler measure is invariant under the action of nonsingular integer matrices.

Proposition 2.9 *For an n-variable polynomial $P(\mathbf{z}) \in \mathbb{R}[\mathbf{z}]$ and a nonsingular $n \times n$ integer matrix V, we have*

$$M(P(\mathbf{z})) = M(P(\mathbf{z}^V)).$$

Here, as in (2.4), $\mathbf{z}^V$ denotes $(\prod_j z_j^{v_{j1}}, \ldots, \prod_j z_j^{v_{jn}})$, where $\mathbf{z} = (z_1, \ldots, z_n)$, $V = (v_{ij})$.

Proof For $\mathbf{z} = (e^{2\pi i\theta_1}, \ldots, e^{2\pi i\theta_n})$, $\boldsymbol{\theta} = (\theta_1, \ldots, \theta_n)$ and $\boldsymbol{\psi} = \boldsymbol{\theta} V$, we have

$$\mathbf{z}^V = (e^{2\pi i\psi_1}, \ldots, e^{2\pi i\psi_n}).$$

Then the map $\boldsymbol{\theta} \mapsto \boldsymbol{\psi}$ from $\mathbb{R}^n/\mathbb{Z}^n$ to itself is a $|\det V|$-fold linear covering of $\mathbb{R}^n/\mathbb{Z}^n$. On the other hand, this map has Jacobian $|\det V|$, so that

$$
\begin{aligned}
\log(M(P(\mathbf{z}^V))) &= \int_{\mathbb{R}^n/\mathbb{Z}^n} \log |P(e^{2\pi i(\boldsymbol{\theta} V)_1}, \ldots, e^{2\pi i(\boldsymbol{\theta} V)_n})| \, d\theta_1 \ldots d\theta_n \\
&= |\det V| \int_{\mathbb{R}^n/\mathbb{Z}^n} \log |P(e^{2\pi i\psi_1}, \ldots, e^{2\pi i\psi_n})| \, \frac{d\psi_1 \ldots d\psi_n}{|\det V|} \\
&= \log(M(P(\mathbf{z}))).
\end{aligned}
$$

$\square$

The next result, [Boy81b, Theorem 1], describes all integer polynomials in several variables having Mahler measure 1. These are the several-variable analogues of cyclotomic polynomials.

Proposition 2.10 *Suppose that $F \in \mathbb{Z}[\mathbf{z}_\ell]$ for some $\ell \in \mathbb{N}$. Then $M(F) = 1$ if and only if F belongs to $\mathcal{P}(S)$ for some polynomial S of the form $S(\mathbf{z}_k) = \pm z_1 C_2(z_2) C_3(z_3) \cdots C_\ell(z_k)$ for some k, where $C_2, \ldots, C_k$ are cyclotomic polynomials.*

Lemma 2.11 *If* $B \in \mathbb{Z}^{\ell' \times \ell}$ *and* $A \in \mathbb{Z}^{\ell \times k}$, *then* $\left(\mathbf{z}_\ell^B\right)^A = \mathbf{z}_{\ell'}^{(BA)}$. *Further, if* $G = F_A \in \mathcal{P}(F)$ *for some polynomial* F, *then* $G_B = F_{BA}$.

Proof The first result is easily checked. For the second result, we have

$$G_B(\mathbf{z}_{\ell'}) = (F_A)_B(\mathbf{z}_{\ell'}) = F_A(\mathbf{z}_{\ell'}^B) = F((\mathbf{z}_{\ell'}^B)^A) = F(\mathbf{z}_{\ell'}^{BA}) = F_{BA}(\mathbf{z}_{\ell'}),$$

as claimed. $\qquad\square$

Proposition 2.12 *For a polynomial* $G(\mathbf{z}_\ell)$ *and nonsingular* $V \in \mathbb{Z}^{\ell \times \ell}$, *we have* $M(G_V) = M(G)$. *Further, for a polynomial* $F(\mathbf{z}_k)$ *and any* $A \in \mathbb{Z}^{\ell \times k}$, *we have* $M(F_{VA}) = M(F_A)$.

Proof Firstly $M(G_V) = M(G)$ follows from Proposition 2.9 and the definition of G_V. Next, using also Lemma 2.11, we have

$$M(F_{VA}(\mathbf{z}_\ell)) = M(F(\mathbf{z}_\ell^{VA})) = M(F((\mathbf{z}_\ell)^V)^A)) = M(F_A(\mathbf{z}_\ell^V)) = M(F_A(\mathbf{z}_\ell)).$$

$\qquad\square$

Lemma 2.13 *If* $G \in \mathcal{P}(F)$, *then* $\mathcal{P}(G) \subseteq \mathcal{P}(F)$ *and* $\mathcal{M}(G) \subseteq \mathcal{M}(F)$.

Proof Suppose that $F = F(\mathbf{z}_k)$ and $G = G(\mathbf{z}_\ell) \in \mathcal{P}(F)$. Then $G = F_A$ for some $A \in \mathbb{Z}^{\ell \times k}$, and for any $B \in \mathbb{Z}^{\ell' \times \ell}$ with $0 \leq \ell' \leq \ell$, we have $G_B = F_{BA}$ by Lemma 2.11. Note that $BA \in \mathbb{Z}^{\ell' \times k}$ with $0 \leq \ell' \leq \ell \leq k$. This proves the first assertion, from which the second assertion follows immediately. $\qquad\square$

Lemma 2.14 *For any two multivariable Laurent polynomials* F *and* G, *we have that* $\mathcal{P}(FG) \subseteq \mathcal{P}(F)\mathcal{P}(G)$ *and* $\mathcal{M}(FG) \subseteq \mathcal{M}(F)\mathcal{M}(G)$.

Proof For F, G polynomials in $\mathbf{z}_k$ and $A \in \mathbb{Z}^{\ell \times k}$ and some ℓ with $0 \leq \ell \leq k$, we have

$$(FG)_A = F_A G_A \in \mathcal{P}(F)\mathcal{P}(G),$$

and hence, $M((FG)_A) = M(F_A)M(G_A) \in \mathcal{M}(F)\mathcal{M}(G)$. $\qquad\square$

This immediately implies the following.

Corollary 2.15 *If* $\mathcal{M}(G) = \{1\}$ *(see Proposition 2.10), then* $\mathcal{M}(FG) = \mathcal{M}(F)$.

Next, given $\ell \geq 2$ and $\mathbf{r} = (r_1, \ldots, r_\ell) \in \mathbb{Z}^\ell$, define, following Boyd [Boy81a, Boy81b]

$$q(\mathbf{r}) := \min_{\substack{0 \neq \mathbf{s} = (s_1, \ldots, s_\ell) \in \mathbb{Z}^\ell \\ \mathbf{r} \cdot \mathbf{s} = 0}} \max_{i=1}^\ell |s_i|.$$

The function q measures, in some sense, how different in magnitude the r_i are.

Exercise 2.16 Let $n \in \mathbb{N}$ and $\mathbf{r}_n = (1, n, n^2, \ldots, n^{\ell-1})$. Show that $q(\mathbf{r}_n) = n$ (and so goes to ∞ as $n \to \infty$).

For a Laurent polynomial $F(z_1, \ldots, z_k)$ and $\mathbf{r} = (r_1, \ldots, r_k) \in \mathbb{Z}^k$ note that, from (2.4), $F_{\mathbf{r}}(z)$ denotes the Laurent polynomial $F(z^{r_1}, \ldots, z^{r_k})$.

The next result was first conjectured by Boyd [Boy81a], who also proved in [Boy81b] some partial results in the direction of his conjecture, including essentially the result for $k = 2$.

Proposition 2.17 *Let* $F(z_1, \ldots, z_k)$ *be a Laurent polynomial with complex coefficients, and suppose that* $\mathbf{r}^{(1)}, \mathbf{r}^{(2)}, \ldots, \mathbf{r}^{(n)}, \ldots$ *is a sequence of vectors in* $\mathbb{Z}^k$ *with* $q(\mathbf{r}^{(n)}) \to \infty$ *as* $n \to \infty$. *Then*

$$\lim_{n \to \infty} M\left(F_{\mathbf{r}^{(n)}}(z)\right) = M(F). \tag{2.7}$$

The following two examples show that the limit result (2.7) does not guarantee that $M(F)$ is a genuine limit point of $\mathcal{L}$, as defined in Sect. 2.6.2.

Exercise 2.18 Let $S(\mathbf{z}_k) = \pm z_1 C_2(z_2) C_3(z_3) \cdots C_k(z_k)$ for some k, where C_2, ..., C_k are cyclotomic polynomials, as in Proposition 2.10. Show that for any $\mathbf{r} = (r_1, r_2, \ldots, r_k)$ in $\mathbb{Z}^k$ with all $r_i \neq 0$, the measure $M(S_{\mathbf{r}}(z))$ equals 1.

A *Pisot number* is a real algebraic integer greater than 1, all of whose conjugates except itself lie in the open disc $|z| < 1$.

Exercise 2.19 Let $F(z_1, z_2) := z_1 + z_2 - 2$ and suppose that the vectors $\mathbf{r}^{(n)} \in \mathbb{Z}^2$ satisfy the condition $q(\mathbf{r}^{(n)}) \to \infty$ as $n \to \infty$.

(a) Show that $M(F) = 2$ and that if $\mathbf{r}^{(n)} \in \mathbb{Z}^2$ has positive components for all n, then $M(F_{\mathbf{r}^{(n)}}(z)) = 2$ for all n.
(b) Show, with the help of Rouché's Theorem (Theorem C.1), that for $\mathbf{r}^{(n)} = (1, -n)$ the sequence $\{M(F_{\mathbf{r}^{(n)}}(z))\}_{n \in \mathbb{N}}$ is strictly increasing, with limit 2.
(c) Show that $M(F_{\mathbf{r}^{(n)}}(z))$ from part (b) is a Pisot number.

Exercise 2.20 Show that for any $\ell \times k$ integer matrix $A = (a_{ij})$ and $M(F_A)$ in $\mathcal{M}(F)$, there is a sequence $\{\mathbf{r}^{(n)}\}_{n \in \mathbb{N}}$ of vectors in $\mathbb{Z}^k$ such that $\lim_{n \to \infty} M(F_{\mathbf{r}^{(n)}}) = M(F_A)$.

Proposition 2.21 ([Smy81a, Corollary 2])   *For a nonzero Laurent polynomial* $F(\mathbf{z}_k) = \sum_{\mathbf{j} \in J} c(\mathbf{j}) \mathbf{z}_k^{\mathbf{j}} \in \mathbb{C}[z_1, \ldots, z_k]$, *where* $J \subseteq \mathbb{Z}^k$, *let the polytope* $\mathcal{C}(F) \in \mathbb{R}^k$ *be the convex hull of those* $\mathbf{j} \in J$ *with* $c(\mathbf{j}) \neq 0$. *Then*

$$\max_{\substack{\mathbf{j} \text{ an extreme point of } \mathcal{C}(F)}} |c(\mathbf{j})| \leq M(F) \leq \sum_{\mathbf{j} \in J} |c(\mathbf{j})|.$$

In particular, $M(F) \geq 1$ *when* F *has integer coefficients.*

Here J is a set of column vectors, so that $\mathbf{z}_k^{\mathbf{j}}$, defined by (2.4), is a monomial. The polytope $\mathcal{C}(F)$ is called the *exponent polytope of* F ([Boy81b, p. 460]). Let $\dim(F)$ denote its dimension, the *dimension of* F, which is clearly at most k. Given a

Mahler measure, i.e., a value $M = M(P)$ for some nonzero integer polynomial P in k variables, we can define its **dimension** $\dim(M)$ as the least k for which $M = M(P)$ for such a P in k variables. For instance, all positive integers m have $\dim(M) = 0$, while all values $M = M(P)$ that are not positive integers and for which P is a one-variable integer polynomial have dimension 1.

Exercise 2.22 Let M be the Mahler measure of a k-variable polynomial P. Show that

$$\dim(M) \le \dim(P) \le k,$$

and give examples to show that either inequality may be strict.

Before moving to the proofs of the main theorems, we note that, armed with these preliminaries, we can also now prove Proposition 2.8.

Proof (of Proposition 2.8) For a given $F(\mathbf{z}_k) = \sum_{\mathbf{j} \in J} c(\mathbf{j}) \mathbf{z}_k^{\mathbf{j}} \in \mathbb{Z}[z_1, \ldots, z_k]$, we know that $\mathcal{M}((z_1 - 1)F(\mathbf{z}_k)) = \mathcal{M}(F(\mathbf{z}_k))$, by Corollary 2.15. So, replacing F by $(z_1 - 1)F$, if necessary, we can assume that $\sum_{\mathbf{j} \in J} c(\mathbf{j}) = F(1, \ldots, 1) = 0$. Choose

$$n := \sum_{\mathbf{j} \in J \text{ with } c(\mathbf{j}) > 0} c(\mathbf{j}),$$

and let $F^{(n)}$ be as in (2.6). Then, for each $\mathbf{j}$ with $c(\mathbf{j}) > 0$, replace $c(\mathbf{j})$ of the z_{2i-1} in (2.6) by $\mathbf{z}_k^{\mathbf{j}}$, and for each $\mathbf{j}$ with $c(\mathbf{j}) < 0$, replace $(-c(\mathbf{j}))$ of the z_{2i} in (2.6) by $\mathbf{z}_k^{\mathbf{j}}$. This gives us the polynomial F in the form $F_A^{(n)}$, where A is a matrix such that for each $\mathbf{j} \in J$, the matrix A has $|c(\mathbf{j})|$ of its columns equal to $\mathbf{j}$. Hence, $M(F) \in \mathcal{M}(F^{(n)})$ for this value of n.

To show that the sequence of sets $\{\mathcal{M}(F^{(n)})\}_{n \in \mathbb{N}}$ are nested, it is enough to observe that

$$F^{(n-1)}(\mathbf{z}_{2n-2}) = F^{(n)}(z_1, z_2, \ldots, z_{2n-3}, z_{2n-2}, z_{2n-2}, z_{2n-2}),$$

so that $F^{(n-1)}$ is of the form $F_A^{(n)}$ for some (easily written down) matrix A, and hence that $F^{(n-1)} \in \mathcal{P}(F^{(n)})$. Applying Lemma 2.13, we see that $\mathcal{M}(F^{(n-1)}) \subseteq \mathcal{M}(F^{(n)})$, as claimed. □

2.3　Proof of Theorem 2.5

Proof We first show that every $M(F_A)$ lies in $\overline{\mathcal{M}_1(F)}$. Then we show, for any $\mathbf{r}^{(1)}, \mathbf{r}^{(2)}, \ldots, \mathbf{r}^{(n)}, \ldots$ in $\mathbb{Z}^k$ with $M(F_{\mathbf{r}^{(n)}})$ converging, that its limit is of the form $M(F_A)$ for some A.

So, first, take any $A \in \mathbb{Z}^{\ell \times k}$, and let $\mathbf{r}^{(n)} = (1, n, n^2, \ldots, n^{\ell-1})$ as in Exercise 2.16. Because $q(\mathbf{r}^{(n)}) \to \infty$ as $n \to \infty$, we can apply Proposition 2.17 to F_A to obtain

$$\lim_{n \to \infty} M\left(F_A(\mathbf{r}^{(n)})\right) = M(F_A).$$

Now, for $\mathbf{r} = \mathbf{r}^{(n)}$, we have $F_A(\mathbf{r}) = F_{\mathbf{r}^A}$, so that $M(F_A(\mathbf{r})) = M(F_{\mathbf{r}^A})$. Hence, for the sequence $\{\mathbf{r}^A = (\mathbf{r}^{(n)})^A\}_{n \in \mathbb{N}} \in \mathbb{Z}^k$, we have

$$\lim_{n \to \infty} M(F_{\mathbf{r}^A}) = M(F_A).$$

Hence, $\mathcal{M}(F) = \{M(F_A) : A \in \mathbb{Z}^{\ell \times k}\} \subseteq \overline{\mathcal{M}_1(F)}$.

To prove that these are the only limit points of $\mathcal{M}_1(F)$, we take any sequence in $\mathbb{Z}^k$

$$\mathbf{r}^{(1)}, \ldots, \mathbf{r}^{(n)}, \ldots$$

for which $M(F_{\mathbf{r}^{(n)}})$ converges. We separate the proof into three cases, doing the trivial case $k = 1$ first and then, for $k \geq 2$, separating the cases where the sequence $\{q(\mathbf{r}^{(n)})\}_{n \in \mathbb{N}}$ is either unbounded or bounded.

Case 1: k = 1. Here $F = F(z_1)$ and our sequence is $\{M(F(z^{r_1}))\}_{n \in \mathbb{N}}$, for some sequence of nonzero integers $\{r_1 = r_1^{(n)}\}_{n \in \mathbb{N}}$. But, applying Proposition 2.9 with $n = 1$ and $V = (r_1)$, we have that the sequence $\{M(F(z^{r_1}))\}_{n \in \mathbb{N}}$ is constant, each term being $M(F(z))$.

Case 2: k ≥ 2 and q(r⁽ⁿ⁾) unbounded. Then there is a subsequence of the $\mathbf{r}^{(n)}$ for which $\lim_{n \to \infty} q(\mathbf{r}^{(n)})$ tends to infinity on that subsequence. Thus, by replacing the sequence of the $\mathbf{r}^{(n)}$ by that subsequence, we can assume that, as $n \to \infty$, both $M(F_{\mathbf{r}^{(n)}})$ converges and $q(\mathbf{r}^{(n)}) \to \infty$. Then we can apply Proposition 2.17 to conclude that $\lim_{n \to \infty} M(F_{\mathbf{r}^{(n)}}) = M(F)$.

Case 3: k ≥ 2 and q(r⁽ⁿ⁾) bounded. Our proof is by induction. From Case 1, we already know that the result is true for $k = 1$. We now assume $k \geq 2$ and that the result is true for all Laurent polynomials F in fewer than k variables.

Take a convergent sequence of real numbers $\{M(F_{\mathbf{r}})\}_{n \in \mathbb{N}}$ for $\mathbf{r} = \mathbf{r}^{(n)}$ ($n = 1, 2, 3, \ldots$) such that the integer sequence $\{q(\mathbf{r})\}_{n \in \mathbb{N}}$ is bounded. Then there are only finitely many possibilities for the nonzero vectors $\mathbf{s} \in \mathbb{Z}^k$ in the definition of q such that $\mathbf{s} \cdot \mathbf{r} = 0$. Hence, by the Pigeonhole Principle, we can find an infinite subsequence of integers n for which the corresponding sequence of vectors $\mathbf{s}$ is constant. On replacing our original sequence $n = 1, 2, 3, \ldots$ by this subsequence, we can assume that *all* $\mathbf{r}$ satisfy $\mathbf{r} \cdot \mathbf{s} = 0$.

Next, take a $(k - 1) \times k$ integer matrix U whose rows are a basis of the sublattice $L_{\mathbf{s}} := \{\mathbf{r} \in \mathbb{Z}^k \mid \mathbf{r} \cdot \mathbf{s} = 0\}$ of $\mathbb{Z}^k$. Then each $\mathbf{r} \in L_{\mathbf{s}}$ can be written as $\mathbf{c}U$ for some $\mathbf{c} \in \mathbb{Z}^{k-1}$. Then writing $G(\mathbf{z}_{k-1}) := F_U(\mathbf{z}_{k-1})$, a Laurent polynomial in at most $k - 1$ variables, we have from Lemma 2.11 that $G_{\mathbf{c}} = F_{\mathbf{c}U} = F_{\mathbf{r}}$. Hence, applying the induction hypothesis to G, or Case 2 if $k - 1 \geq 2$ and the sequence $\{q(\mathbf{c})\}_{n \in \mathbb{N}}$ is unbounded, we see that the sequence $\{M(F_{\mathbf{r}}(z))\}_{n \in \mathbb{N}} = \{M(G_{\mathbf{c}}(z))\}_{n \in \mathbb{N}}$ has a limit of the form $M(G_B)$ for some $B \in \mathbb{Z}^{\ell \times (k-1)}$ and some $\ell \le k - 1$.

Next, we note that, by Lemma 2.11 again, $G_B = F_A$, where $A = BU \in \mathbb{Z}^{\ell \times k}$. Hence, $\overline{\mathcal{M}_1(F)} \subseteq \{M(F_A) : A \in \mathbb{Z}^{\ell \times k}\} = \mathcal{M}(F)$, and so $\overline{\mathcal{M}_1(F)} = \mathcal{M}(F)$, as claimed. $\qquad\square$

2.4 Proof of Theorem 2.6

Proof Suppose that $1 \in \mathcal{M}(F)$, but that it is not isolated. Then, because this set is the closure of the set of measures $M(F_{\mathbf{r}})$ of polynomials $F_{\mathbf{r}}$ for $\mathbf{r} \in \mathbb{Z}^k$, we can take a sequence of such polynomials $\{F_{\mathbf{r}^{(n)}}\}_{n \in \mathbb{N}}$ such that none of the $M(F_{\mathbf{r}^{(n)}})$ are 1, but $\lim_{n \to \infty} M(F_{\mathbf{r}^{(n)}}) = 1$. However, by Proposition 2.17, this limit is $M(F)$, which is, therefore, 1.

As in the proof of Theorem 2.5, we now separate three cases.

Case 1: k = 1. Here $F = F(z_1)$ and our sequence is $\{M(F(z^{r_1}))\}_{n \in \mathbb{N}}$, for some sequence of nonzero integers $\{r_1 = r_1^{(n)}\}_{n \in \mathbb{N}}$. But, as in Case 1 of the proof of Theorem 2.5, we have that the sequence $\{M(F(z^{r_1}))\}_{n \in \mathbb{N}}$ is constant, each term being $M(F)$. Hence, $M(F) = 1$, and so all terms of the converging sequence are 1, contrary to our assumption.

Case 2: k $\ge$ 2 and q($\mathbf{r}^{(n)}$) unbounded. Then, as in the proof of Case 2 of Theorem 2.5, there is a subsequence of the $\mathbf{r}^{(n)}$ for which $\lim_{n \to \infty} q(\mathbf{r}^{(n)})$ tends to infinity on that subsequence. Thus, by replacing the sequence of the $\mathbf{r}^{(n)}$ by that subsequence, we can assume that, as $n \to \infty$, both $M(F_{\mathbf{r}^{(n)}}) \to 1$ and $q(\mathbf{r}^{(n)}) \to \infty$.

By Proposition 2.10, F is of the form $\pm z$ times a product of cyclotomic polynomials $C(z)$, where each occurrence of the variable z is replaced by a (possibly different for each occurrence) monomial in $z_1, \ldots, z_k$. Hence, each $F_{\mathbf{r}}$ is of the form $\pm z$ times a product of cyclotomic polynomials $C(z)$, where each occurrence of the variable z is replaced by a (possibly different for each occurrence) power of z, assumed to be nonzero. So $M(F_{\mathbf{r}^{(n)}}) = 1$, contradicting the fact that these values are all assumed to be not equal to 1.

Case 3: k $\ge$ 2 and q($\mathbf{r}^{(n)}$) bounded. Here, we follow quite closely the induction argument in Case 3 of the proof of Theorem 2.5. Thus, the result is true for $k = 1$ by Case 1, so we assume that $k \ge 2$ and that the result is true for all F in fewer than k variables. Following that argument, we get that our sequence $\{M(F_{\mathbf{r}^{(n)}})\}_{n \in \mathbb{N}}$ has limit $M(F_{A'})$, where $A' \in \mathbb{Z}^{\ell \times k}$ for some $\ell \le k - 1$. Thus, $M(F_{A'}) = 1$, and so, again by Proposition 2.10, F is of the form $\pm z$ times a product of cyclotomic polynomials $C(z)$, where each occurrence of the variable z is replaced by a (possibly different for each occurrence) monomial in $z_1, \ldots, z_k$. From the definition of $F_{A'}$, we then see that F itself has the same property. So, as in Case 2, we conclude that $M(F_{\mathbf{r}^{(n)}}) = 1$ for all n, giving the same contradiction again. $\qquad\square$

Anticipating Chap. 8, we note that the limited computational evidence we have is consistent with the possibility that $\mathcal{L}$ may have a structure similar to that of the set of all Pisot numbers S, so that, like S, it would be a Thue set, as defined in Sect. 8.2. Thus, the sequence of the smallest elements of ℓ will increase towards the first limit point. We can label these points $\ell_{00} < \ell_{01} < \ell_{02} < \cdots < \ell_{0n} < \ldots$, with limit $\ell_{1,0} = 1.255 \cdots$, conjecturally the smallest element of the derived set $\mathcal{L}^{(1)}$ of $\mathcal{L}$. Thus, Table D.1 conjecturally shows ℓ_{0n} for $0 \leq n < 236$.

2.5   Computation of Two-Dimensional Mahler Measures

Our first example shows how to compute the Mahler measure of a polynomial in two variables that is linear in one of its variables, say y.

Exercise 2.23 For a polynomial $P(x, y) = Q(x) + R(x)y$, show that

$$m(P) = \int_0^1 \log \max(|Q(e^{2\pi it})|, |R(e^{2\pi it})|)\, dt \, .$$

This formula could also be used to compute any two-dimensional Mahler measure of a polynomial that can be transformed using Proposition 2.9 into $M(P(x, y))$ that is linear in one variable.

For polynomials $P(x, y)$ that are quadratic in y, the quadratic formula can be used to factorise them, and so produce an explicit formula for $M(P(x, y))$. The following two exercises use this method.

Exercise 2.24 Take the third polynomial $Q_3(x, y)$ in Table D.2 of Appendix D:

$$Q_3(x, y) = y^{-1}x^{-2} + y^{-1}x^{-1} + x^{-2} - 1 + x^2 + yx + yx^2 \, .$$

Show that

$$M(Q_3) = \exp\left(\frac{2}{\pi}\int_0^{\pi/2} \log\left(\sqrt{1+g(t)} + \sqrt{g(t)}\right) dt\right) = 1.30909838\ldots\, ,$$

where $f(t) = \dfrac{2\cos(4t) - 1}{2\cos(t)}$ and $g(t) = \max(0, f(t)^2 - 1)$.

Exercise 2.25 For positive integers a, b, define

$$P_{ab}(z) := (z^a - z^{-a})y^{-1} + (z^b - z^{-b}) + (z^a - z^{-a})y \, ,$$

$$f(t) = \frac{\sin(bt)}{2\sin(at)}, \qquad g(t) = \max(0, f(t)^2 - 1) \, .$$

Prove that

$$\sqrt{1 + g(t)} = \begin{cases} 1 & \text{if } g(t) = 0; \\ |f(t)| & \text{if } g(t) > 0 \end{cases}$$

and

$$M(P_{ab}) = \exp\left(\frac{1}{\pi}\int_0^\pi \log\left(\sqrt{1 + g(t)} + \sqrt{g(t)}\right) dt\right).$$

This technique could be extended, albeit with increasing complication, to polynomials that are cubic or quartic in y, using the formulae for solutions of cubics and quartics by radicals. But the following general method is more practical.

Take a two-variable polynomial $P(x, y)$, and factorise it in the form

$$a_0(y) \prod_j (x - \alpha_j(y)).$$

For y on the unit circle, define a function

$$M_P(y) := |a_0(y)| \prod_j \max(1, |\alpha_j(y)|).$$

Then compute $M(P(x, y))$ by

$$M(P(x, y)) := \exp\left(\int_0^1 \log(M_P(e^{2\pi i t})) dt\right).$$

This is how the Mahler measures in Table D.2 were calculated.

This method, too, can clearly be extended to polynomials in three or more variables. But then multiple integration would be required.

Problem 2.26 Compute $M(J)$, where $J(x, y, z)$ is given by (2.8) below.

2.6 Small Limit Points of $\mathcal{L}$?

2.6.1 Shortness Conjectures Implying Lehmer's Conjecture and Structural Results for $\mathcal{L}$

Suppose that all Mahler measures, less than 1.25, of algebraic integers have bounded shortness. Table D.1 suggests that this may be true, even with a shortness bound 10. Then since all such Mahler measures belong to the closed set

$$\mathcal{M}(z_1 + \cdots + z_{10} - z_{11} - \cdots - z_{20}),$$

Lehmer's Conjecture would be true, with $M(z^{12} - z^7 - z^6 - z^5 + 1) = 1.17628082$ being the smallest element of $\mathcal{L}$.

Conversely, we have the following.

Proposition 2.27 *Suppose that Lehmer's Conjecture is false, so that there is a sequence of integer polynomials* $\{P_n(z)\}_{n \in \mathbb{N}}$ *such that all the* $M(P_n)$ *are distinct and* $\lim_{n \to \infty} M(P_n) = 1$. *Then the shortness* $\mathrm{sh}(P_n)$ *of* P_n *tends to infinity as* $n \to \infty$.

Proof Suppose that we have a sequence of polynomials $\{P_n\}_{n \in \mathbb{N}}$ as in the statement of the Proposition, but that for some integer bound B, the sequence $\{\mathrm{sh}(P_n)\}_{n \in \mathbb{N}}$ has an infinite subsequence, $\{P_{n_k}\}_{k \in \mathbb{N}}$ say, with $\mathrm{sh}(P_{n_k}) \leq B$ ($k \in \mathbb{N}$). Since Mahler measures in this subsequence are all distinct and tend to 1, 1 is a genuine limit point of this subsequence. However, by Proposition 2.8, all the polynomials P_{n_k} belong to the set $\mathcal{P}(F_B)$, where

$$F_B(z_1 + z_2 + \cdots + z_{2B}) := z_1 + z_3 \cdots + z_{2B-1} - (z_2 + z_4 \cdots + z_{2B}).$$

By Theorem 2.6, the corresponding set of measures $\mathcal{M}(F_B)$ does not have 1 as a (genuine) limit point, giving a contradiction. Thus, there is no such B. $\square$

In Table D.1 in Appendix D, all known integer noncyclotomic one-variable polynomial Mahler measures less than 1.25 are shown, with a polynomial of that Mahler measure. They are all self-reciprocal, as indeed they must be, by Theorem 12.1.

We say that two polynomials are **similar** ($\sim$) if their quotient is $\pm$ a monomial in their variables. Similar polynomials have the same Mahler measure.

We observe that all small Mahler measures in Table D.1 can be produced from a tiny number of two-variable or three-variable polynomials via transformations described in this chapter. First, take the third polynomial $Q_3(x, y)$ in Table D.2 in Appendix D:

$$Q_3(x, y) \sim 1 + x + (1 - x^2 + x^4)y + x^3(1 + x)y^2.$$

Note that for the three heptanomials in Table D.1, we have $P_{41}(z) \sim Q_3(z, -z^{12})$, $P_{208}(z) \sim Q_3(-z^7, -z^4)$ and $P_{47}(z) \sim Q_3(-z^7, z^8)$.

Next, put

$$H(x, y, z) = x + \frac{1}{x} + \frac{x}{y} + \frac{y}{x} + xy + \frac{1}{xy} + z + \frac{1}{z}.$$

We have $M(H) = 1.47513258$ (see Exercise 12.6). Then the three octanomials are $P_{125}(z) \sim H(-z^{10}, -z^4, -z^{11})$, $P_{39}(z) \sim H(-z^{11}, -z^5, -z^{13})$ and $P_{68}(z) \sim H(-z^{10}, -z^7, -z^{14})$.

Finally, putting

$$J(x, y, z) = x + \frac{1}{x} + y + \frac{1}{y} + z + \frac{1}{z} + xyz + \frac{1}{xyz} - x^2 y^3 - \frac{1}{x^2 y^3}, \quad (2.8)$$

we have for the only decanomial P_{64} in Table D.1 that $P_{64}(z) \sim J(-z^{23}, -z^{13}, z^{45})$.

Exercise 2.28 We know from Exercise 2.4 that, for $F_1(x, y) = x + 1/x + 1 + y + 1/y$, $F_2(x, y) = x + 1/x - 1 + y + 1/y$ and $F_3(x, y) = x + 1/x + 1 - y - 1/y$, none of the sets $\mathcal{P}(F_1)$, $\mathcal{P}(F_2)$ and $\mathcal{P}(F_3)$ are the same. Show, however, that all of the 14 Mahler measures of pentanomials in Table D.1 belong to one of $\mathcal{M}(F_1)$, $\mathcal{M}(F_2)$ and $\mathcal{M}(F_3)$.

Exercise 2.29 Show that $M(x + x^{-1} + y + y^{-1} + z + z^{-1}) = M(x - x^{-1} + y - y^{-1} + z - z^{-1})$.

Exercise 2.30 Show that all of the hexanomials in Table D.1 lie in one of the three sets $\mathcal{P}(x + x^{-1} + y + y^{-1} + z + z^{-1})$, $\mathcal{P}(x + x^{-1} + y + y^{-1} - (z + z^{-1}))$ and $\mathcal{P}(x - x^{-1} + y - y^{-1} + z - z^{-1})$.

2.6.2 Small Elements of the Set of Two-Variable Mahler Measures

The *derived set* $S^{(1)}$ of a set $S^{(0)}$ of real numbers is defined to be the set of its genuine limit points. A number $s \in S^{(0)}$ is a *genuine limit point* if and only if there is an infinite sequence of distinct elements of $S^{(0)}$ that converges to s. For $k \geq 1$, the kth *derived set* $S^{(k)}$ *of* $S^{(0)}$ is then the derived set of $S^{(k-1)}$. After Proposition 2.17, it is plausible that Mahler measures of irreducible integer polynomials of dimension $k > 1$ are elements of $\mathcal{L}^{(k-1)}$. In some special cases, this is known, but not generally. In particular, we expect that Mahler measures of irreducible integer polynomials of dimension 2 are in the derived set $\mathcal{L}^{(1)}$.

Table D.2 shows the 61 smallest known Mahler measures of irreducible dimension-2 polynomials, which all lie in (1, 1.37). They are all self-reciprocal, as defined in Sect. A.1 in the Appendix. If these are indeed the smallest such measures, they would be labelled $\ell_{1,0} < \dots < \ell_{1,60}$. Very plausibly these are all elements of $\mathcal{L}^{(1)}$. There is a small element that (with the same caveats) is plausibly in the second derived set $\mathcal{L}^{(2)}$ of $\mathcal{L}$, possibly the smallest:

$$M(x + \frac{1}{x} + y + \frac{1}{y} + z + \frac{1}{z}) = 1.38135 \cdots, \quad (2.9)$$

which we could label $\ell_{2,0}$.

Exercise 2.31 Show that $M(x + \frac{1}{x} + y + \frac{1}{y} + z + \frac{1}{z}) = M(1 + x + y)$.

Exercise 2.32 Show that $M(x^3 + x^{-3} + y + y^{-1} + yx^2 + y^{-1}x^{-2}) = M(x + x^{-1} + 1 + y + y^{-1})$.

Problem 2.33 (open problem)

Are all the elements of Table D.2 genuine limit points of the set $\mathcal{L}_1$ of one-variable Mahler measures? More generally, are all Mahler measures of dimension k in the $(k-1)$th derived set of $\mathcal{L}_1$?

2.7   Closed Forms for Mahler Measures of Polynomials of Dimension at Least 2

In this section, it is convenient to use the logarithmic Mahler measure $m(P) := \log(M(P))$ of a polynomial P.

The first nontrivial closed form evaluation of a two-dimensional Mahler measure was for the polynomial $x + y + 1$ ([Smy81c]). We first present a lemma from which $m(x + y + 1)$, and other similar results, can be readily deduced.

Lemma 2.34 *If* $0 \le \theta < \pi/2$, *then*

$$m(x + y + 2\cos\theta) = \left(1 - \frac{2\theta}{\pi}\right)\log(2\cos\theta) + \frac{1}{\pi}\sum_{j=1}^{\infty}\frac{(-1)^{j-1}}{j^2}\sin(2j\theta).$$

$$(2.10)$$

Proof First, we apply Proposition 2.9 and replace x, y by xy^{-1}, $x^{-1}y^{-1}$:

$$m_\theta := m(x + y + 2\cos\theta) = m(xy^{-1} + x^{-1}y^{-1} + 2\cos\theta)$$

$$= \frac{1}{(2\pi)^2}\int_0^{2\pi}\int_0^{2\pi}\log\left|(2\cos\theta)(e^{-is})\left(e^{is} + \frac{e^{it} + e^{-it}}{2\cos\theta}\right)\right|\,ds\,dt$$

$$= \log(2\cos\theta) + m\left(y + \frac{x + x^{-1}}{2\cos\theta}\right).$$

Next, we apply Exercise 2.23 to get

$$m_\theta = \log(2\cos\theta) + \frac{1}{2\pi}\int_0^{2\pi}\log_+\left|\frac{\cos t}{\cos\theta}\right|\,dt$$

$$= \log(2\cos\theta) + \frac{1}{2\pi}\int_{[0,\theta]\cup[\pi-\theta,\pi+\theta]\cup[2\pi-\theta,2\pi]}\log\left|\frac{2\cos t}{2\cos\theta}\right|\,dt$$

$$= \left(1-\frac{2\theta}{\pi}\right)\log(2\cos\theta) + \frac{1}{2\pi}\int_{[0,\theta]\cup[\pi-\theta,\pi+\theta]\cup[2\pi-\theta,2\pi]}\log\left|e^{it}+e^{-it}\right|\,dt$$

$$= \left(1-\frac{2\theta}{\pi}\right)\log(2\cos\theta) + \frac{1}{2\pi}\int_{[0,\theta]\cup[\pi-\theta,\pi+\theta]\cup[2\pi-\theta,2\pi]}\log\left|1+e^{2it}\right|\,dt$$

$$= \left(1-\frac{2\theta}{\pi}\right)\log(2\cos\theta) + \frac{1}{2\pi}\int_{[0,\theta]\cup[\pi-\theta,\pi+\theta]\cup[2\pi-\theta,2\pi]}\sum_{j=1}^{\infty}\frac{(-1)^{j-1}}{j}e^{2ijt}\,dt$$

$$= \left(1-\frac{2\theta}{\pi}\right)\log(2\cos\theta)$$

$$+ \frac{1}{2\pi}\sum_{j=1}^{\infty}\frac{(-1)^{j-1}}{2ij^2}\left\{e^{2ij\theta}-1+e^{2ij\theta}-e^{-2ij\theta}+1-e^{-2ij\theta}\right\}$$

$$= \left(1-\frac{2\theta}{\pi}\right)\log(2\cos\theta) + \frac{1}{\pi}\sum_{j=1}^{\infty}\frac{(-1)^{j-1}}{j^2}\sin(2j\theta).$$

$\square$

Exercise 2.35 Show that for any $a, b, c \in \mathbb{C}$, we have

$$m(ax+by+c) = m(|a|x+|b|y+|c|) = m(a+y(bx+c)) = m(ax+cy+b).$$

We next give a formula for a general two-dimensional polynomial that is linear in one of the variables. We need the classical dilogarithm function defined for $|z| \le 1$ by

$$\mathrm{Li}_2(z) := \sum_{n=1}^{\infty}\frac{z^n}{n^2}. \tag{2.11}$$

Theorem 2.36 *Let* $P(z) = v\prod_j(z-\alpha_j)$, $Q(z) = v'\prod_k(z-\beta_k) \in \mathbb{C}[z]$ *and*

$$0 \le \theta_1 < \theta_2 < \cdots < \theta_r < 2\pi$$

be the roots θ *of odd multiplicity in* $[0, 2\pi)$ *of the equation* $|P(e^{i\theta})| = |Q(e^{i\theta})|$. *Put* $r = 0$ *if there are no such roots. Also, let* λ *be the proportion of the unit circle* $|z| = 1$ *on which* $|P(z)| \ge |Q(z)|$. *Assume that if* $r > 0$, *then*

$$|P(e^{i\theta})| \ge |Q(e^{i\theta})| \text{ for } \theta_1 \le \theta \le \theta_2. \tag{2.12}$$

Then

$$m(P(z)+yQ(z)) = \lambda m(P) + (1 - \lambda)m(Q)$$

$$+ \frac{1}{2\pi} \operatorname{Im} \sum_{\ell=1}^{r} (-1)^{\ell} \left(\sum_{j:|\alpha_j| \leq 1} \operatorname{Li}_2\left(\alpha_j e^{-i\theta_\ell}\right) - \sum_{j:|\alpha_j| > 1} \operatorname{Li}_2\left(\alpha_j^{-1} e^{i\theta_\ell}\right) \right.$$

$$\left. - \sum_{k:|\beta_k| \leq 1} \operatorname{Li}_2\left(\beta_k e^{-i\theta_\ell}\right) + \sum_{k:|\beta_k| > 1} \operatorname{Li}_2\left(\beta_k^{-1} e^{i\theta_\ell}\right) \right).$$

$$\text{(2.13)}$$

Also, r is even, and

$$\lambda = \frac{1}{2\pi} \sum_{\ell=1}^{r} (-1)^{\ell} \theta_\ell . \tag{2.14}$$

Since $m(P(z) + yQ(z)) = m(Q(z) + yP(z))$ by Proposition 2.9, the assumption (2.12) is without loss of generality.

We present the proof as a staged exercise, as follows.

Exercise 2.37 (a) Show that for $\alpha \in \mathbb{C}$

$$\int_{\theta_1}^{\theta_2} \log |e^{i\theta} - \alpha|\, d\theta = \begin{cases} \operatorname{Im}\left(\operatorname{Li}_2(\alpha e^{-i\theta_2}) - \operatorname{Li}_2(\alpha e^{-i\theta_1})\right) & \text{if } |\alpha| \leq 1 \\ \operatorname{Im}\left(-\operatorname{Li}_2(\alpha^{-1} e^{i\theta_2}) + \operatorname{Li}_2(\alpha^{-1} e^{i\theta_1})\right) \\ \quad + (\theta_2 - \theta_1) \log |\alpha| & \text{if } |\alpha| > 1. \end{cases}$$

(b) Let $P(z) = v \prod_j (z - \alpha_j)$. Deduce that

$$\int_{\theta_1}^{\theta_2} \log |P(e^{i\theta})|\, d\theta = \sum_{j:|\alpha_j| \leq 1} \operatorname{Im}(\operatorname{Li}_2(\alpha_j e^{-i\theta_2}) - \operatorname{Li}_2(\alpha_j e^{-i\theta_1}))$$

$$+ \sum_{j:|\alpha_j| > 1} \operatorname{Im}(-\operatorname{Li}_2(\alpha_j^{-1} e^{i\theta_2}) + \operatorname{Li}_2(\alpha_j^{-1} e^{i\theta_1}))$$

$$+ (\theta_2 - \theta_1)(\log |v| + \sum_{j:|\alpha_j| > 1} \log |\alpha_j|) .$$

(c) Apply exercise 2.23 to complete the proof of Theorem 2.36.

Before stating a Corollary to the Theorem, we need to define the **Bloch–Wigner dilogarithm** $D(z)$. It is defined initially for $0 < |z| \leq 1$ by

$$D(z) := \operatorname{Im}(\operatorname{Li}_2(z)) - \log |z| \arg(1 - z) ,$$

and this extends by analytic continuation to $z \in \mathbb{C} \setminus [1, \infty)$.

Corollary 2.38 *Suppose that a, b, c are the lengths of sides of a triangle, with oppo-site angles A, B, C, respectively. Then*

$$m(ax + by + c) = \frac{1}{\pi} \left(A \log a + B \log b + C \log c + D \left(\frac{c}{b} e^{iA} \right) \right). \quad (2.15)$$

Proof Now using Exercise 2.35,

$$m(ax + by + c) = m(a + y(bx + c)),$$

and we can assume that $c \le b$. Thus, we can apply the Theorem with $P(x) = a$, $Q(x) = bx + c$, with $m(P) = a$ and $m(Q) = b$. Then, for x on the unit circle, $a = |bx + c|$ for $x = -e^{\pm iA}$. Put $\theta_1 := \pi - A$ and $\theta_2 := \pi + A$. Since $Q(1) = b + c > a = P(1)$, we have $a \ge |bx + c|$ for $x = e^{i\theta}$, where $\theta_1 \le \theta \le \theta_2$.

We can now apply the theorem. By (2.14), we have $\lambda = A/\pi$, no α_i and just one β_k, namely $\beta_1 = -c/b \in [-1, 0)$. Thus, by (2.13), we have

$$2\pi m(ax + by + c) = 2A \log a + 2(\pi - A) \log b + \mathrm{Im} \left(\mathrm{Li}_2 \left(\frac{c}{b} e^{iA} \right) - \mathrm{Li}_2 \left(\frac{c}{b} e^{-iA} \right) \right)$$

$$= 2A \log a + 2(B + C) \log b + 2 \, \mathrm{Im} \left(\mathrm{Li}_2 \left(\frac{c}{b} e^{iA} \right) \right).$$

Now

$$\arg \left(1 - \frac{c}{b} e^{iA} \right) = \arg(b - ce^{iA}) = \arg(ae^{-iC}) = -C,$$

so that

$$D \left(\frac{c}{b} e^{iA} \right) = \mathrm{Im} \left(\mathrm{Li}_2 \left(\frac{c}{b} e^{iA} \right) \right) + (\log c - \log b) \arg \left(1 - \frac{c}{b} e^{iA} \right)$$

$$= \mathrm{Im} \left(\mathrm{Li}_2 \left(\frac{c}{b} e^{iA} \right) \right) - (\log c - \log b) C.$$

Hence,

$$m(ax + by + c) = \frac{1}{\pi} \left(A \log a + (B + C) \log b + (\log c - \log b) C + D \left(\frac{c}{b} e^{iA} \right) \right),$$

which gives (2.15). □

We note in passing that, by symmetry, (2.15) implies that

$$D \left(\frac{c}{b} e^{iA} \right) = D \left(\frac{a}{c} e^{iB} \right) = D \left(\frac{b}{a} e^{iC} \right).$$

Exercise 2.39 (computational exercise)

Write a program to compute $m(P(x) + yQ(x))$ in two ways: by applying Theorem 2.36 and also by directly computing

$$\frac{1}{2\pi} \int_0^{2\pi} \log\max(|P(e^{i\theta})|, |Q(e^{i\theta})|)\, d\theta \;.$$

2.7.1 Dirichlet L-Functions

For any undefined terms used in this subsection, we refer to Apostol [Apo76]. Given a Dirichlet character $\chi : \mathbb{Z} \to \mathbb{C}$, the Dirichlet L-function $L(\chi, s)$ is defined for $s \in \mathbb{C}$, $\operatorname{Re} s > 1$ by

$$L(\chi, s) := \sum_{n=1}^{\infty} \frac{\chi(n)}{n^s} \;.$$

It can be analytically continued to the whole complex plane (except that in the case that χ is a principal character, there is a simple pole at $s = 1$). For a quadratic field K of discriminant D, there is a Dirichlet character χ_D associated with K by

$$\zeta_K(s) := \sum_{\mathfrak{a}} \frac{1}{(N\mathfrak{a})^s} = \zeta(s) L(\chi_D, s).$$

Here, the sum is over the ideals $\mathfrak{a}$ of the ring of integers of K, and $\zeta(s)$ is the Riemann zeta function. For instance, for $K = \mathbb{Q}(\sqrt{-3})$, we have $\chi_{-3}(n)$ equal to 1, -1 or 0 according to whether $n \equiv 1, 2$ or 3 (mod 3). Also, for $K = \mathbb{Q}(\sqrt{-1})$, of discriminant -4, we have that $\chi_{-4}(n)$ is equal to $1, 0, -1$ or 0 according to whether $n \equiv 1, 2, 3$ or 4 (mod 4).

A Dirichlet character χ is *real* if all its values are real, *even* if $\chi(-1) = 1$ and *odd* if $\chi(-1) = -1$. Then [Apo76, Theorem 12.11] tells us that if χ is real and *primitive* [Apo76, p. 168] with period k, then $L(\chi, s)$ satisfies the functional equation

$$L(\chi, 1 - s) = \frac{k^{s-1}\Gamma(s)}{(2\pi)^s} \left(e^{-\pi i s/2} + \chi(-1)e^{\pi i s/2} \right) \left(\sum_{j=1}^{k} \chi(j)e^{2\pi i j/k} \right) L(\chi, s).$$

$$\tag{2.16}$$

It is known [BS66, p. 348] that χ_D is primitive and even or odd according to whether D is positive or negative.

Exercise 2.40 By applying the functional equation (2.16) to $L(\chi_{-3}, 2 + \delta)$ and $L(\chi_{-4}, 2 + \delta)$ as $\delta \to 0$, deduce that

$$L'(\chi_{-3}, -1) = \frac{3\sqrt{3}}{4\pi} L(\chi_{-3}, 2) \quad \text{and} \quad L'(\chi_{-4}, -1) = \frac{2}{\pi} L(\chi_{-4}, 2).$$

Exercise 2.41 More generally, show that for $j = 0, 1, 2 \ldots$

$$L'(\chi_{-4}, -1 - 2j) = (-1)^j \frac{2^{2j+1}(2j+1)!}{\pi^{2j+1}} L(\chi_{-4}, 2j+2).$$

2.7.2 Some Explicit Formulae for Two-Dimensional Mahler Measures

We can now prove the following.

Proposition 2.42 *We have*

$$m(x + y + 1) = L'(\chi_{-3}, -1), \tag{2.17}$$

$$m((x + y)^2 - 2) = 2m(x + y + \sqrt{2}) = \tfrac{1}{2} \log 2 + L'(\chi_{-4}, -1), \tag{2.18}$$

$$m((x + y)^2 - 3) = 2m(x + y + \sqrt{3}) = \tfrac{2}{3} \log 3 + \tfrac{4}{3} L'(\chi_{-3}, -1). \tag{2.19}$$

Proof Taking $\theta = \pi/3$, Lemma 2.34 gives

$$m(x + y + 1) = \frac{1}{\pi} \sum_{j=1}^{\infty} \frac{(-1)^{j-1}}{j^2} \sin\left(\frac{2\pi j}{3}\right)$$

$$= \frac{\sqrt{3}}{2\pi} \sum_{j=1}^{\infty} \frac{(-1)^{j-1}\chi_{-3}(j)}{j^2}$$

$$= \frac{\sqrt{3}}{2\pi} \left(\sum_{j=1}^{\infty} \frac{\chi_{-3}(j)}{j^2} - 2 \sum_{j=1}^{\infty} \frac{\chi_{-3}(2j)}{(2j)^2} \right)$$

$$= \frac{\sqrt{3}}{2\pi} \left(\sum_{j=1}^{\infty} \frac{\chi_{-3}(j)}{j^2} \right) \left(1 - \frac{2\chi_{-3}(2)}{4} \right)$$

$$= \frac{3\sqrt{3}}{4\pi} L(\chi_{-3}, 2)$$

$$= L'(\chi_{-3}, -1),$$

the last line coming from Exercise 2.40. Next, using $\theta = \pi/4$,

$$m(x + y + \sqrt{2}) = \frac{1}{4}\log 2 + \frac{1}{\pi}\sum_{j=1}^{\infty}\frac{(-1)^{j-1}}{j^2}\sin\left(\frac{\pi j}{2}\right)$$

$$= \frac{1}{4}\log 2 + \frac{1}{\pi}\sum_{j=1}^{\infty}\frac{\chi_{-4}(j)}{j^2}$$

$$= \frac{1}{4}\log 2 + \frac{1}{\pi}L(\chi_{-4}, 2)$$

$$= \frac{1}{4}\log 2 + \frac{1}{2}L'(\chi_{-4}, -1).$$

Hence, by Exercise 2.2,

$$m((x+y)^2 + 2) = m((x+y)^2 - 2) = m(x+y+\sqrt{2}) + m(x+y-\sqrt{2}) = 2m(x+y+\sqrt{2}),$$

from which (2.18) follows. Finally, for $\theta = \pi/6$,

$$m(x + y + \sqrt{3}) = \frac{1}{3}\log 3 + \frac{1}{\pi}\sum_{j=1}^{\infty}\frac{(-1)^{j-1}}{j^2}\sin\left(\frac{\pi j}{3}\right)$$

$$= \frac{1}{3}\log 3 + \frac{\sqrt{3}}{\pi}\sum_{j=1}^{\infty}\frac{\chi_{-3}(j)}{j^2}$$

$$= \frac{1}{3}\log 3 + \frac{\sqrt{3}}{\pi}L(\chi_{-3}, 2)$$

$$= \frac{1}{3}\log 3 + \frac{1}{2}L'(\chi_{-3}, -1).$$

Hence,

$$m((x+y)^2 + 3) = 2m(x+y+\sqrt{3}) = \frac{2}{3}\log 3 + L'(\chi_{-3}, -1).$$

$\square$

Exercise 2.43 Use Theorem 2.36 to show that

$$m(x + y + 1) = m((x+1)y + 1) = \frac{1}{\pi}\operatorname{Im}\operatorname{Li}_2(e^{i\pi/3}),$$

$$m(x + y + \sqrt{2}) = \frac{3}{8}\log 2 + \frac{1}{\pi}\operatorname{Im}\operatorname{Li}_2((1+i)/2),$$

$$m(x + y + \sqrt{3}) = \frac{5}{12}\log 3 + \frac{1}{\pi}\operatorname{Im}\operatorname{Li}_2(e^{\pi i/6}/\sqrt{3}).$$

Reconcile these with Proposition 2.42.

Exercise 2.44 (Boyd [Boy98]) Show that

$$m((x+1)y + x - 1) = \operatorname{Re} \frac{1}{\pi} \int_{-\pi/2}^{\pi/2} \log(1 + e^{it}) \, dt = L'(\chi_{-4}, -1).$$

Use this result to show that also

$$m(x + \frac{1}{x} + y - \frac{1}{y}) = m(\frac{u}{v} + \frac{v}{u} + uv - \frac{1}{uv}) = L'(\chi_{-4}, -1).$$

A similar result, albeit with a more complicated proof, was given by Ray [Ray87], who showed that

$$m\left((y - y^{-1})^2 \left(\frac{x^7 - x^{-7}}{x - x^{-1}}\right) + 7\left(x + x^{-1}\right)^2\right) = \tfrac{8}{7} L'(\chi_{-7}, -1).$$

Ray has other identities involving $L'(\chi_D, -1)$ for $D = -8, -20$ and -24:

$$m((x^4 + 1)(y - 1)^2 + 8x^2 y) = L'(\chi_{-8}, -1), \quad (2.20)$$

$$m((x^8 - x^6 + x^4 - x^2 + 1)(y - 1)^2 + 20x^2(x^2 - 1)^2 y) = \frac{4}{5} L'(\chi_{-20}, -1), \tag{2.21}$$

$$m((x^8 - x^4 + 1)(y - 1)^2 + 24x^2(x^2 - 1)^2 y) = \frac{1}{3} L'(\chi_{-24}, -1). \tag{2.22}$$

Conjecture 2.45 (Chinburg's Conjecture) For every odd character χ_D associated with a quadratic field of negative discriminant D, there is a two-dimensional polynomial $P(x, y) \in \mathbb{Z}[x, y]$ such that $L'(\chi_D, -1)$ is a rational multiple of $m(P)$.

Towards this conjecture, Boyd and Rodriguez-Villegas [BRV02, BRV05] have, for $f = 11, 15, 19, 35, 39, 40, 55, 84$ and 120, constructed polynomials $P_f \in \mathbb{Z}[x, y]$ such that $m(P_f) = r_f L'(\chi_{-f}, -1)$ for some rational number r_f. As before, χ_{-f} is the odd character associated with the imaginary quadratic field of discriminant $-f \equiv 0$ or $1 \pmod 4$. Typically, these r_f can be computed to 50 decimal places, indicating that they are rationals with small denominators. However, because there is no known upper bound on their denominators, these values r_f can be proved to be correct only in special cases. One example where exact evaluation proved possible was [BRV05, Ex. 3]

$$m((x^2 + x + 1)(y^2 + x) + 3x(x + 1)y) = \tfrac{1}{6} L'(\chi_{-15}, -1).$$

As pointed out by Rodriguez-Villegas [RV99], we can interpret Dirichlet's class number formula for real quadratic fields as a formula of the same type. Let K be a real quadratic field of discriminant D, with $u > 1$ generating its unit group, and class number h. Note that u is a quadratic Pisot number, so that $m(u) = \log u$. Then ([BS66])

$$L(\chi_D, 1) = \frac{h \log u}{\sqrt{D}}.$$

Then, using the functional equation (2.16) for $\chi = \chi_D$ and the fact [BS66, p. 349] that for $D > 0$

$$\sum_{j=1}^{D} \chi_D(j) e^{2\pi i j/D} = \sqrt{D},$$

we obtain

$$L'(\chi_D, 0) = m(u)h.$$

In two very interesting papers [BRV02, BRV05], Boyd and Rodriguez-Villegas extend the above results considerably, finding more evidence in support of Chinburg's Conjecture.

2.7.3 Mahler Measures of Elliptic Curves

For some two-dimensional integer polynomials $P(x, y)$ for which $P(x, y) = 0$ defines an elliptic curve E over $\mathbb{Q}$, the Mahler measure $m(P)$ turns out to be a rational multiple of $L'(E, 0)$. Here $L(E, s)$ is the L-function of the projective version of the curve $P = 0$. It is a Dirichlet series that encodes the number of points on this curve in the finite fields. See for instance Knapp [Kna92, p. 294] for details. Results of this kind were predicted in an important paper of Deninger [Den97]. For an outline of Deninger's work in this area and its connections with the Bloch–Beilinson Conjectures, see Boyd [Boy98, p. 46] and Rodriguez-Villegas [RV99]. These papers have been, and continue to be, a valuable and stimulating source of relevant theory, challenging conjectural results and further references. They have stimulated much further research. The recent book of Brunault and Zudilin [BZ20] is also an excellent source for material in this area.

In particular, Deninger conjectured that, for $k = 1$,

$$m\left(x + \frac{1}{x} + y + \frac{1}{y} + k\right) = r_k L'(E, 2) \qquad (2.23)$$

for some rational number r_k. One of many computations by Boyd [Boy98] showed that $r_1 = 1$, correct to 50 decimal places. In 2014, Rogers and Zudilin [RZ14] proved that indeed $r_1 = 1$. To date, values of k where the result has been proved include $k = 1$, $r_1 = 1$, $k = 2$, $r_2 = 1$, $k = 3$, $r_3 = 2$, $k = 5$, $r_5 = 6$, $k = 8$, $r_8 = 6$, $k = 12$, $r_{12} = 2$, $k = 16$, $r_{16} = 11$, as well as for some irrational k with $k^2 \in \mathbb{Z}$. See Samart's table in [Sam20] for the complete list, as well as their sources, and the conductors of the corresponding elliptic curves.

It might be tempting to expect that if $P(x, y) = 0$ defines an elliptic curve E, then $m(P)$ is always a rational multiple of $L'(E, 0)$. However, by Exercise 2.46 below,

$$m(y^2 - x^3 - k) = \log |k| \tag{2.24}$$

for $|k| \geq 2$, while by Proposition 2.9 and (2.17) we have

$$m(y^2 - x^3 \pm 1) = m(x + y + 1) = L'(\chi_{-3}, -1). \tag{2.25}$$

Furthermore, polynomials defining birationally equivalent elliptic curves need not have Mahler measures of the same kind. For instance, for $P(x, y) = y^2 - 6xy + y - x^3$, and E being the elliptic curve $P = 0$, we have

$$m(P) = 3L'(E, 0). \tag{2.26}$$

However, for $Q(x, Y) := P(x, Y + 3x) = Y^2 + Y - x^3 - 9x^2 + 3x$, we have

$$m(Q) = \log \left(\tfrac{1}{2}(9 + \sqrt{93}) \right), \tag{2.27}$$

by Exercise 2.46. Furthermore, for $R(X, Y) := Q(X - 3, Y) = Y^2 + Y - X^3 + 30X - 63$ we have

$$m(R) = \log 63, \tag{2.28}$$

again by Exercise 2.46. These examples, due to Boyd [Boy98], suggest that when the Mahler measure $m(P)$ of a two-dimensional polynomial P that defines an elliptic curve E satisfies $m(P) = rm(Q)$ for some Q of lower genus or dimension, and rational r, then $m(P)$ will not be a rational multiple of $L'(E, 0)$.

Motivated by some of Boyd's computations, Bertin and Zudilin [BZ16] found examples of genus 2 curves $Q = 0$ whose Mahler measures were of the form $rm(P)$ for some rational number r and elliptic curve $P = 0$. This turned out to be because the Jacobian of $Q = 0$ was a product of two elliptic curves, one of which was $P = 0$. This led to the evaluation of $m(Q)$ for cases where r and $m(P)$ could be evaluated. One example they gave was for

$$Q(x, y) = y^2 + (x^4 + x^3 + 2x^2 + x + 1)y + x^4 \text{ and } P(x, y) = (x + 1)y^2 + (x^2 + x + 1)y + x^2 + x,$$

for which

$$m(Q) = 2m(P) = 2L'(E, 0),$$

where E is an elliptic curve of conductor 14.

Exercise 2.46 Suppose that we are given polynomials $P(x) \in \mathbb{Z}[x]$ and $Q(y) \in \mathbb{Z}[y]$, where Q is monic with $Q(0) = 0$, and

$$|P(x)| \geq |Q(y)| \quad \text{for} \quad |x| = |y| = 1.$$

Show that then

$$m(P(x) + Q(y)) = m(P(x)).$$

2.7.4   Mahler Measure of Three-Dimensional Polynomials

We start with the first nontrivial closed form evaluation of a three-dimensional Mahler measure.

Proposition 2.47 *We have*

$$m(1 + x + y + z) = \frac{7}{2\pi^2}\zeta(3).$$

Here, of course, $\zeta(s)$ is the Riemann zeta function.

Proof Using Proposition 2.9, Exercise 2.23 and the fact that $m(1 + x) = 0$, we have

$$
\begin{aligned}
m(1 + x + y + z) &= m(x(1 + y) + 1 + z) \\
&= m(1 + y) + 4\int_0^{1/2}\int_0^{1/2} \log\max\left(|1 + e^{2\pi iu}|, |1 + e^{2\pi it}|\right) dt\, du \\
&= 8\int_0^{1/2} \log|1 + e^{2\pi it}|\, dt \int_t^{1/2} du \\
&= 8\int_0^{1/2} (\tfrac{1}{2} - t) \log|1 + e^{2\pi it}|\, dt \\
&= -8\,\mathrm{Re}\int_0^{1/2} t \log(1 + e^{2\pi it})\, dt \\
&= 8\sum_{n=1}^{\infty} \frac{(-1)^n}{n}\int_0^{1/2} t \cos(2\pi nt)\, dt \\
&= \frac{2}{\pi^2}\sum_{n=1}^{\infty} \frac{(-1)^n}{n}\left(\frac{(-1)^n - 1}{n^2}\right) \\
&= \frac{4}{\pi^2}\sum_{k=1}^{\infty} \frac{1}{(2k + 1)^3} \\
&= \frac{4}{\pi^2}\zeta(3)\left(1 - \frac{1}{2^3}\right) \\
&= \frac{7}{2\pi^2}\zeta(3).
\end{aligned}
$$

$\square$

Exercise 2.48 Use the functional equation

$$\zeta(s) = 2(2\pi)^{s-1}\Gamma(1-s)\sin\left(\frac{\pi s}{2}\right)\zeta(1-s)$$

[Apo76, p. 259] for $\zeta(s)$ to show that for $k \geq 1$

$$\frac{\zeta(2k+1)}{\pi^{2k}} = (-1)^k \frac{2^{2k+1}}{(2k)!}\zeta'(-2k),$$

so that also

$$m(1 + x + y + z) = -14\zeta'(-2).$$

We now turn our attention to a generalisation of Proposition 2.47. To state this result succinctly, we first need some layers of notation. First of all, we let $\mathrm{Log}\, z$ denote the principal value of the complex logarithm, so that

$$\mathrm{Log}\, z = \log|z| + i\arg z,$$

where $\arg z \in (-\pi, \pi]$. Following Zagier [Zag90], we define

$$L_2(z) := \mathrm{Li}_2(z) + \log|z| \cdot \mathrm{Log}(1-z);$$
$$L_3(z) := \mathrm{Li}_3(z) - \log|z| \cdot \mathrm{Li}_2(z) - \tfrac{1}{2}\log^2|z| \cdot \mathrm{Log}(1-z), \qquad (2.29)$$

where Li_2 is given by (2.11), and similarly Li_3 is the trilogarithm function, defined for $|z| \leq 1$ by

$$\mathrm{Li}_3(z) := \sum_{n=1}^{\infty}\frac{z^n}{n^3}.$$

Next, for $x, y \neq 0$, put

$$g_2(x, y) := L_2\left(\frac{1+x+y}{x}\right),$$

$$g_3(x, y) := L_3\left(\frac{1+x+y}{x}\right) + L_3\left(\frac{1+x+y}{y}\right) - L_3\left(\frac{1+x+y}{-xy}\right),$$

and for $i = 2$ and 3, let

$$G_i(x, y) := g_i(x, y) - g_i(x, -y) - g_i(-x, y) + g_i(-x, -y).$$

The polynomials we are going to consider are

$$P_{a,b,c}(x, y, z) := a + bx^{-1} + cy + (a + bx + cy)z,$$

where a, b, c are real parameters. Then because

$$m(P_{a,b,c}) = \begin{cases} \log|b| + m(P_{0,1,\,|c/b|}) & \text{for } a = 0, b \neq 0; \\ \log|a| + m(P_{1,\,|b/a|,\,|c/a|}) & \text{for } a \neq 0, \end{cases}$$

it is enough to consider the following special cases.

Theorem 2.49 (Smyth [Smy02])

(a) Let $0 \leq c \leq 1$. Then

$$m(P_{0,1,c}) = \frac{2}{\pi^2}(\text{Li}_3(c) - \text{Li}_3(-c)).$$

(b) Let $b > 0$ and $c > 0$. Then

$$m(P_{1,b,c}) = \frac{1}{\pi^2}(G_3(b, c) + \log c \cdot G_2(b, c) + \log b \cdot G_2(c, b)).$$

The proof is based on obtaining formulae for

$$\iint \frac{\text{Log}(x + cy)}{xy} \, dx \, dy \quad \text{and} \quad \iint \frac{\text{Log}(1 + bx + cy)}{xy} \, dx \, dy, \qquad (2.30)$$

as follows. The second of these formulae is somewhat complicated. We need to define

$$\Lambda_2(z) := \text{Li}_2(z) + \text{Log}(z) \cdot \text{Log}(1 - z) \quad \text{and}$$
$$\Lambda_3(z) := \text{Li}_3(z) - \text{Log}(z) \cdot \text{Li}_2(z) - \tfrac{1}{2}\text{Log}^2 z \, \text{Log}(1 - z),$$

and then put

$$f_3(x, y) := \Lambda_3\left(\frac{1 + x + y}{x}\right) + \Lambda_3\left(\frac{1 + x + y}{y}\right) - \Lambda_3\left(\frac{1 + x + y}{-xy}\right)$$
$$+ (\text{Log}\, y - i\pi)\Lambda_2\left(\frac{1 + x + y}{x}\right) + (\text{Log}\, x - i\pi)\Lambda_2\left(\frac{1 + x + y}{y}\right)$$
$$- i\pi \, \text{Log}\, x \cdot \text{Log}\, y.$$

(Note that Λ_2 and Λ_3 are not identical to L_2 and L_3 in (2.29): they are analytic versions of them.) Also, let $\mathcal{H}$ denote the upper half-plane $\text{Im}\, z > 0$ of the complex plane $\mathbb{C}$.

Proposition 2.50 ([Smy02, Prop. 2])

(a) For $x, y \in \mathcal{H}$ with $|x| < |y|$, we have

$$\frac{\partial^2}{\partial x \partial y}(\text{Li}_3(-x/y) + \tfrac{1}{2}\text{Log}\, x \cdot \text{Log}^2 y) = \frac{\text{Log}(x + y)}{xy}. \qquad (2.31)$$

(b) *The function $f_3(x, y)$ is analytic both x and y, for $x, y \in \mathcal{H}$, where*

$$\frac{\partial^2}{\partial x \partial y} f_3(x, y) = -\frac{\operatorname{Log}(1 + x + y)}{xy}. \tag{2.32}$$

From this result, the indefinite integrals (2.30) can be readily calculated. A second ingredient in the proof is to observe that, for $b > 0, c > 0$ for instance,

$$
\begin{aligned}
m(P_{1,b,c}) &= \frac{1}{(2\pi)^2} \int_{-\pi}^{\pi} \int_{-\pi}^{\pi} \log \max \left(|1 + be^{i\theta} + ce^{i\psi}|, |1 + be^{-i\theta} + ce^{i\psi}| \right) d\theta \, d\psi \\
&= \frac{1}{2\pi^2} \int_{0}^{\pi} \left(\int_{-\pi}^{\pi} \log \max \left(|1 + be^{i\theta} + ce^{i\psi}|, |1 + be^{-i\theta} + ce^{i\psi}| \right) d\psi \right) d\theta \\
&= \frac{1}{\pi^2} \int_{0}^{\pi} \int_{0}^{\pi} \log |1 + be^{i\theta} + ce^{i\psi}| \, d\psi \, d\theta \,.
\end{aligned}
$$

This last line follows from the fact that $|1 + be^{i\theta} + ce^{i\psi}|$ and $|1 + be^{-i\theta} + ce^{i\psi}|$ have the same real part and that

- for $0 \le \psi \le \pi$, $|1 + be^{i\theta} + ce^{i\psi}|$ has the larger imaginary part;
- for $-\pi \le \psi \le 0$, $|1 + be^{i\theta} + ce^{-i\psi}|$ has the larger imaginary part, but equals $|1 + be^{i\theta} + ce^{i(\psi + \pi)}|$.

Thus, the integrals required to compute the Mahler measure of $P_{a,b,c}$ are

$$m(P_{a,b,c}) = \operatorname{Re} \frac{1}{(i\pi)^2} \iint \frac{\log(a + bx + cy)}{xy} \, dx \, dy \,,$$

where the integral is over the 'half-tori' $x = e^{i\theta}$ ($0 \le \theta \le \pi$) and $y = e^{i\psi}$ ($0 \le \theta \le \pi$). These are the integrals evaluated in Proposition 2.50.

Exercise 2.51 Deduce Proposition 2.47 from Theorem 2.49.

Exercise 2.52 Use Theorem 2.49 and the identity

$$2L_3(3) - L_3(-3) = \frac{13}{6} \zeta(3)$$

([Smy02, Lemma 6]) to show that

$$m(1 + x^{-1} + y + (1 + x + y)z) = \frac{14}{3\pi^2} \zeta(3) \,.$$

2.7.5 Mahler Measure Formulae for Some Polynomials of Dimension at Least 4

The following beautiful results have been proved by Lalín.

Proposition 2.53 (Lalín [Lal06b]) *We have*

$$m((1+w)(1+x) + (1-w)(1+y)z) = \frac{24}{\pi^3} L(\chi_{-4}, 4) = -\frac{1}{2} L'(\chi_{-4}, -3).$$

Proposition 2.54 (Lalín [Lal03]) *We have*

$$m((1+v)(1+w)(1+x) + (1-v)(1-w)(1+y)z) = \frac{93}{\pi^4} \zeta(5) = 124\, \zeta'(-4).$$

From these results, combined with (2.17), Exercise 2.44 and Proposition 2.47, it is tempting to ask the following questions.

Problem 2.55 (open problem)

- For a given odd integer $k \geq 7$, does there exist a k-dimensional integer polynomial P for which $m(P)$ is a rational multiple of $\zeta'(1-k)$?
- For a given even integer $k \geq 6$, does there exist a k-dimensional integer polynomial P for which $m(P)$ is a rational multiple of $L'(\chi, 1-k)$ for some odd character χ and L-function associated with a quadratic field?

Towards answering such questions, Lalín has proved the formulae below (among others). To state her result, we need to recall the elementary symmetric functions: given numbers $b_1, b_2, \ldots, b_n$, for $j > 0$ denote by $e_j(b_1, b_2, \ldots, b_n)$ the sum of all $\binom{d}{j}$ possible products of j numbers with distinct indices chosen from $b_1, b_2, \ldots, b_n$. Also, $e_0 = 1$.

Theorem 2.56 (Lalín [Lal06a, Theorem 1(i)]) *Let* $n \geq 1$, *and*

$$P_n(z_0, z_1, \ldots, z_n) = (1+z_1)(1+z_2) \cdots (1+z_n) + (1-z_1)(1-z_2) \cdots (1-z_n)z_0.$$

Then for $n = 2k$ *even*

$$m(P_{2k}) = \frac{1}{(2k-1)!} \sum_{j=1}^{k} (-1)^j \binom{2^{2j+1}}{2} e_{k-j}\left(2^2, \ldots, (2k-2)^2\right) \zeta'(-2j),$$

while for $n = 2k+1$ *odd*

$$m(P_{2k+1}) = \frac{1}{(2k)!} \sum_{j=1}^{k} (-1)^j e_{k-j}\left(1^2, \ldots, (2k-1)^2\right) L'(\chi_{-4}, -2j-1).$$

These results can be translated into expressions involving $\zeta(s)$ and $L(\chi_{-4}, s)$ using Exercises 2.48 and 2.41. That was how Lalín originally stated them. The proof of the theorem, too long to give here, uses hyperlogarithms, a special kind of iterated integral, in an extended but clear sequence of integral and sum evaluations.

2.7.6 An Asymptotic Mahler Measure Result

We record here the following.

Theorem 2.57 (Myerson and Smyth [Smy81b]; Rodriguez-Villegas, Toledano and Vaaler [RVTV04]) *As $n \to \infty$, we have*

$$m(z_1 + z_2 + \cdots + z_n) = \frac{1}{2} \log n - \frac{\gamma}{2} + O\left(\frac{1}{n}\right).$$

Here, γ is the Euler–Mascheroni constant. Rodriguez-Villegas, Toledano and Vaaler [RVTV04] improved the error term in [Smy81b]; that improvement is given here. In fact, they proved the following more general result.

Theorem 2.58 *Let $P(z) \in \mathbb{C}[z]$. Then as $n \to \infty$,*

$$\int_{(\mathbb{R}/\mathbb{Z})^n} \log |P(1)e^{2\pi i\theta_1} + P(2)e^{2\pi i\theta_2} + \cdots + P(n)e^{2\pi i\theta_n}| \, d\boldsymbol{\theta} =$$

$$\frac{1}{2} \log \left(|P(1)|^2 + |P(2)|^2 + \cdots + |P(n)|^2\right) - \frac{\gamma}{2} + O_P\left(\frac{1}{n}\right),$$

where the constant implied by O_P depends on P.

2.8 Notes

The Mahler measure dates back at least to a paper of Szegő [Sze15] in 1915, where he showed that

$$M(f) = \inf \int_0^1 \left|f(e^{2\pi it})g(e^{2\pi it})\right|^2 \, dt,$$

the infimum being taken over all polynomials g with $g(0) = 1$. Szegő's result was generalised to polynomials in several variables and to general p-norms (not just the 2-norm) by Ruzsa [Ruz99].

Jensen stated and proved his theorem in [Jen99]. The integral representation (1.5) of $M(f)$ for one-variable polynomials f follows immediately from a more general result (for certain meromorphic functions) in [Jen99]. It appears explicitly in [Mah60], with the generalisation to several variables in Mahler [Mah62].

The study of the set $\mathcal{L}$ defined in (2.2) was initiated by Boyd [Boy81b]. The definition of the function q and Exercise 2.16 are also due to Boyd [Boy81a, Boy81b]. For Exercise 2.7, see Boyd [Boy81b].

Theorems 2.5 and 2.6 come from [Smy18]. The case of Theorem 2.6 where $F(\mathbf{z}_k)$ is a linear form was proved in 1977 by Lawton [Law77].

Proposition 2.8 is due to Boyd [Boy81a, Theorem 1]—see also Schinzel [Sch00, Section 3.4, Cor.17, p. 260], Smyth [Smy81a] and [Smy18, Cor. 1].

Proposition 2.9 is due to Smyth [Smy02, Lemma 7]. See also Schinzel [Sch00, Section 3.4, Cor. 8, p. 226] for the case $\det(V) = \pm 1$.

Lemma 2.34 comes from [Smy81b, Lemma 1].

Theorem 2.36 is a variant of a result of Boyd and Rodriguez-Villegas [BRV02, Proposition 1].

Corollary 2.38 is due to Maillot [Mai00]. Vandervelde [Van03] proved a similar result for $P(x, y) = axy + bx + cy + d$.

Ray's formulae (2.20), (2.21) and (2.22) are quoted from Boyd [Boy98, p. 76].

Proposition 2.47 is due to Smyth—see [Boy81b, Appendix 1].

Exercise 2.16 is due to Boyd [Boy81a, p. 118].

Proposition 2.17 is due to Lawton [Law83]. Earlier Boyd had proved this result when F does not vanish on the k-torus T^k, this being a special case of [Boy81a, Lemma 1], which states that for a continuous function $f : T^k \to \mathbb{C}$

$$\lim_{n \to \infty} \int_T f\left(z^{\mathbf{r}^{(n)}}\right) \mathrm{d}z = \int_{T^k} f(\mathbf{z}_k)\,\mathrm{d}\mathbf{z}_k$$

for the same sequence of vectors $\{\mathbf{r}^{(n)}\}_{n \in \mathbb{N}}$.

Proposition 2.21 comes from Smyth [Smy81a, Cor. 2].

Exercise 2.31 comes from Boyd and Mossinghoff [BM05].

Amoroso [Amo08] conjectured that there is a constant $C > 1$ such that for integer polynomials F of dimension at least 2, then $M(F) \geq C$. This is clearly a slightly weakened form of Lehmer's Conjecture. As evidence for this, he showed that any such F with $M(F) < \exp(1/23)$ must have a very large degree.

As noted in (2.3), the set of Mahler measures of one-variable integer polynomials forms a semigroup. Dixon and Dubickas [DD04, Theorem 10] showed that if $\alpha_1 > 1$, $\alpha_2 > 1$ are quadratic units, so that they belong to this set, their product $\alpha_1\alpha_2$ does not. Thus, the subset of such Mahler measures of *irreducible* polynomials is not a semigroup. Also, we know (Exercise 1.10) that all these Mahler measures are Perron numbers. But the products $\alpha_1\alpha_2$ are easily seen to be Perron numbers, so show that not all Perron numbers are Mahler measures.

The Mahler measures in Table D.1 were originally computed by Lehmer, Boyd and Mossinghoff. The list was shown to be complete up to degree 44 by Mossinghoff et al. [MRW08] .

Let Q_1 be the first Laurent polynomial in Table D.2. Presumably, there are infinitely many elements of $\mathcal{L}$ that are small in the sense of Boyd, i.e., less than 1.3. However, if $M(Q_1)$ is indeed the smallest limit point of $\mathcal{L}$, then there are only finitely many elements of $\mathcal{L}$ that are less than 1.25.

Chinburg's Conjecture 2.45 is stated in [Ray87, p. 697]. The Mahler measures in (2.20), (2.21) and (2.22) are given in [Boy98, p. 76].

The examples (2.24), (2.25), (2.26), (2.27) and (2.28) are all due to Boyd [Boy98].

2.9 Glossary

$\mathcal{C}(F)$. The exponent polytope of a Laurent polynomial F.

$F(\mathbf{z}_k)$. A convenient shorthand for $F(z_1, \ldots, z_k)$.

$F_A(\mathbf{z}_\ell)$. Let A be an $\ell \times k$ integer matrix, and let F be a nonzero Laurent polynomial in k variables. Then F_A is the Laurent polynomial in ℓ variables defined by $F_A(\mathbf{z}_\ell) = F(\mathbf{z}_\ell^A)$.

$F^{(n)}(\mathbf{z}_{2n})$. This is the polynomial $(z_1 + z_3 + \cdots + z_{2n-1}) - (z_2 + z_4 + \cdots + z_{2n})$.

$\mathcal{L}$. The set of all Mahler measures of polynomials that have integer coefficients, in any number of variables.

$\mathcal{L}_1$. The set of all Mahler measures of polynomials in one variable, with integer coefficients.

$\mathrm{Li}_2(z)$. The dilogarithm function, namely $\displaystyle\sum_{n=1}^{\infty} \frac{z^n}{n^2}$.

$\mathcal{M}(F)$. The set of all Mahler measures of nonzero Laurent polynomials in $\mathcal{P}(F)$. This is the closure of $\mathcal{M}_1(F)$ in $\mathbb{R}$ (Theorem 2.5).

$\mathcal{M}_1(F)$. Let $F(\mathbf{z}_k)$ be a polynomial in k variables. The set $\mathcal{M}_1(F)$ is the set of Laurent polynomials of the shape $F(z_1^{r_1}, z_2^{r_2}, \ldots, z_k^{r_k})$, where each of $r_1, \ldots, r_k$ ranges over $\mathbb{Z}$.

$\mathcal{P}(F)$. Let F be a nonzero Laurent polynomial in k variables with integer coefficients. Then $\mathcal{P}(F)$ is the set of all Laurent polynomials of the form either F_A or $-F_A$ as A ranges over all $\ell \times k$ integer matrices, for all $\ell \geq 0$.

$S^{(k)}$. The kth derived set of S.

$\mathbf{z}_k$. The vector $(z_1, \ldots, z_k)$, or possibly the transpose of this depending on context, with the subscript of the boldface letter indicating the number of components in the vector.

$\mathbf{z}_\ell^A$. Let $A = (a_{ij})$ be an $\ell \times k$ integer matrix. Then $\mathbf{z}_\ell^A$ is the k-tuple whose ith component is the monomial $z_1^{a_{1i}} z_2^{a_{2i}} \cdots z_\ell^{a_{\ell i}}$. If $\ell = 0$, this is interpreted as the k-tuple whose entries are all equal to 1.

derived set. The derived set $S^{(1)}$ of a set $S = S^{(0)}$ is the set of limit points in S. When we speak of limit points, we always mean 'genuine' limit points, i.e., limits of sequences of distinct elements. The kth derived set $S^{(k)}$ is defined recursively for $k \geq 2$ as the derived set of the $(k-1)$th derived set.

dimension. The dimension of a nonzero Laurent polynomial is defined to be the dimension of its exponent polytope. The dimension of a Mahler measure M is defined to be the least k for which M is the Mahler measure of a k-variable polynomial. See Exercise 2.22.

exponent polytope.   Let F be a nonzero Laurent polynomial in k variables. Each monomial that appears in F (with nonzero coefficient) is viewed as defining a vector in $\mathbb{Z}^k$ containing the exponents of the k variables. The convex hull of these vectors is the exponent polytope of F, written $\mathcal{C}(F)$.

isolated.   Let S be a subset of $\mathbb{C}$. An element x of S is called isolated if it is not a limit point of S. Equivalently, there is some $\varepsilon > 0$ such that all elements y in S distinct from x satisfy $|y - x| > \varepsilon$.

limit point.   Let S be a subset of $\mathbb{C}$. A limit point of S is a number $z \in \mathbb{C}$ such that for all $\varepsilon > 0$ there is some $x \in S$ satisfying $0 < |x - z| < \varepsilon$. Equivalently, there is sequence $\{a_n\}_{n \in \mathbb{N}}$ such that $a_n \to z$ as $n \to \infty$ but with a_n never equal to z. Equivalently, there is a sequence of distinct elements of S converging to z.

Pisot number, Pisot–Vijayaraghavan number, PV number.   Any of these is used for a real algebraic integer greater than 1 whose other conjugates lie in the open disc $|z| < 1$.

similar.   Two Laurent polynomials in $z_1, \ldots, z_k$ are called similar if their quotient is a monomial in $z_1, \ldots, z_k$. Similar polynomials have the same Mahler measure.

Chapter 3
Dobrowolski's Theorem

3.1 The Theorem and Preliminary Lemmas

The following theorem of Dobrowolski is the strongest known result in the direction of proving Lehmer's conjecture. It caused considerable surprise when it appeared, in 1979.

Theorem 3.1 (Dobrowolski [Dob79]) *Let α be an algebraic integer of degree d, not zero or a root of unity, and $\varepsilon > 0$. Then for $d \geq d(\varepsilon)$ we have*

$$M(\alpha) > 1 + (2 - \varepsilon) \left(\frac{\log \log d}{\log d} \right)^3 .$$

For the proof, we first need some preliminary lemmas.

Lemma 3.2 (Hadamard's Inequality) *For any positive integer d and complex $d \times d$ matrix $Z = (z_{ij})$, we have*

$$| \det(Z) |^2 \leq \prod_{j=1}^{d} \left(\sum_{i=1}^{d} |z_{ij}|^2 \right) .$$

Furthermore, equality occurs if and only if the columns of Z are mutually orthogonal.

Proof We can clearly assume that Z is nonsingular. We first prove the result when all the columns of Z are of unit (Euclidean) length. Put $P = Z^*Z$, where X^* denotes the conjugate transpose of the matrix X, and let $\lambda_1, \ldots, \lambda_d$ be (the multiset of) its eigenvalues. Then, since $\mathbf{z}^* P \mathbf{z} = |Z\mathbf{z}|^2 > 0$ for all nonzero $\mathbf{z} \in \mathbb{C}^d$, P is positive definite, all its diagonal entries are 1, its eigenvalues are positive and

© Springer Nature Switzerland AG 2021
J. McKee and C. Smyth, *Around the Unit Circle*, Universitext,
https://doi.org/10.1007/978-3-030-80031-4_3

$$| \det(Z) |^2 = \det(P) = \prod_{i=1}^{d} \lambda_i \leq \left(\frac{1}{d} \sum_{i=1}^{d} \lambda_i \right)^d = \left(\frac{1}{d} \operatorname{trace}(P) \right)^d = 1^d = 1 \, .$$

Here, we have used the classical Arithmetic Mean/Geometric Mean inequality, along with the basic fact that the trace of a square matrix equals the sum of its eigenvalues.

The general result for columns of Z of arbitrary length then follows by applying this special case to the matrix Z with the entries of each column being divided by the length of that column.   □

Exercise 3.3 Prove the 'equality' part of the lemma.

Corollary 3.4 *If the entries z_{ij} of Z all satisfy $|z_{ij}| \leq B$, then $|\det(Z)| \leq B^d d^{d/2}$.*

Lemma 3.5 (The Vandermonde matrix) *For the $d \times d$ complex matrix*

$$V = (z_j^{i-1})_{(i,j=1,\ldots,d)},$$

we have

$$\det(V) = \prod_{i>j} (z_i - z_j) \, .$$

Proof We view the determinant as a polynomial in variables $z_1, \ldots, z_d$. On subtracting the jth column from the ith, we see that $\det(V)$ has a factor $(z_i - z_j)$. There are $\binom{d}{2}$ such factors. This is the same as the total degree of $\det(V)$ in the variables $z_1, \ldots, z_n$, as can be seen by expanding the determinant. Hence, $\det(V)$ is a constant multiple of the polynomial $\prod_{i>j}(z_i - z_j)$. Since the diagonal term of $\det(V)$ is $z_2 z_3^2 \cdots z_d^{d-1}$, we see that this constant must be 1.   □

Corollary 3.6 *We have*

$$\det(V)^2 = (-1)^{\binom{d}{2}} \prod_{i \neq j} (z_i - z_j) \, .$$

Lemma 3.7 *For any $\mathbf{z} = (z_1, \ldots, z_d) \in \mathbb{C}^d$, we have*

$$\prod_{i \neq j} |z_i - z_j| \leq \left(M(\mathbf{z})^2 d \right)^d \, ,$$

where $M(\mathbf{z}) = \prod_i \max(1, |z_i|)$.

Proof Now

$$
\prod_{i \neq j} |z_i - z_j| = \begin{vmatrix} 1 & 1 & \cdots & 1 \\ z_1 & z_2 & \cdots & z_d \\ z_1^2 & z_2^2 & \cdots & z_d^2 \\ \vdots & \vdots & \cdots & \vdots \\ z_1^{d-1} & z_2^{d-1} & \cdots & z_d^{d-1} \end{vmatrix}^2
$$

$$
\leq \prod_{j=1}^{d} \left(\sum_{i=0}^{d-1} |z_j|^{2i} \right) \qquad \text{by Lemma 3.2}
$$

$$
\leq \prod_{j:|z_j| \leq 1} d \cdot \prod_{j:|z_j| > 1} |z_j|^{2(d-1)} \left(1 + |z_j|^{-2} + \cdots + |z_j|^{-2(d-1)} \right)
$$

$$
\leq d^d M(\mathbf{z})^{2(d-1)} \leq \left(M(\mathbf{z})^2 d \right)^d .
$$

$\square$

Lemma 3.8 *Suppose that an algebraic integer α of degree d has the property that there are two distinct conjugates of α whose quotient is a root of unity. Then there is an algebraic number β of degree less than d with $M(\beta) = M(\alpha)$ and such that no quotient of distinct conjugates of β is a root of unity.*

Proof Let F be a finite extension of $\mathbb{Q}$ containing all conjugates of α, where α has the stated property. Define a relation on the set A of conjugates α_i of α by saying that $\alpha_i \sim \alpha_j$ if α_i/α_j is a root of unity. This divides A into equivalence classes $C_1, \ldots, C_k$ say. Let $\alpha_j \in C_j$ $(j = 1, \ldots, k)$ be equivalence class representatives of $C_1, \ldots, C_k$. Note that for any automorphism $\sigma \in \text{Gal}(F/\mathbb{Q})$, if α_i and α_j are in the same equivalence class, so are $\sigma\alpha_i$ and $\sigma\alpha_j$. Furthermore, such an automorphism permutes the equivalence classes. Indeed, it acts transitively on these classes, because any automorphism mapping $\alpha_i \mapsto \alpha_j$ must map the class containing α_i to the class containing α_j. Thus, the transitivity property of A is inherited by the set of equivalence classes. This implies that all equivalence classes are of equal size $r := d/k$ say, where $r > 1$. (Proof: take an equivalence class, say C_1, of maximum size. Then an automorphism taking α_1 to α_j will map distinct elements of C_1 to distinct elements of C_j. Hence, $|C_j| \geq |C_1|$. But C_1 has maximum size, so all $|C_j|$ are equal.)

We now define $\beta := \prod_{\alpha \in C_1} \alpha$ and note that β is an algebraic integer of degree k, whose set of conjugates consists of the numbers $\beta_j := \prod_{\alpha \in C_j} \alpha$ $(j = 1, \ldots, k)$. Then $\beta_j = \alpha_j^r \rho_j$, where ρ_j is a root of unity. We claim that for $j \neq j'$ the quotient $\beta_j/\beta_{j'}$ cannot be a root of unity. For if $\beta_j/\beta_{j'}$ were a root of unity, then $\alpha_j^r/\alpha_{j'}^r$, and hence, $\alpha_j/\alpha_{j'}$, would also be a root of unity. This would contradict the fact that α_j and $\alpha_{j'}$ are in different equivalence classes.

In particular, $\beta_j/\beta_{j'} \neq 1$, i.e., the β_j are distinct. Hence, β has degree k over $\mathbb{Q}$. Also, note that all conjugates of α in the same equivalence class have the same modulus. Thus,

$$M(\alpha) = \prod_{|\alpha_i|>1} |\alpha_i| = \prod_{|\beta_j|>1} |\beta_j| = M(\beta).$$

□

Problem 3.9 Is Lemma 3.8 still true if α is assumed only to be an algebraic *number*?

Corollary 3.10 *Suppose there is a nondecreasing function $f : \mathbb{N} \to \mathbb{R}_+$ such that $M(\alpha) > 1 + 1/f(d)$ for all algebraic integers of degree d satisfying both: (i) $M(\alpha) > 1$ and (ii) $\alpha_i^n \neq \alpha_j^n$ for all distinct pairs of conjugates α_i, α_j and integers $n > 1$. Then the result holds without condition (ii): $M(\alpha) > 1 + 1/f(d)$ whenever $M(\alpha) > 1$.*

Proof Assume the inequality $M(\alpha) > 1/f(d)$ holds for all algebraic integers α of degree d satisfying both (i) and (ii) and take α of degree d that satisfies (i) but fails (ii), so that some ratio of its conjugates is a root of unity. By Lemma 3.8, there is some β with $M(\beta) = M(\alpha)$ with $e := \deg(\beta) < \deg(\alpha) =: d$ and such that $\beta_i^n \neq \beta_j^n$ for all distinct conjugates β_i, β_j of β and all n. Hence, if $M(\beta) > 1 + 1/f(e)$, then

$$M(\alpha) = M(\beta) > 1 + 1/f(e) \geq 1 + 1/f(d).$$

□

We can now deduce a weak form of Dobrowolski's Theorem.

Theorem 3.11 *Let α be a nonzero algebraic integer of degree d, not a root of unity. Then*

$$M(\alpha) > 1 + \frac{1}{22d}.$$

Proof We apply Lemma 3.7 with

$$\mathbf{z} = (\alpha_1^p, \ldots, \alpha_d^p, \alpha_1, \ldots, \alpha_d),$$

where p is some prime number. This gives

$$\prod_{i \neq j} |\alpha_i - \alpha_j| \prod_{i \neq j} |\alpha_i^p - \alpha_j^p| \prod_{i,j} |\alpha_i^p - \alpha_j|^2 \leq \left(2d(M(\alpha))^{2(p+1)}\right)^{2d}.$$

Now both $\prod_{i \neq j} |\alpha_i - \alpha_j|$ and $\prod_{i \neq j} |\alpha_i^p - \alpha_j^p|$ are symmetric functions of the α_i, so are integers. The first one (the modulus of the discriminant of the minimal polynomial of α) is clearly nonzero, and so is at least 1. The second product could be 0. In this case, we use Lemma 3.8 to replace α by an algebraic integer of lower degree having the same Mahler measure as α, but with $\alpha_i^n \neq \alpha_j^n$ for all integers n. This enables us to assume that $\prod_{i \neq j} |\alpha_i^p - \alpha_j^p| \geq 1$. Hence, using also Corollary A.24, we have that

$$p^d \leq \prod_{i,j} |\alpha_i^p - \alpha_j| \leq \left(2d(M(\alpha))^{2(p+1)}\right)^d ,$$

giving

$$M(\alpha) \geq \left(\frac{p}{2d}\right)^{1/(2(p+1))} .$$

By Bertrand's Postulate (actually a theorem of Chebyshev [Che52]), for every $n > 1$, there is a prime between n and $2n$, so we can choose p to be a prime in the range $6d < p < 12d$. Then

$$M(\alpha) \geq 3^{1/(24d)} = e^{\log 3/24d} > 1 + \frac{1}{22d} .$$

$\square$

We next define a generalisation of the Vandermonde matrix, where a single column

$$(1, v, v^2, \ldots, v^{N-1})^\mathsf{T}$$

of the Vandermonde matrix is replaced by a matrix $M(v, N, m)$ of m columns of the same length N, where for $j = 1, \ldots, m$ the jth column of $M(v, N, m)$ is the column of coefficients of t^{j-1} in the binomial expansion of $(1, v + t, (v + t)^2, \ldots, (v + t)^{N-1})^\mathsf{T}$, namely $(0, 0, \ldots, 0, 1, \binom{j}{j-1}v, \ldots, \binom{N-1}{j-1}v^{N-j})^\mathsf{T}$. A **confluent Vandermonde matrix** is an $N \times N$ block matrix

$$W = M(v_1, N, m_1) | M(v_2, N, m_2) | \cdots | M(v_k, N, m_k) , \tag{3.1}$$

where $v_1, \ldots, v_k$ are distinct complex numbers, and $N = m_1 + \cdots + m_k$.

Lemma 3.12 (The confluent Vandermonde matrix (Méray [Mér99])) *For the $N \times N$ complex matrix W given by (3.1), we have*

$$\det(W) = \prod_{i > j} (v_i - v_j)^{m_i m_j} . \tag{3.2}$$

Proof We fix N (and thus the size of the matrix W). Our induction hypothesis is that for any k with $1 \leq k \leq N$ and positive integers m_j ($k = 1, \ldots, n$) whose sum is N and distinct complex numbers v_j ($k = 1, \ldots, n$) that $\det(W)$ is given by (3.2). We use *downward* induction on k, starting at $k = N$. For this 'base' case, all the m_j must be 1, and so the result is precisely Lemma 3.5. Next, we take a positive integer k with $k < N$ and assume as our induction hypothesis that the result holds for $k + 1$. Thus, we have a matrix W with k blocks of sizes $m_1, \ldots, m_k$. Then we know that at least one of the block sizes m_j must be at least 2. Let us suppose that $m_1 \geq 2$. Then by the induction hypothesis we can assume that the result is true when the block $M(v_1, N, m_1)$ in W is replaced by the two blocks $M(v_1, N, m_1 - 1)|M(v_1 + t, N, 1)$, giving an $N \times N$ matrix, W' say, having $k + 1$ blocks. Here $v_1 + t$ is a complex number not equal to any v_j $(j = 1, \ldots, k)$. Thus, we have that

$$\det(W') = \det(M(v_1, N, m_1 - 1)|M(v_1 + t, N, 1)|M(v_2, N, m_2)|\cdots|M(v_k, N, m_k))$$
$$= \Pi_1\Pi_2\Pi_3, \tag{3.3}$$

say, where

$$\Pi_1 = \prod_{i>j\geq 2}(v_i - v_j)^{m_i m_j}, \quad \Pi_2 = \prod_{i>1}(v_i - v_1)^{(m_1-1)m_i}(v_i - v_1 - t)^{m_i}, \quad \Pi_3 = t^{m_1-1}.$$

Note that in (3.3) we can use the columns of $M(v_1, N, m_1 - 1)$ to remove all terms up to and including terms in t^{m_1-2} from $M(v_1 + t, N, 1)$, without changing $\det(W')$. This column then becomes

$$t^{m_1-1}\left(0, 0, \ldots, 0, 1, \binom{m_1}{m_1-1}v_1, \ldots, \binom{N-1}{m_1-1}v_1^{N-m_1}\right)^{\mathsf{T}}$$

plus terms in higher powers of t. We now divide both sides of (3.3) by t^{m_1-1}, and then let $t \to 0$. Then $M(v_1, N, m_1 - 1)|M(v_1 + t, N, 1) \to M(v_1, N, m_1)$, so that

$$\lim_{t\to 0}\det(W')/t^{m_1-1} = \det(W).$$

From (3.3), we see that this limit also equals $\prod_{i>j}(v_i - v_j)^{m_i m_j}$. This verifies the result for k, and the induction goes through. $\qquad\qquad\square$

For our application, we need to apply the inequality of Lemma 3.7 to the following particular kind of confluent Vandermonde matrix, where many of the multiplicities m_i are equal.

Corollary 3.13 *For $i = 1, \ldots, T$, let β_{i1} be an algebraic integer which has conjugates β_{ij} $(j = 1, \ldots, d)$. Then for positive integers $m_1, \ldots, m_T$, $N = d\sum_{i=1}^d m_i$ and $N_2 = d\sum_{i=1}^d m_i^2$, we have*

$$\prod_{\substack{i,k=1 \\ (i,j)\neq(k,\ell)}}^{T}\left(\prod_{j,\ell=1}^{d}|\beta_{ij} - \beta_{k\ell}|\right)^{m_i m_k} \leq N^{N_2}\left(\prod_{i=1}^{T}M(\beta_{i1})^{m_i}\right)^{2N}. \tag{3.4}$$

Proof Now a typical column c_r of a confluent Vandermonde matrix (3.1) is of the form

$$\left(0, 0, \ldots, 0, 1, \binom{r}{r-1}v, \ldots, \binom{N-1}{r-1}v^{N-r}\right)^{\mathsf{T}} \tag{3.5}$$

for a positive integer r and complex number v. Such a column has squared length $|c_r|^2$ bounded above by

$$N \left(\binom{N-1}{r-1} \max(1, |v|)^{N-r} \right)^2 \leq N^{2r-1} \max(1, |v|)^{2N}. \tag{3.6}$$

Hence,

$$\prod_{r=1}^{m} |c_r|^2 \leq N^{m^2} \max(1, |v|)^{2Nm}. \tag{3.7}$$

We now write the left-hand side of (3.4) as a confluent Vandermonde determinant, using Lemma 3.12. We then apply Hadamard's Inequality (Lemma 3.2) to this determinant and make use of (3.7). $\qquad\square$

3.2 Proof of Theorem 3.1: Dobrowolski's Lower Bound for $M(\alpha)$

Proof The idea of the proof is to consider a confluent Vandermonde determinant (3.1), call it V, where the v_i are the algebraic integers

$$\alpha, \alpha^2, \alpha^3, \alpha^5, \ldots, \alpha^{p_j}, \ldots, \alpha^{p_T}$$

and their conjugates, and the multiplicities are S for α and its conjugates and 1 for the α^p and their conjugates. (So the powers of α used are 1 and the first T primes p_i ($i \leq T$). The parameters S and T will be chosen later.) The determinant thus has $N := d(S + T)$ columns of length N, the first Sd of which are say c_{ir} ($i = 1, \ldots, d; r = 1, \ldots, S$) as in (3.5), with $v = \alpha_i$. Then, from (3.6),

$$|c_{ir}|^2 \leq N^{2r-1}(\max(1, |\alpha_i|))^{2N},$$

so that

$$\prod_{i=1}^{d} \prod_{r=1}^{S} |c_{ir}|^2 \leq N^{d(\sum_{r=1}^{S} 2r-1)} (\prod_i \max(1, |\alpha_i|))^{2NS} = N^{dS^2} M(\alpha)^{2NS}.$$

The other columns are of the form

$$c'_{ij} = \left(1, \alpha_i^{p_j}, \ldots, \alpha_i^{p_j(N-1)} \right)^{\mathsf{T}} \quad (i = 1, \ldots, d; j = 1, \ldots, T),$$

having squared length $|c'_{ij}|^2$ bounded by $N(\max(1, |\alpha_i|))^{2Np_j}$. Thus,

$$\prod_{i=1}^{d} \prod_{j \leq T}^{S} |c'_{ip_j}|^2 \leq N^{dT} M(\alpha)^{2N \sum_{j \leq T} p_j} .$$

Hence, using these two upper bounds in Hadamard's Inequality (Lemma 3.2), we have that

$$|V|^2 \leq \prod_{i=1}^{d} \prod_{m=1}^{S} |c_{im}|^2 \cdot \prod_{i=1}^{d} \prod_{j \leq T}^{S} |c'_{ip_j}|^2$$

$$\leq N^{d(S^2+T)} M(\alpha)^{2N(S+\sum_{j \leq T} p_j)} .$$

We shall need the following estimates, which are consequences of the Prime Number Theorem:

$$\sum_{j \leq T} p_j = \frac{1}{2}(T^2 \log T)(1 + o(1)) \tag{3.8}$$

and

$$\sum_{j \leq T} \log p_j = T \log T (1 + o(1)) . \tag{3.9}$$

These are stated explicitly, for instance, in [Axl19, (1.8)] and [Dus18, p. 246], respectively, (3.8) giving

$$V^2 \leq N^{d(S^2+T)} M(\alpha)^{2N\left(S+\frac{1}{2}(T^2 \log T)(1+o(1))\right)} .$$

On the other hand, we have from Lemma 3.12 that

$$V^2 = \prod_{\substack{1 \leq i,i' \leq d \\ i \neq i'}} |\alpha_i - \alpha_{i'}|^{S^2} \cdot \prod_{\substack{1 \leq i,i' \leq d \\ j,j' \leq T \\ (i,j) \neq (i',j')}} \left|\alpha_i^{p_j} - \alpha_{i'}^{p_{j'}}\right| \cdot \prod_{\substack{1 \leq i,i' \leq d \\ j \leq T}} \left|\alpha_i^{p_j} - \alpha_{i'}\right|^{2S} .$$

Now it could be that for some j and $i \neq i'$ we have $\alpha_i^{p_j} = \alpha_{i'}^{p_j}$, making $V = 0$. However, for the moment, we assume that this does not happen. Then we have that the symmetric function

$$\prod_{\substack{1 \leq i,i' \leq d \\ i \neq i'}} |\alpha_i - \alpha_{i'}|^{S^2} \cdot \prod_{\substack{1 \leq i,i' \leq d \\ j,j' \leq T \\ (i,j) \neq (i',j')}} \left|\alpha_i^{p_j} - \alpha_{i'}^{p_{j'}}\right|$$

is at least 1 and, from Lemma A.23,

$$\prod_{\substack{1 \le i, i' \le d \\ j \le T}} |\alpha_i^{p_j} - \alpha_{i'}|^{2S} \ge \left(\prod_{j \le T} p_j \right)^{2Sd} = e^{2Sd \sum_{j \le T} \log(p_j)}.$$

Now from (3.9)

$$e^{2Sd(T \log T)(1+o(1))} \le V^2 \le N^{d(S^2+T)} M(\alpha)^{2N(S+\frac{1}{2}(T^2 \log T)(1+o(1)))}.$$

Recalling that $N = d(S + T)$ and taking logs, we have

$$\log(M(\alpha)) \ge \frac{2S(T \log T)(1 + o(1)) - (S^2 + T) \log(d(S + T))}{2(S + T) \left(S + \frac{1}{2}(T^2 \log T)(1 + o(1))\right)}. \tag{3.10}$$

Now put $u := \log d$, $\ell := \log \log d$,

$$S := 2 \left\lfloor \frac{u}{2\ell} \right\rfloor \quad \text{and} \quad T := \frac{S^2}{2}.$$

Then the numerator of the lower bound in (3.10) is asymptotically

$$\frac{2u^3}{\ell^2}(1 + o(1)) - \frac{3u^2}{2\ell^2} u(1 + o(1)),$$

while its denominator is asymptotically

$$\frac{u^6}{4\ell^5}(1 + o(1)).$$

Thus, their quotient is $2S^{-3}(1 + o(1))$. This gives our result when $\alpha_i^{p_j} \ne \alpha_{i'}^{p_j}$ for all j and $i \ne i'$. But if we do have $\alpha_i^{p_j} = \alpha_{i'}^{p_j}$ for some j and $i \ne i'$, then we can invoke Lemma 3.10 to show that the result still holds. □

Exercise 3.14 Suppose that instead of setting $T = S^2/2$ in the above proof, T is defined as $\lfloor \lambda S^2 \rfloor$, for some constant $\lambda > 0$. Show that the value of λ to take to give the best lower bound in the theorem is $\lambda = 1/2$.

3.3 Notes

The proof of Theorem 3.1 presented is essentially that given by Cantor and Straus [CS83], based on the confluent Vandermonde determinant. Dobrowolski in fact had $1 - \varepsilon$ instead of $2 - \varepsilon$ in his result, while Cantor and Straus had $2 - \varepsilon$. Louboutin [Lou83] improved this constant to $9/4 - \varepsilon$. Dobrowolski also proved a version of

Theorem 3.1 valid for all $d \geq 2$, but with $1 - \varepsilon$ replaced by $1/1200$. For $d \geq 2$, Voutier [Vou96] improved this constant to $1/4$.

The first results that gave an unconditional lower bound for $M(\alpha)$, for α nonzero and not a root of unity, all used the arithmetic fact that $\prod_{j=1}^{d}(\alpha_j^k - 1)$ is a nonzero integer (the α_j the conjugates of α, and k any positive integer). These include the results of Schinzel and Zassenhaus [SZ65], Blanksby and Montgomery [BM71] and Stewart [Ste78a, Ste78b]. Dobrowolski's breakthrough was his use of the extra arithmetic information in his Lemma A.23. Note that in his proof he uses only the bound $|\prod_{i \neq j}(\alpha_i - \alpha_j)| \geq 1$ for the discriminant of the minimal polynomial of α. Better such discriminant bounds are known and indeed were used by Voutier in the proof of his result mentioned above. See Brunault and Zudilin [BZ20, Sect. 2.3] for an interesting discussion of discriminant bounds. It may be that stronger lower bounds for this discriminant, nearer to best possible, could lead to a proof of Lehmer's Conjecture.

The proof of Hadamard's Inequality essentially follows the second proof in [MS99a]—see also Wikipedia (Hadamard's Inequality). Lemma 3.7 is due to Mahler [Mah64].

Lemmas 3.2 and 3.7 were combined and generalised by Dubickas [Dub93] as follows.

Lemma 3.15 *If* $Z := |a_{ij} z_j^{i-1}|_{i,j=1,\ldots,n}$, *where* $a_{ij}, z_j \in \mathbb{C}$ *and* $|z_1| \geq |z_2| \geq \cdots \geq |z_n|$, *then*

$$|Z| \leq |z_1|^{n-1} |z_2|^{n-2} \cdots |z_{n-1}| \prod_{j=1}^{n} \left(\sum_{i=1}^{n} |a_{ij}|^2 \right)^{1/2}.$$

3.4   Glossary

X^*.   The conjugate transpose of the matrix X, namely $\overline{X}^{\mathsf{T}} = \overline{X^{\mathsf{T}}}$.

confluent Vandermonde matrix.   For a complex number v and natural numbers N and m, let $M(v, N, m)$ be the $m \times N$ matrix whose jth column contains as its ith entry the coefficient of t^{j-1} in $(v + t)^{i-1}$. A confluent Vandermonde matrix is constructed using some of these $M(v, N, m)$ matrices as blocks, as follows. Given N, let $m_1, \ldots, m_k$ be natural numbers that sum to N and let $v_1, \ldots, v_k$ be distinct complex numbers. Then the $N \times N$ matrix obtained by stacking the k matrices $M(v_1, N, m_1), \ldots, M(v_k, N, m_k)$ next to each other, as in (3.1), is a confluent Vandermonde matrix.

orthogonal.   Two complex vectors $\mathbf{v}$ and $\mathbf{w}$ are orthogonal if their usual inner product $\mathbf{v} \cdot \overline{\mathbf{w}} = 0$, noting the complex conjugation of one of the two vectors.

Vandermonde matrix. A matrix of the shape

$$
\begin{pmatrix}
1 & 1 & 1 & \cdots & 1 \\
\alpha_1 & \alpha_2 & \alpha_3 & \cdots & \alpha_n \\
\alpha_1^2 & \alpha_2^2 & \alpha_3^2 & \cdots & \alpha_n^2 \\
\vdots & \vdots & \vdots & \ddots & \vdots \\
\alpha_1^{n-1} & \alpha_2^{n-1} & \alpha_3^{n-1} & \cdots & \alpha_n^{n-1}
\end{pmatrix}
$$

for some numbers $\alpha_1, \ldots, \alpha_n$, so that the (i, j) entry is α_j^{i-1}. The transpose of this matrix is often used instead in the literature. If any two of the α_i are equal, the matrix is singular.

Chapter 4
The Schinzel–Zassenhaus Conjecture

4.1 Introduction

Let α be a nonzero algebraic integer, not a root of unity, with conjugates $\alpha_1 = \alpha, \alpha_2, \ldots, \alpha_d$. In 1965, Schinzel and Zassenhaus [SZ65] stated that '…we cannot disprove the inequality

$$\max_{1 \leq j \leq d} |\alpha_i| > 1 + \frac{c}{d} \tag{4.1}$$

for some absolute constant $c > 0$'. The truth of this inequality has since been routinely referred to as the 'Schinzel–Zassenhaus Conjecture'. In their paper, Schinzel and Zassenhaus proved the weaker inequality

$$\max_{1 \leq j \leq d} |\alpha_i| > 1 + 2^{-2s-4}$$

when α has $2s > 0$ nonreal conjugates. Recently, however, the conjecture has been proved by Vesselin Dimitrov. Using $\boxed{\alpha} := \max_{1 \leq j \leq d} |\alpha_j|$, the **house** of α, we can state his result as follows.

Theorem 4.1 (Dimitrov [Dim19]) *Let α be a nonzero algebraic integer of degree d, not a root of unity, whose conjugates have d' different phases between them. Then $\boxed{\alpha} \geq 2^{1/(4d')}$.*

Thus, from $d' \leq d$, we have the inequality $\boxed{\alpha} \geq 2^{1/(4d)}$ and so the following.

Corollary 4.2 *The Schinzel–Zassenhaus Conjecture holds with $c = \frac{1}{4}\log 2$.*

Corollary 4.3 *If α is a reciprocal algebraic integer of degree $d > 1$ with $d'' > 0$ conjugates not on the unit circle, then $\boxed{\alpha} \geq 2^{1/(4d-2d'')}$. In particular, if $d'' = d$, then $\boxed{\alpha} \geq 2^{1/(2d)}$.*

For then the conjugates of α have at most $d - \frac{1}{2}d''$ different phases.

© Springer Nature Switzerland AG 2021
J. McKee and C. Smyth, *Around the Unit Circle*, Universitext,
https://doi.org/10.1007/978-3-030-80031-4_4

4.1.1 A Simple Proof of a Weaker Result

Before proving the full strength of the Schinzel–Zassenhaus Conjecture, we first present a simple argument, proving a slightly weaker result.

Proposition 4.4 *If a nonzero algebraic integer α of degree d satisfies*

$$\lceil \alpha \rceil \leq 1 + \frac{1}{4ed^2}, \tag{4.2}$$

then α is a root of unity.

Proof As usual let $s_k = \alpha_1^k + \cdots + \alpha_d^k$, where $\alpha = \alpha_1, \ldots, \alpha_d$ are the conjugates of α. Let p be a prime between $2ed$ and $4ed$. If (4.2) holds, then if $k \leq d$

$$|s_k| \leq d \left(1 + \frac{1}{4ed^2}\right)^d < de$$

and

$$|s_{kp}| \leq d \left(1 + \frac{1}{4ed^2}\right)^{4ed^2} < de.$$

Hence,

$$|s_{kp} - s_k| \leq 2ed < p.$$

But we know from Theorem A.33 that $s_{kp} \equiv s_k \pmod{p}$, so that $s_{kp} = s_k$ for $k = 1, \ldots, d$. Since these values determine the coefficients of the minimal polynomials of α and α^p (using Theorem A.31), we see that they have the same minimal polynomial. Hence, α^p is a zero of the minimal polynomial of α. Thus, α is conjugate to α^p and so, applying appropriate automorphisms, also to α^{p^m} for all m. So two of these powers must be equal.

4.2 Proof of Dimitrov's Theorem

For the proof of Theorem 4.1, we need the following results.

Proposition 4.5 *A necessary and sufficient condition for the series*

$$f(z) = \sum_{j=0}^{\infty} f_j z^j \in \mathbb{C}[[z]]$$

to represent a rational function is that for the matrices

$$\Delta_k = \begin{pmatrix} f_0 & f_1 & \cdots & f_k \\ f_1 & f_2 & \cdots & f_{k+1} \\ \vdots & \vdots & \vdots & \vdots \\ f_k & f_{k+1} & \cdots & f_{2k} \end{pmatrix}, \tag{4.3}$$

their determinants $\det \Delta_k$ *are* 0 *for all k sufficiently large.*

Proof Suppose first that $f(z)$ is a rational function, $P(z)/Q(z)$ say, with $Q(z) = q_r + q_{r-1}z + \cdots + q_0 z^r$ and the q_j not all 0. Then, from $Qf = P$, we see that

$$q_0 f_k + q_1 f_{k+1} + \cdots + q_r f_{k+r} = 0 \text{ for } k > \deg P. \tag{4.4}$$

Thus, for $k > r + \deg P$, the rightmost $r + 1$ columns of Δ_k are linearly dependent, so $\det \Delta_k = 0$.

Conversely, suppose that $\det \Delta_k = 0$ for all $k \geq p$, where we can assume that p is the smallest integer with this property. Then the rightmost column of Δ_p is a linear combination of the first p columns of Δ_p, so that for some $q_0, q_1, \ldots, q_{p-1}$, we have

$$L_{p+j} := q_0 f_j + q_1 f_{j+1} + \cdots + q_{p-1} f_{j+p-1} + f_{p+j} = 0 \quad (j = 0, 1, \ldots, p).$$

We now show by induction that $L_{p+j} = 0$ for all $j \geq 0$. Assume that for some $m > p$ we have $L_{p+j} = 0$ for $j = 0, 1, 2, \ldots, m - 1$. We need to show that $L_{p+m} = 0$.

Let us write Δ_m as

$$\Delta_m = \left(\begin{array}{ccc|ccc} & & & f_p & \cdots & f_m \\ & \Delta_{p-1} & & \vdots & \vdots & \vdots \\ & & & f_{2p-1} & \cdots & f_{p+m-1} \\ \hline f_p & \cdots & f_{2p-1} & f_{2p} & \cdots & f_{p+m} \\ \vdots & \vdots & \vdots & \vdots & \ddots & \vdots \\ f_m & \cdots & f_{p+m-1} & f_{p+m} & \cdots & f_{2m} \end{array} \right).$$

Now, starting at the $(p + 1)$th column, add to each column a linear combination of the previous p columns with coefficients $q_0, q_1, \ldots, q_{p-1}$. This gives

$$\det \Delta_m = \begin{vmatrix} & & & L_p & \cdots & L_m \\ & \Delta_{p-1} & & \vdots & \vdots & \vdots \\ & & & L_{2p-1} & \cdots & L_{p+m-1} \\ f_p & \cdots & f_{2p-1} & L_{2p} & \cdots & L_{p+m} \\ \vdots & \vdots & \vdots & \vdots & \ddots & \vdots \\ f_m & \cdots & f_{p+m-1} & L_{p+m} & \cdots & L_{2m} \end{vmatrix}$$

$$= \begin{vmatrix} & \Delta_{p-1} & & & O & \\ f_p & \cdots & f_{2p-1} & O & & L_{p+m} \\ \vdots & \vdots & \vdots & & \ddots & \vdots \\ f_m & \cdots & f_{p+m-1} & L_{p+m} & \cdots & L_{2m} \end{vmatrix}$$

$$= \pm (L_{p+m})^{m-p+1} \det \Delta_{p-1} .$$

Then since $\det \Delta_m = 0$ and $\det \Delta_{p-1} \neq 0$, we have $L_{p+m} = 0$, completing the induction step. Thus, $f(z)$ represents a rational function with numerator of degree $p - 1$ and denominator of degree p. □

The following result, along with the remainder of this chapter, requires some knowledge of the transfinite diameter of a set E, written $\tau(E)$. This is covered in Chap. 10, and those unfamiliar with this theory may wish to skim the remainder of the current chapter and return to it later. The proof of Dimitrov's Theorem is motivation enough to study the theory of transfinite diameters!

We write $\widehat{\mathbb{C}}$ for $\mathbb{C} \cup \{\infty\}$.

Theorem 4.6 *Let E be a compact subset of $\mathbb{C}$ that is symmetric about the real axis, so that $\overline{E} = E$, where $\overline{E}$ is the set of complex conjugates of elements of E. Suppose that the transfinite diameter $\tau(E)$ of E is less than 1. Suppose too that*

$$f(z) = \sum_{j=0}^{\infty} \frac{a_j}{z^j}$$

is regular on the complement $\widehat{\mathbb{C}} \setminus E$ of E and that all of the a_j are integers. Then $f(z)$ is a rational function.

Proof Let $\varepsilon > 0$, and E_ε be an ε-thickening of E, as defined in Proposition 10.1(d). From there we can, by choosing ε sufficiently small, ensure that E_ε also has transfinite diameter, τ_ε say, less than 1. Then also $f(1/z)$ is regular in $\widehat{\mathbb{C}} \setminus E_\varepsilon$.

Next, consider the $k \times k$ matrices

$$\Delta_k := \begin{pmatrix} a_1 & a_2 & \cdots & a_k \\ a_2 & a_3 & \cdots & a_{k+1} \\ \vdots & \vdots & \vdots & \vdots \\ a_k & a_{k+1} & \cdots & a_{2k} \end{pmatrix}$$

and $C_k := (c_{ij})_{i,j=0,1,\ldots,k-1}$, where c_{ij} is the residue of $f(1/z)T_i(z)T_j(z)$ at $z = 0$. Here, the polynomials $T_i(z)$ of degree i are chosen as in Proposition 10.1(b), but for the set E_ε; they have maximum modulus m_i on E_ε. Now define the $k \times k$ upper triangular matrix $B_k := (b_{ij})_{i,j=0,1,\ldots,k-1}$ where for a given j and $i \le j$ its entries are defined by $T_j(z) = b_{jj}z^j + b_{j-1,j}z^{j-1} + \cdots + b_{0j}$. Note that, as the polynomials T_j are all monic (so $b_{jj} = 1$), the matrix B_k has all its diagonal entries equal to 1. Then we have the identity

$$B_k^\mathsf{T} \Delta_k B_k = C,$$

because the matrices on both sides of this equation have (i, j)th entry

$$\sum_{0 \le \ell \le i} \sum_{0 \le m \le j} b_{\ell i} a_{\ell+m+1} b_{mj} \quad \text{for} \quad 0 \le i \le k-1, \ 0 \le j \le k-1.$$

Hence, $\det \Delta_k = \det C$. This identity gives us a way of bounding $|\det \Delta_k|$ above, as we shall see.

We note that $f(1/z)$ is bounded on the boundary of E_ε, where ε has been chosen as above. Hence, by Cauchy's Integral Theorem, $|c_{ij}|$ is bounded above by a constant times $m_i m_j$. We can choose a number τ_2 with $\tau_\varepsilon < \tau_2 < 1$ and then an integer I such that for all $i \ge I$ we have $m_i < \tau_2^i$. Next, we apply Hadamard's Inequality (Lemma 3.2) to $\det C$. It is not difficult to check that for $i < I$ there is a constant, c_1 say, with $\sum_{j=1}^\infty |c_{ij}|^2 \le c_1$, independent of k, while for $i \ge I$ there is a constant c_2 with $\sum_{j=1}^\infty |c_{ij}|^2 \le c_2\tau_2^{2i}$. Then Hadamard gives that for some constants c_3, c_4 we have that $|\det \Delta_k|^2 \le c_3(c_4\tau_2^k)^k$. Because each $\det \Delta_k$ is an integer, they must be 0 for all k sufficiently large. Then Proposition 4.5 tells us that $f(z)$ is rational. $\square$

To complete the preliminaries for the proof of Dimitrov's Theorem, we need one final auxiliary result. This is due to Dubinin [Dub84], [Dub14]. We state it without proof here. See also Baernstein [Bae87] for an alternative proof.

Proposition 4.7 *Let* $z_1, \ldots, z_N \in \mathbb{C}$, *and* $\mathcal{K}(z_1, \ldots, z_N)$ *be the 'hedgehog' formed by the union of the line segments joining 0 to* z_j ($j = 1, \ldots, N$). *Then* $\mathcal{K}(z_1, \ldots, z_N)$ *has transfinite diameter at most* $4^{-1/N} \max_{1 \le j \le N} |z_j|$.

Now taking $h(z) := z^N$ and $E := [0, L]$, we see from Proposition 10.1(c) that the hedgehog $h^{-1}(E)$ has transfinite diameter $(L/4)^{1/N}$ which, for $L := (\max_{1 \le j \le N} |z_j|)^N$, gives $4^{-1/N} \max_{1 \le j \le N} |z_j|$. Thus, the upper bound in Proposition 4.7 is attained. The difficult part of the proof, which we do not cover, is to show that if the 'spikes' of the hedgehog are not equally spaced, then its transfinite diameter is strictly smaller.

Proof (of Theorem 4.1) In what follows, let α be a nonzero algebraic integer of degree d having minimal polynomial $P(z)$, and $P_m(z)$ be the polynomial whose zeros are the mth powers of the conjugates of α, as in Chap. 2.

Since the Maclaurin series of $(1 + 4Y)^{-1/2} = \sum_{j=0}^{\infty} (-1)^j \binom{2j}{j} Y^j$ has integer coefficients, so does that of $\sqrt{1 + 4Y}$. (Its sequence of coefficients is, up to sign, $2\times$ the sequence of Catalan numbers.) Now from Corollary A.35 we have that $P_4(z) = P_2(z) + 4R(z)$ for some $R(z) \in \mathbb{Z}[z]$ of degree at most $d - 1$. Then we obtain

$$\sqrt{P_2(z)P_4(z)} = \sqrt{P_2(z)(P_2(z) + 4R(z))} = P_2(z)\sqrt{1 + 4R(z)/P_2(z)}.$$

We now rewrite this identity in terms of reciprocal polynomials, defined in Sect. A.1 of Appendix A. So, on dividing by z^d, rewriting this identity in terms of the reciprocal polynomials P_2^*, P_4^* and R^* and using the fact that $P_2^*(0) = 1$, we see that

$$F(z) := \sqrt{P_2^*(\tfrac{1}{z})P_4^*(\tfrac{1}{z})} = P_2^*(\tfrac{1}{z})\sqrt{1 + 4z^{-(d-\deg R)}R^*(\tfrac{1}{z})/P_2^*(\tfrac{1}{z})}$$

lies in $\mathbb{Z}[[1/z]]$ and has zeros precisely at α_i^2, α_i^4 $(i = 1, \ldots, d)$, where the α_i are the conjugates of α.

The next step in Dimitrov's proof is to construct a set E of transfinite diameter less than 1 containing all the nonregular points z_j of $F(z)$. This is done using Proposition 4.7. First take $N = 2d$ and the z_j to be $\alpha_1^2, \ldots, \alpha_d^2$ and $\alpha_1^4, \ldots, \alpha_d^4$. Because these numbers z_j have at most $2d'$ different phases, at most $2d'$ of them are needed to define $\mathcal{K}$. So, by Proposition 4.7, $\mathcal{K}$ has transfinite diameter at most $2^{-1/d'}\overline{|\alpha|}^4$. So if $\overline{|\alpha|} < 2^{1/(4d')}$, then $\mathcal{K}$ has transfinite diameter less than 1. Hence, $F(z)$ is rational, by Theorem 4.6, and, having no poles except at $z = 0$, is in fact a polynomial in $1/z$.

Finally, assume that Theorem 4.1 is true for all nonzero algebraic integers of degree less than d, not a root of unity and that α is such a number, but of degree d. If $P_2(z)$ is irreducible, then, as $z^d F(z)$ is a polynomial, $P_4(z) = P_2(z)$. However, the maximal modulus of a zero of $P_2(z)$ is $\overline{|\alpha|}^2$, while the maximal modulus of a zero of $P_4(z)$ is $\overline{|\alpha|}^4$. So α is a root of unity, by Kronecker's First Theorem 1.3. But this is contrary to our assumption.

We, therefore, have that $P_2(z)$ is reducible, with α^2 one of its zeros. Hence, the degree of α^2 is less than that of α so, by assumption, the conclusion of the theorem holds for α^2. Also, since a pair $\alpha_j, -\alpha_j$ must occur among the conjugates of α, it follows from Galois theory that the phases of the conjugates of α are in pairs $\theta, \theta + \pi$, say. Hence, the numbers α_j^2 have at most $\frac{1}{2}d'$ phases between them. Thus, $\overline{|\alpha^2|} \geq 2^{1/(4(d'/2))}$, or $\overline{|\alpha|} \geq 2^{1/(4d')}$. Thus, the Theorem is true for α of degree d. $\square$

4.3 Notes

Proposition 4.4 is due to Dobrowolski (unpublished, 1977). Soon afterwards he showed [Dob78], by only a slightly more complicated argument, that the proposition was true with (4.2) replaced by the weaker hypothesis

$$\lceil\alpha\rceil \le 1 + \frac{\log d}{6d^2}. \tag{4.5}$$

Proposition 4.5 is due to Kronecker [Kro81, pp. 566–567]. Our proof follows Salem [Sal63, pp. 5–7]. Theorem 4.6 is due to Pólya [Pól28]. Our proof is based on the outline given by Dimitrov in [Dim19].

The Schinzel–Zassenhaus Conjecture would also follow from the truth of Lehmer's Conjecture, were the latter to be proved. Suppose that for α a nonzero algebraic integer, not a root of unity, we had that $M(\alpha) \ge C > 1$, say. Then if $\lceil\alpha\rceil \le 1 + \frac{\log C}{d}$, we would have $M(\alpha) < (1 + \frac{\log C}{d})^d < C$. Therefore, the Schinzel–Zassenhaus Conjecture (4.1) would hold with $c = \log C$. Results in the direction of Lehmer's Conjecture give corresponding results for the Schinzel–Zassenhaus Conjecture. For instance, Dobrowolski's Theorem 3.1 implies that there is a $c > 1$ such that for nonzero α of degree d, not a root of unity, and $d > d_1(\varepsilon)$ that

$$\lceil\alpha\rceil > 1 + \frac{c - \varepsilon}{d}\left(\frac{\log\log d}{\log d}\right)^3.$$

The best such c calculated was $64/\pi^2$ by Dubickas [Dub93]. These results are of course all superseded by Theorem 4.1.

In the case of α nonreciprocal, we can conclude from Theorem 12.1 that (4.1) holds for such α with $c = \log\theta_0 = 0.2811\cdots$, where $\theta_0 = 1.3247\cdots$ is the real zero of $x^3 - x - 1$. Dubickas [Dub97] has improved the value of c for such α to $c = 0.3096\cdots$. Bazylewicz [Baz88] generalised the result in the nonreciprocal case to integer polynomials in several variables.

Pritsker [Pri21, Theorem 1] has recently improved the value of c for reciprocal α having no conjugates on the unit circle from $\frac{1}{2}\log 2 = 0.3465\cdots$ in Corollary 4.3 to $\frac{1}{2}\log(1 + \sqrt{2}) = 0.44068\cdots$.

4.4 Glossary

$P_m(z)$. The polynomial whose zeros are the mth powers of the zeros of P.

house. Let α be an algebraic integer. The house of α, written $\lceil\alpha\rceil$, is the maximum of the absolute values of the conjugates of α.

Chapter 5
Roots of Unity and Cyclotomic Polynomials

5.1 Introduction

A *root of unity* ω is a root in $\mathbb{C}$ of $z^n = 1$ for some n. For a given ω, choose n to be as small as possible; then, ω is a *primitive* nth root of unity. In this chapter, ω_n will denote a primitive nth root of unity. Denote by $\Phi_n(z)$ the *nth cyclotomic polynomial*

$$\Phi_n(z) = \prod_{\substack{j=1 \\ \gcd(j,n)=1}}^{n} (z - \omega_n^j). \tag{5.1}$$

We shall show below that $\Phi_n(z)$ is irreducible, and thus is the minimal polynomial of all primitive nth roots of unity. Thus, its degree is the Euler function $\varphi(n)$, defined as the number of positive integers at most n and coprime to n.

Exercise 5.1 Show that every conjugate of a primitive nth root of unity is again a primitive nth root of unity.

We also need the Möbius function $\mu(n)$, defined as $(-1)^k$ if n is square-free with k prime factors, and 0 otherwise. The following facts are well known.

Proposition 5.2 *(a) If ω_n is a primitive nth root of unity, then the set of all primitive nth roots of unity is given by ω_n^j for $j = 1, \ldots, n$ and coprime to n. Furthermore, their sum is $\mu(n)$.*

(b) The polynomial $\Phi_n(z)$ has integer coefficients, is irreducible, and is therefore the minimal polynomial of all primitive nth roots of unity.

(c) (See [Neu99, Proposition 10.2]) The roots of unity $1, \omega_n, \omega_n^2, \ldots, \omega_n^{\varphi(n)-1}$ form an integral basis for the ring of integers of $\mathbb{Q}(\omega_n)$.

(d) The Galois group $\mathrm{Gal}(\mathbb{Q}(\omega_n)/\mathbb{Q})$ is isomorphic to the abelian group $(\mathbb{Z}/n\mathbb{Z})^{\times}$ of residue classes $j \pmod n$ that are coprime to n, under the bijection defined by $(\omega_n \mapsto \omega_n^j) \leftrightarrow j$.

© Springer Nature Switzerland AG 2021
J. McKee and C. Smyth, *Around the Unit Circle*, Universitext,
https://doi.org/10.1007/978-3-030-80031-4_5

(e) **(Kronecker–Weber theorem)** *Let K be an abelian extension of $\mathbb{Q}$ of degree d and discriminant Δ. Define*

$$2^* := \begin{cases} 2 \ \textit{if } d \textit{ and } \Delta \textit{ are both even} \\ 1 \ \textit{otherwise} \end{cases} \quad \textit{and} \quad m := 2^* \prod_{\substack{\textit{primes } p: \\ p | \Delta \textit{ and } p^a \| d}} p^{a+1}.$$

Then K is a subfield of $\mathbb{Q}(\omega_m)$.

In (e), the smallest m with $K \in \mathbb{Q}(\omega_m)$ is called the **conductor** of K.

Proof We shall prove only parts (a), (b) and (d). The first sentence of (a) is immediate. Then we see that

$$\prod_{j|n} \Phi_j(z) = z^n - 1.$$

Taking the reciprocal polynomials of both sides of this identity gives

$$(1 - z) \prod_{1 < j | n} \Phi_j(z) = 1 - z^n.$$

Here, we have used the fact that $z^{\deg \Phi_j} \Phi_j(1/z) = \Phi_j(z)$ for $j > 1$. Now apply the multiplicative version of Möbius inversion [Apo76, Theorem 2.9] to this identity ('take logs, apply Möbius inversion, exponentiate'), to obtain, for $n > 1$,

$$\Phi_n(z) = \prod_{j|n}(1 - z^{n/j})^{\mu(j)}. \tag{5.2}$$

Then for $n > 1$, the coefficient of z of $\Phi_n(z)$ is $-\sum_{j : \gcd(j,n)=1} \omega_n^j$, while the right-hand side of (5.2), expanded (mod z^2) as a series in z, is $(1 - z)^{\mu(n)} \equiv 1 - \mu(n)z$.

For part (b), there are many proofs—see the notes at the end of the chapter. We give one here, depending on Dirichlet's theorem on primes in arithmetic progressions [Dir37]—see also [Sel49].

Let $n > 2$ and ω, ω^k be two primitive nth roots of unity. Then k is prime to n, with $kk' \equiv 1 \pmod{n}$, say. Next, note that since $\omega^k \in \mathbb{Q}(\omega)$ and $\omega = (\omega^k)^{k'} \in \mathbb{Q}(\omega^k)$, the degrees $[\mathbb{Q}(\omega) : \mathbb{Q}]$ and $[\mathbb{Q}(\omega^k) : \mathbb{Q}]$ are the same, so that the minimal polynomials $f(z)$ of ω over $\mathbb{Q}$ and $f_k(z)$ of ω^k over $\mathbb{Q}$ have the same degree, N say, where $N \le \varphi(n)$.

Now Dirichlet's theorem tells us that because $\gcd(k, n) = 1$, there are infinitely many primes p of the form $p = k + rn$. Thus, we can choose such a prime with $p > 2^{\varphi(n)}$.

Next, we know from the Multinomial theorem and Fermat's little theorem that

$$f(z^p) \equiv f(z)^p \pmod{p},$$

so that, if $f(z) = \sum_\ell a_\ell z^l$, then

$$f(z^p) = \sum_\ell a_\ell z^{\ell p} = \left(\sum_\ell a_\ell z^l\right)^p + g(z)p\,,$$

say, where $g(z) \in \mathbb{Z}[z]$. Hence, since $\omega^p = \omega^k$ we have

$$f(\omega^k) = g(\omega)p\,.$$

If $g(\omega) \neq 0$, then by Kronecker's first theorem (Theorem 1.3) $g(\omega)$, being an algebraic integer, has a conjugate $g(\omega_*)$ say, of modulus at least 1. Then

$$p \leq |g(\omega_*)|p = |f(\omega_*^k)| = \prod_{\omega'}(\omega_*^k - \omega')| \leq 2^{\varphi(n)}\,,$$

a contradiction. Here, the product is taken over the (at most $\varphi(n)$) conjugates ω' of ω. Hence,

$$f(\omega^k) = g(\omega)p = 0\,,$$

so that ω^k is a conjugate of ω.

For part (d), let $\omega := \omega_n$ and for any j with $\gcd(j, n) = 1$ define the map $\sigma_j \in \mathrm{Gal}(\mathbb{Q}(\omega)/\mathbb{Q})$ by $\sigma_j(\omega) := \omega^j$. Then for all k, we have $\sigma_j(\omega^k) = \omega^{jk}$. Also, all $\sigma \in \mathrm{Gal}(\mathbb{Q}(\omega)/\mathbb{Q})$ are equal to σ_j for some j, since $\mathrm{Gal}(\mathbb{Q}(\omega)/\mathbb{Q})$ acts transitively on the conjugates of ω, as given by parts (a) and (b). Then if $\gcd(k, n) = 1$ and $F : \mathrm{Gal}(\mathbb{Q}(\omega)/\mathbb{Q}) \to (\mathbb{Z}/n\mathbb{Z})^\times$ is defined as $F(\sigma_j) = j$, then

$$\sigma_j(\sigma_k(\omega)) = \sigma_j(\omega^k) = \omega^{jk} = \sigma_{jk}(\omega)$$

so that

$$F(\sigma_j \circ \sigma_k) = jk = F(\sigma_j)F(\sigma_k)\,.$$

$\square$

As a consequence, every conjugate set of roots of unity consists of a set of all primitive nth roots of unity, for some n. The minimal polynomial of the primitive nth roots of unity is the nth cyclotomic polynomial, $\Phi_n(z)$, as in (5.1). While these polynomials are traditionally called the cyclotomic polynomials, recall that we use the term more generally to denote any monic integer polynomial having all its zeros on $|z| = 1$. Such a polynomial is of course a product of polynomials $\Phi_n(z)$, taken over some multiset of values of n.

5.2   Solving Polynomial Equations in Roots of Unity

We first prove a result which turns out to be very useful for finding zeros of integer polynomials that are roots of unity. We call such zeros **cyclotomic zeros**. Clearly, this is equivalent to finding the cyclotomic factors of the polynomial.

Proposition 5.3 *Every root of unity ω is conjugate to exactly one of $-\omega$, ω^2 or $-\omega^2$.*

Conversely, suppose that α is a nonzero algebraic number that is conjugate to $-\alpha$, α^2 or $-\alpha^2$. Then in the first case, the minimal polynomial of α is a polynomial in z^2, in the second case $\alpha = \omega_n$ for some primitive root of unity ω_n with n odd, while in the third case $\alpha = \omega_n$ for some primitive root of unity ω_n with $n = 2n'$, n' odd.

Proof Suppose that $\omega = \omega_n$, a primitive nth root of unity. If $n = 4n'$ then $\omega_n^{2n'} = -1$ and $\omega_n^{2n'+1} = -\omega_n$ is a conjugate of ω_n, since $\gcd(2n' + 1, n) = 1$. If n is odd, then ω^2 is a conjugate of ω_n, since $\gcd(2, n) = 1$. If $n = 2n'$ with n' odd, then $\omega_n^{n'} = -1$ and $\omega_n^{n'+2} = -\omega^2$ is a conjugate of ω_n, since $\gcd(n' + 2, n) = 1$.

For the first converse part, write the minimal polynomial of α as

$$P(z) = P_1(z^2) + zP_2(z^2),$$

as we clearly can. Note that $P_1(z) \neq 0$. But then $0 = P(\alpha) = P_1(\alpha^2) + \alpha P_2(\alpha^2)$ and $0 = P(-\alpha) = P_1((-\alpha)^2) - \alpha P_2((-\alpha)^2)$ gives $P_1(\alpha^2) = 0$ and $P_2(\alpha^2) = 0$. Since α^2 has exactly half as many conjugates as α, the degree of α must be even. Hence $\deg P = 2 \deg P_1$. Thus, the degree of P_2 must be less than that of P_1. Since $P_2(\alpha^2) = 0$, P_2 must be identically 0, and $P(z) = P_1(z^2)$.

For the second and third converse parts, it is enough to show that α must be a root of unity. Indeed, if it is not, by Corollary A.17 in Appendix A there is some valuation $|\ |_p$ on the field $\overline{\mathbb{Q}}$ of algebraic numbers such that $|\alpha|_p > 1$. Choosing such an α with $|\alpha|_p$ maximal, we have that α has a conjugate $\pm\alpha^2$ with $|\pm\alpha^2|_p = |\alpha|_p^2 > |\alpha|_p$, contradicting maximality. Hence, $|\alpha|_p = 1$ for all p, and so α is a root of unity. ∎

Exercise 5.4 Suppose that $P(z) \in \mathbb{C}[z]$ is a polynomial with the property that for every zero α of P, at least one of $\pm\alpha^2$ is also a zero, and $P(0) \neq 0$. Show that then all zeros of P are roots of unity.

Exercise 5.5 For a primitive nth root of unity ω_n, show that

$$1 + \omega_n^k + \omega_n^{2k} + \cdots + \omega_n^{(n-1)k} = \begin{cases} n \text{ if } n \mid k; \\ 0 \text{ otherwise.} \end{cases}$$

Suppose that we want to find all the cyclotomic factors of a given polynomial $P(z) \in \mathbb{Z}[z]$, with or without multiplicities. We now give algorithms to perform these tasks. We assume that $P(z)$ is monic, and not divisible by z. Firstly, define

$$(CP)(z) = \prod_{\substack{P(\omega)=0 \\ \omega \text{ a root of } 1}} (z - \omega),$$

where the zeros all appear with multiplicity 1. Then define GP by

$$(GP)(z) := \gcd(P(z), P(z^2) \cdot P(-z^2)).$$

If $\gcd(P(z), P(-z)) = 1$, then we have the simple formula $(CP)(z) = (GP)(z)$. In general, we claim that CP is given recursively in pseudocode by the following algorithm.

Algorithm 1 (CycFacs)

```
Function C
Input P, Output CP
[P(z) is a monic integer polynomial not divisible by z;
 CP(z) is the product of all distinct cyclotomic polynomials
 Φₙ dividing P(z)]
Local P₂, Q, H
if GP = P then H := P
else
  P₂(z) := gcd(P(z), P(−z))
  Q(z) := P₂(√z)
  H(z) := (C(GP))(z) · (C(Q))(z²))
fi
Return (H/gcd(H, H′))
end
```

Proof *(of termination and correctness of the algorithm)* We work by induction on the degree of $P(z)$. When $P = 1$, the output is $GP = P$, which is correct.

Assume now that P has positive degree and suppose that the algorithm works for all polynomials of lower degree. First note that if $GP = P$, then P has only roots of unity as zeros, by Exercise 5.4. In that case, we also have $CP = P / \gcd(P, P')$ and thus CP is the correct output. Let us now assume that GP has strictly lower degree than P. Suppose that ω is a cyclotomic zero of P. Then, by Proposition 5.3, ω is a zero of at least one of $P_2(z) = \gcd(P(z), P(-z))$ or $(GP)(z)$. Hence, $(CP)(z)$ has the same zero set as $(CGP)(z) \cdot (CP_2)(z)$. (The latter may contain zeros of higher multiplicity.) Note that $P_2(z)$ is a polynomial in z^2, say $P_2(z) = Q(z^2)$, so finding the common cyclotomic zeros of $P(z)$ and $P(-z)$ comes down to finding the cyclotomic zeros of G. Hence $(CP)(z)$ also has the same zero set as $H := (CGP)(z) \cdot (CPg)(z^2)$, giving $(CP)(z) = H / \gcd(H, H')$. Since, by assumption, the degrees of GP and Q are strictly smaller than the degree of P, our inductive hypothesis guarantees that the algorithm works on GP and Q. So it works on P. □

Let us call the product of the cyclotomic polynomials dividing P, with multiplicities, the **cyclotomic part** of P. Its calculation is straightforward in principle: it is $\gcd(P(z), (CP)(z)^{\deg P})$. However, it can also be calculated directly by the algorithm MultiCycFacs below (Algorithm 2).

The idea of the algorithm is first to strip out the largest factor P_2 of P that is a polynomial in z^2, say $P_2(z) = P_3(z^2)$. Denoting the cyclotomic part of $P_3(z)$ by $P_4(z)$, the cyclotomic part of $P_2(z)$ is then $P_4(z^2)$. Thus, the cyclotomic part of $P_2(z)$ can be computed recursively. The cyclotomic part of P/P_2 is then readily computed by the next lemma.

Lemma 5.6 *Suppose that $P(z) \in \mathbb{Z}[z]$ is monic with $P(0) \neq 0$, and is not divisible by any nonconstant polynomial of the form $v(z^2)$. Then $\gcd(P(z), P(z^2)P(-z^2))$ is the cyclotomic part of $P(z)$.*

Proof Let $u(z)$ be an irreducible factor of $P(z)$ of largest Mahler measure that divides $\gcd(P(z), P(z^2)P(-z^2))$. We can suppose that $u \neq 1$. Then $u(z) \mid P(\varepsilon z^2)$ for $\varepsilon = 1$ or -1. Also $u(z) \neq u(-z)$ or else $u(z)$ would equal $v(z^2)$ for some nonconstant polynomial v. So $u(z)u(-z) \mid P(\varepsilon z^2)$. But $u(z)u(-z) = w(z^2)$ for some polynomial w. Since $w(z^2) \mid P(\varepsilon z^2)$, we have $w(z) \mid P(\varepsilon z)$ and so $w(\varepsilon z) \mid P(z)$. But then

$$M(w(\varepsilon z)) = M(w(z)) = M(w(z^2)) = M(u(z))^2 \geq M(w(\varepsilon z))^2,$$

since $M(u)$ is maximal in the sense above. Hence $M(w(\varepsilon z)) = 1$, giving $M(u) = 1$, so that u is cyclotomic.

For the nth cyclotomic polynomial $\Phi_n(z)$, we know that $\Phi_{4k}(z)$ is a polynomial in z^2, so no such polynomials are factors of $P(z)$. On the other hand, for n odd or $n = 2n'$ with n' odd, we see from Proposition 5.3 that any factor Φ_n^j of P for such an n is also a factor of $\gcd(P(z), P(z^2)P(-z^2))$.   □

These considerations lead to the following.

Algorithm 2 (MultiCycFacs)
```
Function C′
Input P, Output C′P
[P(z) is a monic integer polynomial not divisible by z;
 C′P(z) is the product of all cyclotomic polynomials
 Φₙ dividing P(z), with multiplicities]
Local P₂, Q, H
if P = 1 then return(1)
else
  P₂(z) := gcd(P(z), P(−z))
  Q(z) := P₂(√z)
  P₁(z) := P/P₂
  if GP₁ = P₁ then H := P₁
  else H := C′(G(P₁))
  fi
  return(H(z) · (C′Q)(z²))
fi
end
```

Remark It is easy to extend the algorithm CycFacs (Algorithm 1) to find the cyclotomic zeros of polynomials $P(z)$ having coefficients in some number field K. To do this, one simply computes $\gcd\left(P(z), (C\operatorname{Norm}_{K/\mathbb{Q}} P)(z)\right)$.

Exercise 5.7 Show that if a monic integer polynomial $P(z)$ has at most one irreducible noncyclotomic factor and no factors that are polynomials in z^4, then

$$(CP)(z) = \gcd(P(z), P(-z)P(z^2)P(-z^2)).$$

We define a **Salem polynomial** to be a monic integer self-reciprocal polynomial having exactly one zero in $|z| > 1$, and at least one zero on $|z| = 1$. Also, a **Salem number** is a real algebraic integer greater than 1 whose minimal polynomial is a Salem polynomial.

Exercise 5.8 Show that every reducible Salem polynomial $P(z)$ is the product of a cyclotomic polynomial and the minimal polynomial, $S(z)$ say, of either a Salem number or a self-reciprocal quadratic Pisot number.

See Sect. 2.2 for the definition of a Pisot number.

5.3 Cyclotomic Points on Curves

5.3.1 Definitions

The method of the previous section will also work for polynomials in more than one variable. Suppose that we have a Laurent polynomial $f(x, y) \in \mathbb{Z}[x, y, x^{-1}, y^{-1}]$, and we want to find all points (x, y) on the plane curve $f(x, y) = 0$ for which x and y are roots of unity. We call such a point a **cyclotomic point**.

For any Laurent polynomial $f(x, y) = \sum_{i,j} a_{ij} x^i y^j \in \mathbb{C}[x, y]$, we define its **exponent lattice** $\mathcal{L}(f)$ to be the lattice spanned by all the differences $(i, j)^\mathsf{T} - (i', j')^\mathsf{T}$ where $a_{ij} \neq 0$, $a_{i'j'} \neq 0$. Clearly, if f has at least two terms, $\mathcal{L}(f)$ has rank 1 or 2. If $\mathcal{L}(f)$ has rank 2, we say that it is **full** if it equals the whole of $\mathbb{Z}^2$. Also, say that two polynomials $f_1(x, y)$ and $f_2(x, y)$ are **equivalent** if their quotient f_1/f_2 is a nonzero scalar multiple of a monomial $x^a y^b$.

5.3.2 $\mathcal{L}(f)$ of Rank 1

In this case, it is clear that f is equivalent to a Laurent polynomial $P(x^k y^\ell)$ for some one-variable polynomial P. If f has a cyclotomic point (x, y), then P must have a zero ω that is a root of unity. We can use Algorithm 1 above to find the zeros of P that are roots of unity. Thus, to find all cyclotomic points on f, we need only solve the equations $x^k y^\ell = \omega$ for every zero ω of P. If $\gcd(k, \ell) = 1$ then from $kk_1 + \ell\ell_1 = 1$ we obtain the solutions $(x, y) = (\omega^{k_1} \zeta^\ell, \omega^{\ell_1} \zeta^{-k})$, where ζ is any root of unity. If $g := \gcd(k, \ell) > 1$ we can consider $P(x^{k/g} y^{\ell/g})$, to reduce to the coprime case. Thus, f will have no cyclotomic points when P has no zeros that are roots of unity, and infinitely many such points otherwise.

5.3.3 $\mathcal{L}(f)$ Full of Rank 2

Proposition 5.9 *Suppose that* $f(x, y) \in \mathbb{Z}[x, y]$ *is irreducible, and* $\mathcal{L}(f)$ *is full of rank 2. Then every cyclotomic point* (x, y) *on* f *also lies on at least one of the 7 curves* $f_i(x, y) = 0$ $(i = 1, \ldots, 7)$, *where*

$$f_1(x, y) = f(x, -y), \quad f_2(x, y) = f(-x, y),$$

$$f_3(x, y) = f(-x, -y), \quad f_4(x, y) = f(x^2, y^2)$$

and

$$f_i(x, y) = f_{i-4}(x^2, y^2) \quad \text{for } i = 5, 6, 7.$$

Furthermore f *does not divide any of the* f_i.

Proof We can suppose that x and y are powers of a common root of unity, say $x = \omega^a$, $y = \omega^b$, where we can clearly assume that $\gcd(a, b) = 1$. In particular, a and b are not both even. Thus, $f(\omega^a, \omega^b) = 0$ and so, by Proposition 5.3, at least one of $f(\omega^a, -\omega^b)$, $f(-\omega^a, \omega^b)$, $f(-\omega^a, -\omega^b)$, $f(\omega^{2a}, \omega^{2b})$, $f(-\omega^{2a}, -\omega^{2b})$, $f(-\omega^{2a}, \omega^{2b})$ and $f(-\omega^{2a}, -\omega^{2b})$ must also be 0. Thus, our cyclotomic point (x, y) also lies on at least one of the 7 curves $f_i(x, y) = 0$ $(i = 1, \ldots, 7)$ defined above.

It remains to show that f does not divide any of the f_i. It is easy to see that if $f \mid f_1$ then f is equivalent to a polynomial in $\mathbb{Q}[x, x^{-1}, y^2, y^{-2}]$, in which case $\mathcal{L}(f)$ would not be full. Similarly, if $f \mid f_2$ then f is equivalent to a polynomial in $\mathbb{Q}[x^2, x^{-2}, y, y^{-1}]$, while if $f \mid f_3$ then f is equivalent to a polynomial in $\mathbb{Q}[xy, 1/xy, x/y, y/x]$, so that again $\mathcal{L}(f)$ would not be full. Also note that if any of f_1, f_2, f_3 (all irreducible, since f is) were to divide any other one of f_1, f_2, f_3 then the same contradiction would apply.

Next, suppose that $f \mid f_4$. Then, as $f_4(x, y) \in \mathbb{Z}[x^2, y^2]$, we have that each of f_1, f_2, f_3 also divides f_4. Hence, $ff_1f_2f_3 \mid f_4$, clearly impossible on degree grounds (say, in x). Exactly the same argument applies to f_5, f_6 and f_7. □

This result gives the following algorithm for finding all the cyclotomic points on $f(x, y) \in \mathbb{Z}[x, y, x^{-1}, y^{-1}]$, irreducible, and full of rank 2.

Algorithm 3

1. For $i = 1, 2, \ldots, 7$ compute the y-resultant

$$r_i(x) := \text{res}_y(f, f_i)$$

   and

$$(Cr_i)(x) = \prod_{\substack{r_i(\omega)=0 \\ \omega \text{ a root of } 1}} (x - \omega).$$

2. For $i = 1, 2, \ldots, 7$ and each zero ω of $(Cr_i)(x)$ compute

$$\gcd(f(\omega, y), (C \operatorname{Norm}_{\mathbb{Q}(\omega)/\mathbb{Q}} f(\omega, y))),$$

as in the Remark after Algorithm 2. Each of the zeros ω' of these polynomials gives a cyclotomic point (ω, ω') on f.

5.3.4 $\mathcal{L}(f)$ of Rank 2, but Not Full

Suppose that $\mathcal{L}(f)$ has a basis $(a, b)^\mathsf{T}$, $(c, d)^\mathsf{T}$ with index $N = |ad - bc|$ in $\mathbb{Z}^2$. Put $u := x^a y^b$, $v = x^c y^d$. We have, taking any i_0, j_0 such that the coefficient of $x^{i_0} y^{j_0}$ in f is nonzero:

$$f(x, y) = a_{i_0 j_0} x^{i_0} y^{j_0} + \sum_{r,s} c_{rs} x^{i_0+ra+sc} y^{j_0+rb+sd}$$

$$= x^{i_0} y^{j_0} \left(a_{i_0 j_0} + \sum_{r,s} c_{rs} u^r v^s \right),$$

so that f is equivalent to $f^*(u, v)$ for some Laurent polynomial f^* with $\mathcal{L}(f^*)$ full. Put $A = \begin{pmatrix} a & c \\ b & d \end{pmatrix}$. As in (2.4), we have $(x, y)^A = (x^a y^b, x^c y^d) = (u, v)$. Putting A into Smith normal form [New72, p. 26] yields two matrices U and W in $\mathrm{GL}_2(\mathbb{Z})$ with $WAU = D$ say, where $D = \operatorname{diag}(d_1, d_2)$ and d_1 and d_2 are positive integers with $d_1 \mid d_2$. Also $d_1 d_2 = N$. Hence, using Lemma 2.11,

$$(x, y)^{W^{-1}D} = (x, y)^{AU} = ((x, y)^A)^U = (u, v)^U = (u^*, v^*),$$

say. Now, as $\mathcal{L}(f^*)$ is full, we can find all cyclotomic points (u^*, v^*) of F^*. Then, letting u_1, u_2 be all possible d_1th, d_2th roots of u^*, v^*, respectively, each cyclotomic point (u^*, v^*) of f^* gives N points $(x, y) = (u_1, u_2)^W$ of f. Hence, the cyclotomic points of f can be obtained from those of f^*. Also, we see that f has N times as many cyclotomic points as f^*.

5.3.5 The Case of f Reducible

If f is reducible, we can clearly apply the algorithm to each factor of f separately. If any such factor has a lattice of rank 1, we apply the method of Sect. 5.3.2. If the lattice of a factor is full of rank 2, apply Algorithm 3. If this lattice is of rank 2 but not full, apply the method of Sect. 5.3.4.

5.3.6 An Example

Example 5.10 We study the following problem. Find all algebraic integers α with the property that both α and α^{-1} are the sums of two roots of unity, but neither is a root of unity.

It is clearly enough to find all solutions of the equation $(1 + x)(1 + y) = z$, where x, y and z are roots of unity. For then, all solutions are of the form $\alpha = w(1 + x)$, $\alpha^{-1} = (1 + y)(wz)^{-1}$.

Starting with $(1 + x)(1 + y) = z$, we know that if x, y, z are roots of unity then (complex conjugation) also $(1 + x^{-1})(1 + y^{-1}) = z^{-1}$. Eliminating z from these equations, we obtain $f(x, y) := (1 + x)^2(1 + y)^2 - xy = 0$. We must solve this equation in roots of unity, and then retain only the solutions where z is also a root of unity, and also neither $1 + x$ nor $1 + y$ is a root of unity.

We need consider only 5 cases, because $f(x, y) = f(y, x)$. Assume $f(x, y) = 0$ in each case.

1. $f(x, -y) = 0$. Here the y-resultant gives $x = -1$, but then $y = 0$, so there are no cyclotomic points.
2. $f(-x, -y) = 0$. The y-resultant is $16\Phi_{12}(x)^2$.
3. $f(x^2, y^2) = 0$. Here, the y-resultant is $16\Phi_5(x)^2\Phi_3(x)^4$. But x cannot equal ω_3, as $1 + \omega_3$ is a root of unity.
4. $f(-x^2, y^2) = 0$. The y-resultant has no cyclotomic factors.
5. $f(-x^2, -y^2) = 0$. The y-resultant is $16x^2(1 + x)^4\Phi_{12}(x)^2$.

Thus, we see that x, and by symmetry y, must be a fifth or twelfth root of unity. It is easy (especially by computer) to check which pairs (x, y) actually occur, and what their corresponding value of z is. Thus we obtain the following result.

Proposition 5.11 *The solutions to the equation $(1 + x)(1 + y) = z$ in roots of unity with neither $1 + x$ nor $1 + y$ being a root of unity are*

- $(x, y, z) = (\omega_5, \omega_5^2, -\omega_5^4)$ *(4 solutions);*
- $(x, y, z) = (\omega_5, \omega_5^3, -\omega_5^2)$ *(4 solutions);*
- $(x, y, z) = (\omega_{12}, -\omega_{12}, -\omega_{12}^4)$ *(4 solutions);*
- $(x, y, z) = (\omega_{12}, -\omega_{12}^{-1}, \omega_{12}^3)$ *(4 solutions).*

Exercise 5.12 Suppose that for some $f(x, y) \in \mathbb{Z}[x, y]$, the cyclotomic point (ω, ω'), where ω is an odd power of ω', lies on the curve $f(x, y) = 0$. Show that precisely one of the points $(-\omega, -\omega')$, (ω^2, ω'^2) and $(-\omega^2, -\omega'^2)$ also lies on $f(x, y) = 0$.

5.4 Cyclotomic Integers

5.4.1 Introduction to Cyclotomic Integers

A *cyclotomic integer* is a finite sum $\beta := \sum_j a_j \zeta_j$, where the $a_j \in \mathbb{Z}$ and the ζ_j are roots of unity. Since all roots of unity are algebraic integers, and the algebraic integers form a ring (Proposition A.22 in Appendix A), β is an algebraic integer. If $\zeta_j = \omega_{n_j}$ say, then $\beta \in \mathbb{Q}(\omega_n)$, where $n := \text{lcm}_j \, n_j$. Further, from Proposition 5.2(c) we know that all algebraic integers in $\mathbb{Q}(\omega_n)$ are of this form.

We are interested in how β and its conjugates are distributed on the complex plane. In particular, we would like to find all 'small' β. We consider three different height functions of β, which will give us three different definitions of 'small'.

- $\overline{M}(\beta) = M(\beta)^{1/\deg \beta}$, the *absolute Mahler measure* of β;
- $\lceil \beta \rceil$, the maximum modulus of β and its conjugates, called the *house* of β;
- $\mathcal{M}(\beta)$, the *Cassels height* of β, defined as

$$\mathcal{M}(\beta) := \frac{1}{\deg \beta} \sum_j |\beta_j|^2 \,,$$

where the β_j are the conjugates of $\beta = \beta_1$.

Exercise 5.13 Show that

$$M(\beta)^{1/\deg \beta} \le \lceil \beta \rceil \quad \text{and} \quad \mathcal{M}(\beta) \le \lceil \beta \rceil^2$$

for any cyclotomic integer β.

Exercise 5.14 Show that

$$\mathcal{M}(\beta_1 + \beta_2) + \mathcal{M}(\beta_1 - \beta_2) = 2(\mathcal{M}(\beta_1) + \mathcal{M}(\beta_2))$$

for any two cyclotomic integers β_1, β_2.

Exercise 5.15 Show that if $\beta \in \mathbb{Q}(\omega_n)$ and $\beta' \in \mathbb{Q}(\omega_{n'})$, where $\gcd(n, n') = 1$, then

$$\mathcal{M}(\beta\beta') = \mathcal{M}(\beta)\mathcal{M}(\beta') \,.$$

Exercise 5.16 Using the inequality of arithmetic and geometric means, show that if β is a nonzero algebraic integer then $\mathcal{M}(\beta) \ge 1$.

Lemma 5.17 *For any cyclotomic integer β with conjugates β_j, the conjugates of $|\beta|^2$ are the $|\beta_j|^2$.*

Proof Let τ be the complex conjugation automorphism $\alpha \mapsto \overline{\alpha}$ of the field $\mathbb{Q}(\beta)$, and let σ_j be an automorphism of this field that maps $\beta \mapsto \beta_j$. Then because $\mathrm{Gal}(\mathbb{Q}(\beta)/\mathbb{Q})$ is abelian (Proposition 5.2(d)), we have

$$|\beta_j|^2 = \beta_j\overline{\beta_j} = \sigma_j(\beta)\tau(\sigma_j(\beta)) = \sigma_j(\beta)(\sigma_j(\tau(\beta))) = \sigma_j(\beta\,\tau(\beta)) = \sigma_j(|\beta|^2).$$

$\square$

Corollary 5.18 *For any cyclotomic integer β, we have that $|\beta|$ is totally real, and*

$$\mathscr{M}(\beta) = \frac{1}{\deg \gamma} \sum_j \gamma_j = \overline{\mathrm{tr}}\,\gamma,$$

where the sum is over all conjugates γ_j of $\gamma := |\beta|^2$, and $\overline{\mathrm{tr}}$ denotes the mean trace, defined in (A.9) in Appendix A.

Corollary 5.19 *If β is a cyclotomic integer with $|\beta|^2 \in \mathbb{Z}$ then all the conjugates β_j of β lie on the circle $|z| = |\beta|$.*

Corollary 5.20 *As β' runs through the conjugates of the cyclotomic integer β, the conjugates of $|\beta|^2$ each occur the same number of times.*

Exercise 5.21 Show that for any cyclotomic integer β, the heights $M(\beta)^{1/\deg\beta}, \overline{|\beta|}$ and $\mathscr{M}(\beta)$ are in fact functions only of $|\beta|$. (Thus, if β' is another cyclotomic integer with $|\beta'| = |\beta|$ then $M(\beta')^{1/\deg\beta'} = M(\beta)^{1/\deg\beta}, \overline{|\beta'|} = \overline{|\beta|}$ and $\mathscr{M}(\beta') = \mathscr{M}(\beta)$.)

Exercise 5.22 Given a finite set of cyclotomic integers, show how to express each one of them as the sum of the same number of roots of unity.

Definition 5.23 We say that two cyclotomic integers, say β and β^*, are **equivalent** ($\beta \sim \beta^*$) if for some conjugate β' of β we have that β^*/β' is a root of unity. Further, define $q_{\min}(\beta)$ to be the least integer q such that $\beta \in \mathbb{Q}(\omega_q)$ and call β **minimal** if $q_{\min}(\beta^*) \geq q_{\min}(\beta)$ for every cyclotomic integer β^* equivalent to β.

Exercise 5.24 Show that $q_{\min}(\beta)$ is the greatest common divisor of all q for which $\beta \in \mathbb{Q}(\omega_q)$. In particular, if $\beta \in \mathbb{Q}(\omega_q)$ then $q_{\min}(\beta) \mid q$.

It follows from Exercise 5.21 that the absolute Mahler measure, house and Cassels height are the same for equivalent cyclotomic integers. Thus they are functions defined on these equivalence classes.

The following elementary facts, given in the next two exercises, turn out to be very useful.

Exercise 5.25 (Loxton [Lox75, Lemma 1]) Suppose that $q = pq_1$, where p is prime and $p \nmid q_1$.

(a) Show that there are integers a, b such that $\omega_{pq_1} = \omega_p^a \omega_{q_1}^b$.

(b) Show that every $\beta \in \mathbb{Q}(\omega_q)$ can be written as

$$\beta = \sum_{j=0}^{p-1} \alpha_j \omega_p^j, \tag{5.3}$$

with the $\alpha_j \in \mathbb{Q}(\omega_{q_1})$.

(c) Show further that any such representation of β is of the form

$$\beta = \sum_{j=0}^{p-1} (\alpha_j + \alpha) \omega_p^j \tag{5.4}$$

for some $\alpha \in \mathbb{Q}(\omega_{q_1})$.

(d) Show that the conjugates of β in (5.3) over the field $\mathbb{Q}(\omega_{q_1})$ are the

$$\beta' = \sum_{j=0}^{p-1} \alpha_j \omega_p^{ij} \text{ for } 1 \leq i \leq p-1. \tag{5.5}$$

(e) Show that if β in (5.3) is an integer then all the α_j can be chosen to be integers. (Consider forcing $\alpha_{p-1} = 0$, and use Proposition 5.2(c).)

Exercise 5.26 (Loxton [Lox75, Lemma 2]) Suppose that $q = p^N q_2$, where p is prime and $p \nmid q_2$. Let L be a positive integer with $L < N$ and define q_1 by $q = p^L q_1$.

(a) Show that $\varphi(q) = p^L \varphi(q_1)$.
(b) Show that there are integers r, s, b with $0 \leq s \leq p^L - 1$ such that we have $1 = (rp^L + s)q_2 + bp^N$.
(c) Show that every $\beta \in \mathbb{Q}(\omega_q)$ can be written *uniquely* as

$$\beta = \sum_{j=0}^{p^L-1} \alpha_j \omega_{p^N}^j, \tag{5.6}$$

with the $\alpha_j \in \mathbb{Q}(\omega_{q_1})$.

(d) Show that the automorphisms of $\mathbb{Q}(\omega_q)$ that fix ω_{q_1} are given by $\omega_q \mapsto \omega_q^j$ where $\gcd(j, q) = 1$ and $j \equiv 1 \pmod{q_1}$.
(e) Show that the conjugates of β in (5.6) over the field $\mathbb{Q}(\omega_{q_1})$ are the

$$\beta = \sum_{j=0}^{p^L-1} \alpha_j (\omega \omega_{p^N})^j, \tag{5.7}$$

where ω runs through all the p^Lth roots of unity.

(f) Show that if β in (5.6) is an integer, then so are all the α_j.

Lemma 5.27 *For β a nonzero cyclotomic integer, with say $\beta \in \mathbb{Z}[\omega_n]$, there is some power ω_n^i of ω_n such that $\omega_n^i \beta$ has nonzero trace.*

Proof We can write $\beta = \sum_{k=0}^{d-1} a_k \omega_n^k$, where the a_k are integers, and $d = \varphi(n)$. Suppose that the trace of $\omega_n^i \beta$ is 0 for all $i = 0, \ldots, n - 1$. Then the traces of all $a_k \omega_n^{-k} \beta$ would be 0, and so the same would be true for the mean traces of all of these numbers. But by Lemma A.27 in Appendix A, the mean trace is an additive function. Hence, the mean trace of $\sum_{k=0}^{d-1} a_k \omega_n^{-k} \beta = |\beta|^2$ would also be 0. But, by Lemma 5.17, the conjugates of $|\beta|^2$ are all positive, so its mean trace is positive, a contradiction. $\square$

We now define

$$\mu_\varphi(n) := \frac{\mu(n)}{\varphi(n)}. \tag{5.8}$$

From Proposition 5.2, we see that $\mu_\varphi(n)$ is the mean trace $\overline{\mathrm{tr}}(\omega_n)$ of ω_n. In particular, it is 0 when n is not square-free.

Exercise 5.28 (a) ([CMS11, p. 883]) Show that

$$\mathcal{M}(1 + \omega_n) = 2(1 + \mu_\varphi(n)). \tag{5.9}$$

(b) More generally, for $\beta := \sum_{j=1}^k \omega_n^{e_j}$, show that

$$\mathcal{M}(\beta) = \sum_{i=1}^k \sum_{j=1}^k \mu_\varphi \left(\frac{n}{\gcd(n, e_i - e_j)} \right). \tag{5.10}$$

(c) In particular, if $n = p$ is prime and β is the sum of k distinct powers of ω_p, show that then

$$\mathcal{M}(\beta) = \frac{k(p - k)}{p - 1}.$$

5.4.2 The Function $\mathcal{N}(\beta)$

Cyclotomic integers are sums of roots of unity. For a cyclotomic integer β, we define $\mathcal{N}(\beta)$ to be the smallest number of roots of unity whose sum is β. Thus, β can be written as a sum of $\mathcal{N}(\beta)$ roots of unity, but no fewer.

Suppose that a cyclotomic integer β lies in a cyclotomic field $\mathbb{Q}(\omega)$. Loxton showed that if β is the sum of n roots of unity, then it is the sum of at most n roots of unity in $\mathbb{Q}(\omega)$. This follows straight from the following.

Theorem 5.29 *Take a cyclotomic integer β, with (as above) $\mathbb{Q}(\omega_{q_{\min}})$ being the smallest cyclotomic field containing it. Suppose that β is the sum of n roots of unity, and that $\mathbb{Q}(\omega_{q^*})$ is the smallest cyclotomic field containing them. Then*

(a) If $n = \mathcal{N}(\beta)$ then $q^* = q_{\min}$.
(b) If $n = \mathcal{N}(\beta) + 1$ then $q^* = q_{\min}$ or $3q_{\min}$, where the latter case occurs only if $3 \nmid q_{\min}$.

Proof Suppose that $n \leq \mathcal{N}(\beta) + 1$. Clearly $q_{\min} \mid q^*$ (Exercise 5.24). If $q_{\min} \neq q^*$, we have that $p^N \| q^*$ for some prime p, but $p^N \nmid q_{\min}$. Putting $q_1 = q^*/p$ we have that $\beta \in \mathbb{Q}(\omega_{q_1})$. By hypothesis we can write β as a sum of n roots of unity, which we can write in the form $\gamma_j \omega^{r_j}$ ($j = 1, \ldots, n$), where $\omega := \omega_{p^N}$, γ_j is a root of unity in $\mathbb{Q}(\omega_{q_1})$ and $0 \leq r_j \leq p - 1$. Putting $\alpha_j = \sum_{r_i = j} \gamma_i \in \mathbb{Q}(\omega_{q_1})$, we have

$$\beta = \sum_{j=0}^{p-1} \alpha_j \omega^j . \tag{5.11}$$

For the exponent N above, we now separate the two cases $N > 1$ and $N = 1$.

The case N > 1. Then Exercise 5.26 tells us that the representation (5.11) for β is unique. But since $\beta \in \mathbb{Q}(\omega_{q_1})$, we have simply $\beta = \alpha_0$ and $\alpha_i = 0$ for $i > 0$. However, if all the γ_i with $r_i = j > 0$ were zero, then the n roots of unity summing to β would lie in $\mathbb{Q}(\omega_{q_1})$, contradicting the definition of q^*. Hence, for some $j > 0$, $\alpha_j = 0$ is the sum of at least two nonzero γ_i. But then the equation $\beta = \alpha_0$ expresses β as the sum of at most $n - 2$ roots of unity, contradicting $n \leq \mathcal{N}(\beta) + 1$. Hence $q_{\min} = q^*$ and $\mathcal{N}(\beta) = n$, so that β is a sum of $\mathcal{N}(\beta)$ roots of unity in $\mathbb{Q}(\omega_{q_{\min}})$.

The case N = 1. First note that $p \neq 2$ here, because if N is odd then $\mathbb{Q}(\omega_{2N}) = \mathbb{Q}(\omega_N)$. Then from Exercise 5.25, we have

$$\alpha_0 - \beta = \alpha_1 = \alpha_2 = \cdots = \alpha_{p-1} = \alpha , \tag{5.12}$$

say. If $\alpha = 0$ the argument follows as in the $N > 1$ case. So suppose $\alpha \neq 0$. Then the equation $\beta = \alpha_0 - \alpha$ expresses β as a sum of at most $n - (p - 2)\mathcal{N}(\alpha)$ roots of unity. Since $p \geq 3$ and $\mathcal{N}(\alpha) \geq 1$, this will give a contradiction unless $p = 3$ and $\mathcal{N}(\alpha) = 1$. Thus $n = \mathcal{N}(\beta) + 1$. $\qquad \square$

Exercise 5.30 (Loxton [Lox75, p. 163]) Show that all the possibilities described in the theorem can actually occur.

Exercise 5.31 (Loxton [Lox75, Theorem 1(iii)]) Show that if $n \geq \mathcal{N}(\beta) + 2$, then q^* can be any integer divisible by both $q_{\min}$ and 3. Deduce that β is a sum n roots of unity in infinitely many ways if and only if it is a sum of n' roots of unity for some $n' \leq n - 2$.

Exercise 5.32

(a) Show that $\mathcal{N}(\omega\beta) = \mathcal{N}(\beta)$ for any cyclotomic integer β and root of unity ω.
(b) Show that $\mathcal{N}(\beta\beta') \leq \mathcal{N}(\beta)\mathcal{N}(\beta')$ for any cyclotomic integers β, β'.
(c) Show that $\mathcal{N}(k\beta) \leq k\mathcal{N}(\beta)$ for any integer k and cyclotomic integer β.

5.4.3   *Evaluating or Estimating* $\mathcal{N}\left(\sqrt{d}\right)$

For an odd prime p and an integer k, as usual let $\left(\frac{k}{p}\right)$ be the Legendre symbol, so that

$$\left(\frac{k}{p}\right) = \begin{cases} 0 & \text{if } p \mid k; \\ 1 & \text{if } k \text{ is a quadratic residue} \quad (\text{mod } p); \\ -1 & \text{if } k \text{ is a quadratic nonresidue} \quad (\text{mod } p). \end{cases}$$

For our odd prime p, define the special **Gauss sum**

$$S := \sum_{k=1}^{p-1} \left(\frac{k}{p}\right) \omega_p^k. \tag{5.13}$$

Exercise 5.33 (See, for instance, [ME05, Th. 7.2.1]) Use the standard properties of the Legendre symbol to show that $S^2 = \left(\frac{-1}{p}\right) p$. Deduce that all $\sqrt{d}$ for $d \in \mathbb{Z}$ are cyclotomic integers.

Proposition 5.34 *We have* $\mathcal{N}\left(\sqrt{2}\right) = 2$ *and furthermore, for an odd prime p, that* $\mathcal{N}\left(\sqrt{p}\right) = p - 1$. *Also, for integers q and r with q square-free, we have that*

$$\mathcal{N}\left(r\sqrt{q}\right) \leq \begin{cases} r\varphi(q) & \text{if } q \text{ is odd}; \\ 2r\varphi(q) & \text{if } q \text{ is even}. \end{cases}$$

Proof Firstly, since $\omega_8 + \omega_8^{-1} = \sqrt{2}$, we have $\mathcal{N}\left(\sqrt{2}\right) = 2$. Consider next the case $p \equiv 1 \pmod 4$, where $S = \pm\sqrt{p}$. (In fact, after Proposition 5.35 below we see that $S = \sqrt{p}$, but we do not need this extra information here.) From Exercise 5.33 we have $N := \mathcal{N}\left(\sqrt{p}\right) \leq p - 1$, with all the roots of unity $\pm\omega_p^k$ in the corresponding sum being in $\mathbb{Q}(\omega_p)$. Since $\mathbb{Q}(\omega_n) \cap \mathbb{Q}(\omega_p)$ is either $\mathbb{Q}(\omega_p)$ or $\mathbb{Q}$, and $\sqrt{p} \notin \mathbb{Q}$, the field $\mathbb{Q}(\omega_p)$ is the smallest cyclotomic field containing $\sqrt{p}$. So by Theorem 5.29(a), $\sqrt{p}$ is the sum of N roots of unity in $\mathbb{Q}(\omega_p)$. Since $1, \omega_p, \omega_p^2, \ldots, \omega_p^{p-2}$ are a $\mathbb{Z}$-basis for the ring of integers of $\mathbb{Q}(\omega_p)$ and

$$1 + \omega_p + \omega_p^2 + \cdots + \omega_p^{p-2} + \omega_p^{p-1} = 0, \tag{5.14}$$

any $p - 1$ of $1, \omega_p, \omega_p^2, \ldots, \omega_p^{p-2}, \omega_p^{p-1}$ can be chosen as such a $\mathbb{Z}$-basis. Hence, $\sqrt{p}$ has a unique representation as a sum of integer multiples of any such $p - 1$ roots of unity. By adding (5.14) to (5.13), we obtain $\sqrt{p}$ as a sum of multiples of 1 and the $(p - 1)/2$ numbers ω_p^k where k is a quadratic residue. This gives $\sqrt{p}$ as a sum of p roots of unity. Similarly, by subtracting (5.14) from (5.13), we obtain $\sqrt{p}$ as a sum of multiples of 1 and the $(p - 1)/2$ numbers ω_p^k where k is a quadratic nonresidue. This again gives $\sqrt{p}$ as a sum of p roots of unity. Every ω^k ($k = 0, 1, \ldots, p - 1$)

is missing from one of the three representations of $\sqrt{p}$ as a sum of roots of unity that we have described. Hence, they are the only representations of $\sqrt{p}$ as a sum of roots of unity in $\mathbb{Q}(\omega_p)$ using at most $p - 1$ different roots of unity. All other representations of $\sqrt{p}$ will use all p roots of unity. Thus $N = p - 1$, from (5.13).

For $p \equiv 3 \pmod 4$, the above argument gives $\mathcal{N}\left(\sqrt{-p}\right) = p - 1$. Thus by Exercise 5.32(a), we again have $\mathcal{N}\left(\sqrt{p}\right) = p - 1$.

The final inequality follows straight from Exercise 5.32(b),(c). $\qquad\square$

5.4.4 Evaluation of the Gauss Sum

Gauss famously evaluated the sign of S in (5.13), as follows.

Proposition 5.35 *For an odd prime p, we have*

$$
S = \begin{cases} \sqrt{p} & \text{if } p \equiv 1 \pmod 4, \\ i\sqrt{p} & \text{if } p \equiv 3 \pmod 4. \end{cases}
$$

The proof we present starts, following Gauss, by proving the following, which we present as an exercise. Here $k/2$ is the integer $k' \in \{0, 1, \ldots, p - 1\}$ with $2k' \equiv k \pmod p$, and $-k/2$ is the integer $k'' \in \{0, 1, \ldots, p - 1\}$ with $-2k'' \equiv k \pmod p$.

Exercise 5.36 Let $H := \prod_{k=1}^{(p-1)/2}(\omega_p^{-k/2} - \omega_p^{k/2})$. Show that H is positive and real if $p \equiv 1 \pmod 4$, and i times a positive real if $p \equiv 3 \pmod 4$. Show also that $H^2 = (-1)^{(p-1)/2}p$.

With the same interpretations of $\pm k/2$, we define

$$
H(x) := \prod_{k=1}^{(p-1)/2} (x^{-k/2} - x^{k/2}).
$$

Also, put

$$
S(x) := \sum_{j=1}^{p-1} \left(\frac{j}{p}\right) x^j.
$$

We know that $S^2 = S(\omega_p)^2 = H^2 = H(\omega_p)^2$, so that $S(\omega_p) = \varepsilon H(\omega_p)$, where $\varepsilon = \pm 1$. We need to prove that $\varepsilon = 1$.

We know that ω_p is a zero of $S(x) - \varepsilon H(x)$. By Proposition 5.2(b), the minimal polynomial of ω_p is $\Phi_p(x) = x^{p-1} + \cdots + x + 1$ and so $\Phi_p(x)$ is a factor of $S(x) - \varepsilon H(x)$. Since $S(x) - \varepsilon H(x)$ has integer coefficients, we have say

$$
\Phi_p(x)D(x) = S(x) - \varepsilon H(x) \qquad \text{for some } D(x) \in \mathbb{Z}[x].
$$

Now, working modulo p we have

$$\Phi_p(x) = \frac{x^p - 1}{x - 1} \equiv (x - 1)^{p-1} \quad (\text{mod } p).$$

Hence, in $\mathbb{Z}/(p)[x]$ we have

$$S(x) \equiv \varepsilon H(x) \quad (\text{mod } (x - 1)^{p-1}).$$

Next, put $x = 1 + u$ and expand both $H(1 + u)$ and $S(1 + u)$ in powers of u, up to $u^{(p-1)/2}$. First, we note that in $\mathbb{Z}/(p)[u]$

$$(1 + u)^{-k/2} - (1 + u)^{k/2} \equiv (-k)u \quad (\text{mod } u^2).$$

So, since $p - 1 \geq (p + 1)/2$,

$$S(1 + u) \equiv \varepsilon H(1 + u) \equiv \varepsilon(-1)(-2) \cdots (-(p - 1)/2)u^{(p-1)/2} \quad (\text{mod } u^{(p+1)/2}). \tag{5.15}$$

To calculate $S(1 + u)$ (mod $u^{(p+1)/2}$), we need the following.

Lemma 5.37 *We have*

$$\sum_{j=0}^{p-1} j^d \left(\frac{j}{p}\right) \equiv \begin{cases} 0 & (\text{mod } p) & \text{for } 0 \leq d < (p - 1)/2; \\ -1 & (\text{mod } p) & \text{for } d = (p - 1)/2. \end{cases}$$

Proof Define $J_e := \sum_{j=0}^{p-1} j^e$. Using Euler's criterion $\left(\frac{j}{p}\right) \equiv j^{(p-1)/2}$ (mod p), the sum we wish to evaluate is $J_{d+(p-1)/2}$ mod p. If $d = (p - 1)/2$, then by Fermat's little theorem this is just $J_{p-1} \equiv -1$ (mod p). If $0 \leq d < (p - 1)/2$, then putting $e = d + (p - 1)/2$ we can choose r such that $r^e \not\equiv 0$ or 1 (mod p). Then on the one hand, $J_e \equiv \sum_{j=0}^{p-1}(rj)^e$ (mod p), and on the other hand, this sum is $r^e J_e$. Hence $J_e \equiv 0$ (mod p). □

To complete the evaluation of ε, note that we have

$$S(1 + u) = \sum_{j=1}^{p-1} \left(\frac{j}{p}\right)(1 + u)^j = \sum_{j=0}^{p-1} \sum_{k=0}^{j} \frac{j(j - 1) \cdots (j - k + 1)}{k!} \left(\frac{j}{p}\right) u^k,$$

so that by Lemma 5.37 we have (in $\mathbb{Z}/(p)[u]$)

$$S(1 + u) \equiv \frac{-u^{(p-1)/2}}{((p - 1)/2)!} \quad (\text{mod } u^{(p+1)/2}).$$

Then from (5.15), on comparing coefficients of $u^{(p-1)/2}$ we see that

$$\varepsilon(-1)(-2)\cdots(-(p-1)/2) \equiv \frac{-1}{((p-1)/2)!} \pmod{p},$$

so that

$$\varepsilon \equiv \frac{-1}{(p-1)!} \equiv 1 \pmod{p},$$

using Wilson's theorem. Thus Proposition 5.35 is proved.

5.4.5 The Absolute Mahler Measure of Cyclotomic Integers

Lemma 5.38 *Suppose that the cyclotomic integer β has exactly k conjugates of modulus $|\beta|$. Then for $\gamma := |\beta|^2$ and any conjugate γ' of γ, β has k conjugates of modulus $\sqrt{\gamma'}$. Furthermore $\deg \beta = k \deg \gamma$ and $\overline{M}(\beta) = \overline{M}(\gamma)$.*

Proof Suppose that $\beta_j \overline{\beta_j} = \gamma$ for k distinct conjugates β_j of β. For a conjugate γ' of γ, take an automorphism σ' of a field containing all the conjugates of β that maps $\gamma \mapsto \gamma'$. Then σ' maps $\beta_1, \ldots, \beta_k$ to $\beta'_1, \ldots, \beta'_k$ say, and hence, because σ' commutes with complex conjugation, it maps $\overline{\beta}_1, \ldots, \overline{\beta}_k$ to $\overline{\beta}'_1, \ldots, \overline{\beta}'_k$. Hence

$$\beta'_1 \overline{\beta}'_1 = \beta'_2 \overline{\beta}'_2 = \cdots \beta'_k \overline{\beta}'_k = \gamma' > 0.$$

The last statement of the lemma follows immediately. □

With the help of Lemma 5.38, we can now apply Exercise 14.10 and Corollary 14.13 from Chap. 14.

Theorem 5.39 *Suppose that β is a cyclotomic integer. Then the only $\overline{M}(\beta)$ in the interval $(0, 1.311703]$ are for β equal to 1, $2\cos(2\pi/5)$, $2\cos(2\pi/7)$ and $2\cos(2\pi/60)$.*

Note that other cyclotomic integers can give one of the values stated in the theorem. For instance, equivalent cyclotomic integers will give the same value.

Of course, this result does not cover cyclotomic non-integers. For the general case of cyclotomic numbers (algebraic numbers in cyclotomic extensions), see Theorem 11.8 of Chap. 11.

5.5 Robinson's Problems and Conjectures

The study of cyclotomic integers began in earnest with the paper of Raphael Robinson in 1965 [Rob65]. In it, he proposed two problems and five conjectures concerning cyclotomic integers β. Most relate either to $|\beta|$ or $\lceil \beta \rceil$. (This was before the invention of the Cassels height $\mathscr{M}$.) The first conjecture remains open, but the other conjectures and both problems have been resolved. In this section, we present the problems and conjectures; progress is discussed in Sects. 5.7 and 5.8 after presenting some key lemmas in Sect. 5.6.

Robinson's Problem 1.

(a) *How can we decide whether there are any cyclotomic integers of a given absolute value? How can we find all the cyclotomic integers with this absolute value?*

(b) *Can there ever be infinitely many inequivalent cyclotomic integers of the same absolute value?*

Robinson's Problem 2.

How can we tell whether a given cyclotomic integer can be expressed as a sum of a prescribed number of roots of unity?

The conjectures were more specific.

Robinson's Conjecture 1. *Any cyclotomic integer β with $\lceil \beta \rceil < 2$ is either*

- *a sum of at most two roots of unity; or*
- *equivalent to a number of the form $\frac{1}{2}(\sqrt{a} + i\sqrt{b})$ where a and b are positive integers; or*
- *equivalent to one of*

$$\frac{1}{4}\left(3 + \sqrt{13} + i\sqrt{26 - 6\sqrt{13}}\right) \text{ or } 1 + \frac{1}{2}(1 + \sqrt{5})i \text{ or } 2\cos\frac{2\pi}{7} + \frac{1}{2}(1 + i\sqrt{3}).$$

$$(5.16)$$

Exercise 5.40 For which positive integers a, b is $\beta := \frac{1}{2}(\sqrt{a} + i\sqrt{b})$ a cyclotomic integer satisfying $\lceil \beta \rceil < 2$?

Exercise 5.41 Show that the three algebraic numbers in (5.16) are all cyclotomic integers with house less than 2.

Robinson's Conjecture 2.

There are only a finite number of inequivalent cyclotomic integers β with $\lceil \beta \rceil^2 \leq 5$ which are not equivalent to a cyclotomic integer of one of the following forms, for N a positive integer:

$$2\cos\frac{\pi}{N} \quad \text{or} \quad 1 + 2i\cos\frac{\pi}{N} \quad \text{or} \quad \sqrt{5}\cos\frac{\pi}{N} + i\sin\frac{\pi}{N}.$$

Robinson's Conjecture 3.

The cyclotomic integers $1 + 2i\cos(\pi/N)$ and $\sqrt{5}\cos(\pi/N) + i\sin(\pi/N)$ are equivalent only for $N = 2, 10$ and 30.

Robinson's Conjecture 4.

If β is a cyclotomic integer with $\lceil \beta \rceil \leq \sqrt{5}$, then either $\lceil \beta \rceil$ has one of the forms

$$2 \cos \frac{\pi}{N} \quad \text{or} \quad \sqrt{1 + 4 \cos^2 \frac{\pi}{N}},$$

where N is a positive integer, or else is equal to one of

$$\sqrt{\tfrac{1}{2}(5 + \sqrt{13})} \quad \text{or} \quad \tfrac{1}{2}(\sqrt{7} + \sqrt{3}).$$

Robinson's Conjecture 5.

Let β be a cyclotomic integer, not 0 or a root of unity, with $\lceil \beta \rceil \leq \sqrt{5}$. Then the only such β that are expressible as sums of at most three roots of unity are those equivalent to one of

$$2 \cos \frac{\pi}{N} \quad \text{or} \quad 1 + 2i \cos \frac{\pi}{N},$$

or to one of the five numbers

$$\tfrac{1}{2}(1 + i\sqrt{7}) \quad \text{or} \quad \tfrac{1}{2}(\sqrt{5} + i\sqrt{3}) \quad \text{or} \quad 2 \cos \tfrac{2\pi}{7} + \tfrac{1}{2}(1 + i\sqrt{3})$$
$$\text{or} \quad 1 + \omega_{13} + \omega_{13}^4 \quad \text{or} \quad 1 + \omega_{24} + \omega_{24}^7.$$

Exercise 5.42 For each of the algebraic numbers listed in Robinson's conjectures, verify that they are cyclotomic integers: write them as sums of roots of unity.

5.6 Cassels' Lemmas for $\mathcal{M}(\beta)$

The lemmas in this section appear as preliminaries in Cassels' proof of Robinson's second conjecture.

Lemma 5.43 *Let N be a positive integer having a prime factor p with $p \| N$. Put $N' = N/p$. Then every algebraic number $\beta \in \mathbb{Q}(\omega_N)$ can be written (non-uniquely) in the form*

$$\beta = \sum_{j=0}^{p-1} \alpha_j \omega_p^j, \tag{5.17}$$

where the $\alpha_j \in \mathbb{Q}(\omega_{N'})$. When β is an algebraic integer, the α_j can be chosen to be algebraic integers. Also

$$(p-1)\mathcal{M}(\beta) = \sum_{0 \leq i < j \leq p-1} \mathcal{M}(\alpha_i - \alpha_j). \tag{5.18}$$

Furthermore, if exactly X of the α_j are nonzero, then

$$\mathcal{M}(\beta) \geq \frac{X(p-X)}{p-1}. \tag{5.19}$$

Equality occurs when the nonzero α_j are equal, their common value being a root of unity.

Proof We can write any $\beta \in \mathbb{Q}(\omega_N)$ as $\beta = \sum_{i=0}^{\varphi(N)} a_i \omega_N^i$, with $a_0, \ldots, a_{\varphi(N)} \in \mathbb{Q}$. Next write $1 = aN' + bp$ for integers a, and b, so that $\omega_N = \omega_N^1 = \omega_p^a \omega_{N'}^b$. Substituting this into our expression for β, reducing each exponent of ω_p mod p and gathering like terms gives (5.17). Moreover, if β is an integer, then by Proposition 5.2(c), the a_i are in $\mathbb{Z}$, and so the α_j produced by the above process are algebraic integers. (The representation is not unique, as any fixed $\alpha \in \mathbb{Q}(\omega_{N'})$ could be added to each α_j.)

Denote by $\mathcal{M}'(\beta)$ the mean of $|\beta'|^2$ over all the conjugates β' of β over $\mathbb{Q}(\omega_{N'})$. The β' are obtained from β by letting ω_p in (5.17) range over all the primitive pth roots of 1. Hence

$$(p-1)\mathcal{M}'(\beta) = \sum_{\omega_p} \left| \sum_{0 \leq j \leq p-1} \alpha_j \omega_p^j \right|^2$$
$$= (p-1) \sum_j |\alpha_j|^2 - \sum_{i \neq j} \overline{\alpha}_i \alpha_j$$
$$= \sum_{0 \leq i < j \leq p-1} |\alpha_i - \alpha_j|^2.$$

Here, $\sum_{\omega_p}$ denotes the sum taken over all primitive pth roots of unity. Now, on taking the mean over all the conjugates over $\mathbb{Q}$ of the $\alpha_j \in \mathbb{Q}(\omega_{N'})$, we obtain (5.18).

If X of the α_j are nonzero, then (5.18) gives

$$(p-1)\mathcal{M}(\beta) \geq (p-X) \sum_{\alpha_j \neq 0} \mathcal{M}(\alpha_j). \tag{5.20}$$

Since $\mathcal{M}(\alpha_j) \geq 1$ when $\alpha_j \neq 0$, we obtain (5.19). $\square$

Lemma 5.44 *Let β, N, p and N' be as in Lemma 5.43, with p odd and β an algebraic integer. Suppose that $\mathcal{M}(\beta) < \frac{1}{4}(p+3)$. Then there is a representation (5.17) of β (with the α_j being algebraic integers) in which at most $\frac{1}{2}(p-1)$ of the α_j are nonzero.*

Proof It is enough to show that in any representation (5.17) (with the α_j algebraic integers), at least $(p+1)/2$ of the α_j are equal: we can then subtract this common value from all of the α_j to obtain a representation with at most $(p-1)/2$ of them nonzero.

From (5.18), and using Exercise 5.16, we have

$$\tfrac{1}{4}(p-1)(p+3) > \sum_{0 \le i < j \le p-1} \mathcal{M}(\alpha_i - \alpha_j)$$

$$\ge \sum_{\alpha_i \ne \alpha_j} 1$$

$$= \binom{p}{2} - \sum_{\ell} \binom{K_\ell}{2}, \tag{5.21}$$

if the p values of the α_j fall into multisets containing $K_1, K_2, \ldots, K_\ell, \ldots$ equal values. If $K_\ell \le \frac{1}{2}(p-1)$ for all ℓ, then it is an easy exercise to see that the minimum of (5.21) occurs for $K_1 = K_2 = \frac{1}{2}(p-1)$ and $K_3 = 1$. This gives $\binom{p}{2} - \sum_{\ell} \binom{K_\ell}{2} = \frac{1}{4}(p-1)(p+3)$, producing a contradiction. $\qquad \square$

Exercise 5.45 Verify the claim in the above proof: for p odd, show that the maximum of $\sum_{\ell=1}^{t} \binom{K_\ell}{2}$, subject to $K_1 + \cdots + K_t = p$ and $1 \le K_\ell \le (p-1)/2$ for all ℓ, is attained when $t = 3$, $K_1 = K_2 = (p-1)/2$ and $K_3 = 1$.

Lemma 5.46 *Let N be a positive integer having a prime factor p with $p^r \| N$, where $r \ge 2$. Put $N_1 = N/p$. Then every algebraic number $\beta \in \mathbb{Q}(\omega_N)$ can be written uniquely in the form*

$$\beta = \sum_{j=0}^{p-1} \alpha_j \omega_{p^r}^j, \tag{5.22}$$

where the $\alpha_j \in \mathbb{Q}(\omega_{N_1})$. Furthermore, when β is an algebraic integer, the α_j are also algebraic integers. Also

$$\mathcal{M}(\beta) = \sum_{j=0}^{p-1} \mathcal{M}(\alpha_j). \tag{5.23}$$

Proof Writing $p^{r-1} = aN_1 + bp^r$ for some $a, b \in \mathbb{Z}$, we can write $\omega_N = \omega_{p^r}^a \omega_{N_1}^b$. Substituting the above expression for ω_N in the unique representation of β as a $\mathbb{Q}$-linear combination of $1, \omega_N, \ldots, \omega_N^{\varphi(N)-1}$, and reducing exponents of ω_{p^r} to lie between 0 and $p-1$ by writing $\omega_{p^r}^p$ as a power of ω_{N_1}, we get β as a $\mathbb{Q}$-linear combination of $\{\omega_{N_1}^i \omega_{p^r}^j\}_{0 \le i < \varphi(N_1), \, 0 \le j < p}$. This latter spanning set has $\varphi(N)$ elements, so is a basis: we have found a representation as in (5.22) and the coefficients α_j are unique. Moreover, if β is an algebraic integer then it is a $\mathbb{Z}$-linear combination of $1, \omega_N, \ldots, \omega_N^{\varphi(N)-1}$, and the above process yields each α_j as a $\mathbb{Z}$-linear combination of $1, \omega_{N_1}, \ldots, \omega_{N_1}^{\varphi(N_1)-1}$, and the α_j are algebraic integers.

The conjugates of ω_{p^r} over $\mathbb{Q}(\omega_{N_1})$ are just the $\zeta_1 = \zeta \omega_{p^r}$, where ζ ranges over all the pth roots of unity. Note that $\zeta \in \mathbb{Q}(\omega_{N_1})$. Denoting as before by $\mathcal{M}'(\beta)$ the mean of $|\beta'|^2$ over all the conjugates β' of β over $\mathbb{Q}(\omega_{N_1})$, this time we have

$$p\mathcal{M}'(\beta) = \sum_{\zeta_1} \left| \sum_j \alpha_j \zeta_1^j \right|^2 = p \sum_j |\alpha_j|^2,$$

giving

$$\mathcal{M}'(\beta) = \sum_j |\alpha_j|^2 \, .$$

On averaging over the conjugates of $\mathcal{M}'(\beta)$ and the α_j over $\mathbb{Q}$, we obtain (5.23). $\square$

Lemma 5.47 *Suppose that β is a cyclotomic integer with $\mathcal{M}(\beta) > 1$. Then $\mathcal{M}(\beta) \geq 3/2$, with equality precisely when β is equivalent to $1 + \omega_5$. Furthermore, if β is not representable as a sum of two roots of unity then $\mathcal{M}(\beta) \geq 2$, with equality if and only if β is equivalent to one of*

$$1 + \omega_7 + \omega_7^j \text{ for } j = 2, 3, 4, 5 \text{ or } 6, \tag{5.24}$$

or to one of

$$\omega_3 - \omega_5 - \omega_5^j \text{ for } j = 2, 3 \text{ or } 4. \tag{5.25}$$

Proof Suppose that $\beta \in \mathbb{Q}(\omega_N)$, where β is chosen up to equivalence so that N is as small as possible. We shall refer to this as *minimality*. Assume too that β is not a root of unity, so that $\mathcal{M}(\beta) > 1$.

If N is not square-free, then Lemma 5.46 applies, and (5.23) tells us that $\mathcal{M}(\beta) \geq 2$, with equality occurring if and only if exactly two of the α_j are nonzero, and both are roots of unity. Thus, β is then a sum of two roots of unity.

We can therefore assume that N is square-free. Let p be the largest prime dividing N. If $p = 3$ then

$$\mathcal{M}(\beta) = \text{Norm}(\beta) \geq 3, \tag{5.26}$$

so we can take $p \geq 5$. Let X be the number of nonzero α_j in (5.17). By minimality, $X \geq 2$, and so

$$\mathcal{M}(\beta) \geq \frac{X(p - X)}{p - 1} \geq \frac{3}{2}, \tag{5.27}$$

with equality when $p = 5$ and $X = 2$, with the two nonzero α_j being equal roots of unity. Hence β is then equivalent to $1 + \omega_5$.

We now separate the argument into two cases, and each case into two subcases.

(a) $p \geq 7$. Then we can assume that $\mathcal{M}(\beta) < \frac{1}{4}(p + 3)$, so that we can apply Lemma 5.44 to conclude that $2 \leq X \leq \frac{1}{2}(p - 1)$.

(a1) $X \geq 3$. Then from (5.19)

$$\mathcal{M}(\beta) \geq \frac{3(p - 3)}{p - 1} \geq 2, \tag{5.28}$$

with equality when $p = 7$ and $X = 3$, with all the nonzero α_j equal to the same root of unity. This produces the possibility in (5.24).

(a2) $X = 2$. For β not representable as the sum of two roots of unity, the two nonzero α_j cannot both be roots of unity. So at least one α_j has $\mathcal{M}(\alpha_j) \geq 3/2$, by (5.27), so that (5.20) gives

$$\mathcal{M}(\beta) \geq \frac{\left(1 + \frac{3}{2}\right)(p - 2)}{p - 1} > 2. \tag{5.29}$$

(b) $p = 5$. If $\mathcal{M}(\beta) \leq 2$ then (5.18) gives

$$4 \times 2 \geq \sum_{0 \leq i < j \leq p-1} \mathcal{M}(\alpha_i - \alpha_j) \geq \binom{5}{2} - \sum_{\ell} \binom{K_\ell}{2}, \tag{5.30}$$

similarly to (5.21), yielding

$$\sum_{\ell} \binom{K_\ell}{2} \geq 2. \tag{5.31}$$

Of course we also have $\sum_{\ell} K_\ell = 5$. We now have two possibilities.

(b1) Some $K_\ell \geq 3$, so that at least three of the α_j are equal, say to α. On subtracting $\alpha(1 + \omega_5 + \omega_5^2 + \omega_5^3 + \omega_5^4) = 0$ from (5.17), we have $X \leq 2$. But $X = 1$ contradicts minimality, so $X = 2$, with say $\beta = \alpha_1 + \alpha_2 \omega_5$, with $\alpha_1, \alpha_2 \in \mathbb{Z}[\omega_3]$. Since we have excluded β being the sum of two roots of unity, we see using (5.26) and (5.18) that $\mathcal{M}(\beta) \geq (5 - 2)(3 + 1)/4 = 3$.

(b2) The only other case is $\{K_\ell\} = \{1, 2, 2\}$. We can assume that two of the α_j are 0. Then on multiplying β by some fifth root of unity and conjugating, we can further assume that $\beta = \alpha_0 + \alpha_1 \omega_5 + \alpha_2 \omega_5^r$ for some $r = 2, 3$ or 4. But then, since $\alpha_0, \alpha_1 \in \mathbb{Z}[\omega_3]$ and we have equality in (5.18), all three of α_0, α_1 and $\alpha_0 - \alpha_1$ must be roots of unity. Hence β is equivalent to one of the cyclotomic integers of (5.25).

$\square$

5.7 Discussion of Robinson's Problems

5.7.1 Robinson's First Problem

We now give Loxton's solution to Robinson's first problem. He shows how to decide whether there exist cyclotomic integers β with $|\beta|$ given. He also shows that there are only finitely many such β, and that they can be effectively determined.

Theorem 5.48 ([Lox75, Theorem 5]) *For a given positive real number h, up to equivalence there are only finitely many cyclotomic integers β with $|\beta| = h$. Further, all these β can be effectively determined.*

The bulk of the proof is contained in the following important subcase.

Proposition 5.49 *Theorem 5.48 holds when h^2 is a positive integer.*

Proof (of Proposition 5.49) Take β with $\overline{|\beta|}^2 = h^2 = R$ say, where $R \in \mathbb{Z}$. By Lemma 5.19, we know that all conjugates of β also lie on the circle $|z|^2 = R$. Also, we can assume that β is minimal (recall that this means that $q_{\min}(\beta^*) \geq q_{\min}(\beta)$ for every cyclotomic integer β^* equivalent to β). Put $q := q_{\min}(\beta)$ and $n := \mathcal{N}(\beta)$, and take a prime divisor p of q with say $p^N \| q$.

Now apply Theorem 5.51 below with $k = 1$. Then there is an effectively determined constant c_1 such that

$$R = \overline{|\beta|}^2 \geq c_1 n^{1 - \frac{1}{\log \log n}} \, ,$$

showing that n is bounded above by a number depending only on R. If we can show that q has the same property, then there will only be finitely many choices for β. (How many at most? Easy exercise.) To do this, it is enough to show that

$$p \leq 4^n + 1 \quad \text{and} \quad p^{N-1} \leq 4^n + 1 \tag{5.32}$$

for every prime divisor p of q. We divide the argument into two cases.

First Case: $N = 1$. Suppose that $p > 4^n + 1$. So certainly $n < p/2$. Write $q = pq_1$. Since every root of unity in $\mathbb{Q}(\omega_q)$ is of the form $\omega \omega_p^{r_j}$ for some root of unity $\omega \in \mathbb{Q}(\omega_{q_1})$ and r_j with $0 \leq r_j \leq p - 1$, we can write β as

$$\beta = \sum_{j=1}^{X} \gamma_j \omega_p^{r_j} \, , \tag{5.33}$$

where the γ_j are nonzero integers in $\mathbb{Q}(\omega_{q_1})$, and the r_j are incongruent (mod p). Furthermore, since $X \leq n < p/2$, this representation is unique, by Exercise 5.25(b). By Dirichlet's theorem on simultaneous rational approximation [Cas65, p. 13, Theorem VI], there is an integer u between 1 and $p - 1$ such that

$$\left\| \frac{u r_j}{p} \right\| \leq (p - 1)^{-1/X} \text{ for } 1 \leq j \leq X$$

where, as usual, $\|\theta\|$ denotes the distance of θ from the nearest integer. Also, by our hypothesis that $p > 4^n + 1$, we have $(p - 1)^{-1/X} \leq (p - 1)^{-1/n} < \frac{1}{4}$. So, after replacing β by one of its conjugates over $\mathbb{Q}(\omega_{q_1})$ and multiplying it by a suitable root of unity, we can assume that $r_1 = 0$ and $0 < r_j < \frac{1}{2}p$ $(2 \leq j \leq X)$.

We now rewrite (5.33) as

$$\beta = \sum_{i=1}^{\ell} \alpha_i \omega_p^i,$$

where $1 \le \ell < \frac{1}{2} p$ and each α_i is either 0 or one of the γ_j. Choosing ℓ minimally we have $\alpha_0 \ne 0$ and $\alpha_\ell \ne 0$. Then for the integer R, we have

$$R = |\beta|^2 = \sum_{k=0}^{p-1} \theta_k \omega_p^k, \tag{5.34}$$

where

$$\theta_k = \sum_{i-j \equiv k \pmod{p}} \alpha_i \overline{\alpha_j}.$$

From our hypothesis $p > 4^n + 1$, we see that the number of nonzero θ_k is at most $n^2 < p - 1$, so on applying Exercise 5.25(c) to (5.34) we have that θ_0 is the only nonzero θ_k (since $R \in \mathbb{Z}$ there is a representation with only the coefficient of 1 nonzero, and any other representation would have $p - 1$ nonzero coefficients). However, $\theta_\ell = \alpha_\ell \overline{\alpha_0} \ne 0$, a contradiction. So $p \le 4^n + 1$.

Second Case: $N > 1$. The argument in this case runs parallel to the first case. Put $L := N - 1$, assume $p^L > 4^n + 1$ and $q := p^L q_1$. Let $\omega := \omega_{p^N}$. By Exercise 5.26(a), there is a unique representation of β as

$$\beta = \sum_{j=1}^{X} \gamma_j \omega^{r_j}, \tag{5.35}$$

where, as in the First case, the γ_j are nonzero integers in $\mathbb{Q}(\omega_{q_1})$, the r_j are incongruent (mod p) and $X \le n$. By Dirichlet's theorem again, there is an integer u between 1 and $p^L - 1$ such that

$$\left\| \frac{u r_j}{p^L} \right\| \le (p^L - 1)^{-1/X} \text{ for } 1 \le j \le X.$$

Write $u = p^r v$, where $p \nmid v$, and put $M = L - r$. Then

$$\left\| \frac{v r_j}{p^M} \right\| \le (p^L - 1)^{-1/X} \text{ for } 1 \le j \le X \text{ and } 1 \le v \le p^M - 1.$$

After applying the automorphism $\omega \mapsto \omega^v$ of $\mathbb{Q}(\omega_q)/\mathbb{Q}$ and replacing β by one of its conjugates, we can assume that in the representations (5.35)

$$\left\| \frac{r_j}{p^M} \right\| \le (p^L - 1)^{-1/X} \text{ for } 1 \le j \le X. \tag{5.36}$$

We now rewrite (5.35) as

$$\beta = \sum_{i=0}^{p^M - 1} \alpha_i \omega^i ,$$

$$\alpha_i = \sum_{r_j \equiv i \pmod{p^M}} \gamma_j \omega^{r_j - i} .$$

Also set $q = p^M q_2$. Then all the α_i lie in $\mathbb{Q}(\omega_{q_2})$. Since all r_j / p^M are, by (5.36), close to integers, so also must be i / p^M if α_i is to be nonzero. More precisely, $\alpha_i = 0$ unless $\|i/p^M\| \le (p^L - 1)^{-1/X}$. By hypothesis, $(p^L - 1)^{-1/X} \le (p^L - 1)^{-1/n} < 1/4$, so these $\|i/p^M\|$ belong to $[0, 1/4)$, which happens for less than half of the values of $i \in \{0, 1, 2, \ldots, , p^M - 1\}$. So, by multiplying β by a suitable p^Nth root of unity we can assume that $\alpha_0 \ne 0$ and that $\alpha_i = 0$ for $i \ge \frac{1}{2} p^M$. Let ℓ with $1 \le \ell < \frac{1}{2} p^M$ be the largest integer for which $\alpha_\ell \ne 0$. Again, for the integer R we have

$$R = |\beta|^2 = \sum_{k=0}^{p^M - 1} \theta_k \omega^k , \tag{5.37}$$

where

$$\theta_k = \sum_{i - j \equiv k \pmod{p^M}} \alpha_i \overline{\alpha_j} \omega^{i - j - k} \ (0 \le k \le p^M - 1) .$$

On applying Exercise 5.26(c) to (5.37), we have that θ_0 is the only nonzero θ_k. However, $\theta_\ell = \alpha_\ell \overline{\alpha_0} \ne 0$, a contradiction. So $p^{N-1} \le 4^n + 1$. □

We can now prove Theorem 5.48. This extends the proposition by removing the restriction that R is an integer.

Proof (of Theorem 5.48*)* Let β be a cyclotomic integer with $|\beta|^2 = R$, where $[\mathbb{Q}(R) : \mathbb{Q}] = d$. By Exercise 5.50 below, we know that the conjugates β' of β lie on d circles centred at the origin, with the same number of conjugates on each circle. Pick one conjugate from each circle, say $\beta_1 = \beta$, $\beta_2, \ldots, \beta_d$, and let θ be their product. Then

$$|\theta|^2 = |\beta_1 \cdots \beta_d|^2 = \mathrm{Norm}_{\mathbb{Q}(R)/\mathbb{Q}}(R) \in \mathbb{Z} .$$

So, by Proposition 5.49, θ is equivalent to an element of a certain finite set, depending on R. The same is true for $\theta^* = \beta_1 \overline{\beta_2} \cdots \overline{\beta_d}$ and hence for

$$\theta \theta^* = \beta^2 |\beta_2 \cdots \beta_d|^2 = \beta^2 \, \mathrm{Norm}_{\mathbb{Q}(R)/\mathbb{Q}}(R)/R .$$

Thus, β^2 and thus also β is equivalent to an element of a certain finite set. □

Exercise 5.50 Let β be a cyclotomic integer, put $R := |\beta|^2$ and $d := [\mathbb{Q}(R) : \mathbb{Q}]$. Show that the conjugates of R are all of the form $|\beta'|^2$ for some conjugate β' of β, and that the conjugates of β lie on d circles, with the same number on each circle.

5.7.2 Robinson's Second Problem

Since $-\omega$ is a root of unity for any root of unity ω, we know from Proposition 5.2(c) that every cyclotomic integer β is a sum of roots of unity. Recall that $\mathcal{N}(\beta)$ is the smallest number of roots of unity whose sum is β. A natural question is: if $\mathcal{N}(\beta) = n$, clearly $\lceil \beta \rceil \le n$. But how small can $\lceil \beta \rceil$ be? This question was answered by Loxton [Lox72], in the following results. The theorems also solve Robinson's second problem.

Theorem 5.51 *For every $k > \log 2$, there is a positive number c depending only on k such that*

$$\lceil \beta \rceil^2 \ge cn^{1 - \frac{k}{\log\log n}}$$

for all cyclotomic integers β with $\mathcal{N}(\beta) = n$.

Loxton showed too that Theorem 5.51 fails for $k = \log 2$:

Theorem 5.52 *Suppose that $c > 0$. Then there are infinitely many positive integers n with the following property: there are infinitely many inequivalent cyclotomic integers β with $\mathcal{N}(\beta) = n$ and*

$$\lceil \beta \rceil^2 < cn^{1 - \frac{\log 2}{\log\log n}} .$$

5.8 Discussion of Robinson's Conjectures

5.8.1 The First Conjecture

We want to find all cyclotomic integers β with $\lceil \beta \rceil < 2$. Now $\lceil \beta \rceil^2 \in (0, 4)$ so, by Kronecker's second theorem (Theorem 1.4),

$$\lceil \beta \rceil = 2 \cos \frac{2\pi}{N}$$

for some $N \ge 4$. We can always take $\beta = \omega_N + \omega_N^{-1}$. For $N = 6$ we can take β to be any root of unity. From Cassels' solution to the second conjecture (see below), we know that there are only finitely many others. A good place to search for such β is in quadratic extensions of $\mathbb{Q}$, and their quadratic extensions. Another possibility is to look in the totally real fields $\mathbb{Q}(\omega_N + \omega_N^{-1})$ and their quadratic extensions. Such fields are of course Galois and abelian, and therefore subfields of cyclotomic fields. This is what Robinson seems to have done to find all his examples that are not a sum of at most two roots of unity. These are $1 + \omega_7 + \omega_7^3$, $1 + \omega_8 + \omega_8^3$, $1 + \omega_{11} + \omega_{11}^2 + \omega_{11}^4 + \omega_{11}^7$, $1 + \omega_{13} + \omega_{13}^3 + \omega_{13}^9$, $1 + \omega_{20} + \omega_{20}^9$, $1 + \omega_{20} + \omega_{20}^4 + \omega_{20}^{13}$, $1 + \omega_{30} + \omega_{30}^{12}$, $1 + \omega_{42} + \omega_{42}^{13}$ and $1 + \omega_{70} + \omega_{70}^{11} + \omega_{70}^{42} + \omega_{70}^{51}$.

This conjecture seems not to have been proved, in spite of the assertion in [RW13] that it had been. However, the calculations in that paper may prove useful for constructing a proof.

Exercise 5.53 Show that all these examples are equivalent to one of the forms given in Robinson's Conjecture 1.

Problem 5.54 (*open problem*)
Prove Robinson's Conjecture 1.

5.8.2 The Second Conjecture

Cassels [Cas69] proved a stronger version of Robinson's second conjecture: the bound $\lceil \beta \rceil^2 \leq 5$ in the hypothesis of the conjecture is weakened to $\lceil \beta \rceil^2 \leq 5.01$.

Theorem 5.55 (Cassels [Cas69]) *Suppose that β is a cyclotomic integer. A necessary and sufficient condition that $\lceil \beta \rceil^2 < 5.01$ is that one of the following conditions hold:*

I. *β can be expressed as the sum of not more than two roots of unity;*
II. *β is equivalent to $1 + \omega - \omega^{-1}$ for some root of unity ω;*
III. *β is equivalent to $\omega_5 + \omega_5^4 + \rho(\omega_5^2 + \omega_5^3)$ for some root of unity ρ;*
    or
IV. *β is equivalent to an element of a certain finite set.*

Thus, Cassels' theorem also shows that, in the set of all houses of cyclotomic integers, $\sqrt{5}$ is not a limit point from the right. It is, however, such a limit point from the left, as for instance, $\left| \omega_5 + \omega_5^4 + \omega_{2n-1}^n(\omega_5^2 + \omega_5^3) \right| \to \sqrt{5}$ as $n \to \infty$. We know from Proposition 5.35 that $\omega_5 + \omega_5^4 - (\omega_5^2 + \omega_5^3) = \sqrt{5}$.

The finite number of exceptions (IV) in the theorem remain a mystery. In [CMS11] Calegari, Morrison and Snyder say that 'A careful study of Cassels' analysis shows that any exceptions must lie in the field $\mathbb{Q}(\omega_N)$ with $N = 144 \prod_{\text{primes } p \leq 53} p$'. They also show that there are no exceptions where β is equivalent to a real number.

5.8.3 The Third Conjecture

We can prove Robinson's third conjecture of Sect. 5.5 using Algorithm 3 for finding cyclotomic points on plane algebraic curves. The conjecture says that the equation

$$1 + i(\omega + 1/\omega) = ((\omega + 1/\omega)\theta + \omega)\zeta,$$

where $i^2 = -1$ and $\theta = (\sqrt{5} - 1)/2$, has a solution (ω, ζ) in roots of unity if and only if $\omega := \omega_{2N}$, where $N = 2, 10$ or 30. Note that $\theta^2 + \theta = 1$.

So we put

$$g(w, x) := w(1 + i(w + 1/w) - x(\theta(w + 1/w) + w))$$

and regarding i and θ temporarily as indeterminates, we put

$$r := \text{res}_i(g, i^2 + 1) \quad \text{and} \quad r' := \text{res}_\theta(r, \theta^2 + \theta - 1),$$

which is a polynomial in w^2 and x. Replacing w^2 by u, and discarding the factor $u^2 + 3u + 1$ (which clearly has no cyclotomic zeros), r' becomes

$$f(u, x) := u^4 x^4 + (3x^2 + 2x + 3)x^2 u^3 + (x^4 - 2x^3 - 2x + 1)u^2 + (3x^2 + 2x + 3)u + 1.$$

Since this polynomial is full, we can apply Algorithm 3. Thus, we form the 7 resultants

$$r_i := \text{res}_x(f, f_i) \quad (i = 1, \ldots, 7),$$

where the f_i are defined in Proposition 5.9, and the r_i are polynomials in u. We then use the algorithm CycFacs above (Algorithm 1) to find their cyclotomic zeros. Only r_1, r_6 and r_7 have cyclotomic factors; these give the possible root-of-unity values for u. However, on computing $r_i' := \text{res}_u(f, f_i)$ $(i = 1, \ldots, 7)$ we find that r_1' has no cyclotomic factors, so there are no corresponding root-of-unity values for x in this case. The results are that $\text{CycFacs}(r_6) = \Phi_{30}(w)$, $\text{CycFacs}(r_6') = \Phi_{30}(x)$, $\text{CycFacs}(r_7) = \Phi_2(w)\Phi_{10}(w)$ and $\text{CycFacs}(r_7') = \Phi_2(x)\Phi_{10}(x)$. Thus indeed $N = 2$, 10 or 30. It is then easy, by direct substitution of the possibilities, to find all cyclotomic solutions (u, x) of $f(u, x) = 0$. These are $(u, x) = (-1, -1)$, $(\omega_{10}, \omega_{10})$, $(\omega_{10}, \omega_{10}^3)$, $(\omega_{30}, -\omega_{30}^5)$ and $(\omega_{30}, \omega_{30}^{-6})$.

Exercise 5.56 Which of these solutions corresponds to a solution of the original problem (i.e. with the 'positive' square roots of -1 and 5)?

5.8.4 The Fourth Conjecture

This was proved by F. Robinson and M. Wurtz [RW13] in 2013. In fact they found all possible values $\lceil \beta \rceil$ for cyclotomic integers β with $\lceil \beta \rceil^2 \leq 5.04$. They gave one value of $\lceil \beta \rceil$ with $\beta = 1 + \omega_{70} + \omega_{70}^{10} + \omega_{70}^{29}$ additional to the values of $\lceil \beta \rceil$ mentioned in the fourth conjecture.

5.8.5 The Fifth Conjecture

This was proved by A. Jones [Jon68] in 1968.

5.9 Multiplicative Relations Between Conjugate Roots of Unity

Proposition 5.57 *Let α be an algebraic number with conjugates $\alpha_1 = \alpha$, α_2, ..., α_d that satisfy*

$$\alpha_1^{b_1} \alpha_2^{b_2} \cdots \alpha_d^{b_d} = 1 \tag{5.38}$$

for some integers $b_1, \ldots, b_d$. Suppose further that either

$$|b_1| > |b_2| + \cdots + |b_d| \tag{5.39}$$

or, for some prime p,

$$p \nmid b_1 \quad but \quad p \mid b_j \, (j = 2, \ldots, d). \tag{5.40}$$

Then α is a root of unity.

Proof Let $K = \mathbb{Q}(\alpha_1, \ldots, \alpha_d)$, and assume that α is not a root of unity. By Corollary A.17 there is a prime q, possibly $q = \infty$, such that $|\alpha_j|_q > 1$. Choose d automorphisms $\sigma_i \in \text{Gal}(K/\mathbb{Q})$ such that $\sigma_i(\alpha_1) = \alpha_i$ ($i = 1, \ldots, d$). This is possible by the transitivity of the action of $\text{Gal}(K/\mathbb{Q})$ on conjugate sets in K, as described in Sect. A.4. Apply the σ_i to (5.38), to give d identities

$$\alpha_1^{b_{i1}} \cdots \alpha_d^{b_{id}} = 1,$$

say, with $b_{ii} = b_1$, and the other exponents in the ith identity being $b_2, \ldots, b_d$ in some order. Put $B := (b_{ij})$ and $\mathbf{b} := (\log |\alpha_1|_q, \ldots, \log |\alpha_d|_q)^\mathsf{T}$. We have $A\mathbf{b} = \mathbf{0}$ and, by the choice of q, that $\mathbf{b}$ is nonzero. But also $\det B \neq 0$, as its diagonal term dominates, giving a contradiction. (If (5.39) holds and $B\mathbf{x} = \mathbf{0}$, $\mathbf{x} \neq \mathbf{0}$ and the jth component of $\mathbf{x}$ has maximum modulus, then the jth component of $B\mathbf{x}$ is nonzero. If (5.40) holds, then the diagonal term in the expansion of $\det B$ is the only term not divisible by p.) $\qquad\square$

Problem 5.58 (*open problem*)
Show that, conversely, if under every relabelling of the b_i in (5.38), neither (5.39) nor (5.40) holds for any prime p, then (5.38) *does* have a solution in conjugates of an algebraic number α that is not a root of unity.

The following result centres around a special case of (5.38) in Proposition 5.57, where *every* conjugate set of roots of unity gives a solution to the equation.

Proposition 5.59 *Let α be a nonzero algebraic number. Then α has two conjugates α' and α'', not necessarily distinct from α or from each other, satisfying $\alpha^4 = \alpha'\alpha''$ if and only if α is a root of unity.*

Proof Take such α, α' and α''. Suppose that there is a valuation $|\ |_p$ for which $|\alpha|_p > 1$. By applying an appropriate automorphism of $\overline{\mathbb{Q}}_p/\mathbb{Q}_p$ and replacing α by a conjugate, we can assume that $|\alpha|_p \geq \max(|\alpha'|_p, |\alpha''|_p)$. Then

$$|\alpha^4|_p > |\alpha|_p^2 \geq |\alpha'|_p|\alpha''|_p = |\alpha'\alpha''|_p,$$

a contradiction. Hence $|\alpha|_p \leq 1$ for all primes p, including for $p = \infty$, so that, by Corollary A.17 in Appendix A, α is a root of unity.

Conversely, suppose that $\alpha^n = 1$, where $n = 2^k\ell$, with ℓ odd. Then

$$\alpha^4 = \alpha^{2+\ell}\alpha^{2-\ell},$$

where

$$\gcd(n, 2 \pm \ell) = \gcd(\ell, 2 \pm \ell) = \gcd(\ell, 2) = 1,$$

showing that both $\alpha^{2\pm\ell}$ are conjugates of α. $\qquad\square$

Exercise 5.60 Show that the exponent 4 cannot be replaced by 3 in Proposition 5.59. Which exponents, if any, *can* replace 4?

Exercise 5.61 Given an irreducible polynomial $P(z) \neq z$ with integer coefficients, define $R_P := \mathrm{res}_z(\mathrm{res}_y(\mathrm{res}_x(x^4 - yz, P(x)), P(y)), P(z))$. Show that R_P is zero if and only if $P(z)$ is a cyclotomic polynomial.

Proposition 5.62 ([Smy87]) *Given a rational number p/q, there are conjugate algebraic integers, α, α' with $|\alpha| \neq 0$ or 1 and $|\alpha|^{p/q} = |\alpha'|$.*

Proof Let $\beta = 1.3247\cdots$, β_2, $\bar{\beta}_2$ be the zeros of $z^3 - z - 1$, whose Galois group is the symmetric group on three symbols. Then

$$\alpha = \beta_2^{q-p}\bar{\beta}_2^{2q+p} \quad \text{and} \quad \alpha' = \beta^{q-p}\bar{\beta}_2^{2q+p}$$

are conjugate. But $|\beta_2| = |\bar{\beta}_2| = \beta^{-1/2}$, so

$$|\alpha| = \beta^{-3q/2} \quad \text{and} \quad |\alpha'| = \beta^{-3p/2} = |\alpha|^{p/q}.$$

$$\square$$

This result shows that (5.38) with $b_1 = b_2 = p$, $b_3 = b_4 = q$, with all other $b_i = 0$, has a solution in conjugate algebraic numbers that are *not* roots of unity.

5.10 Notes

The version of the Kronecker–Weber theorem in Proposition 5.2(e) comes from [Lox75, Lemma 6].

Algorithm 1 (CycFacs) and most of the material on cyclotomic points on curves in Sect. 5.3 come from Beukers and Smyth [BS02]. Algorithm 2 (MultiCycFacs) is new.

Proposition 5.11 was first considered and solved by Cassels [Cas69, Lemma 7]. The second statement of Lemma 5.31 was first proved by Schinzel [Sch66, Corollary 2].

Corollary 5.20 is due to Loxton [Lox75, Lemma 3]. Theorem 5.29 is also due to Loxton [Lox75, Theorem 1(i),(ii)].

The neat proof of Proposition 5.35 (evaluation of the sign of the Gauss sum) given here is due to David Speyer [Spe08].

Lemmas 5.43, 5.44, 5.46 and 5.47 are due to Cassels [Cas69].

Proposition 5.57 comes from Smyth [Smy86], while Proposition 5.62 comes from [Smy87]. Proposition 5.59 is new.

5.11 Glossary

$\lceil \beta \rceil$. The house of β.

$\mathcal{L}(f)$. The exponent lattice of f.

$\left(\frac{k}{p}\right)$. The Legendre symbol for an odd prime p, equal to 0 if $p \mid k$, 1 if k is a nonzero square modulo p, and -1 if k is not a square modulo p.

$||\theta||$. The distance between $\theta \in \mathbb{R}$ and the nearest integer.

$M(\beta)$. The Mahler measure of β.

$\mathcal{M}(\beta)$. The Cassels height of β.

$\mu(n)$. The Möbius function: $\mu(n)$ is the trace of ω_n.

$\mathcal{N}(\beta)$. For a cyclotomic integer β, we define $\mathcal{N}(\beta)$ to be the smallest number of roots of unity whose sum is β. Thus β can be written as a sum of $\mathcal{N}(\beta)$ roots of unity, but no fewer.

$\varphi(n)$. The Euler function: $\varphi(n)$ is the degree of Φ_n.

Φ_n. The nth cyclotomic polynomial, defined by (5.1).

ω, ω_n. Complex roots of unity. We use ω_n for a primitive nth root of unity: $\omega_n = e^{2\pi i j/n}$ for some j with $\gcd(j, n) = 1$.

$q_{\min}(\beta)$. The smallest q such that $\beta \in \mathbb{Q}(\omega_q)$.

absolute Mahler measure. If β has degree d, then the absolute Mahler measure of β is $M(\beta)^{1/d}$.

Cassels height. If β has conjugates $\beta_1, \ldots, \beta_d$, then the Cassels height of β, written as $\mathcal{M}(\beta)$, is the mean of the squares of the $|\beta_j|$. If β is a cyclotomic integer, then this lies in $\mathbb{Q}$.

conductor. If K is a finite abelian extension of $\mathbb{Q}$, then it is a subfield of $\mathbb{Q}(\omega_m)$ for some m, and the minimal such m is the conductor of K.

cyclotomic integer. A sum of roots of unity. Equivalently, a $\mathbb{Z}$-linear combination of roots of unity (after all, if ω is a root of unity, then so is $-\omega$). Equivalently, an element of the ring of integers of $\mathbb{Q}(\omega_n)$ for some n.

cyclotomic point.   A point (x, y) on a plane curve whose coordinates x and y are both roots of unity.

cyclotomic polynomial, nth cyclotomic polynomial.   If we speak simply of a cyclotomic polynomial, then we mean a product of polynomials Φ_n for some multiset of positive integers n. Thus, a cyclotomic polynomial is a monic integer polynomial whose zeros are roots of unity, but we do not require it to be irreducible, and we do not require it to have distinct zeros. If we speak of the nth cyclotomic polynomial, then we mean Φ_n, as in (5.1).

cyclotomic zero.   A zero of a polynomial that is also a root of unity.

equivalent.   This is a much-used word!

In the context of cyclotomic integers, we say that two are equivalent if dividing one of them by some conjugate of the other gives a root of unity. Equivalent cyclotomic integers share the same absolute Mahler measure, the same house and the same Cassels height.

In the context of Laurent polynomials, two are called equivalent if their quotient is a nonzero Laurent monomial $cx^a y^b$.

Euler function.   The function, written $\varphi(n)$, that counts the number of positive integers between 1 and n that have no prime factor in common with n. It is the degree of the algebraic integer ω_n, and hence of its minimal polynomial Φ_n.

exponent lattice.   We can represent a Laurent monomial $x^i y^j$ as a vector $(i, j)^\top \in \mathbb{Z}^2$. Given a Laurent polynomial $f(x, y) = \sum_{i,j} a_{ij} x^i y^j \in \mathbb{C}[x, y, x^{-1}, y^{-1}]$, its exponent lattice is the sublattice of $\mathbb{Z}^2$ spanned by the differences of the vectors corresponding to the monomials that appear, i.e., spanned by those $(i - i', j - j')^T$ for which $a_{ij} a_{i'j'} \neq 0$. The notation used for this lattice is $\mathcal{L}(f)$.

full.   An exponent lattice $\mathcal{L}(f)$ is called full if it equals $\mathbb{Z}^2$.

Gauss sum.   For $n \in \mathbb{N}$ and χ a character modulo n, the corresponding Gauss sums are sums of the form $\sum_{i=1}^n \chi(i) \omega_n^{ai}$ for some a. The only Gauss sum that we will need is the special case where $n = p$ is an odd prime, $a = 1$, and $\chi(i)$ is the Legendre symbol $\left(\frac{i}{p}\right)$.

minimal.   A cyclotomic integer β is called minimal if $q_{\min}(\beta)$ is minimal amongst all equivalent cyclotomic integers: if β' is equivalent to β, then $q_{\min}(\beta) \leq q_{\min}(\beta')$.

Möbius function.   The function, written $\mu(n)$, which is 0 if n is divisible by the square of any prime, and otherwise is $(-1)^k$ if n has k distinct prime factors. A key result concerning this function is the Möbius inversion formula: given a function $f : \mathbb{N} \to \mathbb{C}$, if we define $g : \mathbb{N} \to \mathbb{C}$ by $g(n) := \sum_{d|n} f(d)$, then $f(n) = \sum_{d|n} g(d) \mu(n/d)$, and conversely if f is defined in terms of g by the second formula then the first gives g in terms of f.

primitive nth root of unity.   A root of unity $\omega \in \mathbb{C}$ for which $n \geq 1$ is minimal such that $\omega^n = 1$.

root of unity.   A solution $\omega \in \mathbb{C}$ to $\omega^n = 1$ for some n. If n is specified, then we have an nth root of unity. For a given root of unity ω, the smallest positive n for which $\omega^n = 1$ is the order of ω, and ω is then called a primitive nth root of unity.

Salem number.   A real algebraic integer $\tau > 1$ such that all conjugates of τ (other than τ itself) have modulus at most one and at least one conjugate has modulus exactly one. The minimal polynomial of a Salem number is a Salem polynomial.

Salem polynomial.   A monic self-reciprocal integer polynomial that has exactly one zero in $|z| > 1$ and at least one zero on $|z| = 1$. If a Salem polynomial is irreducible, it is the minimal polynomial of a Salem number (and see Exercise 5.8).

Chapter 6
Cyclotomic Integer Symmetric Matrices I: Tools and Statement of the Classification Theorem

6.1 Introduction

The smallest known Mahler measures greater than 1 are all exhibited by polynomials that have most of their zeros on the unit circle: 'almost cyclotomic' polynomials. We might attempt to find small Mahler measures by considering polynomials attached to combinatorial objects. To do this, a first step is to find the 'cyclotomic' combinatorial objects (i.e., those to which cyclotomic polynomials are attached), and then the search for small Mahler measures can be via combinatorial objects that are 'almost cyclotomic', but now in the sense of being combinatorially close to cyclotomic. Herein is the power of the combinatorial approach, which goes right back to the discovery of Lehmer's number. Lehmer's polynomial (after replacing z by $-z$) is the Alexander polynomial of a certain pretzel knot, known as $P(7, 3, -2)$, which appeared in a book published in 1932 by Reidemeister [Rei74]. The Alexander polynomials of $P(3, 3, -2)$ and $P(5, 3, -2)$ are both cyclotomic.

In this chapter and the next, we start this programme for an important class of combinatorial objects: integer symmetric matrices. We attach reciprocal polynomials to these matrices and classify all those for which the Mahler measure of the reciprocal polynomial is 1. Later, in Chap. 13, by considering examples that are combinatorially close to cyclotomic, we shall identify the spectrum of the smallest Mahler measures greater than 1 for these reciprocal polynomials, solving the analogue of Lehmer's problem in this setting.

We shall need two important tools, included in Appendix B. The Interlacing Theorem (Theorem B.1) will play a prominent role, and we shall frequently use the words 'by interlacing' to mean that we have made the stated deduction by a suitable appeal to that theorem, usually in the form of one of the corollaries (Corollaries B.2 or B.3). We shall also use a little Perron–Frobenius theory (Theorem B.6).

The exposition here is restricted to integer symmetric matrices. There are extensions by Greaves and Taylor to Hermitian matrices whose entries are imaginary quadratic integers and to real matrices whose entries are real quadratic integers. See [Tay10, Tay11, Tay12, Gre12a, Gre12b, Gre12c, GT13] for these extensions.

© Springer Nature Switzerland AG 2021
J. McKee and C. Smyth, *Around the Unit Circle*, Universitext,
https://doi.org/10.1007/978-3-030-80031-4_6

The chapter starts (Sect. 6.2) with a general definition of the Mahler measure of a square matrix (not necessarily symmetric). Matrices that have Mahler measure 1 are called cyclotomic. In Sect. 6.3 some notions of equivalence of integer symmetric matrices will be defined and discussed: our classification goal is considerably simplified by working with a pertinent definition of equivalence. One key reason for our success when we tackle matrix analogues of problems relating to the Mahler measure is that we can grow examples from smaller examples. The theory and practice of efficient growing are developed in Sect. 6.4. One final tool is needed before embarking on the classification: Gram vectors. Representing vertices of charged signed graphs by vectors, such that adjacencies are given by dot products, provides a powerful application of linear algebra to assist our proofs. Gram vectors are introduced in Sect. 6.5 before the classification theorem is stated in Sect. 6.6. The proof is long and is deferred to the next chapter, where it is split into two main sections treating signed graphs (Sect. 7.1) and charged signed graphs (Sect. ??) before wrapping things up in Sect. ??. The chapter is definition-heavy, particularly at the start. The glossary definitions sometimes provide additional comments, beyond those that appear in the main text, with the intention that this section not only provides a convenient reference point while reading the chapter but also provides further help to understanding the text.

6.2   The Mahler Measure of a Matrix and Cyclotomic Matrices

Let $A \in M_n(\mathbb{Z})$, i.e., A is an $n \times n$ matrix with integer entries. The *characteristic polynomial* of A is the polynomial $\chi_A(x)$ defined by

$$\chi_A(x) = \det(xI - A),$$

where I is the $n \times n$ identity matrix. The zeros of χ_A are all algebraic integers, since $\chi_A(x) \in \mathbb{Z}[x]$ and is monic. In our search for small Mahler measures, we know that we can restrict to reciprocal polynomials [Bre51, Smy71] (and see Chap. 12), and this motivates the desire to attach a reciprocal polynomial to our matrix. We define the *reciprocal polynomial* of A, written $R_A(z)$, by

$$R_A(z) = z^n \chi_A(z + 1/z).$$

The *Mahler measure* of A, written $M(A)$, is defined to be the Mahler measure of R_A.

The matrix A is said to be *cyclotomic* if $M(A) = 1$. By Theorem 1.3, this is equivalent to saying that $R_A(z)$ has all its zeros on the unit circle (as 0 is never a zero of a reciprocal polynomial).

Exercise 6.1 Let $A \in M_n(\mathbb{Z})$.

(i) Verify that $R_A(z)$ is indeed a reciprocal polynomial and has degree $2n$.
(ii) If A has eigenvalues $\lambda_1, \ldots, \lambda_n$, establish the formula

$$M(A) = \prod_{i=1}^{n} \max(\theta_i, 1/\theta_i),$$

where $\theta_i = \left| \lambda_i + \sqrt{\lambda_i^2 - 4} \right| /2$ (for $i = 1, \ldots, n$).
(iii) In the special case where all the eigenvalues of A are real, show that

$$M(A) = \prod_{\lambda_i > 2} \left(\left(\lambda_i + \sqrt{\lambda_i^2 - 4} \right) /2 \right) \prod_{\lambda_i < -2} \left(\left(-\lambda_i + \sqrt{\lambda_i^2 - 4} \right) /2 \right).$$

Exercise 6.2 Given $R_A(z)$, show that the resultant of $R_A(z)$ and $z^2 - xz + 1$ (treating z as the variable) is $(\chi_A(x))^2$. This provides a convenient way to recover the characteristic polynomial from the reciprocal polynomial.

Compare with Exercise 1.6 and derive a method to recover χ_A from R_A that does not involve taking square roots of polynomials.

Although the definition of the Mahler measure of a matrix applies to an arbitrary integer matrix, we shall for the next couple of chapters be restricting to integer symmetric matrices: matrices $A \in M_n(\mathbb{Z})$ for which $A = A^\mathsf{T}$ (A^T is the transpose of A). In this chapter, we shall give a complete classification of those integer symmetric matrices for which the Mahler measure equals 1 (the *cyclotomic* integer symmetric matrices), and in the next chapter, we shall consider small Mahler measures greater than 1.

For an integer symmetric matrix A, all eigenvalues are real and the result of Exercise 6.1(iii) applies: A is cyclotomic if and only if all its eigenvalues lie in the interval $[-2, 2]$. We highlight this important condition.

Remark An integer symmetric matrix is cyclotomic if and only if all its eigenvalues lie in the interval $[-2, 2]$.

Suppose that A is an $n \times n$ matrix. Let S be any subset of $\{1, 2, \ldots, n\}$, and let $B = A_S$ be the matrix formed by deleting from A each row and each column corresponding to each element that is *not* in S, so that the rows of $B = A_S$ correspond to the elements of S. If A is symmetric, then so is B. We shall refer to B as an *induced submatrix* of A, taking the language from graph theory, as for cases of interest, the matrix A will correspond to a graph (or signed graph, or charged signed graph), and the matrix B will then correspond to an induced subgraph (see Appendix B).

If A is an integer symmetric matrix that is cyclotomic, then by interlacing (Corollary B.3), any induced submatrix of A is cyclotomic. Note that symmetry of A is essential to make this argument. We record this remark as a Lemma.

Lemma 6.3 *Let A be a cyclotomic integer symmetric matrix. Then any induced submatrix of A is cyclotomic.*

Exercise 6.4   (i) Verify that the matrix $A = \begin{pmatrix} -1 & 1 & 0 & 0 \\ 1 & 1 & 1 & -1 \\ 0 & 1 & 1 & 1 \\ 0 & -1 & 2 & 0 \end{pmatrix}$ is cyclotomic.

(ii) Let B be the submatrix of A obtained by deleting the fourth row and column. Verify that B is *not* cyclotomic. Why does this not conflict with Lemma 6.3?

As detailed in Appendix B, when dealing with integer symmetric matrices whose entries come from $\{-1, 0, 1\}$, it is sometimes convenient to view these as adjacency matrices of charged signed graphs. The charged signed graph for which A is the adjacency matrix is written G_A. It is sometimes convenient to use graph language and matrix language interchangeably. We may speak of row i or vertex i, and sometimes one choice will seem more natural than the other. Two vertices i and j in a charged signed graph G_A corresponding to an integer symmetric matrix $A = (a_{ij})$ are said to be **adjacent** if $a_{ij} \neq 0$. In other words, there is an edge (positive or negative) between the vertices i and j, or $i = j$ and the vertex i is charged. (Here it feels more natural to speak of the vertices as being adjacent rather than the rows.) The **degree** of a vertex i is the number of nonzero entries in row i (including the diagonal entry); it is the ith entry in the diagonal of A^2. (For more general integer symmetric matrices, the degree of vertex i may be defined as the sum of the squares of the entries in the ith row, again corresponding to the ith entry in the diagonal of the squared matrix.) A matrix is called **connected** if the graph G_A is connected, and the **components** of A are the induced submatrices corresponding to the connected components of G_A.

Exercise 6.5 With A satisfying (B.2) of Appendix B, show that the eigenvalues of A are obtained by pooling the eigenvalues of $A_1, \ldots, A_r$. Show that the Mahler measure of A is the product of the Mahler measures of the A_i.

The above extends to arbitrary integer symmetric matrices $A = (a_{ij})$, where adjacency of vertices i and j corresponds to $a_{ij} \neq 0$. We define connectedness (in other words, indecomposability) in the same way, and again we see that if A decomposes into block diagonal form via conjugating by a permutation matrix, then the Mahler measure of A is the product of the Mahler measures of the blocks. In order to describe cyclotomic integer symmetric matrices, it is, therefore, sufficient to restrict to connected ones: an arbitrary cyclotomic matrix is then formed by taking connected cyclotomic matrices for the blocks in (B.2).

We use the adjective **cyclotomic** to apply to graphs, signed graphs or charged signed graphs, to indicate that the Mahler measure of the associated matrix is 1. We also speak of the **Mahler measure** of a graph, signed graph, or charged signed graph to indicate the Mahler measure of the associated matrix. Our task in this chapter is to classify cyclotomic integer symmetric matrices. In fact, the bulk of the work will be in classifying cyclotomic charged signed graphs.

6.3 Flavours of Equivalence: Isomorphism, Equivalence and Strong Equivalence of Matrices

We say that two $n \times n$ matrices A and B are ***isomorphic*** if there exists a permutation matrix P such that $B = P^\mathsf{T} A P = P^{-1} A P$. Isomorphic matrices share the same eigenvalues; we have merely renumbered the rows and columns, and any eigenvector $\mathbf{v}$ for A corresponds to an eigenvector $P^{-1}\mathbf{v} = P^\mathsf{T}\mathbf{v}$ for B with the same eigenvalue.

We write $[A]_{\mathrm{iso}}$ for the set of matrices that are isomorphic to A. The property 'is isomorphic to' is an equivalence relation, and $[A]_{\mathrm{iso}}$ is the equivalence class containing A. Thus, A and B are isomorphic if and only if $B \in [A]_{\mathrm{iso}}$.

In our correspondence between matrices A and weighted digraphs G_A, the class $[A]_{\mathrm{iso}}$ corresponds precisely to the weighted digraph G_A without assigning any ordering to the vertices. Choosing an ordering for the vertices pins down a particular element of the class.

Exercise 6.6 *(computational exercise)* (Most of the exercises in this section are of a computational nature, producing a library of routines that can be used to check the results of computations. The reader who has no desire to engage in such exercises can safely skip them.)

Write and test a program to determine whether or not two integer symmetric matrices A and B are isomorphic. This is essentially a generalised graph isomorphism checker. We shall need this only for relatively small matrices (no larger than 16×16), so the naive approach outlined below should be fast enough.

- Check a list of possible cheap isomorphism invariants for the two matrices. Most nonisomorphic pairs of matrices will reveal their nonisomorphicness cheaply. This list of invariants might include: the size of the matrix; the multiset of diagonal entries; the multiset of row sums (the sum of all the entries in a row); the multiset of row sums of the square of the matrix; or the characteristic polynomial of the matrix.
- For pairs of matrices that pass the cheap necessary tests for the existence of an isomorphism, give each row a 'score' that is some combination of isomorphism invariants: e.g., the row sum plus the sum of the entries in the corresponding row of the square of the matrix. The multisets of row scores for the two matrices must agree if the two matrices are isomorphic. Backtrack through possible assignments of rows of A to rows of B mapping each row to one with the same score, checking as soon as a new row is assigned that the induced submatrix formed by the rows of A that have been assigned so far agrees with the relevant submatrix of B (else backtrack). Either an isomorphism is found by this process, or it is shown that none exists.

If A is an $n \times n$ matrix, then for any $i \in \{1, \ldots, n\}$ if we change the sign of row i of A and then change the sign of column i of A (thereby fixing the diagonal entry on row i), the eigenvalues of A are unchanged. We refer to this process as ***sign switching***. Of course, we can switch at any number of rows (or, in graph language, vertices), and we shall use the phrase 'sign switching' to refer to the cumulative effect of sign switchings at any number of rows/vertices. A sign switching from A to B is a transformation of the form $B = D^\mathsf{T} AD = D^{-1}AD = DAD$, where D is a diagonal matrix with all diagonal entries ± 1 (the negative diagonal entries flag the vertices at which we switch). We use the notation $[A]_{\mathrm{sw}}$ for the set of matrices that may be obtained from A by sign switching and refer to matrices A and B as being *switch-equivalent* if $B \in [A]_{\mathrm{sw}}$.

In the connected case, there is a surprisingly simple test for whether one can obtain $B = (b_{ij})$ from $A = (a_{ij})$ simply by sign switching. Of course, trivial necessary conditions are that $|b_{ij}| = |a_{ij}|$ for all i and j, that $b_{ii} = a_{ii}$ for all i and $\chi_B = \chi_A$. But it may be that all these conditions hold and yet there is no sign switching that transforms A to B.

For example, consider the matrices

$$A = \begin{pmatrix} 0 & -1 & 1 & 1 \\ -1 & 0 & 0 & 1 \\ 1 & 0 & 0 & 1 \\ 1 & 1 & 1 & 0 \end{pmatrix}, \quad B = \begin{pmatrix} 0 & 1 & -1 & 1 \\ 1 & 0 & 0 & 1 \\ -1 & 0 & 0 & 1 \\ 1 & 1 & 1 & 0 \end{pmatrix}.$$

These satisfy $|b_{ij}| = |a_{ij}|$ for all i and j and $b_{ii} = a_{ii}$ for all i. Their characteristic polynomials are equal (indeed they are isomorphic matrices: swap vertices 2 and 3). Yet there is no diagonal matrix $D = \mathrm{diag}(d_1, d_2, d_3, d_4)$ with each $d_i \in \{-1, 1\}$ such that $D^{-1}AD = DAD = B$, for consideration of the $(1, 2)$, $(1, 4)$ and $(2, 4)$ entries would give $d_1 d_2 = -1, d_1 d_4 = 1$ and $d_2 d_4 = 1$, leading to $-1 = d_2^2 d_1 d_4 = d_1 d_4 = 1$, a contradiction. How can we discriminate between the classes of A and B, without the gory detail of chasing such equations for a contradiction? Let C be the matrix whose entries are the component-wise products of those of A and B, and let E be the matrix whose entries are the absolute values of those of C:

$$C = \begin{pmatrix} 0 & -1 & -1 & 1 \\ -1 & 0 & 0 & 1 \\ -1 & 0 & 0 & 1 \\ 1 & 1 & 1 & 0 \end{pmatrix}, \quad E = \begin{pmatrix} 0 & 1 & 1 & 1 \\ 1 & 0 & 0 & 1 \\ 1 & 0 & 0 & 1 \\ 1 & 1 & 1 & 0 \end{pmatrix}.$$

Observe that any switching that takes A to B would take C to E (for $d_i c_{ij} d_j = d_i a_{ij} b_{ij} d_j$). Yet C and E have distinct characteristic polynomials, so no such switching exists.

We shall now prove that if all the off-diagonal entries of one of the matrices A and B are nonnegative, then the trivial necessary conditions that $|b_{ij}| = |a_{ij}|$ for all i and j, $b_{ii} = a_{ii}$ for all i and $\chi_B = \chi_A$ are also sufficient for there to be a sign switching that transforms A to B, at least if the underlying graph is connected.

Theorem 6.7 *Let $A = (a_{ij})$ and $B = (b_{ij})$ be $n \times n$ matrices with integer entries such that*

- $a_{ii} = b_{ii}$ *for* $1 \le i \le n$;

- $a_{ij} = |b_{ij}|$ for $1 \leq i \leq n, 1 \leq j \leq n, i \neq j$ (in particular the underlying graphs of the two matrices are the same, and the off-diagonal entries of A are nonnegative);
- the common underlying graph of A and B is connected.

Then $\chi_A = \chi_B$ if and only if there exists a diagonal matrix $D = \mathrm{diag}(d_1, \ldots, d_n)$ with each d_i either 1 or −1 such that $A = DBD^{-1}$.

Proof One direction is easy: if $A = DBD^{-1}$, then $\chi_A = \chi_B$.

Conversely, suppose that $\chi_A = \chi_B$, and choose an integer k such that $A + kI$ has all entries at least 0 (we know already that this is true for the off-diagonal entries, but the diagonal entries of A might include some negative values). Let λ be the largest modulus of any eigenvalue of $A + kI$; by the penultimate part of Theorem B.6, λ is actually an eigenvalue of $A + kI$. Now $A + kI$ and $B + kI$ share the same eigenvalues, so λ is an eigenvalue of $B + kI$: let $\mathbf{x}$ be an eigenvector of $B + kI$ with eigenvalue λ. Let $D = \mathrm{diag}(\varepsilon_1, \ldots, \varepsilon_n)$, where

$$\varepsilon_i = \begin{cases} -1 & x_i < 0, \\ 1 & x_i \geq 0. \end{cases}$$

Then $D\mathbf{x}$ is an eigenvector for $D(B + kI)D^{-1}$, with eigenvalue λ, and all entries of $D\mathbf{x}$ are at least 0. Now the entries of $DBD^{-1} + kI$ have modulus equal to those of $A + kI$, and so together with $(D\mathbf{x})_i \geq 0$ for each i, we have

$$((A + kI)D\mathbf{x})_i \geq ((DBD^{-1} + kI)D\mathbf{x})_i = \lambda(D\mathbf{x})_i .$$

By the final part of Theorem B.6, we conclude that $D\mathbf{x}$ is an eigenvector of $A + kI$, with eigenvalue λ, and that every entry of $D\mathbf{x}$ is strictly positive. Then from

$$(A - DBD^{-1})D\mathbf{x} = ((A + kI) - (DBD^{-1} + kI))D\mathbf{x} = \mathbf{0} ,$$

we conclude that $A = DBD^{-1}$. □

In light of this theorem, we can describe a simple test to see if B is a sign switching of A, in the case where both share the same connected underlying graph. First one checks that $|b_{ij}| = |a_{ij}|$ for all i and j and that $a_{ii} = b_{ii}$ for all i (diagonal entries are unchanged by sign switching, other entries may change sign). If any of these checks fail, then B cannot be obtained from A by sign switching. Let

$$C = (c_{ij}) = (|a_{ij}|\mathrm{sgn}(a_{ij})\mathrm{sgn}(b_{ij})) ,$$

so that $|c_{ij}| = |a_{ij}| = |b_{ij}|$ for all i and j, and let $E = (|c_{ij}|)$. (We shall generally use this in the case where all the nonzero off-diagonal entries have modulus 1, so that the above can be written more simply as $c_{ij} = a_{ij}b_{ij}$.) Suppose that $D = \mathrm{diag}(d_1, \ldots, d_n)$ with each d_i either 1 or −1. If $D^{-1}AD = B$, then $D^{-1}CD = E$, and conversely. We are, therefore, reduced to testing whether or not C and E are switch-equivalent, and for this we can use Theorem 6.7 (with $A = E$, $B = C$).

Modifying our previous example to illustrate this process, let

$$A_2 = \begin{pmatrix} 0 & -1 & 1 & 1 \\ -1 & -1 & 0 & 1 \\ 1 & 0 & -1 & 1 \\ 1 & 1 & 1 & 0 \end{pmatrix}, \quad B_2 = \begin{pmatrix} 0 & 1 & -1 & 1 \\ 1 & -1 & 0 & 1 \\ -1 & 0 & -1 & 1 \\ 1 & 1 & 1 & 0 \end{pmatrix}.$$

We verify first that the diagonal entries agree and the absolute values of off-diagonal entries agree. (In fact, also the characteristic polynomials agree.) Form C_2 by multiplying the signs of the entries of A_2 and B_2 together while preserving the moduli of the entries, and E_2 by taking absolute values of all entries:

$$C_2 = \begin{pmatrix} 0 & -1 & -1 & 1 \\ -1 & 1 & 0 & 1 \\ -1 & 0 & 1 & 1 \\ 1 & 1 & 1 & 0 \end{pmatrix}, \quad E_2 = \begin{pmatrix} 0 & 1 & 1 & 1 \\ 1 & 1 & 0 & 1 \\ 1 & 0 & 1 & 1 \\ 1 & 1 & 1 & 0 \end{pmatrix}.$$

We compute that C_2 and E_2 have different characteristic polynomials, and hence, A_2 and B_2 are not switch-equivalent (although they are isomorphic as charged signed graphs: there is a bijection between the two sets of vertices that preserves signed adjacencies, including charges). If it had turned out that C_2 and E_2 had the same characteristic polynomial, then we would have concluded that A_2 and B_2 were switch-equivalent.

We can combine sign switching with permutations of the rows and columns, still preserving both the characteristic polynomial and the property of being cyclotomic (or not). Following [MS07], we say that two $n \times n$ matrices A and B are **strongly equivalent** if there exists a **signed permutation matrix** P (i.e., P is a matrix with entries from $\{-1, 0, 1\}$ with a single nonzero entry in each row and a single nonzero entry in each column, i.e., $P \in O_n(\mathbb{Z})$, the group of $n \times n$ orthogonal integer matrices) such that $B = P^T A P = P^{-1} A P$. The matrices A and B share their eigenvalues, so have the same Mahler measure, and in particular, one is cyclotomic if and only if the other is. Moreover, if A is symmetric, then so is B.

We use the notation $[A]_{\text{str}}$ to denote the set of matrices that are strongly equivalent to A.

Exercise 6.8 Show that

$$[A]_{\text{str}} = \bigcup_{B \in [A]_{\text{sw}}} [B]_{\text{iso}} = \bigcup_{B \in [A]_{\text{iso}}} [B]_{\text{sw}}.$$

We say that a charged signed graph G is **bipartite** if $-G$ (changing the signs of all charges and all edges) is strongly equivalent to G, i.e., $-G \in [G]_{\text{str}}$ (interpreting this as meaning that if A is the adjacency matrix then $-A \in [A]_{\text{str}}$).

Exercise 6.9 Show that if G is bipartite (as defined here), then the eigenvalues of G are symmetric about the origin. Verify that the notion of bipartiteness defined here generalises the usual one for graphs. (See [BH12, Proposition 3.4.1(i)] for the spectral characterisation of bipartite graphs.)

Exercise 6.10 For ordinary graphs, one has the result that a connected graph that has largest eigenvalue ρ is bipartite if and only if $-\rho$ is also an eigenvalue ([BH12, Proposition 3.4.1(ii)]). For charged signed graphs, this may be false. Show that the largest eigenvalue of the charged signed graph shown below is 2, the smallest eigenvalue is -2, yet the charged signed graph is not bipartite. (Indeed the number of positive charges in a bipartite charged signed graph must be the same as the number of negative charges.)

Exercise 6.11 *(computational exercise)* Adapt the code from Exercise 6.6 to determine whether or not two connected integer symmetric matrices A and B are strongly equivalent. Perform the cheap tests as before but on the matrices $(|a_{ij}|)$ and $(|b_{ij}|)$. When doing the search with backtracking, the decision to allow a new assignment of a vertex should be based on whether some *switching* of the assignment of the vertices so far gives agreement. For this check, use the test outlined above.

If A is cyclotomic, then so is $-A$. The characteristic polynomials of A and $-A$ are not generally equal, but it will be convenient for our classification to regard A and $-A$ as equivalent; for the purposes of classifying cyclotomic matrices, this reduces the effort. We therefore define A and B to be **equivalent** if either A is strongly equivalent to B or A is strongly equivalent to $-B$. We write $[A]_{\text{eq}}$ for the set of matrices that are equivalent to A. Clearly, we have

$$[A]_{\text{eq}} = [A]_{\text{str}} \cup [-A]_{\text{str}} .$$

Exercise 6.12 *(computational exercise)* Make the trivial adaptation to the code from Exercise 6.11 to determine whether or not two connected integer symmetric matrices A and B are equivalent.

Exercise 6.13 Let

$$A = \begin{pmatrix} 1 & 1 & 0 \\ 1 & -1 & 1 \\ 0 & 1 & 0 \end{pmatrix}, \quad B = \begin{pmatrix} -1 & 1 & 1 \\ 1 & 0 & 0 \\ 1 & 0 & 1 \end{pmatrix}, \quad C = \begin{pmatrix} 1 & 1 & 0 \\ 1 & -1 & -1 \\ 0 & -1 & 0 \end{pmatrix},$$

$$D = \begin{pmatrix} -1 & -1 & 1 \\ -1 & 0 & 0 \\ 1 & 0 & 1 \end{pmatrix} \quad \text{and } E = \begin{pmatrix} -1 & 1 & 0 \\ 1 & 1 & 1 \\ 0 & 1 & 0 \end{pmatrix} .$$

Draw the charged signed graphs corresponding to A, B, C, D and E.

Verify that:

- $B \in [A]_{\mathrm{iso}}$;
- $C \in [A]_{\mathrm{sw}}, C \notin [A]_{\mathrm{iso}}$;
- $D \in [A]_{\mathrm{str}}, D \notin [A]_{\mathrm{iso}}, D \notin [A]_{\mathrm{sw}}$;
- $E \in [A]_{\mathrm{eq}}, E \notin [A]_{\mathrm{str}}$;
- A, B, C, D and E are all cyclotomic;
- $\chi_A = \chi_B = \chi_C = \chi_D \neq \chi_E$.

6.4　Growing Cyclotomic Matrices

Let A be an $n \times n$ integer symmetric matrix, and let S be a subset of $\{1, 2, \ldots, n\}$. Recall that we can form the induced submatrix A_S by restricting to the rows and columns of A that are in S: if $S = \{i_1, \ldots, i_m\}$, then A_S is the $m \times m$ matrix whose (r, s) entry is a_{i_r, i_s}. In a complementary manner, let $S' = \{1, 2, \ldots, n\} - S$, then the induced submatrix $A_{S'}$ is called the matrix obtained by deleting the rows/columns in S. By Lemma 6.3, if A is cyclotomic, then so are both A_S and $A_{S'}$.

Exercise 6.14 Show that if A is connected with at least two rows, then there is at least one *connected* induced submatrix obtained by deleting one of the rows of A (and the corresponding column).

We say that a connected matrix B is ***contained*** in a connected integer symmetric matrix A if B is equivalent to an induced submatrix of A. Note that we work up to equivalence and restrict to connected matrices, and symmetrically we say that A ***contains*** B. A natural notation for this is $B \subseteq A$ or $A \supseteq B$, but it is important to remember that we are working up to equivalence. If A is cyclotomic, then by interlacing so is any matrix contained in A. We say that G_B is (equivalent to) an induced subgraph of G_A, dropping the possible adjectives 'charged' or 'signed' as these will be clear from the context. Again we write $H \subseteq G$ to mean that H is equivalent to an induced subgraph of G, stressing that we work up to equivalence when using this notation.

If we have a complete list L_n of all $n \times n$ connected cyclotomic integer symmetric matrices (up to equivalence), then by Exercise 6.14 every $(n + 1) \times (n + 1)$ connected cyclotomic integer symmetric matrix contains at least one of the elements of L_n. In principle, we can use this process, which we call ***connected growing***, to produce all connected cyclotomic integer symmetric matrices up to any particular size. To do this in full generality requires obtaining bounds on the sizes of the entries of the larger matrices, but if we restrict to charged signed graphs, then that aspect of the process is trivial.

If a connected cyclotomic integer symmetric matrix A cannot be grown connectedly to give any larger connected cyclotomic integer symmetric matrix, then we say that A is ***maximal***. Part of our classification theorem will be to show that every connected cyclotomic integer symmetric matrix is contained in a maximal one. For an integer symmetric matrix A, we define the **degree** of the ith row/vertex to be the ith diagonal entry in the matrix A^2. If A is the adjacency matrix of a graph, then this agrees with the usual definition of the degree of a vertex.

Lemma 6.15 *Let A be a cyclotomic integer symmetric matrix.*

(i) The degree of each row of A is at most 4.
(ii) If A is a connected cyclotomic integer symmetric matrix and each row has degree exactly 4, then A is maximal.

Proof (i) Let d_i be the ith diagonal entry of A^2. By interlacing, A^2 has an eigenvalue $\mu \geq d_i$. Hence, A has an eigenvalue λ with $|\lambda| \geq \sqrt{d_i}$. Given that A is cyclotomic, it follows that $d_i \leq 4$.

(ii) If each row of A has degree exactly 4, then any matrix connectedly grown from A contains a row that has degree at least 5, and so cannot be cyclotomic. □

Exercise 6.16 *(computational exercise)* Use your code for identifying when two matrices are equivalent to write a program that takes as its input a connected cyclotomic charged signed graph G on n vertices and gives as its output all connected cyclotomic charged signed graphs on $n + 1$ vertices, up to equivalence, that contain G.

It turns out that there are arbitrarily large connected cyclotomic charged signed graphs, so one cannot find them all in finite time simply by connected growing. Nevertheless, for some particular cyclotomic graphs G, the process of repeated connected growing terminates in a finite list of maximal cyclotomic examples.

Exercise 6.17 *(computational exercise)* Let

$$A = \begin{pmatrix} 0 & 1 & 1 \\ 1 & 0 & 1 \\ 1 & 1 & 0 \end{pmatrix},$$

with G_A an uncharged triangle. Verify that this is cyclotomic. Grow G_A connectedly to produce all cyclotomic charged signed graphs on four vertices that contain G_A, and then regrow the larger charged signed graphs produced, and repeat (always restricting to connected cyclotomic examples): you should find that starting from G_A this process terminates. Verify that up to equivalence the complete list of connected cyclotomic charged signed graphs containing G_A is

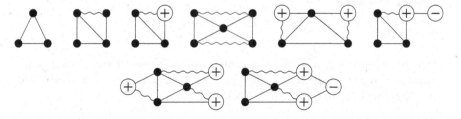

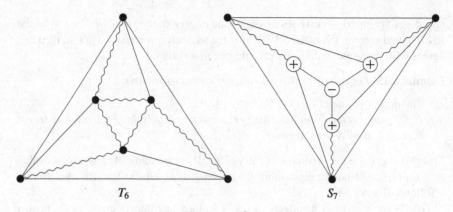

T_6 S_7

(The two maximal examples are named for later reference.)

Reflecting on the result of Exercise 6.17, we have the following remark.

Lemma 6.18 *Let G be a connected cyclotomic charged signed graph that contains a triangle of neutral vertices. Then either $G \subseteq T_6$ or $G \subseteq S_7$, where T_6 and S_7 are drawn in Exercise 6.17.*

(Recall that $H \subseteq G$ means that H is equivalent to a subgraph of G.)

Proof If the triangle has three positive edges, then the claim is immediate from the exercise; if it has exactly one positive edge, then switching at the vertex that is not one of the endpoints of the positive edge reduces to the previous case; if it has exactly two positive edges or none, then negating all the signs of edges and charges (an equivalence operation) reduces to one of the previous cases.   □

This illustrates an important tool which we shall use repeatedly in the classification of connected cyclotomic charged signed graphs: we have found a subgraph (here a neutral triangle) which if contained in a connected cyclotomic charged signed graph G forces G to be contained in one of a finite number of maximal possibilities (here T_6 or S_7). Having established this, we can restrict our search to graphs that do not contain any such subgraph. Exercises 6.19–6.21 produce other examples of such subgraphs: the exclusion of these as subgraphs will greatly help in establishing the full classification of connected cyclotomic charged signed graphs.

Exercise 6.19 *(computational exercise)* Show that the charged signed graph G below is cyclotomic.

$$\oplus\!\!-\!\!\ominus$$
$$G$$

By connected growing, show that any connected cyclotomic charged signed graph containing G is equivalent to one contained in one of the examples S_7 (Exercise 6.17), C_4^{+-} (Exercise 6.20), S_8 (below) or S_8' (below).

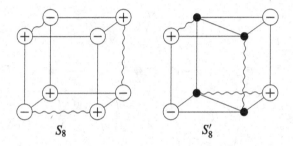

$$S_8 \qquad\qquad S_8'$$

Exercise 6.20 *(computational exercise)* Show that the charged signed graph G below is cyclotomic.

$$G$$

Show by connected growing that the only connected cyclotomic charged signed graphs containing G that have all their vertices charged (so no neutral vertices allowed) are equivalent to subgraphs of one of C_4^{++} or C_4^{+-} shown below.

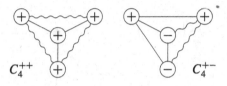

Exercise 6.21 *(computational exercise)* This exercise represents a more substantial computational challenge. Let A, M, B be the three signed graphs shown.

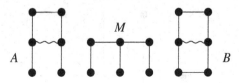

Show by connected growing that the only connected cyclotomic charged signed graphs containing a subgraph equivalent to either A, B or M are equivalent to subgraphs of one of S_{14} (Fig. 6.1) or S_{16} (Fig. 6.2).

Exercise 6.22 Given a connected $n \times n$ matrix $A = (a_{ij})$ $(n > 1)$, define $v(i)$ $(1 \le i \le n)$ to be the smallest j such that $a_{ij} \ne 0$ (connectedness implies that not all a_{ij} are zero). We say that A is in ***echelon form*** if $v(2) \le \cdots \le v(n)$ (note that this starts with $v(2)$, not $v(1)$) and $a_{iv(i)} > 0$ for $2 \le i \le n$.

(i) Show that any connected $n \times n$ matrix $A = (a_{ij})$ $(n > 1)$ is equivalent to one in echelon form.

(ii) Show that if A is connected and in echelon form, then so is the induced submatrix formed by the first r rows and columns $(2 \le r \le n)$.

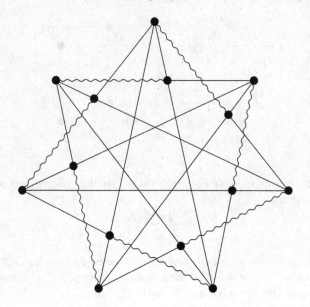

Fig. 6.1 S_{14}

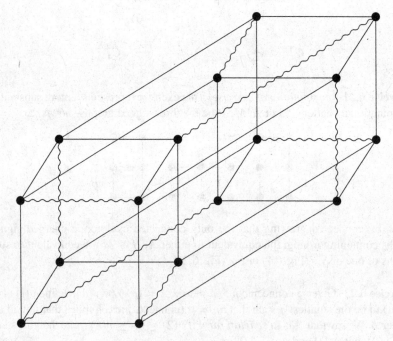

Fig. 6.2 S_{16}

(iii) (This part is a programming exercise.) Write a program that will grow connected cyclotomic matrices while preserving echelon form. When growing from nothing, this will be *much* faster than naive connected growing. Why can it not be used in Exercise 6.21?

We shall refer to the process outlined in Exercise 6.22 as *echelon growing*. There is a disadvantage over connected growing that having grown examples from a list, preserving echelon form, if one wishes to grow further, one cannot reduce the list to one that contains only one representative of each equivalence class; for it might be that to reach a larger example while preserving echelon form, one needs to grow from a different representative of the equivalence class. There is, therefore, the risk that repeated echelon growing might lead to an explosion of the number of matrices being grown. For the examples we shall need, however, the improved speed of each growing operation more than compensates for this cost.

6.5 Gram Vectors

Let $A = (a_{ij})$ be an $n \times n$ integer symmetric matrix that has all its eigenvalues in the interval $[-2, \infty)$. This includes the possibility that A is cyclotomic, but is more general. Then $A + 2I$ has all eigenvalues in the interval $[0, \infty)$, so that if we diagonalise $B = A + 2I$ (over $\mathbb{R}$) using an orthogonal transformation matrix P, we get $P^\mathsf{T} B P = \mathrm{diag}(d_1, \ldots, d_n)$, with all $d_i \geq 0$. Then

$$
\begin{aligned}
B &= P \, \mathrm{diag}(d_1, \ldots, d_n) P^\mathsf{T} \\
&= P \, \mathrm{diag}(\sqrt{d_1}, \ldots, \sqrt{d_n}) \, \mathrm{diag}(\sqrt{d_1}, \ldots, \sqrt{d_n}) P^\mathsf{T} \\
&= M M^\mathsf{T},
\end{aligned}
$$

where $M = P \, \mathrm{diag}(\sqrt{d_1}, \ldots, \sqrt{d_n})$. Let $\mathbf{v}_1, \ldots, \mathbf{v}_n$ be the rows of M, viewed as vectors in $\mathbb{R}^n$. Then for each i and j we have (always using the usual dot product on $\mathbb{R}^n$)

$$
a_{ij} = \begin{cases} \mathbf{v}_i \cdot \mathbf{v}_j & i \neq j, \\ \mathbf{v}_i \cdot \mathbf{v}_i - 2 & i = j. \end{cases} \tag{6.1}
$$

Conversely, let $\mathbf{v}_1, \ldots, \mathbf{v}_n$ be any vectors in $\mathbb{R}^m$ (not necessarily distinct, and perhaps with $m \neq n$), and define $A := (a_{ij})$, where (a_{ij}) is given by (6.1). Let M be the $n \times m$ matrix whose rows are given by the $\mathbf{v}_i$. Then $A = M M^\mathsf{T} - 2I$ is symmetric, and all eigenvalues of A are at least -2; for if $\mathbf{w}$ is any eigenvector of A (so that, in particular, $\mathbf{w}^\mathsf{T} \mathbf{w} > 0$), with eigenvalue λ, then

$$
(\lambda + 2)\mathbf{w}^\mathsf{T}\mathbf{w} = \mathbf{w}^\mathsf{T}(A + 2I)\mathbf{w} = \mathbf{w}^\mathsf{T} M M^\mathsf{T} \mathbf{w} = (M^\mathsf{T}\mathbf{w})^\mathsf{T}(M^\mathsf{T}\mathbf{w}) \geq 0,
$$

which implies that $\lambda \geq -2$.

Thus, we see that $n \times n$ integer symmetric matrices A with all eigenvalues at least -2 are precisely those for which $A + 2I$ can be *represented by Gram vectors*, i.e., for some m there exist $\mathbf{v}_1, \ldots, \mathbf{v}_n \in \mathbb{R}^m$ such that (6.1) holds. When all entries of A are from $\{-1, 0, 1\}$, we speak of such a set of Gram vectors either as a *Gram representation* of the matrix $A + 2I$ or as a Gram representation of the associated charged signed graph G_A. Note the language: it is convenient to speak of the vectors giving a Gram representation of G_A rather than using the more cumbersome G_{A+2I}. Given a set of Gram vectors, the corresponding matrix is called a *Gram matrix*.

Exercise 6.23 Show that an integer symmetric matrix A is cyclotomic if and only if both $A + 2I$ and $-A + 2I$ can be represented by Gram vectors.

As we shall see, Gram vectors provide an extraordinarily powerful tool in our classification of all cyclotomic integer symmetric matrices. For charged signed graphs, all diagonal entries are -1, 0 or 1, so the Gram vectors representing a cyclotomic charged signed graph have length 1, $\sqrt{2}$ or $\sqrt{3}$, according as the corresponding vertex has charge -1, is neutral or has charge $+1$. It may be convenient when drawing cyclotomic charged signed graphs to label the vertices with corresponding Gram vectors (which must have the right length, as detailed above, and correct dot products with the vectors that label other vertices, be they adjacent or not).

For example, let $\mathbf{e}_1, \mathbf{e}_2, \mathbf{e}_3$ be an orthonormal basis of $\mathbb{R}^3$. Then the triangle graph of Exercise 6.17 can be redrawn with appropriate vectors labelling the vertices:

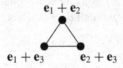

Here each vertex vector has squared length 2, as required for neutral vertices, and the dot product of each (adjacent) pair of vertex vectors is $+1$.

Exercise 6.24 Let $\mathbf{e}_1, \mathbf{e}_2, \mathbf{e}_3$ be an orthonormal basis for $\mathbb{R}^3$. Show that the six vectors $\pm\mathbf{e}_1 + \mathbf{e}_2$, $\mathbf{e}_1 \pm \mathbf{e}_3$, $\pm\mathbf{e}_2 + \mathbf{e}_3$ give a Gram representation of T_6 in Exercise 6.17, thereby reproving that all its eigenvalues are at least -2. Verify that $-T_6$ (changing the signs of all edges and the signs of any charged vertices, although here all vertices are neutral) is isomorphic to T_6, and hence reprove that T_6 is cyclotomic.

Exercise 6.25 Let A be an $n \times n$ integer symmetric matrix that is cyclotomic, and let $\mathbf{v}_1, \ldots, \mathbf{v}_n \in \mathbb{R}^n$ be a set of Gram vectors that represents $A + 2I$. Show that a sign switching at vertex i corresponds to changing the sign of the ith Gram vector, $\mathbf{v}_i$.

A charged signed graph G is called a *path* if the vertices of G can be labelled $v_1, \ldots, v_n$ such that the only edges between distinct vertices in G are $n-1$ signed edges which occur between each consecutive pair v_i, v_{i+1} ($1 \leq i \leq n-1$). A charged signed graph G is called a *cycle* if the vertices of G can be labelled $v_1, \ldots, v_n$ such that the only edges in G are n signed edges which occur between each consecutive pair v_i, v_{i+1} ($1 \leq i \leq n-1$) and also between v_1 and v_n.

Lemma 6.26 *Let P_r be the path on $r+1$ vertices shown (with the labels being the names of the vertices), where there is one charged vertex v_0 (negatively charged) at one end of the path, and the other r vertices $(v_1, ..., v_r)$ are neutral:*

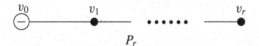

Let $\mathbf{e}_1, ..., \mathbf{e}_{r+1}$ be an orthonormal basis for $\mathbb{R}^{r+1}$. Then the vectors $\mathbf{e}_1, \mathbf{e}_1 + \mathbf{e}_2$, $\mathbf{e}_2 + \mathbf{e}_3, ..., \mathbf{e}_r + \mathbf{e}_{r+1}$ give a Gram representation of P_r.

Moreover, let $\mathbf{v}_0, ..., \mathbf{v}_r$ be any set of Gram vectors that represents P_r, with $\mathbf{v}_i$ associated with the vertex v_i. Then there is a uniquely determined orthonormal basis $\mathbf{e}_1, ..., \mathbf{e}_{r+1}$ for the span of $\mathbf{v}_0, ..., \mathbf{v}_r$ such that $\mathbf{v}_0 = \mathbf{e}_1$ and $\mathbf{v}_i = \mathbf{e}_i + \mathbf{e}_{i+1}$ for $1 \leq i \leq r$.

Proof The first paragraph is clear: the vector $\mathbf{e}_1$ associated with the negatively charged vertex v_0 has length 1, the vectors associated with the neutral vertices all have length $\sqrt{2}$, the dot product of consecutive vectors in the given list is always 1, and the dot product of nonconsecutive vectors is 0.

For the second paragraph, given $\mathbf{v}_0, ..., \mathbf{v}_{r+1}$, we start by defining $\mathbf{e}_1 = \mathbf{v}_0$. Since v_0 has a negative charge, $\mathbf{e}_1 \cdot \mathbf{e}_1 = \mathbf{v}_0 \cdot \mathbf{v}_0 = 1$. Projecting orthogonally onto $\langle \mathbf{e}_1 \rangle$, we write $\mathbf{v}_1 = \lambda_1 \mathbf{e}_1 + \mathbf{e}_2$, where $\mathbf{e}_1 \cdot \mathbf{e}_2 = 0$. Since v_0 and v_1 are joined by a positive edge, we have $\lambda_1 = \mathbf{e}_1 \cdot \mathbf{v}_1 = \mathbf{v}_0 \cdot \mathbf{v}_1 = 1$. Then since v_1 is neutral, $\mathbf{v}_1 \cdot \mathbf{v}_1 = 2$, from which we deduce that $\mathbf{e}_2 \cdot \mathbf{e}_2 = 1$. Now project $\mathbf{v}_2$ orthogonally onto $\langle \mathbf{e}_1, \mathbf{e}_2 \rangle$ and write $\mathbf{v}_2 = \lambda_1 \mathbf{e}_1 + \lambda_2 \mathbf{e}_2 + \mathbf{e}_3$, where $\mathbf{e}_1 \cdot \mathbf{e}_3 = \mathbf{e}_2 \cdot \mathbf{e}_3 = 0$. Since $\mathbf{v}_2 \cdot \mathbf{v}_0 = 0$ and $\mathbf{v}_2 \cdot \mathbf{v}_1 = 1$, we compute $\lambda_1 = 0$, $\lambda_2 = 1$. Then since $\mathbf{v}_2 \cdot \mathbf{v}_2 = 2$, we have $\mathbf{e}_3 \cdot \mathbf{e}_3 = 1$, and so on, producing the required orthonormal list $\mathbf{e}_1, ..., \mathbf{e}_{r+1}$. Uniqueness is clear, as $\mathbf{e}_1 = \mathbf{v}_0$ and $\mathbf{e}_i = \mathbf{v}_{i-1} - \mathbf{e}_{i-1}$ for $i > 1$. $\qquad \square$

Exercise 6.27 Find a Gram representation of $-P_r$ (it may be convenient to work in $r+2$ dimensions), and conclude that P_r is cyclotomic.

Lemma 6.26 will play a crucial role in our classification of those connected cyclotomic charged signed graphs that include at least one charged vertex. For signed graphs without charges, we shall use the Gram representation of a slightly more complicated object.

Lemma 6.28 *Consider the signed graph $P_{\ell,r}$ ($\ell \geq 0$, $r \geq 1$) having $\ell + r + 2 \geq 3$ vertices $v_0, v_1, ..., v_\ell, w_0, w_1, ..., w_r$, and signed adjacencies as shown:*

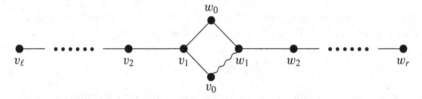

If $\ell = 0$, then this signed graph is simply

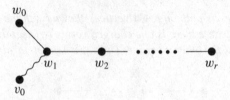

Let $\mathbf{e}_1, \dots, \mathbf{e}_{\ell+1}, \mathbf{f}_1, \dots, \mathbf{f}_{r+1}$ be an orthonormal basis for $\mathbb{R}^{\ell+r+2}$. Put

$$\mathbf{v}_0 = \mathbf{e}_1 - \mathbf{f}_1 \,,$$
$$\mathbf{w}_0 = \mathbf{e}_1 + \mathbf{f}_1 \,,$$
$$\mathbf{v}_i = \mathbf{e}_i + \mathbf{e}_{i+1} \qquad\qquad (1 \le i \le \ell) \,,$$
$$\mathbf{w}_j = \mathbf{f}_j + \mathbf{f}_{j+1} \qquad\qquad (1 \le j \le r) \,. \tag{6.2}$$

By associating each vector $\mathbf{v}_i$ with the vertex v_i $(0 \le i \le \ell)$ and associating each vector $\mathbf{w}_j$ with the vertex w_j $(0 \le j \le r)$, one obtains a Gram vector representation of $P_{\ell,r}$.

Moreover, let $\mathbf{v}_i$ $(0 \le i \le \ell)$ and $\mathbf{w}_j$ $(0 \le j \le r)$ be an arbitrary Gram representation of the signed graph $P_{\ell,r}$, with each $\mathbf{v}_i$ corresponding to the vertex v_i $(0 \le i \le \ell)$ and each $\mathbf{w}_j$ corresponding to the vertex w_j $(0 \le j \le r)$. Then there is a uniquely determined orthonormal basis $\mathbf{e}_1, \dots, \mathbf{e}_{\ell+1}, \mathbf{f}_1, \dots, \mathbf{f}_{r+1}$ for the span of $\mathbf{v}_0, \dots, \mathbf{v}_\ell, \mathbf{w}_0, \dots, \mathbf{w}_r$ such that the formulas (6.2) hold.

Proof The first paragraph is easy: the stated $\mathbf{v}_i$ and $\mathbf{w}_j$ all have length 2 (as they should for neutral vertices), and the dot products between them are easily seen to be precisely as desired.

For the second paragraph, we start by defining

$$\mathbf{e}_1 = (\mathbf{v}_0 + \mathbf{w}_0)/2 \quad \text{and} \quad \mathbf{f}_1 = (\mathbf{v}_0 - \mathbf{w}_0)/2 \,.$$

Using $\mathbf{v}_0 \cdot \mathbf{w}_0 = 0$ and $\mathbf{v}_0 \cdot \mathbf{v}_0 = \mathbf{w}_0 \cdot \mathbf{w}_0 = 2$, we compute

$$\mathbf{e}_1 \cdot \mathbf{e}_1 = \mathbf{f}_1 \cdot \mathbf{f}_1 = 1 \quad \text{and} \quad \mathbf{e}_1 \cdot \mathbf{f}_1 = 0 \,.$$

Write $\mathbf{w}_1 = \lambda_1 \mathbf{e}_1 + \mu_1 \mathbf{f}_1 + \mathbf{f}_2$ with

$$\mathbf{e}_1 \cdot \mathbf{f}_2 = \mathbf{f}_1 \cdot \mathbf{f}_2 = 0 \,.$$

Then $\mathbf{v}_0 \cdot \mathbf{w}_1 = -\mathbf{w}_0 \cdot \mathbf{w}_1 = 1$ gives $1 = (\mathbf{e}_1 + \mathbf{f}_1) \cdot \mathbf{w}_1 = \lambda_1 + \mu_1$ and $-1 = \lambda_1 - \mu_1$, so $\lambda_1 = 0$ and $\mu_1 = 1$. Then from the length of $\mathbf{w}_1$, we have

$$\mathbf{w}_1 = \mathbf{f}_1 + \mathbf{f}_2 \quad \text{and} \quad \mathbf{f}_2 \cdot \mathbf{f}_2 = 1 \,.$$

Then, as in the proof of Lemma 6.26, we repeat this process to conclude that

$$\mathbf{w}_2 = \mathbf{f}_2 + \mathbf{f}_3 \,, \ \ldots \,, \ \mathbf{w}_r = \mathbf{f}_r + \mathbf{f}_{r+1}$$

with $\mathbf{e}_1, \mathbf{f}_1, \ldots, \mathbf{f}_{r+1}$ an orthonormal list.

If $\ell \geq 1$, we write $\mathbf{v}_1 = \lambda_1 \mathbf{e}_1 + \mu_1 \mathbf{f}_1 + \cdots + \mu_{r+1}\mathbf{f}_{r+2} + \mathbf{e}_2$, with

$$\mathbf{e}_2 \cdot \mathbf{e}_1 = \mathbf{e}_2 \cdot \mathbf{f}_2 = \cdots = \mathbf{e}_2 \cdot \mathbf{f}_{r+1} = 0 \,.$$

Adjacencies give $\lambda_1 + \mu_1 = 1, \lambda_1 - \mu_1 = 1$, so $\lambda_1 = 1$ and $\mu_1 = 0$, and then $\mu_1 + \mu_2 = 0$ gives $\mu_2 = 0$, and successively $\mu_3 = \cdots = \mu_{r+1} = 0$, and finally a length computation gives

$$\mathbf{v}_1 = \mathbf{e}_1 + \mathbf{e}_2 \quad \text{and} \quad \mathbf{e}_1 \cdot \mathbf{e}_1 = 1 \,.$$

Then, essentially as before, we repeat this process to conclude that

$$\mathbf{v}_2 = \mathbf{e}_2 + \mathbf{e}_3 \,, \ \ldots \,, \ \mathbf{v}_\ell = \mathbf{e}_\ell + \mathbf{e}_{\ell+1}$$

with $\mathbf{e}_1, \ldots, \mathbf{e}_{\ell+1}, \mathbf{f}_1, \ldots, \mathbf{f}_{r+1}$ an orthonormal list. Uniqueness is clear as we can solve for the $\mathbf{e}_i$ and $\mathbf{f}_i$ in (6.2) in terms of the $\mathbf{v}_i$ and $\mathbf{w}_i$. □

Exercise 6.29 The above Gram representation of $P_{\ell,r}$ shows that all eigenvalues of $P_{\ell,r}$ are at least -2. Check that if one changes the signs of all edges of $P_{\ell,r}$, one obtains a signed graph that is simply a sign switching of $P_{\ell,r}$, and deduce that $P_{\ell,r}$ is cyclotomic.

Exercise 6.30 Use the Gram representation of $P_{\ell,r}$ to show that $P_{\ell,r}$ in fact has all its eigenvalues in the open interval $(-2, 2)$. (Show that the Gram vectors are linearly independent, and deduce that if A is the corresponding integer symmetric matrix, then $A + 2I$ has a nonzero determinant. Also, use the previous exercise.)

6.6 Statement of the Classification Theorem for Cyclotomic Integer Symmetric Matrices

For $k \geq 3$, let T_{2k} be the signed graph on $2k$ vertices shown in Fig. 6.3. The letter T is used to indicate that T_{2k} can be drawn on a torus without any of the edges crossing. The smallest case T_6 can in fact be drawn in the plane without crossings: it is the signed octahedron that was seen in Exercise 6.17.

For $k \geq 2$, let C_{2k}^{++} and C_{2k}^{+-} be the charged signed graphs shown in Fig. 6.4. Each has $2k - 4 \geq 0$ neutral vertices. The cases C_4^{++} and C_4^{+-} were seen in Exercise 6.20.

Recall that a connected (indecomposable) cyclotomic integer symmetric matrix is called maximal if it is not contained in a larger connected cyclotomic integer symmetric matrix. We can now state the main classification theorem.

Theorem 6.31 *Let A be a connected cyclotomic integer symmetric matrix. Then A is contained in a maximal connected cyclotomic integer symmetric matrix B (which*

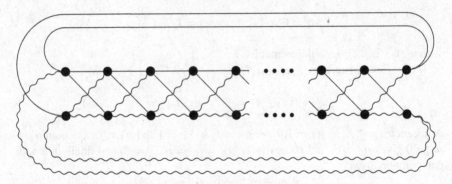

Fig. 6.3 The toroidal tessellation T_{2k} ($2k$ vertices)

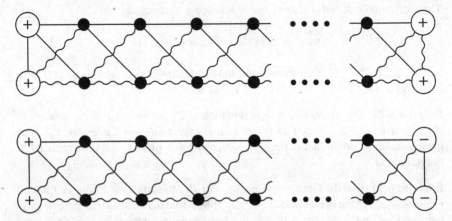

Fig. 6.4 The cylindrical tessellations C_{2k}^{++} and C_{2k}^{+-} (each has $2k$ vertices)

depends on A). Either B is equivalent to one of the two matrices

$$(2), \quad \begin{pmatrix} 0 & 2 \\ 2 & 0 \end{pmatrix},$$

or G_B is equivalent to one of the charged signed graphs:

- T_{2k} *(Fig. 6.3), for some $k \geq 3$;*
- C_{2k}^{++} *(Fig. 6.4), for some $k \geq 2$;*
- C_{2k}^{+-} *(Fig. 6.4), for some $k \geq 2$;*
- S_7, S_8 *or S_8' from Exercises 6.17 and 6.19;*
- S_{14} *or S_{16} (Figs. 6.1 and 6.2).*

We see that nearly all the maximal connected cyclotomic integer symmetric matrices are associated with charged signed graphs. We will deal with signed graphs first in Sect. 7.1, then charged signed graphs in Sect. **??**, and then complete the proof of the theorem in Sect. **??**.

Exercise 6.32 Let A be an integer symmetric matrix such that G_A is one of the charged signed graphs listed in Theorem 6.31. Show that $A^2 = 4I$. Deduce that the only possible eigenvalues of A are ± 2, and hence that A is cyclotomic.

Exercise 6.33 Show that the signed graphs A, M, B of Exercise 6.21 are not equivalent to subgraphs of any T_{2k}, C_{2k}^{++} or C_{2k}^{+-}. (In fact, this follows from Exercise 6.21, but it is easier to prove it directly.)

Exercise 6.34 Let $e_1, \ldots, e_k$ be an orthonormal basis for $\mathbb{R}^k$. Verify that the $2k$ vectors $e_1 \pm e_2, e_2 \pm e_3, \ldots, e_{k-1} \pm e_k, e_k \pm e_1$ give a Gram representation for the signed graph T_{2k} of Fig. 6.3.

Exercise 6.35 Let $e_1, \ldots, e_{k+1}$ be an orthonormal basis for $\mathbb{R}^{k+1}$. Verify that the $2k$ vectors $\pm e_1 + \sqrt{2}e_k, e_1 \pm e_2, e_2 \pm e_3, \ldots, e_{k-2} \pm e_{k-1}, e_{k-1} \pm \sqrt{2}e_{k+1}$ give a Gram representation for the charged signed graph C_{2k}^{++} of Fig. 6.4.

Let $e_1, \ldots, e_k$ be an orthonormal basis for $\mathbb{R}^k$. Verify that the $2k$ vectors $\pm e_1 + \sqrt{2}e_k, e_1 \pm e_2, e_2 \pm e_3, \ldots, e_{k-2} \pm e_{k-1}, e_{k-1}, e_{k-1}$ (yes, using e_{k-1} twice) give a Gram representation for the charged signed graph C_{2k}^{+-} of Fig. 6.4.

In proving the classification theorem, it will be more convenient to work with the equivalent charged signed graph $-C_{2k}^{++}$ (change signs of all charges and edges) rather than C_{2k}^{++}. Let $e_1, \ldots, e_{k-1}$ be an orthonormal basis for $\mathbb{R}^{k-1}$. Verify that the $2k$ vectors e_1, e_1 (again), $\pm e_1 + e_2, \pm e_2 + e_3, \ldots, \pm e_{k-2} + e_{k-1}, e_{k-1}, -e_{k-1}$ give a Gram representation for the charged signed graph $-C_{2k}^{++}$.

6.7 Glossary

$[A]_{\text{eq}}.$ The set of those matrices that are equivalent to A.

$[A]_{\text{iso}}.$ The set of those matrices that are isomorphic to A.

$[A]_{\text{str}}.$ The set of those matrices that are strongly equivalent to A.

$[A]_{\text{sw}}.$ The set of those matrices that are switch-equivalent to A.

$A_S.$ The induced submatrix of A that uses entries only from rows and columns indexed by S.

$\chi_A.$ The characteristic polynomial of A.

$A^\top.$ The transpose of the matrix A. We have $(A^\top)_{ij} = (A)_{ji}$. The matrix A is symmetric if and only if $A = A^\top$.

$C_{2k}^{++}, C_{2k}^{+-}.$ The maximal connected cyclotomic charged signed graphs of Fig. 6.4.

$\text{diag}(d_1, \ldots, d_n).$ The diagonal matrix whose diagonal entries are $d_1, \ldots, d_n$.

$I.$ The $n \times n$ identity matrix, where n is understood from context. Later we shall define a family of charged signed graphs I_n and wish to reduce confusion by not using the subscript for identity matrices.

$G_A.$ If A is an integer symmetric matrix whose entries are all either 0 or ± 1, then A can be viewed as the adjacency matrix of a charged signed graph, which we denote by G_A. More generally, if A is any integer matrix, then it can be viewed as the adjacency matrix of a weighted digraph, which we denote by G_A.

R_A.   The reciprocal polynomial of the matrix A.

$M(A)$.   The Mahler measure of the matrix A.

$M(G)$.   The Mahler measure of the charged signed graph G.

T_{2k}.   The maximal connected cyclotomic charged signed graphs of Fig. 6.3.

adjacent.   Let G be a charged signed graph with adjacency matrix $A = (a_{ij})$. Two vertices i and j in G are said to be adjacent if $a_{ij} \neq 0$. In other words, there is an edge (positive or negative) between the vertices i and j, or $i = j$ and the vertex i is charged. Note in particular that a charged vertex is viewed as being adjacent to itself. One could view charges on vertices as being directed signed loops, but we prefer to use the language of charges, and this simplifies the pictures.

bipartite.   A charged signed graph is bipartite if its adjacency matrix A satisfies $-A \in [A]_{\text{str}}$. This can be seen to generalise the definition for graphs, but knowing the graph definition, one should be careful when treating charged signed graphs, as intuition may be misleading.

characteristic polynomial.   The characteristic polynomial of a matrix, A, is the polynomial $\det(xI - A)$. This polynomial is always monic (top coefficient 1). Its zeros are the eigenvalues of A. If A has integer entries, then χ_A has integer coefficients.

charged signed graph.   A generalisation of a graph where its adjacency matrix A can be any symmetric matrix with entries in $\{-1, 0, 1\}$; for $i \neq j$, the off-diagonal element a_{ij} is the 'sign' of the edge joining vertex i and vertex j, while the diagonal element a_{ii} is the 'charge' on vertex i.

component.   The components of an integer symmetric matrix A are the induced submatrices corresponding to the connected components of G_A.

connected.   An integer symmetric matrix A is called connected if G_A is connected.

connected growing.   The process of finding connected examples of matrices or graphs having some property by growing them from smaller connected examples. Of course, this process can only be applied if it is known that every connected example having the desired property can be grown from a smaller such connected example.

containment of matrices or graphs.   We say that a matrix A is contained in a matrix B if A is equivalent to an induced submatrix of B, and if so we write $A \subseteq B$. Similarly for graphs, we have $H \subseteq G$ if H is equivalent to an induced subgraph of G. Note that in either case we work up to equivalence. Naturally, we may also say that B contains A and write $B \supseteq A$ and so on.

cycle.   A cycle is a connected charged signed graph for which the vertices can be named $v_1, \ldots, v_n$ in such a way that the only edges are: an edge between v_i and v_{i+1} for $1 \leq i \leq n - 1$ and an edge between v_n and v_1. The edges can be of either sign, and the vertices can be charged. Given that vertices may have charges, the degree of each vertex is either 2 (neutral vertices) or 3 (charged vertices). If the cycle has r vertices, then it is called an r-cycle.

cyclotomic.   The adjective cyclotomic is used to indicate that the Mahler measure is 1. It can be applied to monic integer polynomials, integer symmetric matrices or charged signed graphs. For matrices (or graphs via their adjacency matrices),

being cyclotomic is equivalent to the reciprocal polynomial being cyclotomic (a
product of irreducible cyclotomic polynomials).

cylindrical tessellation. One of the two maximal connected cyclotomic charged
signed graphs of Fig. 6.4.

degree. In a charged signed graph, the degree of vertex i is the number of neighbours of that vertex, with the convention that a charged vertex is considered to be
a neighbour of itself. In terms of the adjacency matrix A, the degree of vertex i is
the number of nonzero entries in the row corresponding to that vertex. This also
equals the corresponding diagonal entry in A^2. This latter observation is used to
define the degree of the ith row of an arbitrary integer symmetric matrix A as
the corresponding diagonal entry in A^2. As ever, we speak of rows and vertices
interchangeably (moving between A and G_A without comment), using whichever
seems most natural and helpful in context.

echelon form. A connected $n \times n$ matrix is in echelon form if the first nonzero
entries in rows $i = 2, \ldots, n$ are all positive and move weakly to the right as i
increases. Note that this is not the usual definition of echelon form, but it is what
we need.

echelon growing. A special form of connected growing in which all the matrices
have echelon form.

equivalent. The matrices A and B are equivalent if A is strongly equivalent to at
least one of B and $-B$. Then either $\chi_A = \chi_B$ or $\chi_A(x) = (-1)^n \chi_B(-x)$ (where
n is the number of rows). Equivalent matrices have the same Mahler measure, and
in particular, whether or not a matrix is cyclotomic is an equivalence invariant.

Gram vectors. An $n \times n$ matrix $A = (a_{ij})$ is represented by Gram vectors if there
are vectors $v_1, \ldots, v_n \in \mathbb{R}^m$ (the Gram vectors) such that $a_{ij} = v_i \cdot v_j$ for each
relevant i and j. If a matrix can be represented by Gram vectors, then there
exists such a representation for which $m = n$. An integer symmetric matrix can
be represented by Gram vectors if and only if all eigenvalues are at least 0. We
say that a charged signed graph G is represented by Gram vectors if we have a
Gram vector representation for $A + 2I$, where A is the adjacency matrix. Note
the addition of $2I$ to the adjacency matrix.

induced submatrix. An induced submatrix is one formed by deleting some of the
rows along with the corresponding columns. If A is symmetric, then so is any
induced submatrix. The induced submatrix A_S is formed by deleting all the rows
and columns that are *not* in S. If an integer symmetric matrix A is cyclotomic,
then so is any induced submatrix. The language here is that of graph theory: if A
is the adjacency matrix of a graph G, then induced submatrices of A correspond
to induced subgraphs of G.

isomorphic. Two $n \times n$ matrices A and B are said to be isomorphic if there is
a permutation matrix P such that $B = P^\mathsf{T} A P$. In other words, there is some
bijection $f : \{1, \ldots, n\} \rightarrow \{1, \ldots, n\}$ such that $b_{ij} = a_{f(i)f(j)}$ for all relevant i
and j. In other words, the graphs G_A and G_B are isomorphic as weighted digraphs.
Isomorphic matrices share the same eigenvalues. In particular, the Mahler measure
of a matrix is an isomorphism invariant.

Mahler measure.   The Mahler measure of a matrix A is defined to be the Mahler
measure of its reciprocal polynomial, R_A. This definition applies to arbitrary integer square matrices, although nearly always we will have that A is also symmetric.
The definition is extended to charged signed graphs via their adjacency matrix:
the Mahler measure of a graph G is the Mahler measure of its adjacency matrix
A, which in turn is the Mahler measure of the polynomial R_A.

maximal.   The matrix A (or graph G) is maximal with respect to some property P
if A has property P (or G does), but that any strictly larger matrix containing A (or
strictly larger graph containing G) does not have property P. Usually, the property
P will include connectedness. Note that containment is always up to equivalençe.
The most important instance for us is a maximal connected cyclotomic integer
symmetric matrix.

path.   A path is a connected charged signed graph for which the vertices can be
named $v_1, \ldots, v_n$ in such a way that the only edges are an edge between v_i and
v_{i+1} for $1 \leq i \leq n - 1$. The edges can be of either sign, and the vertices can be
charged.

reciprocal polynomial.   The reciprocal polynomial of a matrix, A, is the polynomial $z^n \chi_A(z + 1/z)$, where n is the degree of χ_A.

sign switching.   The process of sign switching a matrix A is to take some subset
S of the row indices, then to change the sign of each row indexed by S and then
to change the sign of each column indexed by S. Note that this process leaves all
the diagonal entries fixed (they change signs twice or not at all).

signed permutation matrix.   A square matrix P is a signed permutation matrix if
each row and each column contains exactly one nonzero entry, and that nonzero
entry is ± 1. Note that if P is a signed permutation matrix, then $P^{-1} = P^{\mathsf{T}}$.

strongly equivalent.   A square integer matrix B is strongly equivalent to A if there
exists a signed permutation matrix P such that $B = P^{\mathsf{T}} A P$. Strong equivalence
combines sign switching and isomorphism: B is isomorphic to a matrix that is
switch-equivalent to A; B is switch-equivalent to a matrix isomorphic to A.

switch-equivalent.   Two matrices A and B are switch-equivalent if B can be
obtained from A by sign switching. In other words, there exists a diagonal matrix
D with each diagonal entry either 1 or -1 such that $B = DAD = D^{\mathsf{T}} A D = D^{-1} A D$.

toroidal tessellation.   One of the maximal connected cyclotomic signed graphs of
Fig. 6.3.

Chapter 7
Cyclotomic Integer Symmetric Matrices II: Proof of the Classification Theorem

7.1 Cyclotomic Signed Graphs

In this section, we prove the restriction of Theorem 6.31 to signed graphs: all vertices are neutral. Recall that the notation $H_1 \subseteq H_2$ means that H_1 is *equivalent to* a subgraph of H_2.

Theorem 7.1 *Let G be a connected cyclotomic signed graph. Then one of the following holds:*

- $G \subseteq S_{14}$ *(Fig. 6.1);*
- $G \subseteq S_{16}$ *(Fig. 6.2);*
- $G \subseteq T_{2k}$ *for some $k \geq 3$ (Fig. 6.3).*

Proof Let G be a connected cyclotomic signed graph. If G is a path or a cycle, then by sign switching we can arrange for at most one edge of G to be negative, and then can see that $G \subseteq T_{2k}$ for some k. If G contains any sort of triangle (three vertices that are pairwise adjacent, via edges of either sign), then,If G contains a subgraph equivalent to any of the graphs A, M or B of Exercise Thus, we are done unless all of the following hold:

G is neither a path nor a cycle; $\qquad\qquad\qquad\qquad\qquad\qquad\qquad$ (7.1)

G contains no triangles; $\qquad\qquad\qquad\qquad\qquad\qquad\qquad\qquad\qquad$ (7.2)

G does not contain a subgraph equivalent to any of A, M, or B of Exercise 6.21.
$\qquad\qquad\qquad\qquad\qquad\qquad\qquad\qquad\qquad\qquad\qquad\qquad\qquad$ (7.3)

From now on, we assume all of these restrictions on G. Our next step is to show that

$$G \subseteq T_8, \text{ or } P_{0,3} \subseteq G, \text{ or } P_{1,2} \subseteq G \qquad\qquad (7.4)$$

(see Lemma 6.28 for the definition of $P_{\ell,r}$).

© Springer Nature Switzerland AG 2021
J. McKee and C. Smyth, *Around the Unit Circle*, Universitext,
https://doi.org/10.1007/978-3-030-80031-4_7

By (7.1) and (7.2), $P_{0,2} \subseteq G$. If this is the whole of G, then $G \subseteq T_8$ and (7.4) holds. Otherwise, we connectedly grow $P_{0,2}$ to get all possible 5-vertex cyclotomic supergraphs, and find that G must contain a subgraph equivalent to one of $P_{0,3}$, $P_{1,2}$, G_1 or G_2, where the latter two graphs are pictured as follows:

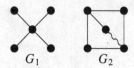

$$G_1 \qquad\qquad G_2$$

If G contains $P_{0,3}$ or $P_{1,2}$, then (7.4) holds. If G equals either G_1 or G_2, then (7.4) holds, as both G_1 and G_2 are subgraphs of T_8. Otherwise, we connectedly grow both G_1 and G_2. We find that either we obtain a signed graph which contains a subgraph equivalent to $P_{1,2}$ (in which case, (7.4) holds), or, up to equivalence, G contains one of G_3 or G_4 pictured as follows:

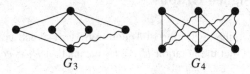

$$G_3 \qquad\qquad\qquad G_4$$

If G equals either G_3 or G_4, then (7.4) holds, since both G_3 and G_4 are contained in T_8. Otherwise, when we connectedly grow we either get T_8 with a single vertex deleted (G_5, say), or a signed graph which contains $P_{1,2}$. In the latter case, (7.4) holds; in the former case, we have (7.4) if G is G_5, and otherwise we connectedly grow G_5 and obtain only T_8, which we then find is maximal (Lemma 6.15).

We have shown using (7.1) and (7.2) that (7.4) holds. If $G \subseteq T_8$ then we are done, otherwise we have imposed several restrictions on G. Not only does it satisfy (7.2) and (7.3), but also it contains some $P_{\ell,r}$ with $\ell + r \geq 3$ (which implies (7.1)).

We now take a subgraph H of G that is equivalent to $P_{\ell,r}$ with $\ell + r \geq 3$ maximal. By sign switching vertices of G, we may suppose that $\ell \leq r$ and that H actually equals $P_{\ell,r}$.

Choose Gram vectors to represent G. Label the vertices of H as in Lemma 6.28. By that lemma, we can choose orthonormal vectors $\mathbf{e}_1, \ldots, \mathbf{e}_{\ell+1}, \mathbf{f}_1, \ldots, \mathbf{f}_{r+1}$ such that the Gram vectors $\mathbf{v}_i$ and $\mathbf{w}_j$ for the vertices v_i and w_j of H are given by the formulas (6.2). For convenience, we redraw $P_{\ell,r} = H$ here, with the Gram vectors written in terms of the $\mathbf{e}_i$ and $\mathbf{f}_j$:

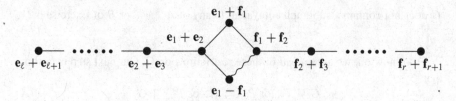

Recall that $\mathbf{v}_0 = \mathbf{e}_1 - \mathbf{f}_1$, $\mathbf{v}_i = \mathbf{e}_i + \mathbf{e}_{i+1}$ ($1 \le i \le \ell$), $\mathbf{w}_0 = \mathbf{e}_1 + \mathbf{f}_1$, $\mathbf{w}_j = \mathbf{f}_j + \mathbf{f}_{j+1}$ ($1 \le j \le r$).

If $G = H$ we are done, as $H \subseteq T_{2(\ell+r+2)}$.

Otherwise, let v be any vertex of G that is not in H but that is adjacent to at least one vertex of H, and let $\mathbf{v}$ be its Gram vector.

We write

$$\mathbf{v} = \lambda_1 \mathbf{e}_1 + \cdots + \lambda_{\ell+1} \mathbf{e}_{\ell+1} + \mu_1 \mathbf{f}_1 + \cdots + \mu_{r+1} \mathbf{f}_{r+1} + \mathbf{g},$$

where $\mathbf{g}$ is orthogonal to all the $\mathbf{e}_i$ and $\mathbf{f}_j$. The story now diverges according to how many of v_0 and w_0 are adjacent to v.

First suppose that v is adjacent to both v_0 and w_0. By switching, we may suppose that $\mathbf{v} \cdot \mathbf{w}_0 = 1$. By (7.2), $\mathbf{v} \cdot \mathbf{w}_1 = 0$. We therefore have $\lambda_1 + \mu_1 = 1$, $\lambda_1 - \mu_1 = \pm 1$ and $\mu_1 + \mu_2 = 0$. If we are in the case $\lambda_1 - \mu_1 = -1$, then we have $\lambda_1 = 0$, $\mu_1 = 1$ and $\mu_2 = -1$, and by the length of $\mathbf{v}$ the only possibility is $\mathbf{v} = \mathbf{f}_1 - \mathbf{f}_2$. If we are in the case $\lambda_1 - \mu_1 = 1$, then we have $\lambda_1 = 1$, $\mu_1 = 0$. In the case $\ell > 0$, we have (by (7.2)) $\mathbf{v}_1 \cdot \mathbf{v} = 0$, giving $\lambda_1 + \lambda_2 = 0$, and by the length of $\mathbf{v}$ the only possibility is $\mathbf{v} = \mathbf{e}_1 - \mathbf{e}_2$. In the case $\ell = 0$, we must have v adjacent to some w_i for $i \ge 2$, else we would have $P_{1,r}$ as a subgraph of G, contradicting maximality of $\ell + r$. If i is minimal in the range $2 \le i \le r$ with v adjacent to w_i, then chasing adjacencies and lengths gives $\mu_2 = \cdots = \mu_i = 0$, $\mu_{i+1} = \pm 1$ and $\mathbf{v} = \mathbf{e}_1 \pm \mathbf{f}_{i+1}$. If $i < r$, this is not possible, as it implies that v is also adjacent to v_{i+1}, contradicting (7.2). To sum up, if v is adjacent to both v_0 and w_0, then there are three possibilities:

$$\mathbf{v} = \mathbf{f}_1 - \mathbf{f}_2, \quad \mathbf{v} = \mathbf{e}_1 - \mathbf{e}_2, \quad \text{or } (\ell = 0 \text{ and } \mathbf{v} = \mathbf{e}_1 \pm \mathbf{f}_{r+1}).$$

Next, suppose that v is adjacent to just one of v_0 and w_0. If $\ell > 0$, then in order to avoid graph A (as required by (7.3)) on vertices v, v_1, v_0, w_0, w_1 and w_2, we must have v adjacent to w_2. Depending on the sign of this adjacency, the subgraph on vertices v, v_1, v_0, w_0, w_1 and w_2 is either equivalent to graph B (excluded by (7.3)) or is not cyclotomic (excluded by Lemma 6.3). So we must have $\ell = 0$, although we shall soon see that this too is impossible. With $\ell = 0$, in order to avoid graph M (as required by (7.3)) on vertices v, v_0, w_0, w_1, w_2 and w_3 ($r \ge 3$ since $\ell = 0$), we must have v adjacent to some w_i with $i = 2$ or 3 (but not both by (7.2)). In the former case, the vertices v, v_0, w_0, w_1, w_2 and w_3 give a subgraph equivalent to the excluded graph A, and the latter case gives a subgraph equivalent to the noncyclotomic graph

which is excluded by Lemma 6.3. To sum up, we cannot have v adjacent to just one of v_0 and w_0.

Next, we consider the possibility that v is adjacent to neither v_0 nor w_0, so that $\lambda_1 = \mu_1 = 0$. If v is adjacent to (at least) one of $w_1, \ldots, w_{r-1}$, then let w_i be the first such. Since v cannot be adjacent to w_{i+1} (by (7.2)), we solve in a now familiar fashion to find $\mathbf{v} = \mathbf{f}_{i+1} - \mathbf{f}_{i+2}$. Similarly, if v is adjacent to one of $v_1, \ldots, v_{\ell-1}$, then we compute that $\mathbf{v} = \mathbf{e}_{j+1} - \mathbf{e}_{j+2}$. If v is not adjacent to any of $v_0, \ldots, v_{\ell-1}, w_0, \ldots, w_{r-1}$, then it must be adjacent to *both* v_ℓ and w_r, else we could find a subgraph of G equivalent to either $P_{\ell+1,r}$ or to $P_{\ell,r+1}$, contradicting maximality of $\ell + r$. In this case, we find that $\mathbf{v} = \pm\mathbf{e}_{\ell+1} \pm \mathbf{f}_{r+1}$. To sum up: if v is adjacent to neither v_0 nor w_0, then we have one of three cases, namely

$$\mathbf{v} = \mathbf{f}_{i+1} - \mathbf{f}_{i+2} \text{ (for some } 1 \le i \le r - 1),$$
$$\mathbf{v} = \mathbf{e}_{j+1} - \mathbf{e}_{j+2} \text{ (for some } 1 \le j \le \ell - 1),$$
$$\text{or } \mathbf{v} = \pm\mathbf{e}_{\ell+1} \pm \mathbf{f}_{r+1}.$$

In all cases, we see that (after appropriate switching) $\mathbf{v}$ is one of the vectors $\mathbf{f}_i - \mathbf{f}_{i+1}$ $(1 \le i \le r)$, $\mathbf{e}_j - \mathbf{e}_{j+1}$ $(1 \le j \le \ell)$ and $\mathbf{e}_{\ell+1} \pm \mathbf{f}_{r+1}$. We cannot have two distinct vertices of G giving the same Gram vector $\mathbf{v}$, nor one giving the negative of another, else their dot product would be ± 2. Now the vectors $\mathbf{f}_i \pm \mathbf{f}_{i+1}$ $(1 \le i \le r)$, $\mathbf{e}_j \pm \mathbf{e}_{j+1}$ $(1 \le j \le \ell)$ and $\mathbf{e}_{\ell+1} \pm \mathbf{f}_{r+1}$ together with $\mathbf{e}_0 \pm \mathbf{f}_0$ give a Gram representation of $T_{2(\ell+r+2)}$, as in Exercise 6.34.

Let K be the subgraph of G formed by $H = P_{\ell,r}$ along with all vertices that are adjacent to one or more vertices of H. We have shown that

- $K \subseteq T_{2(\ell+r+2)}$;
- all Gram vectors used in a Gram representation of K lie in the span of $\mathbf{e}_1, \ldots, \mathbf{e}_{\ell+1}$, $\mathbf{f}_1, \ldots, \mathbf{f}_{r+1}$.

If K is not the whole of G, let w be a vertex of G that is adjacent to a vertex of K but is not in K. Since w is not adjacent to any vertex of H, we see that its Gram vector $\mathbf{w}$ is orthogonal to all of $\mathbf{e}_1, \ldots, \mathbf{e}_{\ell+1}, \mathbf{f}_1, \ldots, \mathbf{f}_{r+1}$ (we can write each of these vectors in terms of the $\mathbf{v}_i$ and $\mathbf{w}_j$: $\mathbf{e}_1 = (\mathbf{v}_0 + \mathbf{w}_0)/2$; $\mathbf{f}_1 = (\mathbf{v}_0 - \mathbf{w}_0)/2$, and recursively $\mathbf{e}_i = \mathbf{v}_{i-1} - \mathbf{e}_{i-1}$ and $\mathbf{f}_i = \mathbf{w}_{i-1} - \mathbf{f}_{i-1}$). Hence, it is orthogonal to every vertex in K, which gives a contradiction. Hence $G = K$ and $G \subseteq T_{2(\ell+r+2)}$. $\qquad\square$

7.2 Cyclotomic Charged Signed Graphs

Now we turn to charged signed graphs that are not signed graphs, and prove the relevant parts of Theorem 6.31. Some of the techniques will be just as in the signed graph case, but the presence of three flavours of vertex complicates things. The Gram vectors may now have length 1, $\sqrt{2}$ or $\sqrt{3}$. It will be convenient to write $G \subseteq C_{2k}^{\bullet\bullet}$ to mean that G is equivalent to a subgraph of one of C_{2k}^{++} or C_{2k}^{+-}.

We shall again use Lemma 6.3 to exclude certain subgraphs, and also will use a charged analogue of Exercise 6.21 to exclude others. The special graph $P_{\ell,r}$ will be replaced by P_r, and we shall again consider the maximal such subgraph, and make heavy use of its Gram vector representation. In the neutral case, we had the striking conclusion that (having excluded certain subgraphs) every vertex of a connected cyclotomic signed graph was either in a maximal $P_{\ell,r}$ or adjacent to one of the vertices in a maximal $P_{\ell,r}$. In the charged case, things will not be quite so straightforward, but the use of excluded subgraphs will rescue us in the end.

We start, then, by recording the subgraphs that we shall need to exclude during the proof.

Lemma 7.2 *Let G be a connected cyclotomic charged signed graph that contains a triangle on three vertices v_1, v_2 and v_3 with v_1 charged and the other two neutral. Then either $G \subseteq S_7$ or $G \subseteq S_8'$.*

Proof By negating all charges and signed edges if necessary, we can assume that v_1 has negative charge. By sign switching, we can assume that our triangle is one of T_1 or T_2 shown as follows:

Now T_2 is not cyclotomic, so our triangle must be T_1. Connectedly growing T_1 while keeping the graphs cyclotomic, one stays inside either S_7 or S_8'; the details are left as an exercise. □

Lemma 7.3 *Let G be a connected cyclotomic charged signed graph. If G contains a subgraph equivalent to the charged signed graph H_1 shown below, then $G \subseteq S_8'$.*

The charged signed graphs H_2 and H_3 shown above are not cyclotomic.

Proof Connectedly growing H_1 while keeping the graphs cyclotomic, one remains inside S_8'. The final sentence is verified by a small computation. □

In addition, we shall need to exclude triangles of negatively charged vertices, using a special case of the following lemma.

Lemma 7.4 *Let G be a connected cyclotomic charged signed graph. If G contains a triangle that consists of three charged vertices, then $G \subseteq C_4^{\bullet\bullet}$.*

Proof If two of the charged vertices have opposite signs, then we are done, by Exercise 6.19. Otherwise, by working up to equivalence we may suppose that all three vertices are negative and either one or no edges are negative. Then a small amount of connected growing completes the proof. □

We now state the classification theorem for connected cyclotomic charged signed graphs that are not signed graphs.

Theorem 7.5 *Let G be a connected cyclotomic charged signed graph that contains at least one charged vertex. Then one of the following holds:*

- $G \subseteq S_7$ *(Exercise 6.17)*;
- $G \subseteq S_8$ *(Exercise 6.19)*;
- $G \subseteq S_8'$ *(Exercise 6.19)*;
- $G \subseteq C_{2k}^{\bullet\bullet}$ *for some $k \geq 2$ (Fig. 6.4).*

Proof If $G \subseteq S_7$, $G \subseteq S_8$ or $G \subseteq S_8'$, then we are done, so we may assume that

$$G \nsubseteq S_7, \ G \nsubseteq S_8, \ G \nsubseteq S_8'. \tag{7.5}$$

If G contains only charged vertices, then either $G \subseteq S_8$ or $G \subseteq C_4^{\bullet\bullet}$ (Exercises 6.19 and 6.20) and we are done. Otherwise, G must contain a charged vertex that is adjacent to a neutral vertex. By negating G if necessary, we can assume that G contains a negative vertex that is adjacent to a neutral vertex, namely the path P_1 of Lemma 6.26. So we can take $r \geq 1$ maximal such that $P_r \subseteq G$. By switching, we can suppose that our maximal P_r is actually an induced subgraph of G.

Let v be any vertex of G not in our maximal P_r with v adjacent to one of the vertices in P_r. Let $\mathbf{v}$ be the vector associated with v in some Gram representation of G; we have seen (Lemma 6.26) that there is an orthonormal list $\mathbf{e}_1, \ldots, \mathbf{e}_{r+1}$ such that the vertices in P_r have Gram vectors $\mathbf{e}_1$ (the negatively charged vertex), $\mathbf{e}_1 + \mathbf{e}_2, \ldots,$ $\mathbf{e}_r + \mathbf{e}_{r+1}$. We write

$$\mathbf{v} = \lambda_1 \mathbf{e}_1 + \cdots + \lambda_{r+1}\mathbf{e}_{r+1} + \mathbf{f},$$

where $\mathbf{f}$ is orthogonal to all the $\mathbf{e}_i$.

For convenience, we redraw P_r with the vertices labelled by their Gram vectors (v_0 corresponds to $\mathbf{v}_0 = \mathbf{e}_1$ and v_i corresponds to $\mathbf{v}_i = \mathbf{e}_i + \mathbf{e}_{i+1}$ for $1 \leq i \leq r$):

$$P_r$$

First, consider the case where v is adjacent to v_0; by switching, we may suppose that $\mathbf{v} \cdot \mathbf{v}_0 = \mathbf{v} \cdot \mathbf{e}_1 = 1$. If v has positive charge, then after Exercise 6.19 either $G \subseteq S_7$ or $G \subseteq S_8'$ contradicting (7.5). If v has a negative charge, then $\mathbf{v} = \mathbf{e}_1$, and we have the same Gram vector for two vertices (which is allowed precisely in this context of two adjacent negatively charged vertices, joined by a positive edge). If v is neutral, then by Lemma 7.2 either $G \subseteq S_7$ or $G \subseteq S_8'$, or v is not adjacent to v_1. In the latter case, $\mathbf{v} = \mathbf{e}_1 - \mathbf{e}_2$. To sum up: if v is adjacent to v_0 and (7.5) holds, then $\mathbf{v}$ is one of $\mathbf{e}_1$ or $\mathbf{e}_1 - \mathbf{e}_2$.

Next consider the case where v is not adjacent to v_0, and let $i \geq 1$ be minimal such that v is adjacent to v_i. If v is negatively charged, then $\mathbf{v} = \mathbf{e}_{r+1}$ (if $i < r$ then v is also adjacent to v_{i+1} and either $G \subseteq S_7$ or $G \subseteq S_8'$, by Lemma 7.2, contradicting (7.5)). If v is positively charged and $i = r$, then $\mathbf{v} = \mathbf{e}_{r+1} + \mathbf{f}$, where $\mathbf{f}$ has length $\sqrt{2}$.

In this case, we note that $\mathbf{v}$ is not in the span of the $\mathbf{e}_i$. If v is positively charged and $i < r$, then by Lemma 7.2 we can assume that v is not adjacent to v_{i+1} (else either $G \subseteq S_7$ or $G \subseteq S_8'$), and depending on whether $i = 1$, $i = 2$ or $i > 2$, we find that $G \subseteq S_8'$ (Lemma 7.3; only $i = 1$ is possible, as the other cases are not cyclotomic). If v is neutral, then $i < r$ or G would contain a path equivalent to P_{r+1}, contradicting maximality of r. Then by Lemma 6.18, v is not adjacent to v_{i+1}, and $\mathbf{v} = \mathbf{e}_{i+1} - \mathbf{e}_{i+2}$. To sum up: if v is not adjacent to v_0 and (7.5) holds, then $\mathbf{v}$ is one of $\mathbf{e}_{r+1} + \mathbf{f}$ or $\mathbf{e}_{i+1} - \mathbf{e}_{i+2}$.

The conclusion of the above two paragraphs is summarised in the following table, showing the possibilities for $\mathbf{v}$ if (7.5) holds (possibly after switching).

	v adjacent to v_0	v not adjacent to v_0
v positively charged	—	$\mathbf{v} = \mathbf{e}_{r+1} + \mathbf{f}$
v negatively charged	$\mathbf{v} = \mathbf{e}_1$	$\mathbf{v} = \mathbf{e}_{r+1}$
v neutral	$\mathbf{v} = \mathbf{e}_1 - \mathbf{e}_2$	$\mathbf{v} = \mathbf{e}_{i+1} - \mathbf{e}_{i+2}$ for some $1 \le i \le r - 1$

Now, unlike the situation for graphs, it is possible for two distinct vertices to share the same Gram vector, or to have Gram vectors of opposite signs. By Lemma 7.4, this can only happen for a pair of negative charges joined by an edge, and one cannot have three Gram vectors that agree up to sign. Hence, if we let H be the subgraph of G comprising P_r and any vertices of G that are adjacent to those in P_r, then comparing the above table (and the Gram representation for P_r) with Exercise 6.35, we see that $H \subseteq C_{2(r+2)}^{\bullet\bullet}$.

The build-up to the punchline is more complicated than in the uncharged case. Suppose that G does not equal H, and let w be a vertex of G that is not in H but that is adjacent to a vertex in H. As before, we have that $\mathbf{w}$ is orthogonal to all the $\mathbf{e}_i$, but we cannot now immediately conclude that it is orthogonal to every Gram vector for H, as H might have a vertex with Gram vector $\mathbf{e}_{r+1} + \mathbf{f}$, where $\mathbf{f}$ has length $\sqrt{2}$ and is orthogonal to all the $\mathbf{e}_i$. Depending on the charge of w, and the value of r, G would then contain a subgraph equivalent to one of

| $r \ge 2$, w neutral | $r = 1$, w neutral | w positive | w negative |

The first three of these are not cyclotomic and so cannot occur, and if the final case occurs then either $G \subseteq S_7$ or $G \subseteq S_8'$ by Exercise 6.19, contradicting (7.5). Hence we must have $G = H$, and $G \subseteq C_{2(r+2)}^{\bullet\bullet}$. $\qquad\square$

7.3   Cyclotomic Integer Symmetric Matrices: Completion of the Classification

Finally, we complete the proof of Theorem 6.31.

Lemma 7.6 *Let A be a connected cyclotomic integer symmetric matrix. If any entry of A has modulus greater than 1, then A is one of*

$$(2), \quad (-2), \quad \begin{pmatrix} 0 & 2 \\ 2 & 0 \end{pmatrix}, \quad \begin{pmatrix} 0 & -2 \\ -2 & 0 \end{pmatrix}.$$

Proof By Lemma 6.15, no row of A can have degree greater than 4, hence no entry has modulus greater than 2. If A has a diagonal entry ± 2, then either $A = (2)$ or $A = (-2)$, since these two matrices are maximal (by Lemma 6.15).

If A has an off-diagonal entry $a_{ij} = \pm 2$, then $a_{ii} = 0$ else the degree of row i would be greater than 4. Then A contains a submatrix

$$\begin{pmatrix} 0 & \pm 2 \\ \pm 2 & 0 \end{pmatrix},$$

and by Lemma 6.15 this is maximal, so equals A. □

This lemma completes the proof of the classification of cyclotomic integer symmetric matrices (Theorem 6.31). Apart from the trivial cases covered by Lemma 7.6, every connected cyclotomic integer symmetric matrix is the adjacency matrix of a charged signed graph, and those cases are covered by Theorems 7.1 and 7.5.

7.4   Further Exercises

Exercise 7.7 Let e_1, e_2, e_3, e_4 be an orthonormal basis of $\mathbb{R}^4$. Show that the vertices of the charged signed graph S_7 of Exercise 6.17 can be labelled using the vectors $e_1 + e_2$, $e_1 + e_3$, $e_2 + e_3$ (for the neutral vertices), e_4 (for the negatively charged vertex), and $-e_1 + e_3 + e_4$, $e_2 - e_3 + e_4$, $e_1 - e_2 + e_4$ (for the positively charged vertices), so as to produce a Gram representation of S_7.

Exercise 7.8 Find Gram vector representations for the two cyclotomic charged signed graphs S_8 and S_8' of Exercise 6.19. (Hint: try orthonormal e_1, e_2, e_3 and e_4 for the negatively charged vertices.)

Exercise 7.9 Find Gram vector representations for the cyclotomic signed graphs S_{14} and S_{16} of Exercise 6.21. (Hint: for S_{16}, observe that there is a set of 8 vertices that are pairwise nonadjacent; label these with orthogonal vectors $\mathbf{f}_1, \ldots, \mathbf{f}_8 \in \mathbb{R}^8$, each of length $\sqrt{2}$; label each other vertex with something of the shape $(\pm \mathbf{f}_i \pm \mathbf{f}_j \pm \mathbf{f}_k \pm \mathbf{f}_\ell)/2$ (length $\sqrt{2}$), where the choices of i, j, k, l and the signs are such that adjacencies with the vertices labelled $\mathbf{f}_1, \ldots, \mathbf{f}_8 \in \mathbb{R}^8$ are correct; check that the eight vectors added are pairwise orthogonal. Do something similar in seven-dimensional space for S_{14}.)

Exercise 7.10 Explore the complexity of connected growing. Suppose that one has a connected cyclotomic $n \times n$ integer symmetric matrix A representing a charged signed graph. If one naively considers all possible ways of extending this to an $(n+1) \times (n+1)$ connected integer symmetric matrix representing a charged signed graph, then how many new rows are considered? It is much better to exploit Lemma 6.15: now how many new rows need to be tried? Estimate the complexity of echelon growing, as described in Exercise 6.22.

Exercise 7.11 Complete the proofs of Lemmas 7.2, 7.3 and 7.4 by connectedly growing T_1, H_1 and the relevant charged triangles.

Exercise 7.12 Let G be a charged signed graph. Let $\mathfrak{S}$ be the set of all charged signed graphs that share the same underlying graph as G, with the same charges on the vertices: if G has m edges, then $|\mathfrak{S}| = 2^m$, as each edge can be positive or negative.

Define a binary operation $*$ on $\mathfrak{S}$ as follows. Given $G_1, G_2 \in \mathfrak{S}$, with adjacency matrices $A = (a_{ij})$ and $B = (b_{ij})$, respectively, define $G_1 * G_2 := G_3$, where G_3 has adjacency matrix (c_{ij}) with c_{ij} defined by

$$c_{ij} = \begin{cases} a_{ii} & i = j, \\ a_{ij}b_{ij} & i \neq j. \end{cases}$$

Show that this operation gives $\mathfrak{S}$ the structure of a group. What is the identity element? Show that there is a subgroup $\mathfrak{T}$ of $\mathfrak{S}$ such that the sign switchings of G correspond precisely to multiplication by elements of $\mathfrak{T}$. Show that the sign switching classes of G correspond to the cosets of $\mathfrak{T}$ in $\mathfrak{S}$. In particular, this shows that each sign switching class in $\mathfrak{S}$ has the same number of elements.

Exercise 7.13 Let G be the (charged signed) graph shown as follows:

Consider the set $\mathfrak{S}$ as in the previous exercise, for this particular G. Show that $\mathfrak{S}$ splits into four different equivalence classes under the relation $H_1 \sim H_2$ if and only if there is a sign switching that maps H_1 to H_2. Check that the characteristic polynomial cannot distinguish between two of the four classes, but in accordance with Theorem 6.7 the characteristic polynomial of G distinguishes the elements in the class of G from elements in other classes.

For the elements of $\mathfrak{S}$ that are in the same class as G, verify that the construction in the proof of Theorem 6.7 correctly produces the desired switchings.

7.5 Notes on Chaps. 6 and 7

Theorem 6.31 appeared first in [MS07]. There is a beautiful connection between signed graphs that have all eigenvalues at least -2 and signed graphs represented in the root system E_8 or one of the root systems D_n, and this machinery was used in the original proof (although there was still the complication of dealing with charges). The more direct approach in this chapter is close to that of Greaves in [Gre12c], where the same approach was used in a more general setting to classify cyclotomic matrices that have entries from certain quadratic rings. Other quadratic rings were treated by Taylor in [Tay10].

Studying Theorem 6.31 and identifying which of the charged signed graphs are actually equivalent to graphs, one can recover Smith's classification of cyclotomic graphs [Smi70]. A direct proof of this classification is preferable and instructive, and is outlined here as an exercise.

Exercise 7.14 Let G be a connected cyclotomic graph. Following the steps indicated, show that G is a subgraph of one of the maximal examples shown here.

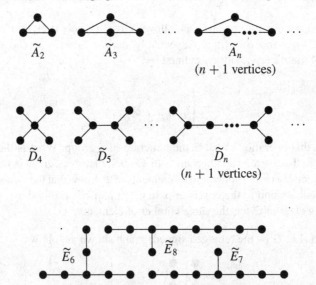

(i) For each of the graphs listed above, find an eigenvector that has all entries strictly positive and eigenvalue 2. Use Theorem B.6 to deduce that each of the graphs listed is a maximal connected cyclotomic graph.

(ii) If G contains a cycle as a subgraph, then it contains some $\tilde{A}_n$ as a subgraph. Use Theorem B.6 to deduce that $G = \tilde{A}_n$ in this case. From now on, we suppose that G contains no cycles.

(iii) If G contains a vertex of degree at least 4, then it contains $\tilde{D}_4$ as a subgraph. Use Theorem B.6 to deduce that $G = \tilde{D}_4$ in this case. From now on, we suppose that G contains no cycles and has no vertex of degree 4 or higher.

(iv) Show that if G contains only vertices of degree 1 or 2, then it is a path (given that it cannot be a cycle), and hence is a subgraph of some $\widetilde{A}_n$. From now on, we suppose that G contains no cycles and that the maximal degree of its vertices is exactly 3.

(v) If G contains two vertices v and w that both have degree 3, then show that G contains some $\widetilde{D}_n$ as a subgraph. Use Theorem B.6 to deduce that $G = \widetilde{D}_n$ in this case. From now on, we suppose that G contains no cycles, that the maximal degree of its vertices is exactly 3 and that it contains exactly one vertex that has degree 3.

(vi) Let v be the unique vertex of G that has degree 3. Show that deleting v leaves three components that are paths, say with a_1, a_2 and a_3 vertices. Show that

- if all the a_i are at least 2, then $G = \widetilde{E}_6$;
- if two of the a_i equal 1, then G is a subgraph of some $\widetilde{D}_n$;
- if one of the a_i is 1, and the others are both at least 3, then $G = \widetilde{E}_7$;
- if one of the a_i is 1, one is 2 and the other is at least 2, then G is a subgraph of $\widetilde{E}_8$.

7.6 Glossary

Most of the notation and terminology appeared in Chap. 6, and we refer the reader to the Glossary at the end of that chapter. There is one new piece of notation for this chapter.

$C_{2k}^{\bullet\bullet}$ Either C_{2k}^{++} or C_{2k}^{+-}; this notation is convenient in situations where it does not matter which of the cylindrical tessellations is meant.

Chapter 8
The Set of Cassels Heights

8.1 Cassels Height and the Set $\mathscr{C}$

Let β be a cyclotomic integer, i.e., a sum of roots of unity. If $\beta_1 = \beta, \beta_2, \ldots, \beta_d$ are the Galois conjugates of β (or indeed a list that includes each Galois conjugate the same number of times), we recall from Chap. 5 the definition of the Cassels height $\mathscr{M}(\beta)$ of β as

$$\mathscr{M}(\beta) = \frac{1}{d} \sum_{j=1}^{d} |\beta_j|^2 .$$

Because, as first noted by Robinson [Rob65], the $|\beta_j|^2$ are the conjugates of $|\beta|^2$ (Lemma 5.17; something that is not true for algebraic integers generally), $\mathscr{M}(\beta)$ is rational. From the inequality of arithmetic and geometric means, it follows immediately that $\mathscr{M}(\beta) \geq 1$ for $\beta \neq 0$ (Exercise 5.16). As before, we say that two nonzero cyclotomic integers are **equivalent** if dividing the first by some conjugate of the second gives a root of unity. Equivalent cyclotomic integers have the same Cassels height.

The aim of this chapter is to study the set

$$\mathscr{C} = \{\mathscr{M}(\beta) \mid \beta \text{ a nonzero cyclotomic integer}\} .$$

This set has an interesting structure. We start by noting that it is closed under addition. As ever we let ω_n denote a primitive nth root of unity.

Lemma 8.1 *The set $\mathscr{C}$ is closed under addition.*

Proof Take nonzero cyclotomic integers α, β. Then $\alpha, \beta \in \mathbb{Q}(\omega_n)$ for some n, and we choose a prime p dividing n, say with $p^r || n$. Put $\omega = \omega_{p^{r+1}}$ and let $\gamma = \alpha + \omega\beta \in \mathbb{Q}(\omega_{pn})$. By Lemma 5.46, there is a unique representation

$$\gamma = \gamma_0 + \gamma_1 \omega + \cdots + \gamma_{p-1} \omega^{p-1},$$

© Springer Nature Switzerland AG 2021
J. McKee and C. Smyth, *Around the Unit Circle*, Universitext,
https://doi.org/10.1007/978-3-030-80031-4_8

where $\gamma_0, \ldots, \gamma_{p-1} \in \mathbb{Q}(\omega_{pn/p}) = \mathbb{Q}(\omega_n)$. By uniqueness, we have $\gamma_0 = \alpha$, $\gamma_1 = \beta$, $\gamma_2 = \cdots = \gamma_{p-1} = 0$. Then Lemma 5.46 gives

$$\mathscr{M}(\gamma) = \mathscr{M}(\gamma_0) + \cdots + \mathscr{M}(\gamma_{p-1}) = \mathscr{M}(\alpha) + \mathscr{M}(\beta).$$

Thus $\mathscr{M}(\alpha) + \mathscr{M}(\beta) \in \mathscr{C}$. $\square$

Exercise 8.2 Let $c \geq 1$ be half an integer. Show that $c \in \mathscr{C}$. (For $c \in \mathbb{Z}$, this is trivial. Otherwise, use Lemma 8.1 along with computing the Cassels height of $1 + \omega_5$.)

Our next structural results for $\mathscr{C}$ are the following.

Theorem 8.3 (i) *The set $\mathscr{C}$ is a closed subset of $\mathbb{Q}$.*
(ii) *For every rational number $r \in [0, 1)$, there is an integer n_0 such that $r + n \in \mathscr{C}$ for all $n \geq n_0$.*

Proof (i) We write $\mathscr{C}$ as a union

$$\mathscr{C} = \bigcup_{n=0}^{\infty} \left(\mathscr{C} \cap \left[\tfrac{n}{2}, \tfrac{n+1}{2} \right] \right),$$

and prove by induction on n that for all $n \geq 0$

$$\mathscr{C} \cap \left[\tfrac{n}{2}, \tfrac{n+1}{2} \right] \quad \text{is closed.} \tag{8.1}$$

Since $\mathscr{M}(\beta) \geq 1$ for $\beta \neq 0$, the result is true for $n = 0$. In order to prove (8.1) for all $n \geq 0$, we take any $n \geq 0$ and assume that it holds for n. We take a convergent sequence $c_1, c_2, \ldots, c_m, \cdots \in \mathscr{C}$, with limit $c \in \left[\tfrac{n+1}{2}, \tfrac{n+2}{2} \right]$. We must show that $c \in \mathscr{C}$. First of all, if c is an integer or half-integer, then by Exercise 8.2 $c \in \mathscr{C}$. So we can assume that both the sequence $c_1, c_2, \ldots, c_m, \ldots$ and c itself lie in the open interval $\left(\tfrac{n+1}{2}, \tfrac{n+2}{2} \right)$. For each m, choose q_m as small as possible so that there is some $\beta_m \in \mathbb{Z}[\omega_{q_m}]$ for which $\mathscr{M}(\beta_m) = c_m$. We now separate three cases.

Case I: The sequence $\{q_m\}_{m \in \mathbb{N}}$ is bounded. By replacing $\{c_m\}_{m \in \mathbb{N}}$ by a suitable subsequence, we can assume that q_m is constant, say equal to q; then $\beta_m \in \mathbb{Z}[\omega_q]$ for all m, and, since $\deg \mathbb{Q}(\omega_q) = \varphi(q)$, we have

$$\mathscr{M}(\beta_m) = c_m = \frac{b_m}{\varphi(q)}$$

for some integer b_m. Thus, the sequence $\{b_m\}_{m \in \mathbb{N}}$ converges, so it is constant for all m sufficiently large. Hence $c = c_m \in \mathscr{C}$ for such an m.

Case II: The sequence $\{q_m\}_{m \in \mathbb{N}}$ is unbounded, and there are infinitely many m for which q_m is not square-free. For the moment, fix some m for which q_m is not square-free. Let p be a prime for which $p^r \| q_m$, where $r \geq 2$, and put $\omega := \omega_{p^r}$. Using Lemma 5.46, write β_m in the form

$$\beta_m = \alpha_0 + \alpha_1 \omega + \cdots + \alpha_{p-1} \omega^{p-1},$$

where the $\alpha_i \in \mathbb{Z}[\omega_{q_m/p}]$. Then (Lemma 5.46 again)

$$\mathscr{M}(\beta_m) = \mathscr{M}(\alpha_0) + \mathscr{M}(\alpha_1) + \cdots + \mathscr{M}(\alpha_{p-1}).$$

Define the set $J := \{j \in \{0, 1, , \ldots, p-1\} \mid \alpha_j \neq 0\}$. Since $\mathscr{M}(\beta_m) = c_m > 0$, clearly J is nonempty. Moreover, writing $\#J$ for the number of elements in the set J, we have $\#J \geq 2$. Indeed, if J consisted of only one element, j say, then we would have $\beta = \alpha_j \omega^j$, implying $\mathscr{M}(\alpha_j) = \mathscr{M}(\beta_j) = c_m$. However, since $\alpha_j \in \mathbb{Z}[\omega_{q_m/p}]$, this would contradict the choice of q_m. Hence $\#J \geq 2$.

Next, since $\mathscr{M}(\alpha_j) \geq 1$ for each $j \in J$, it follows that

$$\#J \leq \sum_{j \in J} \mathscr{M}(\alpha_j) = c_m < \frac{n+2}{2},$$

and, for all $j \in J$,

$$\mathscr{M}(\alpha_j) = \mathscr{M}(\beta_m) - \sum_{i \in J, i \neq j} \mathscr{M}(\alpha_i) \leq \mathscr{M}(\beta_m) - 1 = c_m - 1 < \frac{n}{2}.$$

Now let m vary, and write J as J_m. Replace $\{c_m\}_{m \in \mathbb{N}}$ by a subsequence where the q_m are not square-free, and the $\#J_m$ are constant, equal to k say, where we know that $2 \leq k < (n+2)/2$. For m in this new sequence,

$$c_m = \mathscr{M}(\beta_m) = \sum_{j \in J_m} \mathscr{M}(\alpha_j) = c_{m,1} + \cdots + c_{m,k}$$

say, where $\mathscr{M}(\alpha_j) := c_{m,j}$ and we know that the $c_{m,j} \in [1, \frac{n}{2}]$. Hence, the sequence of k-tuples $\{(c_{m,1}, c_{m,2} \cdots, c_{m,k})\}_{m \in \mathbb{N}}$ has a convergent subsequence, with limit

$$(c_{\infty,1}, c_{\infty,2}, \cdots, c_{\infty,k})$$

say. Then, by the induction hypothesis,

$$c_{\infty,j} = \lim_{m \to \infty} c_{m,j} \in \mathscr{C} \qquad (j = 1, \ldots, k),$$

so that, by Lemma 8.1

$$c = c_{\infty,1} + c_{\infty,2} + \cdots + c_{\infty,k} \in \mathscr{C},$$

as required.

Case III: The sequence $\{q_m\}_{m \in \mathbb{N}}$ is unbounded, and all but finitely many q_m are square-free. Firstly, replace $\{c_m\}_{m \in \mathbb{N}}$ by a subsequence where all the q_m are square-free, and where the largest prime factor of q_m, say p_m, tends to infinity with m. Fix an m, put $\omega := \omega_{p_m}$, $p := p_m$ and (Lemma 5.43) write β_m as

$$\beta_m = \alpha_0 + \alpha_1 \omega + \cdots + \alpha_{p-1} \omega^{p-1}, \tag{8.2}$$

with $\alpha_0, \alpha_i, \ldots, \alpha_{p-1} \in \mathbb{Z}[\omega_{q_m/p}]$. By Lemma 5.43, we have

$$\mathcal{M}(\beta_m) = \frac{1}{2(p-1)} \sum_{0 \le i \le p-1} \sum_{0 \le j \le p-1} \mathcal{M}(\alpha_i - \alpha_j). \tag{8.3}$$

Let r be the largest number of equal coefficients among $\alpha_0, \alpha_i, \ldots, \alpha_{p-1}$. So for each $i \in \{0, 1, \ldots, p-1\}$, there are at least $p - r$ values of j for which $\alpha_j \ne \alpha_i$. For each such pair, we have $\mathcal{M}(\alpha_i - \alpha_j) \ge 1$, giving

$$\mathcal{M}(\beta_m) \ge \frac{1}{2(p-1)} \sum_{0 \le i \le p-1} (p-r) = \frac{p(p-r)}{2(p-1)}.$$

Hence,

$$\frac{p(p-r)}{2(p-1)} \le \mathcal{M}(\beta_m) = c_m < \frac{n+2}{2},$$

and so

$$p - r \le \left\lfloor (n+2)\frac{p-1}{p} \right\rfloor \le n+1,$$

giving $r \ge p - (n+1)$. Say that the r equal α_i are equal to α. Then by subtracting $(1 + \omega + \cdots + \omega^{p-1})\alpha = 0$ from (8.2) and amending the α_i appropriately, we can assume that in (8.2) at most $n + 1$ of the α_i are nonzero. Thus, once more by taking a suitable subsequence, we can assume that the number of nonzero α_i is k, where $k \le n + 1$. Also, as in Case II, $k \ge 2$. Then (8.2) becomes

$$\beta_m = \alpha_{i_1} \omega^{i_1} + \alpha_{i_2} \omega^{i_2} + \cdots + \alpha_{i_k} \omega^{i_k}.$$

Then, setting

$$S := \mathcal{M}(\alpha_{i_1}) + \mathcal{M}(\alpha_{i_2}) + \cdots + \mathcal{M}(\alpha_{i_k})$$

we have from (8.3) that

$$\mathcal{M}(\beta_m) = \frac{1}{p-1}\left(\mathcal{M}(\alpha_{i_1} - \alpha_{i_2}) + \mathcal{M}(\alpha_{i_1} - \alpha_{i_3}) + \cdots + \mathcal{M}(\alpha_{i_{k-1}} - \alpha_{i_k}) \right) + \frac{p-k}{p-1} S.$$

Now from Exercise 5.14, we have $0 \le \mathcal{M}(\alpha_i - \alpha_j) \le 2(\mathcal{M}(\alpha_i) + \mathcal{M}(\alpha_j))$, so that

$$\mathscr{M}(\alpha_{i_1} - \alpha_{i_2}) + \mathscr{M}(\alpha_{i_1} - \alpha_{i_3}) + \cdots + \mathscr{M}(\alpha_{i_{k-1}} - \alpha_{i_k}) \le 2(k-1)S.$$

Hence

$$\frac{p-k}{p-1} S \le \mathscr{M}(\beta_m) \le \frac{p-k}{p-1} S + \frac{2(k-1)}{p-1} S$$

which we can rewrite as

$$\left(1 - \frac{k-1}{p-1}\right) S \le \mathscr{M}(\beta_m) \le \left(1 + \frac{k-1}{p-1}\right) S.$$

So

$$S \in \left[\frac{c_m}{1 + \frac{k-1}{p-1}}, \frac{c_m}{1 - \frac{k-1}{p-1}}\right].$$

Hence each $\mathscr{M}(\alpha_{i_j})$ satisfies

$$1 \le \mathscr{M}(\alpha_{i_j}) \le S - (k-1) \le S - 1 \le \frac{c_m}{1 - \frac{k-1}{p-1}} - 1.$$

Now let m go to infinity, whence also $p = p_m \to \infty$. Since $c_m \le (n+2)/2$, we see that the $\mathscr{M}(\alpha_{i_j})$ are all less than $(n+1)/2$ for m (and thus p) sufficiently large. Furthermore, on some subsequence of the integers m, the k-tuple $(\mathscr{M}(\alpha_{i_1}), \mathscr{M}(\alpha_{i_2}), \ldots, \mathscr{M}(\alpha_{i_k}))$ converges, say to $(u_1, u_2, \ldots, u_k)$. Since all the $\mathscr{M}(\alpha_{i_j})$ are in $[1, \frac{1}{2}(n+1)]$, so are all the u_j, with $\lim_{m \to \infty} \mathscr{M}(\alpha_{i_j}) = u_j$. Thus, by the induction hypothesis, all the u_j are in $\mathscr{C}$, and hence, by Lemma 8.1, so is $u_1 + u_2 + \cdots + u_k$, which equals c.

(ii) Write $r = a/q$, and let p be the smallest prime number greater than or equal to 5 such that $p \equiv 1 \pmod{2q}$. Then we can write $r = m/(p-1)$ for some even integer m. We show first that $\mathscr{C} \cap (r + \mathbb{Z})$ is nonempty. To do this, we need to find a cyclotomic integer β for which the fractional part of $\mathscr{M}(\beta)$ is $m/(p-1)$.

Write m as a sum of four integer squares, say $m = a^2 + b^2 + c^2 + d^2$, and put $x = -(a+b+c+d)/2$. Since $(a+b+c+d)^2 \equiv m \equiv 0 \pmod 2$, we see that x is an integer. Take $\beta := x + a\omega_p + b\omega_p^2 + c\omega_p^3 + d\omega_p^4$. Then we verify that

$$\mathscr{M}(\beta) = \frac{a^2 + b^2 + c^2 + d^2}{p-1} + x^2 + a^2 + b^2 + c^2 + d^2,$$

which has fractional part $m/(p-1) = r$. So we can take $n_0 = \lfloor \mathscr{M}(\beta) \rfloor$, and then use Lemma 8.1, which allows us to add any positive integer to $\mathscr{M}(\beta)$, while remaining in $\mathscr{C}$. $\square$

8.2   The Derived Sets and the Sumsets of $\mathscr{C}$

We extend Lemma 8.1 to obtain the following results, connecting the kth derived set $\mathscr{C}^{(k)}$ of $\mathscr{C}$ and the Minkowski sumset

$$k\mathscr{C} = \{c_1 + c_2 + \cdots + c_k \mid c_1, c_2, \ldots, c_k \in \mathscr{C}\}. \tag{8.4}$$

For the definition of derived sets, see Sect. 2.6.2.

Theorem 8.4 *For $k \geq 1$, the kth derived set $\mathscr{C}^{(k)}$ of $\mathscr{C}$ is the sumset $(k+1)\mathscr{C}$. Furthermore, every element of $\mathscr{C}^{(k)}$ is a limit from both sides of elements of $\mathscr{C}^{(k-1)}$.*

The following is an immediate consequence.

Corollary 8.5 *The smallest element of $\mathscr{C}^{(k)}$ $(k \geq 0)$ is $k+1$. Furthermore, $\mathscr{C}$ is closed under addition. Indeed, a stronger version of additivity holds, namely that $\mathscr{C}^{(k)} + \mathscr{C}^{(\ell)} = \mathscr{C}^{(k+\ell+1)}$ $(k, \ell \geq 0)$.*

Sets having similar topological (though not algebraic) structure to $\mathscr{C}$ have been found before. Salem [Sal44] proved that the set S of all Pisot numbers is closed in $\mathbb{R}$. For $k \geq 0$, the sets $S^{(k)}$ are known to be nonempty, with the smallest element being at least $\sqrt{k}$—see [Boy79]. Also, Boyd and Mauldin in 1996 [BM96] proved that for $k \geq 1$ every member of $S^{(k)}$ is a limit from both sides of elements of $S^{(k-1)}$. This enabled them to specify the order type of S. With this in mind, and recalling that Axel Thue [Thu12] was the discoverer of the Pisot numbers, we define a ***Thue set*** T to be a subset of the positive real line with the following properties:

(i)   The set T is a closed subset of $\mathbb{R}_+$;
(ii) For $k \geq 1$, the kth derived set $T^{(k)}$ is nonempty, and every element of it is a limit from both sides of elements of $T^{(k-1)}$;
(iii) $t_k := \min\{t \mid t \in T^{(k)}\} \to \infty$ as $k \to \infty$.

So S is a Thue set.

Corollary 8.6 *The set $\mathscr{C}$ is a Thue set.*

It is immediately clear that all derived sets of a Thue set are again Thue sets. Thus, all the derived sets $\mathscr{C}^{(k)}$ for $k \geq 1$ are also Thue sets.

8.3   Proof of Theorem 8.4

For the proof, we need a qualitative version of a very precise theorem of Loxton.

Theorem 8.7 ([Lox72, Eq. (6.1)]) *There is a strictly increasing (concave) function g such that for every cyclotomic integer β, we have $\mathscr{M}(\beta) \geq g(\mathscr{N}(\beta))$.*

Here $\mathcal{N}(\beta)$ is the function (introduced in Sect. 5.4.2) which equals the smallest number of roots of unity whose sum is β. Thus, given $B > 0$, there is a constant $B' > 0$) such that $\mathcal{N}(\beta) \leq B'$ whenever $\mathcal{M}(\beta) \leq B$.

Recall that for any algebraic integer α, its mean trace is defined in (A.9) by $\overline{\text{tr}}(\alpha) = (\text{trace } (\alpha))/[\mathbb{Q}(\alpha) : \mathbb{Q}]$. This is the mean of the conjugates of α. In particular $\mathcal{M}(\beta) = \overline{\text{tr}}(|\beta|^2)$.

Recall from (5.8) that $\mu_\varphi(n)$ denotes $\mu(n)/\varphi(n)$, where μ is the Möbius μ-function, and φ is the Euler φ-function. It is the mean trace $\overline{\text{tr}}(\omega_n)$ of ω_n.

Lemma A.27 states that the mean trace is additive. Of course, it is not generally multiplicative, but there is a special case where this property does in fact hold.

Lemma 8.8 *Let m, n be coprime integers and let $\alpha \in \mathbb{Q}(\omega_n)$. Then*

(i) $\overline{\text{tr}}(\omega_m \alpha) = \overline{\text{tr}}(\omega_m)\overline{\text{tr}}(\alpha) = \mu_\varphi(m)\overline{\text{tr}}(\alpha)$;
(ii) If also m is odd, one still has $\overline{\text{tr}}(\omega_{2m}\alpha) = \overline{\text{tr}}(\omega_{2m})\overline{\text{tr}}(\alpha) = -\mu_\varphi(m)\overline{\text{tr}}(\alpha)$, regardless of the parity of n.

Proof Since m and n are coprime, $\omega_m \omega_n$ is a primitive mnth root of unity, and the $\varphi(mn)$ automorphisms of $\mathbb{Q}(\omega_m \omega_n)$ are defined by $\omega_m \omega_n \mapsto \omega_m^a \omega_n^b$ where a is coprime to m and b is coprime to n. From this the formula in (i) is immediate. For (ii), given m is odd we have that $-\omega_{2m}$ is a primitive mth root of unity and since $\overline{\text{tr}}(-\beta) = -\overline{\text{tr}}(\beta)$, one deduces (ii) from (i). $\square$

Proposition 8.9 *Let $\mathcal{L}$ be an infinite increasing sequence of positive integers, and γ_1 and γ_2 be nonzero cyclotomic integers. Then*

$$\lim_{\substack{\ell \to \infty \\ \ell \in \mathcal{L}}} \mathcal{M}(\gamma_1 + \omega_\ell \gamma_2) = \mathcal{M}(\gamma_1) + \mathcal{M}(\gamma_2).$$

Also, $\mathcal{L}$ can be chosen so that infinitely many of the values $\mathcal{M}(\gamma_1 + \omega_\ell \gamma_2)$ are distinct, so that $\mathcal{M}(\gamma_1) + \mathcal{M}(\gamma_2)$ is a genuine limit point of the sequence $\{\mathcal{M}(\gamma_1 + \omega_\ell \gamma_2)\}_{\ell \in \mathcal{L}}$. Furthermore, $\mathcal{L}$ can be chosen so that the limit is approached either from above or from below.

Proof Now from Lemma A.27

$$\begin{aligned}
\mathcal{M}(\gamma_1 + \omega_\ell \gamma_2) &= \overline{\text{tr}}(|\gamma_1 + \omega_\ell \gamma_2|^2) \\
&= \overline{\text{tr}}(|\gamma_1|^2) + \overline{\text{tr}}(|\gamma_2|^2) + \overline{\text{tr}}(\omega_{-\ell}\gamma_1\overline{\gamma_2}) + \overline{\text{tr}}(\omega_\ell \overline{\gamma_1}\gamma_2) \\
&= \mathcal{M}(\gamma_1) + \mathcal{M}(\gamma_2) + \overline{\text{tr}}(\omega_\ell^{-1}\gamma_1\overline{\gamma_2}) + \overline{\text{tr}}(\omega_\ell \overline{\gamma_1}\gamma_2). \quad (8.5)
\end{aligned}$$

Choosing n so that γ_1, $\gamma_2 \in \mathbb{Q}(\omega_n)$, with say

$$\gamma_1 \overline{\gamma_2} = \sum_k a_k \omega_n^k,$$

we see that

$$\overline{\mathrm{tr}}(\omega_\ell^{-1}\gamma_1\overline{\gamma_2}) = \overline{\mathrm{tr}}(\omega_\ell\overline{\gamma_1}\gamma_2) = \sum_k a_k\overline{\mathrm{tr}}(\omega_\ell\omega_n^{-k}) = \sum_k a_k\overline{\mathrm{tr}}(\omega_{\ell'}) = \sum_k a_k\mu_\varphi(\ell'),$$

where $\omega_\ell\omega_n^{-k} = \omega_{\ell'}$, say, where ℓ' depends on k. Since $\ell' \to \infty$ as $\ell \to \infty$, and $\mu_\varphi(\ell') \to 0$ as $\ell' \to \infty$, we see that as $\ell \to \infty$

$$\mathcal{M}(\gamma_1 + \omega_\ell\gamma_2) \to \mathcal{M}(\gamma_1) + \mathcal{M}(\gamma_2),$$

as claimed.

To ensure that this is a genuine limiting process, we need to have $\mathcal{M}(\gamma_1 + \omega_\ell\gamma_2) \neq \mathcal{M}(\gamma_1) + \mathcal{M}(\gamma_2)$ for infinitely many values of ℓ. We now show that $\mathcal{L}$ can be chosen so that this is true.

From Lemma 5.27, we can choose an integer i such that $\overline{\mathrm{tr}}(\omega_n^i\overline{\gamma_1}\gamma_2) \neq 0$. Then also $\overline{\mathrm{tr}}(\omega_n^{-i}\gamma_1\overline{\gamma_2}) \neq 0$. Next, define the numbers ℓ in $\mathcal{L}$ by $\omega_\ell = \omega_{\ell^*}\omega_n^i$, where the ℓ^* are odd primes not dividing n. Then, using Lemma 8.8(i),

$$\overline{\mathrm{tr}}(\omega_\ell\overline{\gamma_1}\gamma_2) = \overline{\mathrm{tr}}(\omega_{\ell^*}\omega_n^i\overline{\gamma_1}\gamma_2) = \overline{\mathrm{tr}}(\omega_{\ell^*})\overline{\mathrm{tr}}(\omega_n^i\overline{\gamma_1}\gamma_2) = -\frac{1}{\ell^* - 1}\overline{\mathrm{tr}}(\omega_n^i\overline{\gamma_1}\gamma_2),$$

which is nonzero for all ℓ. Hence, from (8.5), $\mathcal{M}(\gamma_1 + \omega_\ell\gamma_2)$ tends to $\mathcal{M}(\gamma_1) + \mathcal{M}(\gamma_2)$ from either above or below (say, above), depending on the sign of $\overline{\mathrm{tr}}(\omega_n^i\overline{\gamma_1}\gamma_2)$; it never equals $\mathcal{M}(\gamma_1) + \mathcal{M}(\gamma_2)$.

Finally, if we replace ℓ^* by $2\ell^*$ in the argument (and see Lemma 8.8(ii)), then $-\frac{1}{\ell^*-1}$ is replaced by $\frac{1}{\ell^*-1}$, so that $\mathcal{M}(\gamma_1 + \omega_\ell\gamma_2)$ tends to $\mathcal{M}(\gamma_1) + \mathcal{M}(\gamma_2)$ from below. $\qquad\square$

Note that Proposition 8.9 tells us that $2\mathscr{C} \subseteq \mathscr{C}^{(1)}$.

Proposition 8.10 *Let $\gamma_0, \gamma_1, \ldots, \gamma_r$ be fixed cyclotomic integers, and for all $n \geq 1$ define*

$$\beta_n := \gamma_0 + \gamma_1\omega_{n_1} + \gamma_2\omega_{n_2} + \cdots + \gamma_r\omega_{n_r},$$

where $n_1, \ldots, n_r$ are integers each tending to infinity as $n \to \infty$, and such that for all k, ℓ with $1 \leq k < \ell \leq r$ the order of $\omega_{n_\ell}/\omega_{n_k}$ also tends to infinity as $n \to \infty$. Then the sequence $\{\mathcal{M}(\beta_n)\}$ converges, say to $\mathcal{M}(\beta)$, with

$$\mathcal{M}(\beta) = \mathcal{M}(\gamma_0) + \mathcal{M}(\gamma_1) + \cdots + \mathcal{M}(\gamma_r).$$

Proof Now putting $n_0 = 1$, we have

$$|\beta_n|^2 = \sum_{k=0}^r |\gamma_k|^2 + \sum_{\substack{k,\ell=0\\k\neq\ell}}^r \gamma_k\overline{\gamma_\ell}\frac{\omega_{n_k}}{\omega_{n_\ell}}.$$

Choose an integer t so that all the γ_k belong to $\mathbb{Q}(\omega_t)$. Then taking the mean trace of this expression, we obtain $\mathcal{M}(\beta_n) = \sum_{k=0}^r \mathcal{M}(\gamma_k)$ plus a sum of terms of the form

$\overline{\text{tr}}(a\omega_t^h \omega_{n_k}/\omega_{n_\ell})$, where a and h are integers. Putting $\omega_t^h \omega_{n_k}/\omega_{n_\ell} = \omega_N$ say, we have

$$\overline{\text{tr}}(a\omega_t^h \omega_{n_k}/\omega_{n_\ell}) = a\mu_\varphi(N).$$

Since $N \to \infty$ as $n \to \infty$ we see that as $n \to \infty$ these terms all tend to 0, so that $\mathcal{M}(\beta_n) \to \sum_{k=0}^r \mathcal{M}(\gamma_k)$. $\square$

Proposition 8.11 Let $k \geq 1$. Every element of $(k+1)\mathcal{C}$ belongs to $\mathcal{C}^{(k)}$ and is a limit from both sides of elements of $k\mathcal{C}$.

Proof The case $k = 1$ has been done in Proposition 8.9. So take $k \geq 2$ and assume the result is true for $k - 1$. For cyclotomic integers $\gamma_1, \ldots, \gamma_k, \gamma'_{k+1}$, consider

$$m_{k+1} := \mathcal{M}(\gamma_1) + \cdots + \mathcal{M}(\gamma_k) + \mathcal{M}(\gamma_{k+1}) \in (k+1)\mathcal{C}.$$

By the induction hypothesis, for fixed ℓ the value

$$m_{k,\ell} := \mathcal{M}(\gamma_1) + \cdots + \mathcal{M}(\gamma_{k-1}) + \mathcal{M}(\gamma_k + \omega_\ell \gamma_{k+1})$$

belongs to $\mathcal{C}^{(k-1)}$, and is a limit from above of elements of $(k-1)\mathcal{C}$. Using Proposition 8.9 again, we see that m_{k+1} is a limit from above of elements of $k\mathcal{C} \subseteq \mathcal{C}^{(k-1)}$, namely the $m_{k,\ell}$, as $\ell \to \infty$, for ℓ in some sequence $\mathcal{L}$. Hence $m_{k+1} \in \mathcal{C}^{(k)}$. Since we can replace 'above' with 'below' in the two previous sentences, this proves the result for k. $\square$

So certainly the kth derived set $\mathcal{C}^{(k)}$ of $\mathcal{C}$ contains $(k+1)\mathcal{C}$. We need to show that in fact equality holds.

Proof (of Theorem 8.4) The theorem holds trivially for $k = 0$. So take $k \geq 1$ and assume that it holds for $k - 1$. We need to prove that $\mathcal{C}^{(k)} \subseteq (k+1)\mathcal{C}$. Take $\mathcal{M}(\beta) \in \mathcal{C}^{(k)}$. Then $\mathcal{M}(\beta)$ is a genuine limit of a convergent sequence $\{\mathcal{M}(\beta_n)\}_{n \in \mathbb{N}}$ say, in $\mathcal{C}^{(k-1)}$. By induction hypothesis, $\mathcal{C}^{(k-1)} \subseteq k\mathcal{C}$, so that for each β_n there are cyclotomic integers γ_{in} $(i = 1, \ldots, k)$ such that

$$\mathcal{M}(\beta_n) = \mathcal{M}(\gamma_{1n}) + \mathcal{M}(\gamma_{2n}) + \cdots + \mathcal{M}(\gamma_{kn}). \tag{8.6}$$

Now the sequence $\{\mathcal{M}(\beta_n)\}$ is bounded, so the k sequences $\{\mathcal{M}(\gamma_{in})\}$ $(i = 1, \ldots, k)$ are also bounded, with the same bound, B say. Thus, by replacing $\{\mathcal{M}(\gamma_{in})\}$ by an appropriate subsequence, we can assume that for each $i = 1, \ldots, k$ the sequence $\{\mathcal{M}(\gamma_{in})\}$ converges. Because the set $\mathcal{C}$ is closed, the limit will be $\mathcal{M}(\gamma_{i\infty})$, say, for some cyclotomic integer $\gamma_{i\infty}$. Note too that $\mathcal{M}(\gamma_{i\infty})$ must be a genuine limit point of $\{\mathcal{M}(\gamma_{in})\}$ for at least one value of i.

Further, by Loxton's Theorem 8.7, there is an integer N' such that all γ_{in} can be expressed as the sum of at most N' roots of unity. Hence, by replacing $\{\mathcal{M}(\beta_n)\}$ by a suitable subsequence, we may assume that for each i the numbers γ_{in} can be expressed as the sum of the same minimal number, N_i say, of roots of unity. By writing each γ_{in} as a sum of a minimal number $\mathscr{N}(\gamma_{in})$ of roots of unity, we will have $\mathscr{N}(\gamma_{in}) = N_i$ for each n.

We now study one of these sequences $\{\mathscr{M}(\gamma_{in})\}$. For this purpose, we temporarily drop the 'i' subscript and study the convergent sequence $\{\mathscr{M}(\gamma_n)\}$, where each γ_n is the sum of the same number, N say, of roots of unity. By replacing γ_n by an equivalent cyclotomic integer, we can assume that

$$\gamma_n = 1 + \sum_{j=2}^{N} \rho_{jn}, \tag{8.7}$$

say. By reordering the roots of unity, if necessary, we can also assume that the orders of these roots of unity increase nonstrictly monotonically with j. Consider the sequence $\{\rho_{2n}\}_{n\in\mathbb{N}}$. If infinitely many of these roots of unity are equal, then we can replace $\{\mathscr{M}(\beta_n)\}$ by an infinite subsequence so that all the ρ_{2n} are equal. We do the same for $\{\rho_{3n}\}$, $\{\rho_{4n}\}$, ..., until we find a j_1 for which $\{\rho_{j_1n}\}$ contains only finitely many copies of every root of unity. In this situation, the order of ρ_{j_1n} tends to infinity with n. We can then rewrite (8.7) as

$$\gamma_n = s_0 + \rho_{j_1n} + \rho_{j_1+1,n} + \cdots \tag{8.8}$$

where s_0 is a sum of roots of unity, all independent of n. Note that such a term $\rho_{j_1,n}$ must exist for all those i for which $\mathscr{M}(\gamma_{i\infty})$ is a genuine limit point of $\{\mathscr{M}(\gamma_{in})\}$.

We now temporarily modify (8.8) to

$$\gamma_n = s_0 + \rho_{j_1n}(1 + \rho'_{j_1+1,n} + \rho'_{j_1+2,n} + \cdots), \tag{8.9}$$

say. We then reorder the sequence $\rho'_{j_1+1,n}, \rho'_{j_1+2,n}, \ldots, \rho'_{N,n}$ so that their orders as roots of unity are (nonstrictly) monotonically increasing. If the sequence $\{\rho'_{j_1+1,n}\}_{n\in\mathbb{N}}$ has infinitely many equal terms, then we can take an infinite subsequence of $\{\mathscr{M}(\beta_n)\}$ where $\{\rho'_{j_1+2,n}\}$ is constant. We do the same for $\{\rho'_{j_1+2,n}\}$, if possible. We continue in this way until we encounter a sequence, $\{\rho_{j_2n}\}$ say, that contains only finitely many copies of each root of unity; we then define

$$s_1 := 1 + \sum_{j=j_1+1}^{j_2-1} \rho'_{jn}$$

so that we can rewrite (8.9) as

$$\gamma_n = s_0 + \rho_{j_1n}s_1 + \rho_{j_2n}(1 + \rho'_{j_2+1,n} + \rho'_{j_2+2,n} + \cdots). \tag{8.10}$$

Note that the order of ρ_{j_2n}/ρ_{j_1n} tends to infinity with n.

Continuing in this way, we can finally write γ_n as

$$\gamma_n = s_0 + \rho_{j_1 n} s_1 + \rho_{j_2 n} s_2 + \cdots + \rho_{j_r n} s_r \,, \tag{8.11}$$

where $r \geq 1$ for at least one value of i, and $s_0, s_0, \ldots, s_r$ are sums of roots of unity, all independent of n. In general they will, of course, depend on the (dropped) subscript i. Also, all of the s_k must be nonzero, as γ_n has been written as the sum of a minimal number of roots of unity. Furthermore, for $k = 1, \ldots, r$ and $\ell = k + 1, \ldots, r$ the order of $\rho_{j_\ell n} / \rho_{j_k n}$ tends to infinity with n. For if the sequence {order of $\rho_{j_\ell n} / \rho_{j_k n}$} were bounded, then we could assume, by the above subsequence argument, that it would be constant. Then the term $\rho_{j_\ell n} = \rho_{j_k n} (\rho_{j_\ell n} / \rho_{j_k n})$ would already have contributed a root of unity to s_k.

From (8.11) and Proposition 8.10, we see that $\mathcal{M}(\gamma_n) \to \mathcal{M}(s_0) + \mathcal{M}(s_1) + \cdots + \mathcal{M}(s_r)$ as $n \to \infty$. On reinstating the dropped subscript i, and applying this result to each sequence $\{\mathcal{M}(\gamma_{in})\}$, we see that for each i the limit of this sequence is a sum of $r_i := 1 + r$ elements of $\mathscr{C}$. We have seen above that $r_i \geq 2$ for at least one value of i, so that from (8.6) that $\mathcal{M}(\beta) = \lim_{n \to \infty} \mathcal{M}(\beta_n)$ is a sum of $k + t$ elements of $\mathscr{C}$, where $t \geq 1$. Since by additivity we can express a sum of t elements of $\mathscr{C}$ as a single element of $\mathscr{C}$, we have $\mathcal{M}(\beta) \in (k + 1)\mathscr{C}$, as required. $\square$

This completes the proof of Theorem 8.4. We now know that $\mathscr{C}$ is a countable closed set, having nonempty derived sets of all orders k, with every element of $\mathscr{C}^{(k)}$ being a two-sided limit of elements of $\mathscr{C}^{(k-1)}$, and with the smallest element of the kth derived set tending to infinity as k goes to infinity. Thus $\mathscr{C}$ is a Thue set, proving Corollary 8.6.

8.3.1 Structure and Labelling of Thue Sets

Two totally ordered sets are said to have the **same order type** if there is an order-preserving bijection between them. This defines an equivalence relation on such sets, and their equivalence classes are the **ordinals**. By abuse of notation, we identify an ordinal with a set representative of this equivalence class. The **order type** of a set is then the ordinal of its equivalence class.

We can add and multiply ordinals as follows: To add two ordinals θ and θ', we define $\theta + \theta'$ to be the disjoint union of (sets) θ and θ', ordered so that every element of θ is less than every element of θ'. This definition readily extends to countable sums of ordinals. To multiply these two ordinals, we define $\theta\theta'$ as the order type of the Cartesian product of θ and θ', ordered by reverse lexicographic order, so that for $s_1, s_2 \in \theta$ and $s_1', s_2' \in \theta'$ we have $(s_1, s_1') < (s_2, s_2')$ if $s_1' < s_2'$ or $s_1' = s_2'$ and $s_1 < s_2$.

Now take ψ to be the order type of the positive integers, and put $a_1 = \psi + 1 + \psi^*$, and for $n \geq 1$

$$a_{n+1} = a_n \psi + 1 + (a_n \psi)^* .$$

Here $(\ )^*$ denotes the reverse order. Boyd and Mauldin [BM96] showed that the order type of the set of Pisot numbers is $\sum_{n=1}^{\infty} a_n$. The number 1 in the centre of this formula corresponds to the smallest element of the $(n + 1)$th derived set of the set of Pisot numbers. (See Sect. 2.6.2 for the definition of derived set.)

Let T be any Thue set. We will now build a finite string of integers to label a given element t of T. We proceed as follows. If $t < t_1$ then t is an element of the increasing sequence of all members of T that are less than t_1, which we label $\ell_{00}, \ell_{01}, \ell_{02}, \ldots$. For $t \geq t_1$, choose the largest k such that $t \geq t_k$. Take k as the first element of our string. Then there are no limit points of $T^{(k)}$ (i.e., elements of $T^{(k+1)}$) that are less than t, so that $T^{(k)}$ is discrete in the interval $[t_k, t_{k+1})$, which must contain t. We label the elements of $[t_k, t_{k+1}) \cap T^{(k)}$ in ascending order by $\ell_{k0}, \ell_{k1}, \ell_{k2}, \ldots$. Then t is in one of the half-open intervals $[\ell_{kr}, \ell_{k,r+1})$ say; we take r to be the second element of our string. If $t = \ell_{kr}$, end the string. Otherwise, we note that the elements of $T^{(k-1)}$ in the interval $(\ell_{kr}, \ell_{k,r+1})$ form a countable set with limit points precisely at both endpoints of the interval. For definiteness, we label those in $[\frac{1}{2}(\ell_{kr} + \ell_{k,r+1}), \ell_{k,r+1})$ by $\ell_{kr0}, \ell_{kr1}, \ell_{kr2}, \ldots$ in ascending order, and those in $[\ell_{kr}, \frac{1}{2}(\ell_{kr} + \ell_{k,r+1}))$ by $\ell_{kr,-1}, \ell_{kr,-2}, \ell_{kr,-3}, \ldots$, in descending order. Again, t is in one of the half-open intervals defined by these points, so we label it by the left endpoint. Again, if t is equal to this endpoint, the label ends. Otherwise, we note that in the open interval there is a countable ascending string of elements of $T^{(k-2)}$ with limit points precisely at both endpoints of the interval. So we can proceed as before. Continuing in this way, the string ends by t being a left endpoint of an interval (the elements with the longest strings will be those t in an interval whose endpoints are in $T \setminus T^{(1)}$). Then t must equal the left endpoint of such an interval. Thus, in the end, every element of T is of the form ℓ_s, where s is a string of integers, which we call the *label* of ℓ_s; we have seen that s is of the form $s = kr_1 \cdots r_j$, where $k \geq 0$ and $1 \leq j \leq k + 1$. This tells us that $t_k \leq t < t_{k+1}$ and that $t \in T^{(k-j+1)} \setminus T^{(k-j+2)}$.

The labelling described is ordered by the most significant digits, with the added rule that if two strings are of different lengths, but agree for the whole length of the shorter one, then this shorter one comes first in the ordering. Then this ordering coincides with the ordering on the real line.

Note that the allowable integer string labels are subject to the following constraints:

- The first term, k, is nonnegative;
- If $k = 0$, then the second term is nonnegative;
- The string must contain between 2 and $k + 2$ terms.

Proposition 8.12 *Any two Thue sets have the same order type.*

Proof Label the two sets as above and produce an order-preserving bijection by matching elements whose labels are the same.   □

8.4 Cassels Heights of Cyclotomic Integers in $\mathbb{Q}(\omega_p)$

Let p be an odd prime. Our next result concerns the set of those $\mathcal{M}(\beta)$ where β is a sum of $(2p)$th roots of unity. We denote this set by $\mathscr{C}_p$, so that

$$\mathscr{C}_p = \{\mathcal{M}(\beta) \mid \beta \in \mathbb{Z}[\omega_p]\},$$

where ω_p is a primitive pth root of unity. Note that we have $0 \in \mathscr{C}_p$, whereas $0 \notin \mathscr{C}$.

Theorem 8.13 *For all primes $p \geq 5$, the set $\mathscr{C}_p$ is given by*

$$\mathscr{C}_p = \left\{ \frac{1}{p'} \left(\tfrac{1}{2}s(p-s) + rp \right) \mid s = 0, 1, \ldots, p' \text{ and } r \geq 0 \right\}. \tag{8.12}$$

Here $p' := (p-1)/2$.

It is easy to check that the elements specified by (8.12) are all distinct.

For $p = 3$, the set $\mathscr{C}_3$ is a proper subset of the set given by the right-hand side of (8.12). Indeed $\mathscr{C}_3$ is easily seen to be the set of integers of the form $(a + b\omega_3)(a + b\omega_3^2) = a^2 - ab + b^2$, namely all integers N with prime factorisation of the form $N = \prod_q q^{e_q}$, where e_q is even for all primes $q \equiv 2 \pmod 3$. However for $p = 3$, the set on the right-hand side of (8.12) consists of all integers $N \not\equiv 2 \pmod 3$. So, for instance, 6, 10, 15 and 18 belong to this set, but do not belong to $\mathscr{C}_3$.

For the proof in the case $p = 5$, we need to prove the universality of two ternary quadratic polynomials.

Theorem 8.14 *Both of the quadratic polynomials*

$$a^2 + ab + b^2 + c^2 + a + b + c \tag{8.13}$$

and

$$a^2 + b^2 + c^2 + ab + bc + ca + a + b + c \tag{8.14}$$

*represent all positive integers for integer values of their variables (i.e., they are **universal**).*

Of course, it would be interesting to study $\mathscr{C}_n := \{\mathcal{M}(\beta) \mid \beta \in \mathbb{Z}[\omega_n]\}$ for n composite, too.

Problem 8.15 (*open problem*) Give a complete description of $\mathscr{C}_n$ for all n.

For primes $p \geq 5$, the following are the elements of $\mathscr{C}_p$ less than 5.01, as given by Theorem 8.13:

p	$\mathscr{C}_p \cap [0, 5.01]$
5	$\{1, \frac{3}{2}, \frac{5}{2}, \frac{7}{2}, 4, 5\}$
7	$\{1, \frac{5}{3}, 2, \frac{7}{3}, \frac{10}{3}, 4, \frac{13}{3}, \frac{14}{3}\}$
11	$\{1, \frac{9}{5}, \frac{11}{5}, \frac{12}{5}, \frac{14}{5}, 3, \frac{16}{5}, 4, \frac{22}{5}, \frac{23}{5}, 5\}$
13	$\{1, \frac{11}{6}, \frac{13}{6}, \frac{5}{2}, 3, \frac{19}{6}, \frac{10}{3}, \frac{7}{2}, 4, \frac{13}{3}, \frac{14}{3}\}$
17	$\{1, \frac{15}{8}, \frac{17}{8}, \frac{21}{8}, \frac{25}{8}, \frac{13}{4}, \frac{15}{4}, 4, \frac{33}{8}, \frac{17}{4}, \frac{35}{8}, \frac{9}{2}, \frac{19}{4}\}$

while for primes $p \geq 19$, the values are $M/(p-1)$, where

$$M = p - 1, 2p - 4, 2p, 3p - 9, 3p - 1, 4p - 16, 4p - 4, 5p - 25, 5p - 9$$

and also $6p - 36$ for $p = 19, 23, 29$.

There is nothing special about the choice of bound 5.01 above. But, because $\mathscr{M}(\beta) \leq \lceil \beta \rceil^2$, this bound does at least tell us that for any of the finite number of exceptions β in Cassels' Theorem 5.55:IV, we know all possible values of $\mathscr{M}(\beta)$ for those β lying in some $\mathbb{Q}(\omega_p)$.

8.5 Proof of Theorem 8.14

Proof First, note that for any integer $m \geq 0$

$$a^2 + ab + b^2 + c^2 + a + b + c = m$$

has an integer solution if and only if

$$3(2a + b + 1)^2 + (3b + 1)^2 + 3(2c + 1)^2 = 12m + 7$$

has an integer solution. Note that the class number of $x^2 + 3y^2 + 3z^2$ is one by [Jon35], and by using [O'M63, Sect. 102.5] and [O'M58], one may easily check that there are integers x, y and z such that

$$x^2 + 3y^2 + 3z^2 = 12m + 7.$$

Since x is not divisible by 3, by changing, if necessary, the sign of x, there is an integer b such that $3b + 1 = x$. Assume that x is even. Then b is odd. In this case, since $y - z$ is odd, without loss of generality we may assume that y is odd. Therefore, there are integers a and c such that

$$2a + b + 1 = z, \quad 2c + 1 = y.$$

Now, assume that x is odd. Then b is even and both y and z are odd. Therefore, there are integers a and c satisfying the above. Thus (8.13) is universal, as claimed.

Next, note that for any integer $m \geq 0$, the equation

$$a^2 + b^2 + c^2 + ab + bc + ca + a + b + c = m$$

has an integer solution if and only if

$$6(2a + b + c + 1)^2 + 2(3b + c + 1)^2 + (4c + 1)^2 = 24m + 9$$

has an integer solution. Note that the class number of $x^2 + 2y^2 + 6z^2$ is one by [Jon35], and by again using [O'M63, Sect. 102.5], one may easily check that there are integers x, y and z such that

$$x^2 + 2y^2 + 6z^2 = 24m + 9.$$

Since x is odd, by changing the sign of x, if necessary, there is an integer c such that $4c + 1 = x$. Note that x is divisible by 3 if and only if y is divisible by 3. Hence, there is an integer b such that $3b + c + 1 = y$ by changing, if necessary, the sign of y. Finally, since $y \equiv z \pmod{2}$, there is an integer a such that $2a + b + c + 1 = z$. Thus (8.14) is universal. $\qquad\square$

8.6 Proof of Theorem 8.13

Proof For p an odd prime, let $\beta = \sum_{i=0}^{p-1} a_i \omega_p^i \in \mathbb{Z}[\omega_p]$. These coefficients a_i are not uniquely determined by β: we can replace each a_i by $a_i + t$ for any $t \in \mathbb{Z}$. Thus, we can assume that $s := \sum_{j=0}^{p-1} a_j \in [-p', p']$, where, as before, $p' := (p-1)/2$. In fact, since $\mathcal{M}(-\beta) = \mathcal{M}(\beta)$, we can assume for the study of $\mathcal{C}_p$ that s is an integer in $[0, p']$. Also write $\mathrm{var}(a_0, \ldots, a_{p-1})$ for the variance of $a_0, \ldots, a_{p-1}$. We need the following. $\qquad\square$

Lemma 8.16 *We have*

$$p' \mathcal{M}(\beta) = \frac{1}{2} \left(p \sum_{j=0}^{p-1} a_j^2 - s^2 \right) \tag{8.15}$$

$$= \frac{p^2}{2} \mathrm{var}(a_0, \ldots, a_{p-1}). \tag{8.16}$$

Proof (of Lemma 8.16) We have

$$p'\mathscr{M}(\beta) = \tfrac{1}{2}\sum_{i=1}^{p-1}\left(\sum_{j=0}^{p-1}a_j\omega_p^{ij}\sum_{k=0}^{p-1}a_k\omega_p^{-ik}\right)$$

$$= \tfrac{1}{2}\sum_{j=0}^{p-1}\sum_{k=0}^{p-1}a_ja_k\left(\sum_{i=1}^{p-1}\omega_p^{i(j-k)}\right)$$

$$= \tfrac{1}{2}\sum_{j=0}^{p-1}\sum_{k=0}^{p-1}a_ja_k\left(\sum_{i=0}^{p-1}\omega_p^{i(j-k)}-1\right)$$

$$= \tfrac{1}{2}\left(p\sum_{j=0}^{p-1}a_j^2-s^2\right),$$

giving (8.15). This also equals

$$\frac{p^2}{2}\left(\frac{1}{p}\sum_{j=0}^{p-1}a_j^2-\left(\frac{s}{p}\right)^2\right)=\frac{p^2}{2}\,\mathrm{var}(a_0,\ldots,a_{p-1})\,.$$

□

Thus, we can interpret the Cassels height $\mathscr{M}(\beta)$ as a fixed multiple (depending on p) of the variance of the sequence of coefficients a_i of β. Thus, for p and s given, $p'\mathscr{M}(\beta)$ is minimised when the a_i are as close as possible to each other (and to their mean, which lies in $[0, 1/2)$), and the minimum occurs precisely when s of the a_i equal 1, while the remaining $p - s$ are 0. From the formula (8.15), we see that this minimum of $p'\mathscr{M}(\beta)$ is $\frac{s(p-s)}{2}$. Furthermore, up to permutation of the a_k, this is the only sequence for which the minimum occurs.

We must now show that $p'\mathscr{M}(\beta)$ can take all integer values $\frac{s(p-s)}{2} + rp$, for all integers $r \geq 0$. We separate three cases.

- $p \geq 11$. From the s ones and $p - s$ zeros in $a_0, a_1, \ldots, a_{p-1}$, we can choose $\lfloor\frac{s}{2}\rfloor + \lfloor\frac{p-s}{2}\rfloor > 4$ pairs of equal values (both 1 or both 0). Taking four of these pairs (a, a) and replacing each by $(a + n, a - n)$ for some integer n, we see from (8.15) that $p'\mathscr{M}(\beta)$ is increased by p times the sum of four squares of integers. Since every nonnegative integer r is the sum of four squares [Lag70], we indeed have that $p'\mathscr{M}(\beta)$ can take every value $\frac{s(p-s)}{2} + rp$.
- $p = 7$. We have $s = 0, 1, 2$ or 3. Let $\underline{a} = (0, 0, 0, 0, 0, a_5, a_6)$, so that $s = a_5 + a_6$. Now change $\underline{a}$ to $\underline{a} = (-a, -b, -c, -d, a + b + c + d, a_5, a_6)$. Then s remains unchanged, while $\sum_{j=0}^{6}a_j^2$ increases by 7/2 times

$$a^2 + b^2 + c^2 + d^2 + (a+b+c+d)^2 = 2(a^2+b^2+c^2+d^2+a(b+c+d)+b(c+d)+cd)\,.$$
(8.17)

This quadratic form, with root lattice A_4, has class number 1 (see Nipp [Nip91]), and locally represents all even integers. Hence by [O'M63, Sect. 102.5], it represents all even positive integers. By choosing $a_5, a_6 = 0$ or 1, and so $s = 0, 1$ or 2 we see from (8.15) that $p'\mathcal{M}(\beta)$ can take all values $\frac{1}{2}s(7-s)+7r$ for every integer $r \geq 0$, for these values of s. For $s = 3$ and $\underline{a} = (0, 0, 0, 0, 1, 1, 1)$, we change $\underline{a}$ to $\underline{a} = (a, b, c, -(a+b+c), d+1, -d+1, 1)$. Here still $s = 3$, while $\sum_{j=0}^{p-1} a_j^2$ increases by $7/2$ times

$$a^2 + b^2 + c^2 + (a+b+c)^2 + (d+1)^2 + (-d+1)^2 = 2(a^2 + (b+c)a + b^2 + bc + c^2 + d^2). \tag{8.18}$$

This quadratic form, with root lattice $A_3 \perp A_1$, has class number 1 (see [Nip91]), and locally represents all even integers. Hence by [O'M63, Sect. 102.5], it represents all even positive integers. Thus, $p'\mathcal{M}(\beta)$ can take all values $\frac{1}{2}s(7-s)+7r$ for every integer $r \geq 0$ for $s = 3$ also. .

- $p = 5$. We have $s = 0, 1$ or 2. The case $s = 0$ is essentially the same as for $p = 7$: take $\underline{a} = (-a, -b, -c, -d, a+b+c+d)$, with again $\sum_{j=0}^{4} a_j^2$ given by (8.17). For $s = 1$, start with $\underline{a} = (0, 0, 0, 0, 1)$ and change it to $(0, -a, -b, -c, 1+a+b+c)$. Then $\sum_{j=0}^{4} a_j^2$ increases by $5/2$ times

$$a^2 + b^2 + c^2 + (1+a+b+c)^2 - 1 = 2(a^2 + b^2 + c^2 + ab + bc + ca + a + b + c).$$

Hence, by Theorem 8.14, $p'\mathcal{M}(\beta)$ can take all values $2 + 5r$ for every integer $r \geq 0$. For $s = 2$, start with $\underline{a} = (0, 0, 0, 1, 1)$ and change it to $(-a, -b, -c, 1 + a + b, 1 + c)$. Then $\sum_{j=0}^{4} a_j^2$ increases by $5/2$ times

$$a^2 + b^2 + c^2 + (1+a+b)^2 + (1+c)^2 - 2 = 2(a^2 + ab + b^2 + c^2 + a + b + c).$$

Hence, again by Theorem 8.14, $p'\mathcal{M}(\beta)$ can take all values $3 + 5r$ for every integer $r \geq 0$.

$\square$

Note that it follows that, for $\beta \in \mathbb{Z}[\omega_p]$, $\mathcal{M}(\beta)$ depends only on the set $\{a_k\}$ of coefficients of β, and not on their order. (In fact, it depends only on $\sum_k a_k$ and $\sum_k a_k^2$.) Thus in general, there are many inequivalent $\beta \in \mathbb{Q}(\omega_p)$ with the same value of $\mathcal{M}(\beta)$. Note too that, having established Theorem 8.13, Lemma 8.16 provides a description of the possible values of the variance of a sequence of p integers.

Exercise 8.17 (*computational exercise*) A beautiful result, due to Bhargava and Hanke [BH05], states that a positive definite integral quadratic form represents all positive integers for integer values of its variables (i.e., is universal) if and only if it represents each of the integers

$$1, 2, 3, 5, 6, 7, 10, 13, 14, 15, 17, 19, 21, 22, 23, 26,$$
$$29, 30, 31, 34, 35, 37, 42, 58, 93, 110, 145, 203, 290.$$

Use this result to prove that both the quadratic form $a^2 + b^2 + c^2 + d^2 + a(b+c+d) + b(c+d) + cd$, coming from (8.17), and the quadratic form $a^2 + (b+c)a + b^2 + bc + c^2 + d^2$, coming from (8.18), are universal.

Exercise 8.18 Show that the six smallest elements of $\mathscr{C}_{15}$ and $\mathscr{C}_{21}$ are

For $\mathscr{C}_{15}$: $1, \frac{3}{2}, \frac{7}{4}, 2, \frac{9}{4}, \frac{5}{2}$.

For $\mathscr{C}_{21}$: $1, \frac{5}{3}, \frac{11}{6}, 2, \frac{13}{6}, \frac{7}{3}$.

The next result, a strengthening of Lemma 5.47, gives all points of $\mathscr{C}$ in $(0, 9/4)$.

Theorem 8.19 *The elements of $\mathscr{C}$ that are less than $9/4$ are those $\mathscr{M}(\beta)$ with β equivalent to one of the following:*

- *1, with $\mathscr{M}(\beta) = 1$.*
- *$1 + \omega_7 + \omega_7^i$, where $i \in \{2, 3, 4, 5, 6\}$, with $\mathscr{M}(\beta) = 2$.*
- *$\omega_3 - \omega_5 - \omega_5^i$, where $i \in \{2, 3, 4\}$, with $\mathscr{M}(\beta) = 2$.*
- *$1 + \omega_n$, for all square-free $n \neq 1, 3, 6, 10, 14$ or 15, with $\mathscr{M}(\beta) = 2 + \frac{2\mu(n)}{\varphi(n)}$.*

Thus, we can see that the nine smallest elements of $\mathscr{C}$ are

$$1, \frac{3}{2}, \frac{5}{3}, \frac{7}{4}, \frac{9}{5}, \frac{11}{6}, \frac{15}{8}, \frac{17}{9}, \frac{19}{10} \, .$$

Proof Assume that β is a cyclotomic integer with $1 < v := \mathscr{M}(\beta) < 9/4$, with $\beta \in \mathbb{Q}(\omega_m)$, where m is the conductor of $\mathbb{Q}(\beta)$ and, for given v, β is chosen up to equivalence so that m is as small as possible.

Assume first that $p^r \| m$ for some prime p, where $r \geq 2$. Then on writing $\beta = \sum_i \alpha_i \omega_{p^r}^i$, we have by Lemma 5.46 that $\mathscr{M}(\beta) = \sum_i \mathscr{M}(\alpha_i)$. So if this sum contains at least three nonzero terms, then $\mathscr{M}(\beta) \geq 3$. If this sum contains at least two nonzero terms and at least one of the α_i is not a root of unity, then, by Lemma 5.46, $\mathscr{M}(\beta) \geq 1 + 3/2 > 9/4$. Hence β must be equivalent to $1 + \omega_m$. Thus $\mathscr{M}(\beta) = 2(1 + \mu(m)/\varphi(m)) = 2$, as $\mu(m) = 0$.

We can now assume that m is square-free. Suppose first that $p \mid m$ for some $p > 7$. Since $\mathscr{M}(\beta) < 9/4 < (11 + 1)/4$, we can deduce from Lemma 5.44 that we can write β as a sum of $X \leq (p - 1)/2$ nonzero terms. Next, because at least $X(p - X)$ terms $\alpha_i - \alpha_j$ with $i < j$ will be nonzero, we have from Lemma 5.43 that if $X \geq 3$ then

$$(p - 1)\mathscr{M}(\beta) \geq X(p - X) \geq 3(p - 3) \,,$$

from which it follows that $\mathscr{M}(\beta) \geq 24/10 > 9/4$. So we can assume that $X = 2$. If β is a sum of two roots of unity, then it is easily checked that the possible β are as given in the theorem statement. Otherwise $\beta = \alpha_1 + \alpha_2 \omega_p$, where at least one of $\mathscr{M}(\alpha_1), \mathscr{M}(\alpha_2) \geq 3/2$, so that

$$(p - 1)\mathscr{M}(\beta) \geq (p - 2)(1 + 3/2) \,,$$

giving $\mathscr{M}(\beta) \geq 9/4$, a contradiction.

We can assume from now on that $\beta \in \mathbb{Q}(\omega_{105})$. Write $\beta = \sum_i \alpha_i \omega_5^i$, where now exactly X of the α_i are nonzero, and all lie in $\mathbb{Q}(\omega_{21})$. We consider the possibilities for this new X.

- $X = 1$. Then we can assume that $\beta \in \mathbb{Z}[\omega_{21}]$. We work up to equivalence, so that we can replace β by $\beta\omega$ for a root of unity ω, or by one of its conjugates, without further comment. We can write β as $\beta = \alpha + \gamma\omega_3$, where $\alpha, \gamma \in \mathbb{Z}[\omega_7]$, and then from (5.18) we have

$$2\mathcal{M}(\beta) = \mathcal{M}(\alpha) + \mathcal{M}(\gamma) + \mathcal{M}(\alpha - \gamma). \qquad (8.19)$$

Clearly at least one of $\mathcal{M}(\alpha)$, $\mathcal{M}(\gamma)$ and $\mathcal{M}(\alpha - \gamma)$ must be less than $3/2$. Hence by Lemma 5.47, at least one of α, γ or $\alpha - \gamma$ is 0 or a root of unity. If one of the three is 0, then we can assume that $\beta \in \mathbb{Z}[\omega_7]$. (We use $1 + \omega_3 = -\omega_3^2$ in the $\alpha = \gamma$ case.) But then, again by Lemma 5.47, β is equivalent to one of $1 + \omega_7 + \omega_7^j$ for $j = 2, 3, 4, 5$ or 6, with $\mathcal{M}(\beta) = 2$. From Theorem 8.13, all other $\mathcal{M}(\beta)$ for β in $\mathbb{Z}[\omega_7]$ satisfy $\mathcal{M}(\beta) \geq 7/3 > 9/4$. Since we must have $\mathcal{M}(\alpha) + \mathcal{M}(\gamma) + \mathcal{M}(\alpha - \gamma) < 9/2$, at least two of $\mathcal{M}(\alpha)$, $\mathcal{M}(\gamma)$ and $\mathcal{M}(\alpha - \gamma)$ must be 1, so that at least two of α, γ and $\alpha - \gamma$ must be roots of unity. Thus β is equivalent to $1 + \omega_n$ for some root of unity ω_n. The feasible values of n are readily found.
- $X = 2$. Then $\beta = \alpha + \gamma\omega_5$, where $\alpha, \gamma \in \mathbb{Z}[\omega_3, \omega_7]$, and from (5.18) we have

$$4\mathcal{M}(\beta) = 3\mathcal{M}(\alpha) + 3\mathcal{M}(\gamma) + \mathcal{M}(\alpha - \gamma). \qquad (8.20)$$

Again, we can assume that α and γ are not both roots of unity. If α and γ are equal but not a root of unity, then $\mathcal{M}(\alpha) \geq 3/2$ so that (8.20) shows that $\mathcal{M}(\beta) \geq 9/4$. But if they are different then $4\mathcal{M}(\beta) \geq 3(1 + 5/3) + 1$, giving $\mathcal{M}(\beta) \geq 9/4$.
- $X = 3$. Write $\beta = \sum_{i=0}^{4} \alpha_i \omega_5^i$, with three of the α_i nonzero. If these nonzero ones were equal, say to α, we could subtract $\sum_{i=0}^{4} \alpha\omega_5^i = 0$ to obtain the previous case $X = 2$. Thus at least two of the $\alpha_i - \alpha_j$ are nonzero, so that

$$4\mathcal{M}(\beta) \geq 2\sum_i \mathcal{M}(\alpha_i) + 2.$$

If one or more of the α_i is not a root of unity, we have

$$\mathcal{M}(\beta) \geq \tfrac{1}{4}(2(1 + 1 + \tfrac{5}{3}) + 2) = \tfrac{7}{3} > \tfrac{9}{4}.$$

Thus we can assume that all the nonzero α_i are roots of unity. If moreover they are all different then from (8.20) we would have $4\mathcal{M}(\beta) \geq 3 + 3 + 3$, $\mathcal{M}(\beta) \geq 9/4$. So, on multiplying by a root of unity we can assume that $\beta = \omega + \omega_5^i + \omega_5^j$, where $i \neq j$ and ω is some root of unity. Since then $4\mathcal{M}(\beta) = 6 + 2\mathcal{M}(\omega - 1)$, we would have

$$\mathcal{M}(\beta) \geq \tfrac{1}{4}(6 + 2(\tfrac{3}{2})) = \tfrac{9}{4},$$

unless $\omega - 1$ is also a root of unity. But if ω and $\omega - 1$ are both roots of unity then $\omega = -\omega_3^{\pm 1}$. So, up to equivalence we would have $\beta = \omega_3 - \omega_5 - \omega_5^i$ for $i \in \{2, 3, 4\}$.

- $X = 4$. If any two of the α_i were equal to α we could reduce to $X \leq 3$ by subtracting $\sum_{i=0}^{4} \alpha \omega_5^i = 0$, as above. So we can take the α_i as all different, and then

$$\mathcal{M}(\beta) \geq \tfrac{1}{4} \sum_{i,j} \mathcal{M}(\alpha_i - \alpha_j) \geq \tfrac{10}{4} > \tfrac{9}{4}.$$

- $X = 5$. Here we can subtract $\sum_{i=0}^{4} \alpha_0 \omega_5^i = 0$ to reduce the value of X.

The elements of $\mathscr{C}$ in $(0, 9/4)$ that are sums of two roots of unity are clearly equivalent to $1 + \omega_n$ for some n. The relevant values of n are readily calculated from (5.9). $\square$

Problem 8.20 Find all β with $\mathcal{M}(\beta) = 9/4$, up to equivalence.

8.7 Notes

Cassels [Cas69] showed that the only $\mathcal{M}(\beta) < 2$ were for β that can be written as a sum of at most two roots of unity; this implies that 2 is the smallest limit point of $\mathscr{C}$.

In 2011 Calegari, Morrison and Snyder [CMS11] studied cyclotomic integers β with a view to applications to fusion categories and subfactors. As part of this study (their Theorem 9.0.1) they found all β with $\mathcal{M}(\beta) < 9/4$. Our Theorem 8.19 reproduces this result, and mostly follows their proof.

Theorem 8.3, as well as $\mathscr{C}$ being closed under addition—see Corollary 8.5—are due to Stan and Zaharescu [SZ09, Theorem 4]. We have followed their proofs quite closely. They applied their results to deducing facts about character values of finite groups. Theorems 8.4 and 8.13 and their proofs come from McKee, Oh and Smyth [MOS20].

8.8 Glossary

$\mathscr{C}$.   The set of all Cassels heights of nonzero cyclotomic integers. Having excluded 0, the smallest element of $\mathscr{C}$ is 1.

$\mathscr{C}^{(k)}$.   The kth derived set of $\mathscr{C}$, for which, see Sect. 2.6.2 of Chap. 2.

$\mathscr{C}_n$.   Let n be a positive integer, and ω_n a primitive nth root of unity. The set $\mathscr{C}_n$ is the set of Cassels heights of cyclotomic integers $\beta \in \mathbb{Z}[\omega_n]$; this includes 0.

$k\mathscr{C}$.   The kth sumset of $\mathscr{C}$, whose elements are all possible sums of k elements of $\mathscr{C}$.

$\mathcal{M}(\beta)$.   The Cassels height of a cyclotomic integer β.

$\mu_\varphi(n)$.   This is the function $\mu(n)/\varphi(n)$. It is the mean trace of ω_n.

ω_n.   A primitive nth root of unity.

$\overline{\mathrm{tr}}(\alpha)$.   The mean trace of α.

Cassels height.   As defined in Chap. 5, the Cassels height of a cyclotomic integer β is the mean of the squared absolute values of the conjugates of β.

equivalent.   Two nonzero algebraic integers β and γ are said to be equivalent if β/γ' is a root of unity for some γ', a conjugate of γ.

label.   In our description of Thue sets, we assign a label to each element. The label is a finite string of integers subject to three constraints: (i) the first term, k, is nonnegative; (ii) if $k = 0$, then the second term is nonnegative; (iii) the string contains between 2 and $k + 2$ integers.

mean trace.   The mean trace of an algebraic number is the mean of its conjugates.

(same) order type.   Two totally ordered sets have the same order type if and only if there is an order-preserving bijection between them. The order type of a totally ordered set S is the ordinal that has the same order type as S.

Thue set.   A closed subset T of the positive real numbers such that all its derived sets are nonempty, the smallest element of $T^{(k)}$ tends to infinity as $k \to \infty$, for which every element of $T^{(k)}$ is a two-sided limit of elements of $T^{(k-1)}$ ($k \geq 1$).

universal polynomial.   A polynomial is called universal if it represents all positive integers using integer values for its variables.

Chapter 9
Cyclotomic Integer Symmetric Matrices Embedded in Toroidal and Cylindrical Tessellations

9.1 Introduction

Having classified all cyclotomic integer symmetric matrices, we now proceed to find all the minimal noncyclotomic ones. We say that an integer symmetric matrix M is **minimal noncyclotomic** if it is not cyclotomic, but any proper induced submatrix is cyclotomic. In particular, M must be connected. From the very definition, we see that M can be produced by adding a row and column to a cyclotomic matrix. Our knowledge of the structure of cyclotomic integer symmetric matrices will be crucial. In particular, we see that once these cyclotomic matrices are large enough they must be equivalent to the adjacency matrices of subgraphs of one of T_{2r}, C_{2r}^{++} or C_{2r}^{+-}.

We shall see in Chap. 13 that the smallest Mahler measure of a minimal noncyclotomic integer symmetric matrix is Lehmer's number, and this will settle the analogue of Lehmer's problem in this setting.

This chapter starts by defining some notation and tools that will be invaluable in pinning down the minimal noncyclotomics. It also establishes some useful structure theory concerning cyclotomic integer symmetric matrices that can be embedded in one of the toroidal or cylindrical tessellations of Chap. 6.

9.2 Preliminaries: Notation and Tools

Let H be a charged signed graph, with vertex set $V(H)$ and edge set $E(H)$. As outlined in Appendix B.2, we can view the elements of E as triples (x, y, ε), indicating an edge between x and y that has sign $\varepsilon \in \{-1, 1\}$. We do not distinguish between (x, y, ε) and (y, x, ε). Recall that we regard a charged vertex as being adjacent to itself: the charge can be viewed as an indication of a signed directed loop from the vertex to itself. If x is a charged vertex, with charge ε, then $(x, x, \varepsilon) \in E(H)$. We write $x \sim y$ to indicate that x and y are adjacent, regardless of the sign of the connecting edge (or loop).

© Springer Nature Switzerland AG 2021
J. McKee and C. Smyth, *Around the Unit Circle*, Universitext,
https://doi.org/10.1007/978-3-030-80031-4_9

We shall often be referring to trails and circuits, and it becomes tiresome to keep repeating the words 'trail' or 'circuit', so we use notation from Appendix B.2 that specifies precisely what is needed. We write

$$v_1 \bullet v_2 \bullet \cdots \bullet v_r$$

to indicate the trail $v_1, v_2, \ldots, v_r$ (length $r - 1$), and we write

$$v_1 : v_2 : \cdots : v_r$$

to indicate the circuit $v_1, v_2, \ldots, v_r, v_1$ (length r; an r-circuit). The *edges* in a trail or circuit must be distinct, but with the presence of charged vertices we do *not* insist that the *vertices* in a trail or circuit are distinct. For example, in the charged signed graph shown here, with vertices labelled by their names, we view $c_1 : c_1 : v_1 : c_2$ as a 4-circuit, as it uses 4 distinct edges (one of them is a loop):

$$(9.1)$$

We shall sometimes use the same notation to indicate the set of edges involved in a trail or circuit, rather than the trail or circuit itself. In context, $c_1 : c_1 : v_1 : c_2$ could indicate the set $\{(c_1, c_1, -1), (c_1, v_1, 1), (v_1, c_2, -1), (c_2, c_1, 1)\}$.

If H is a charged signed graph, we write U_H for the **underlying graph** of H, in which all edges become unsigned (in effect, positive). It will be convenient to preserve charges (positive and negative) in U_H: just the edges become unsigned. If H has adjacency matrix (a_{ij}) (for some choice of ordering the vertices), then U_H has adjacency matrix (b_{ij}), where $b_{ij} = |a_{ij}|$ if $i \neq j$, and $b_{ii} = a_{ii}$. For example, if H is the charged signed graph in (9.1), then U_H is

Let H be a charged signed graph. For each vertex $x \in V(H)$, we define the **common neighbour class** (or simply **class**) of x, written as $[x]$, by

$$[x] = \{y \in V(H) \mid \forall z \in V(H), y \sim z \Leftrightarrow x \sim z\}.$$

Thus, $[x]$ contains the vertices that share precisely the same neighbours as x. If x is charged, then $x \sim x$, and one quickly sees that all vertices in $[x]$ must then be adjacent to each other, and all must be charged. If x is uncharged, then no two vertices in $[x]$ are adjacent. We always have $x \in [x]$.

The **class graph** of H, written as C_H, is the graph whose vertices are the classes $[x]$, with adjacency defined by

$$[x] \sim [y] \Leftrightarrow x \sim y .$$

For example, with H as in (9.1), $[c_1] = \{c_1, c_2\}$, $[v_1] = \{v_1, v_2\}$ and C_H is

$$[c_1]\!\!-\!\!-\!\![v_1] \; .$$

Exercise 9.1 Verify that the notion of adjacency in the class graph is well-defined, i.e., independent of the choices of representatives of the classes.

Exercise 9.2 For $k = 3$ and $k \geq 5$, verify that $C_{T_{2k}}$ is a cycle of length k. What happens when $k = 4$?

At times in this chapter, we shall be treating the charged signed graphs C_{2k}^{++} and C_{2k}^{+-} (and their subgraphs) in parallel. With this in mind, as in Chap. 7, we use the convenient notation $C_{2k}^{\bullet\bullet}$ to indicate one or other of C_{2k}^{++} and C_{2k}^{+-} without specifying which.

Exercise 9.3 Show that for $k \geq 3$, the class graph $C_{C_{2k}^{\bullet\bullet}}$ is a path of length $k - 1$. What happens when $k = 2$?

Vertices in the same class as x will be called **conjugates** of x. In the special case when $[x] = \{x, y\}$, we write $y = \overline{x}$, and say that y is *the* conjugate of x. Of course then we have that $\overline{\overline{x}} = x$.

Exercise 9.4 Let H be a signed graph (no charges). Verify that any permutation of the elements of any class $[x]$ induces a graph automorphism of the underlying graph U_H. Show by example that this might not be a charged signed graph automorphism of H.

In the sequel, we shall make heavy use of the parity of the number of positive edges in certain cycles. It is convenient to work with a group that captures precisely the information that we need. We define the **parity check group**, P_H, of a charged signed graph H to be the additive group whose elements are ordered pairs (S, ε), where $S \subseteq E(H)$ (perhaps $S = \emptyset$), and $\varepsilon \in \{0, 1\}$ agrees modulo 2 with the number of *positive* edges (or loops) in S. Addition is defined by

$$(S_1, \varepsilon_1) + (S_2, \varepsilon_2) = (S_1 \triangle S_2, \; \varepsilon_1 + \varepsilon_2 \pmod 2),$$

where

$$S_1 \triangle S_2 = \{e \in E(H) \mid e \in S_1 \cup S_2, e \notin S_1 \cap S_2\}$$

is the symmetric difference of S_1 and S_2. The zero element is $(\emptyset, 0)$.

For example, if H is the charged signed graph

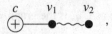

then

$$P_H = \{ \ (\emptyset, 0), \ (\{(c, c, 1)\}, 1), \ (\{(c, v_1, 1)\}, 1), \ (\{(v_1, v_2, -1)\}, 0),$$
$$(\{(c, c, 1), (c, v_1, 1)\}, 0), \ (\{(c, c, 1), (v_1, v_2, -1)\}, 1),$$
$$(\{(c, v_1, 1), (v_1, v_2, -1)\}, 1), \ (\{(c, c, 1), (c, v_1, 1), (v_1, v_2, -1)\}, 0) \ \}.$$

Exercise 9.5 Verify that P_H as described is indeed a group. In particular, the sum of two elements of P_H is indeed in P_H (the parity check works).

What is the abstract structure of this group? (Note that every nonidentity element has order 2.)

The group P_H provides a convenient way of describing certain parity checks that we shall need to do. We illustrate this by using it to prove a couple of lemmas for later use. Recall that we may write $(v_1 : \cdots : v_r, \varepsilon)$ as shorthand for

$$(\{(v_1, v_2, \varepsilon_{1,2}), (v_2, v_3, \varepsilon_{2,3}), \ldots, (v_{r-1}, v_r, \varepsilon_{r-1,r}), (v_r, v_1, \varepsilon_{r,1})\}, \varepsilon),$$

with the $\varepsilon_{i,j}$ being the signs of the relevant edges, i.e., we use the notation for a circuit to indicate the set whose elements are the edges of the circuit, and similarly for trails.

Lemma 9.6 *(i) Let H be a signed graph with vertices v_1, w_1, v_2, w_2, v_3 and w_3, whose underlying graph U_H is drawn as follows:*

Consider the following four 4-cycles in H (there are others):

$$v_1 : v_2 : v_3 : w_2, \quad v_1 : v_2 : w_3 : w_2, \quad w_1 : v_2 : v_3 : w_2, \quad w_1 : v_2 : w_3 : w_2.$$

If any three of these have an odd number of positive edges, then so does the fourth.

(ii) Similarly, suppose that H is a signed graph with vertices $v_1, v_2, \bar{v}_2, v_3, \bar{v}_3$ and v_4, and underlying graph shown as follows:

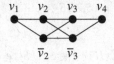

If any three of the four 4-cycles in H that include just one pair of conjugate vertices have an odd number of positive edges, then so does the fourth.

Note that in part (i) of the lemma, we cannot write $w_1 = \bar{v}_1$, as the class $[v_1]$ is $\{v_1, w_1, v_3, w_3\}$. In part (ii), there are five 4-cycles, but one of them, namely $v_2 : v_3 : \bar{v}_2 : \bar{v}_3$, includes two pairs of conjugate vertices.

Proof (i) The four 4-cycles between them cover every edge of H twice, so that for the appropriate values of $\varepsilon_1, \ldots, \varepsilon_4$ we have

$$(v_1 : v_2 : v_3 : w_2, \varepsilon_1) + (v_1 : v_2 : w_3 : w_2, \varepsilon_2)$$
$$+ (w_1 : v_2 : v_3 : w_2, \varepsilon_3) + (w_1 : v_2 : w_3 : w_2, \varepsilon_4) = (\emptyset, 0)$$

in P_H. Hence, if any three of the ε_i equal 1, then so does the fourth.

(ii) Similarly,

$$(v_1 : v_2 : \bar{v}_3 : \bar{v}_2, \varepsilon_1) + (v_1 : v_2 : v_3 : \bar{v}_2, \varepsilon_2)$$
$$+ (v_2 : v_3 : v_4 : \bar{v}_3, \varepsilon_3) + (v_3 : v_4 : \bar{v}_3 : \bar{v}_2, \varepsilon_4) = (\emptyset, 0)$$

in P_H. $\qquad\square$

Lemma 9.7 *Let H be a charged signed graph with vertices $v_1, v_2, v_3, \bar{v}_1, \bar{v}_2$ and $\bar{v}_3$ whose underlying graph U_H is shown below, where the two vertices v_3 and $\bar{v}_3$ are charged with the same sign ε (either positive or negative—it does not matter):*

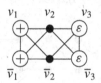

Let $C_1, \ldots, C_8$ in that order be the 4-cycles

$$v_1 : v_1 : v_2 : \bar{v}_1, \quad v_1 : v_1 : \bar{v}_2 : \bar{v}_1, \quad v_2 : v_3 : v_3 : \bar{v}_3, \quad \bar{v}_2 : v_3 : v_3 : \bar{v}_3,$$
$$v_1 : v_2 : v_3 : \bar{v}_2, \quad \bar{v}_1 : v_2 : \bar{v}_3 : \bar{v}_2, \quad v_1 : v_2 : \bar{v}_3 : \bar{v}_2, \quad \bar{v}_1 : v_2 : v_3 : \bar{v}_2.$$

If $(C_1, 1), (C_2, 1), (C_3, 1), (C_4, 1)$ and $(C_5, 1)$ are all in P_H, then so are $(C_6, 1)$, $(C_7, 1)$ and $(C_8, 1)$.

Proof Note that $C_1 \,\Delta\, C_2 \,\Delta\, C_3 \,\Delta\, C_4 = C_5 \,\Delta\, C_6 = C_7 \,\Delta\, C_8$, and $C_3 \,\Delta\, C_4 \,\Delta\, C_5 = C_7$.

Hence, we compute $(C_6, 1) = (C_1, 1) + (C_2, 1) + (C_3, 1) + (C_4, 1) + (C_5, 1)$, $(C_7, 1) = (C_3, 1) + (C_4, 1) + (C_5, 1)$ and $(C_8, 1) = (C_5, 1) + (C_6, 1) + (C_7, 1)$. $\square$

9.3 Cyclotomic Graphs Embedded in T_{2k}

Armed with the tools of the previous section, we now learn more about the structure of cyclotomic signed graphs, and especially the infinite family T_{2k}. Later we shall employ similar techniques to deal with charged signed graphs too, but it eases the exposition if we keep charges out of sight for the moment.

A little more terminology: a cycle of length k in T_{2k} is called a *long cycle*.

Exercise 9.8 Show that for $k = 3$ or $k \geq 5$, there are 2^k long cycles in T_{2k}. How many long cycles are there in T_8?

A path $v_1 \bullet \cdots \bullet v_r$ (length $r - 1$) in a signed graph H is called *chordless* if the only edges between any of the v_i are the edges joining v_i to v_{i+1} for $1 \leq i \leq r - 1$.

Exercise 9.9 Verify that for $k \geq 4$, the longest chordless path in T_{2k} has length $k - 2$. What is the longest chordless path in T_6?

We see from the above that T_6 and T_8 have some unusual properties. We shall quite often state results for T_{2k} that are restricted to $k \geq 5$; the interested reader can work out how much remains true, or what needs to be changed, for $k = 3$ or $k = 4$. For example, when $k \geq 5$ the class graph is a cycle of length k. In T_{2k}, each class $[x]$ contains two vertices, unless $k = 4$. In T_8, the eight vertices split into two common neighbour classes, each of size 4.

In T_{2k}, for $k \geq 5$, there are k pairs of conjugate vertices. If we take any set of representatives of the k classes, they form a long cycle.

If a 4-cycle in T_{2k} ($k \geq 5$) contains precisely one pair of conjugate vertices (so it is of the shape $x : y : \overline{x} : z$, where $z \neq \overline{y}$), then it has an odd number of positive edges. In the standard drawing of T_{2k} (Fig. 6.3), such 4-cycles look like either triangles or parallelograms:

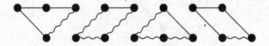

In other words, $(x : y : \overline{x} : z, 1 \in P_{T_{2k}})$ whenever y and z are neighbours of x that are not in the same class. We call this property the 4-*cycle rule*, and we emphasise that it applies only to 4-cycles that contain precisely one pair of conjugate vertices.

We can extend this notion to any signed graph in which each class contains either 1 or 2 vertices. A signed graph is said to satisfy if the 4-*cycle rule* if

- each class $[x]$ contains either 1 or 2 vertices;
- for every 4-cycle of the shape $x : y : \overline{x} : z$ in H for which $z \neq \overline{y}$,   (9.2)
  there holds $(x : y : \overline{x} : z, 1) \in P_H$.

Exercise 9.10 Suppose that H is a connected signed graph containing a chordless path of length 3, and with each common neighbour class of size either 1 or 2. Show that if H satisfies the 4-cycle rule, then any 4-cycle in H that includes *two* pairs of conjugate vertices has an even number of positive edges. (In particular, this is true for T_{2k} itself.) Show that this conclusion may fail if there is no chordless path of length 3.

Lemma 9.11 *Let $k \geq 5$, and let $v_1 : \cdots : v_k$ be a long cycle in $U_{T_{2k}}$, so that the vertices of $U_{T_{2k}}$ are $v_1, \bar{v}_1, \ldots, v_k, \bar{v}_k$. Then the graph automorphisms of $U_{T_{2k}}$ are generated by*

- *the cycle automorphism σ that sends v_i to v_{i+1} and $\bar{v}_i$ to $\bar{v}_{i+1}$ ($1 \leq i \leq k$, with the interpretation that $v_{k+1} = v_1$ and $\bar{v}_{k+1} = \bar{v}_1$);*
- *the orientation reversal ρ which sends v_i to v_{k-i+2} and $\bar{v}_i$ to $\bar{v}_{k-i+2}$ ($1 \leq i \leq k$, with the interpretation that $v_{k+1} = v_1$ and $\bar{v}_{k+1} = \bar{v}_1$);*
- *the conjugations τ_i ($1 \leq i \leq k$), where τ_i swaps v_i and $\bar{v}_i$, and fixes all other vertices.*

Proof Note that σ, ρ and $\tau_1, \ldots, \tau_k$ are all indeed automorphisms of $U_{T_{2k}}$, so we have the containment $\langle \sigma, \rho, \tau_1, \ldots, \tau_k \rangle \subseteq \mathrm{Aut}(U_{T_{2k}})$.

Now take any $\kappa \in \mathrm{Aut}(U_{T_{2k}})$. We have $\kappa(v_1) = v_t$ or $\kappa(v_1) = \bar{v}_t$ for some t, and hence for some choice of $\varepsilon \in \{0, 1\}$ the map $\kappa_1 := \tau_1^\varepsilon \sigma^{-t} \kappa$ fixes v_1; then κ_1 also fixes $\bar{v}_1$, since any automorphism must send conjugate pairs to conjugate pairs.

Now $\kappa_1(v_2)$ is a neighbour of $\kappa_1(v_1) = v_1$, so is one of $v_2, \bar{v}_2, v_k, \bar{v}_k$. Hence, for some choice of $\varepsilon_1, \varepsilon_2 \in \{0, 1\}$, we have that $\kappa_2 := \tau_2^{\varepsilon_1} \rho^{\varepsilon_2} \kappa_1$ fixes v_2 (and hence also fixes $\bar{v}_2$); and κ_2 still fixes v_1 and $\bar{v}_1$ (since ρ and τ_2 fix v_1). Then $\kappa_2(v_3)$ is adjacent to $\kappa_2(v_2) = v_2$ but not to $\kappa_2(v_1) = v_1$, so is one of v_3 or $\bar{v}_3$. Composing with τ_3 if needed, we produce κ_3 fixing $v_1, \bar{v}_1, v_2, \bar{v}_2, v_3$ and $\bar{v}_3$. And so on. Eventually, we express the identity map as the composition of κ with an element of $\langle \sigma, \rho, \tau_1, \ldots, \tau_k \rangle$, and hence κ is in this subgroup.

Thus, we have $\langle \sigma, \rho, \tau_1, \ldots, \tau_k \rangle = \mathrm{Aut}(U_{T_{2k}})$, as claimed. $\qquad \square$

Exercise 9.12 Let $H = \mathrm{Aut}(U_{T_{2k}})$, where $k \geq 5$. Show that $N = \langle \tau_1, \ldots, \tau_k \rangle \cong C_2^k$ is a normal subgroup of H, where C_2 is the group of order 2. Show that H is the semidirect product of N and the dihedral group D_{2k}, with $2k$ elements.

Exercise 9.13 Determine the group of signed graph automorphisms of T_{2k}.

If H is a signed graph, then a graph embedding $\sigma : U_H \hookrightarrow U_{T_{2k}}$ is an embedding of the vertices of H into those of T_{2k} such that adjacencies in the underlying graph are preserved. When such an embedding exists, we shall provide a useful criterion to determine whether or not H is in fact cyclotomic. First we note that such an embedding is unique up to applying automorphisms of $U_{T_{2k}}$, at least if H contains a chordless path of length 3. The necessity of this hypothesis is seen by the chordless path of length 2; for any $k \geq 4$, this path can be embedded in T_{2k} with the endpoints either conjugate or not.

Lemma 9.14 *Let H be a connected signed graph containing a chordless path of length 3.*

If there is an embedding of U_H in $U_{T_{2k}}$, then this embedding is unique up to an automorphism of $U_{T_{2k}}$.

Proof Let κ_1, κ_2 be two embeddings $U_H \hookrightarrow U_{T_{2k}}$.

Let $x_1 \bullet x_2 \bullet x_3 \bullet x_4$ be a chordless path in H. Order the remaining vertices x_5, …, x_n such that for $i > 1$ each x_i is adjacent to at least one earlier vertex in the list.

From Lemma 9.11, it is clear that $\mathrm{Aut}(U_{T_{2k}})$ acts transitively on the vertices of $U_{T_{2k}}$, and the stabiliser of any vertex acts transitively on the four neighbours of that vertex, so we can find $\theta \in \mathrm{Aut}(U_{T_{2k}})$ such that $\theta \kappa_2(x_1) = \kappa_1(x_1)$ and $\theta \kappa_2(x_2) = \kappa_1(x_2)$. Then $\theta \kappa_2(x_3)$ is a neighbour of $\theta \kappa_2(x_2)$, but cannot be either $\theta \kappa_2(x_1)$ or its conjugate, as in the former case $\theta \kappa_2$ would not be injective, and in the latter case $\theta \kappa_2(x_4)$ would be a neighbour of $\theta \kappa_2(x_1)$. We must have that $\theta \kappa_2(x_3)$ is one of $\kappa_1(x_3)$ or its conjugate, and replacing θ by $\tau_{\kappa_1(x_3)}\theta$ if needed we can assume that $\theta \kappa_2(x_3) = \kappa_1(x_3)$.

Having firmly established the orientation of our embedding, we deal in turn with x_4, …, x_n, and in each case we can if necessary replace our automorphism θ by its composition with a conjugation so that κ_1 and $\theta \kappa_2$ agree on the x_i treated so far.

We conclude that κ_1 and κ_2 agree up to composition by an automorphism of $U_{T_{2k}}$. $\qquad\square$

Now we can establish a key criterion for when signed graphs of a particular shape are cyclotomic.

Lemma 9.15 *Let H be a connected signed graph for which U_H can be embedded in $U_{T_{2k}}$, and containing a chordless path of length 3. Then each class of H contains either 1 or 2 vertices, and H is cyclotomic if and only if it satisfies the 4-cycle rule* (9.2).

Proof Suppose that $x \in H$ with $|[x]| \geq 3$. Cyclically permuting three elements of $[x]$ gives an automorphism of U_H (although perhaps not an automorphism of H as a signed graph). Composing this automorphism with the given embedding of U_H into $U_{T_{2k}}$, we get two such embeddings that differ by a cyclic permutation of three vertices. But from Lemma 9.11, we see that no such 3-cycle is an automorphism of $U_{T_{2k}}$ ($k \geq 5$), and we contradict Lemma 9.14. (There might be elements of the automorphism group of $U_{T_{2k}}$ that have order 3, but they would move all the vertices rather than just three of them.) Thus, each class of H contains either 1 or 2 vertices.

Now suppose that H fails the 4-cycle rule: say $(x_1 : x_2 : \overline{x}_1 : x_3, 0) \in P_H$, where $x_3 \neq \overline{x}_2$. Then H must contain vertices other than those in $[x_1] \cup [x_2] \cup [x_3]$, else it would not have a chordless path of length 3. Take $x_4 \in H$ not in $[x_1] \cup [x_2] \cup [x_3]$, but adjacent to a vertex in one of these classes. Note that the preimage under our embedding of a class in $U_{T_{2k}}$ is contained in a class of H. Hence x_4 cannot be adjacent to x_1 (or its conjugate), else its image in $U_{T_{2k}}$ would be in the same class in $U_{T_{2k}}$ as one of the images of x_2 or x_3, giving x_4 in $[x_2] \cup [x_3]$. We have x_4 adjacent to one of x_2 or x_3, but not both, from consideration of the embedding in T_{2k} with $k \geq 5$.

Then H contains a subgraph (induced by $x_1, \overline{x}_1, x_2, x_3, x_4$) equivalent to the noncyclotomic graph

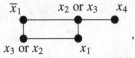

and by interlacing H is not cyclotomic.

Now suppose that H satisfies the 4-cycle rule (9.2). We shall grow H to a signed graph $\widehat{H}$ such that $U_{\widehat{H}} = U_{T_{2k}}$ and $\widehat{H}$ still satisfies the 4-cycle rule. We first reduce to the case where H has a chordless cycle of length k (which maps to a long cycle in T_{2k}, since $k \geq 5$). If not, let $v_1, \ldots, v_\ell$ be a chordless path in H of maximal length. Then the class graph of H is a path on $[v_1], \ldots, [v_\ell]$, with each $[v_i]$ either a singleton or containing one vertex other than v_i, namely $\bar{v}_i$. We have $k > l$, and we see that U_H embeds in $U_{T_{2(l+1)}}$, so we might as well take $k = l + 1$. We shall add a vertex v to H, adjacent to all vertices in $[v_\ell] \cup [v_1]$, and show that the signs of the new edges can be chosen so that the 4-cycle rule still holds in this larger signed graph H_v (we may then replace H by H_v to reduce to the case where H contains a chordless cycle of length k). Consider the following picture of the relevant part of U_H, along with the new vertex v, noting that (i) the vertices $\bar{v}_1, \bar{v}_2, \bar{v}_{\ell-1}$ and $\bar{v}_\ell$ each might not be present, and (ii) if $l = 4$ or $l = 5$, then certain vertices appear more than once in the picture.

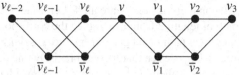

We choose the signs of the edges from v to v_1 and v_ℓ freely. If $\bar{v}_1$ (respectively $\bar{v}_\ell$) is present, then we choose the sign of the edge from v to $\bar{v}_1$ (respectively $\bar{v}_\ell$) to ensure that the 4-cycle rule holds for $v : v_1 : v_2 : \bar{v}_1$ (respectively $v : v_\ell : v_{\ell-1} : \bar{v}_\ell$). If both $\bar{v}_1$ and $\bar{v}_2$ are present (respectively $\bar{v}_\ell$ and $\bar{v}_{\ell-1}$), then we apply part (ii) of Lemma 9.6 to the subgraph induced by $v, v_1, v_2, v_3, \bar{v}_1$ and $\bar{v}_2$ (respectively $v, v_\ell, v_{\ell-1}, v_{\ell-2}, \bar{v}_\ell$ and $\bar{v}_{\ell-1}$) to conclude that the 4-cycle rule holds for $v : v_1 : \bar{v}_2 : \bar{v}_1$ (respectively $v : v_\ell : \bar{v}_{\ell-1} : \bar{v}_\ell$).

We are now reduced to the case where H contains a chordless cycle of length k, the class graph is $[v_1] : \cdots : [v_k]$ (say), with each $[v_i]$ either a singleton or containing the two vertices v_i and $\bar{v}_i$. If $U_H = U_{T_{2k}}$, then we have our desired $\widehat{H} = H$. Otherwise, suppose (after cyclically changing the names) that $\bar{v}_3$ is *not* present in H. We shall show how to add a vertex $v = \bar{v}_3$ to H with signed edges joining it to all vertices in $[v_2] \cup [v_4]$ such that the 4-cycle rule holds for this larger graph H_v. Repeating this construction will eventually produce the desired signed graph $\widehat{H}$. The picture below may help, showing the relevant part of U_H, with the new vertex $v = \bar{v}_3$, and where each of $\bar{v}_1, \bar{v}_2, \bar{v}_4$ and $\bar{v}_5$ might not be present.

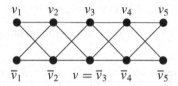

We join $v = \bar{v}_3$ to v_2 with arbitrary sign, and then choose the sign for the edge $(\bar{v}_3, v_4, \varepsilon)$ so that $(v_2 : v_3 : v_4 : \bar{v}_3, 1) \in P_{H_v}$.

If $\bar{v}_4$ (respectively $\bar{v}_2$) is present, choose the sign of $(\bar{v}_3, \bar{v}_4, \varepsilon)$ so that $(\bar{v}_3 : \bar{v}_4 : v_5 : v_4, 1) \in P_{H_v}$ (respectively, choose the sign of $(\bar{v}_3, \bar{v}_2, \varepsilon)$ so that $(\bar{v}_3 : \bar{v}_2 : v_1 : v_2, 1) \in P_{H_v}$). Applying part (ii) of Lemma 9.6 to the subgraph induced by $v_2, v_3, v_4, v_5, \bar{v}_3$ and $\bar{v}_4$ (respectively $v_1, v_2, v_3, v_4, \bar{v}_2$ and $\bar{v}_3$), we deduce that $(v_2 : v_3 : \bar{v}_4 : \bar{v}_3, 1) \in P_{H_v}$ (respectively $(v_4 : v_3 : \bar{v}_3 : \bar{v}_3, 1) \in P_{H_v}$). If both $\bar{v}_4$ and $\bar{v}_5$ are present (respectively $\bar{v}_1$ and $\bar{v}_2$), then applying part (i) of Lemma 9.6 to the subgraph induced by $v_3, v_4, v_5, \bar{v}_3, \bar{v}_4$ and $\bar{v}_5$ (respectively $v_3, v_2, v_1, \bar{v}_3, \bar{v}_2$ and $\bar{v}_1$) shows that $(\bar{v}_3 : \bar{v}_4 : \bar{v}_5 : v_4, 1) \in P_{H_v}$ (respectively $(\bar{v}_3 : \bar{v}_2 : \bar{v}_1 : v_2, 1) \in P_{H_v}$).

Finally, if both $\bar{v}_2$ and $\bar{v}_4$ are present, then applying part (i) of Lemma 9.6 to the subgraph induced by $v_2, v_3, v_4, \bar{v}_2, \bar{v}_3$ and $\bar{v}_4$ shows that $(v_3 : \bar{v}_4 : \bar{v}_3 : \bar{v}_2, 1) \in P_{H_v}$.

We conclude that we can indeed grow H as desired, and eventually reach a signed graph $\widehat{H}$ that has the same underlying graph as T_{2k}, and where $\widehat{H}$ satisfies the 4-cycle rule.

Let A be the adjacency matrix of $\widehat{H}$. Since each vertex of $\widehat{H}$ has four signed edges leading from it, the diagonal entries of A^2 are all 4. Let i and j be distinct vertices in $\widehat{H}$. If there is no path of length 2 from i to j, then the corresponding off-diagonal entry in A^2 is 0. If there is a path of length 2 from i to j, say $i \bullet x \bullet j$, and $j \neq \bar{i}$, then from the structure of $U_{T_{2k}}$ we see that there is precisely one other path of length 2 from i to j, namely $i \bullet \bar{x} \bullet j$; then the cycle $i : x : j : \bar{x}$ must have an odd number of positive edges, and hence the contributions to A^2 from the two paths cancel out and the corresponding off-diagonal entry in A^2 is 0. Finally, consider the paths of length 2 from i to $\bar{i}$; if x and y are neighbours of i that are not conjugate, then there are four such paths, going from i to j via one of $x, \bar{x}, y$ and $\bar{y}$; the contributions to A^2 from the paths via x and y cancel each other, as do the contributions from the paths via $\bar{x}$ and $\bar{y}$. In summary, $A^2 = 4I$.

Thus, $\widehat{H}$ is cyclotomic (its eigenvalues can only be ± 2), and by interlacing so is H. ∎

9.4   Changes for Charges

Much of the work of the previous section adapts painlessly to charged signed graphs. We now consider the families $C_{2k}^{\bullet\bullet}$, usually with $k \geq 3$ to avoid certain pathologies.

We no longer have long cycles, but define a **long path** in $C_{2k}^{\bullet\bullet}$ to be a chordless path of length $k - 1$. The ends of a long path in $C_{2k}^{\bullet\bullet}$ are charged vertices, and any chordless path of maximal length is necessarily a long one.

Exercise 9.16 Show that for $k \geq 3$ there are 2^k long paths in $C_{2k}^{\bullet\bullet}$. How many long paths are there in $C_4^{\bullet\bullet}$?

Our definition of classes is unchanged, but remember that a charged vertex is adjacent to itself, and that if x is charged then all vertices in $[x]$ are charged, and all are adjacent to each other. As before, when $[x] = \{x, y\}$ we write $y = \bar{x}$.

In $C_{2k}^{\bullet\bullet}$ we generally have $|[x]| = 2$, except when $k = 2$ (then all four vertices lie in the same class).

The 4-cycle rule needs to become a 4-circuit rule, to allow circuits that repeat a charged vertex (adjacent to itself). Explicitly, if c and $\overline{c}$ are two charged vertices in the same class in $C_{2k}^{\bullet\bullet}$ ($k \geq 3$), with neighbouring uncharged vertices v and $\overline{v}$, and with the other neighbours of v being w and $\overline{w}$ (charged if $k = 3$, otherwise uncharged) then the 4-circuits that involve the charged vertex c are

$$c : v : w : \overline{v}, \quad c : v : \overline{w} : \overline{v}, \quad c : v : \overline{c} : c, \quad c : v : \overline{c} : \overline{c}, \quad c : \overline{v} : \overline{c} : c, \quad c : \overline{v} : \overline{c} : \overline{c}.$$

We say that a charged signed graph satisfies the 4-*circuit rule* if

- each class $[x]$ contains either 1 or 2 vertices;
- for every 4-circuit of the shape $x : y : \overline{x} : z$ in H for which $z \neq \overline{y}$,   (9.3) there holds $(x : y : \overline{x} : z, 1) \in P_H$.

Recall that if H is a charged signed graph, then the underlying charged graph U_H has vertices that carry the same charges as in H, but with all edges replaced by positive edges. Note that U_H can have negative charges.

The analogues of Lemmas 9.11 and 9.14 can be proved in much the same way, and are given here as exercises.

Exercise 9.17 Let $v_1, \ldots, v_k$ be representatives of the classes of $C_{2k}^{\bullet\bullet}$ with $v_1 \bullet \cdots \bullet v_k$ a long path ($k \geq 3$). The automorphism group of $U_{C_{2k}^{+-}}$ as a charged graph is generated by the k conjugations τ_i ($1 \leq i \leq k$), where τ_i swaps v_i and $\overline{v}_i$, fixing all other vertices. The automorphism group of $U_{C_{2k}^{++}}$ is generated by the conjugations and the orientation-reversing reflection $v_i \mapsto v_{k-i+1}$, $\overline{v}_i \mapsto \overline{v}_{k-i+1}$.

Exercise 9.18 If H is a connected charged signed graph including at least one charged vertex, and there is an embedding of U_H in $U_{C_{2k}^{\bullet\bullet}}$, then this embedding is unique up to an automorphism of $U_{C_{2k}^{\bullet\bullet}}$.

The analogue of our main Lemma for uncharged signed graphs whose underlying graph embeds in T_{2k} is the following.

Lemma 9.19 *Let H be a connected charged signed graph including at least one charged vertex, and with a chordless path of length 3. Suppose there is an embedding of U_H in some $U_{C_{2k}^{\bullet\bullet}}$. Then each class x of H contains at most two vertices, and H is cyclotomic if and only if it satisfies the 4-circuit rule.*

Proof We can eliminate the possibility that any class contains more than two elements as before.

Suppose that H fails the 4-circuit rule. Then H contains a subgraph equivalent to one of the noncyclotomic examples

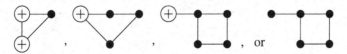

and H is not cyclotomic.

Now suppose that the 4-circuit rule holds in H.

First suppose that the embedding hits only one class of charged vertices in $U_{C_{2k}^{**}}$. There must be at least one uncharged vertex since there is a chordless path of length 3. Let $v_1 \bullet \cdots \bullet v_r$ be a chordless path of maximal length ($r - 1$) in H, with v_1 charged and v_r not. We shall construct a charged signed graph H_v by adding a new charged vertex, with positive charge, adjacent to v_r, and also adjacent to $\bar{v}_r$ if that is present, in such a way that H_v still satisfies the 4-circuit rule, and embeds into C_{2k}^{**} for some $k \geq 3$ (perhaps smaller than the original k). The relevant part of U_{H_v} is shown, with the comments that (i) $\bar{v}_{r-1}$ and $\bar{v}_r$ each might not be present; (ii) v_{r-2} is not charged, since the assumption on there being a chordless path of length 3 forces $r \geq 4$.

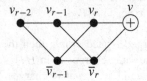

We use an arbitrary sign for the edge from v to v_r. If $\bar{v}_r$ is present, then we choose the sign for the edge from v to $\bar{v}_r$ so that $(v_{r-1} : v_r : v : \bar{v}_r, 1) \in P_{H_v}$. If both $\bar{v}_r$ and $\bar{v}_{r-1}$ are present, then we apply part (ii) of Lemma 9.6 to the subgraph induced by the vertices $v, v_r, \bar{v}_r, v_{r-1}, \bar{v}_{r-1}$ and v_{r-2} to conclude that the 4-circuit rule holds in H_v.

We are now reduced to the case where the embedding of H hits both of the classes of charged vertices in $U_{C_{2k}^{**}}$. We proceed to fill in any missing vertices in the image of the embedding, growing H to a charged signed graph $\widehat{H}$ such that $U_{\widehat{H}}$ is isomorphic to $U_{C_{2k}^{**}}$, and the 4-circuit rule still holds in $\widehat{H}$. As in the uncharged case, we proceed one vertex at a time.

We may now assume that there is a path $v_1 \bullet \cdots \bullet v_k$ in H joining two charged vertices in different classes, with $k \geq 4$.

First suppose that H is missing a charged vertex; say there is no $\bar{v}_k$. We try to form H_v by adding a new vertex $v = \bar{v}_k$; this must have the same charge as v_k, and we join it to v_k with an edge of positive sign (an arbitrary choice). We join $v = \bar{v}_k$ to v_{k-1} with an edge of the correct sign such that $(v_{k-1} : v_k : v_k : \bar{v}_k, 1) \in P_{H_v}$ (and then $(v_{k-1} : v_k : v : v, 1) \in P_{H_v}$ since v and v_k have the same charge). The relevant part of U_{H_v} is shown, where it is understood that $\bar{v}_{k-1}$ and $\bar{v}_{k-2}$ each might not be present, and the sign ε for the charge of v_k is unknown (but we choose the charge of v to be the same).

If $\bar{v}_{k-1}$ is present, we join v to $\bar{v}_{k-1}$ with an edge of the correct sign such that $(\bar{v}_{k-1} : v_k : v_k : v, 1) \in P_{H_v}$. We need then to check the parity condition for the circuit $v_{k-2} : v_{k-1} : v : \bar{v}_{k-1}$, which follows from the computation

$$(v_{k-2} : v_{k-1} : v_k : \bar{v}_{k-1}, 1) + (v_{k-1} : v_k : v_k : v, 1)$$
$$+ (\bar{v}_{k-1} : v_k : v_k : v, 1)$$
$$= (v_{k-2} : v_{k-1} : v : \bar{v}_{k-1}, 1).$$

If both $\bar{v}_{k-1}$ and $\bar{v}_{k-2}$ are present, then we need to check the parity condition for the circuit $\bar{v}_{k-2} : \bar{v}_{k-1} : v : v_{k-1}$. This is done by an equivalent computation, replacing v_{k-2} by $\bar{v}_{k-2}$.

We can now suppose that any missing charged vertices have been filled in, and proceed to deal with any missing uncharged vertices.

Suppose that an uncharged vertex $v \in H$ is in a class of its own. Take x and y adjacent to v in different classes (possible since only the charged vertices in H fail to have this property).

First, we consider the case where a neighbour of v is charged, say x has charge ε. Then $\bar{x} \in H$, with the same charge ε. On the other hand, y must be adjacent to some z that is not in $[v]$; perhaps, z is charged. We shall add $\bar{v}$ to produce H_v. Some of the information about the underlying graph of H_v is pictured here: x and $\bar{x}$ have the same (unknown) charge, the vertex z might be charged or not, and if $\bar{y}$ or $\bar{z}$ is present, then further edges need to be added.

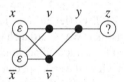

There are up to ten 4-circuits of relevance (those involving $\bar{v}$, and noting that, e.g., $\bar{x} : \bar{x} : v : x$ has the same number of positive edges as $x : x : v : \bar{x}$): let us name them $C_1, \ldots, C_{10}$ in the order shown as follows:

$$x : x : v : \bar{x}, \quad x : x : \bar{v} : \bar{x}, \quad x : v : y : \bar{v}, \quad \bar{x} : v : y : \bar{v}, \quad \bar{x} : \bar{v} : \bar{y} : v,$$
$$v : y : z : \bar{y}, \quad x : v : \bar{y} : \bar{v}, \quad \bar{v} : y : z : \bar{y}, \quad v : y : \bar{z} : \bar{y}, \quad \bar{v} : y : \bar{z} : \bar{y}.$$

We have $(C_1, 1) \in P_{H_v}$, and we choose the signs of the edges from x, $\bar{x}$ and y to $\bar{v}$ such that $(C_2, 1), (C_3, 1) \in P_{H_v}$. Then $(C_4, 1) = (C_1, 1) + (C_2, 1) + (C_3, 1) \in P_{H_v}$. If $\bar{y}$ is present, then $(C_6, 1) \in P_{H_v}$, and we choose the sign of the edge from $\bar{v}$ to $\bar{y}$ such that $(C_5, 1) \in P_{H_v}$. Then $(C_7, 1) = (C_1, 1) + (C_2, 1) + (C_5, 1) \in P_{H_v}$, and $(C_8, 1) = (C_3, 1) + (C_6, 1) + (C_7, 1) \in P_{H_v}$. If $\bar{y}$ and $\bar{z}$ are both present, then $(C_9, 1) \in P_{H_v}$. And then $(C_{10}, 1) = (C_4, 1) + (C_5, 1) + (C_9, 1) \in P_{H_v}$.

If neither x nor y is charged, then x is adjacent to $w \notin [v]$ and y is adjacent to $z \notin [v]$. Either w or z might be charged, but in any event the argument now proceeds as in the uncharged case: we can add $\bar{v}$ adjacent to x, y, $\bar{x}$ (if present), and $\bar{y}$ (if present) such that the 4-circuit rule still holds.

As in the uncharged case, we repeat this process to produce a charged signed graph $\widehat{H}$, with underlying graph equal to that of $C_{2k}^{\bullet\bullet}$, and satisfying the 4-circuit rule. Let A be the adjacency matrix of $\widehat{H}$. The diagonal entries of A all equal 4 (uncharged vertices have four incident edges; charged vertices have three edges to other vertices along with a loop). As in the uncharged case, we check that all off-diagonal entries of A^2 are zero. There are two novel cases coming from a charged vertex x and an uncharged neighbour v: (i) the two paths $x \bullet x \bullet v$ and $x \bullet \bar{x} \bullet v$ add to give the 4-circuit $x : x : \bar{x} : v$, and since $(x : x : \bar{x} : v, 1) \in P_{H_v}$, the contributions from these two paths cancel out; (ii) there are four relevant paths from x to $\bar{x}$, which pair up so as to cancel out ($x \bullet x \bullet \bar{x} \oplus x \bullet \bar{v} \bullet \bar{x} = x : x : \bar{x} : \bar{v}$, and $x \bullet v \bullet \bar{x} \oplus x \bullet \bar{x} \bullet \bar{x} = x : v : \bar{x} : \bar{x}$).

Since $A^2 = 4I$, all eigenvalues of $\widehat{H}$ are ± 2, so $\widehat{H}$ is cyclotomic, and by interlacing so is H.   □

Exercise 9.20 Check that Lemma 9.19 remains true if the longest chordless path in H has length 2. What can go wrong if the longest chordless path is just a single edge?

9.5 Glossary

$v_1 \bullet v_2 \bullet \cdots \bullet v_r$.   The trail $v_1, \ldots, v_r$ (length $r - 1$) in a charged signed graph. Being a trail, the edges are distinct. We regard a charged vertex as being adjacent to itself, and a trail is allowed to include adjacent charged vertices (but any such special edge, which might be viewed as a loop, can only be used once).

$v_1 : v_2 : \cdots : v_r$.   The circuit $v_1, \ldots, v_r, v_1$ (length r) in a charged signed graph. As with a trail, the edges must be distinct, but the vertices do not need to be.

$[x]$.   The (common neighbour) class of x.

4-circuit rule.   A charged signed graph H satisfies the 4-circuit rule if (i) each class $[x]$ contains either one or two vertices, and (ii) $(x : y : \bar{x} : z, 1) \in P_H$ whenever $x : y : \bar{x} : z$ is a 4-circuit for which $y \neq \bar{z}$. One should remember as always that any charge is adjacent to itself so may appear twice as adjacent vertices in a 4-circuit.

4-cycle rule.   A signed graph H satisfies the 4-cycle rule if (i) each class $[x]$ contains either one or two vertices, and (ii) $(x : y : \bar{x} : z, 1) \in P_H$ whenever $x : y : \bar{x} : z$ is a 4-cycle for which $y \neq \bar{z}$.

$\mathrm{Aut}(G)$.   The automorphism group of a charged signed graph G. An automorphism of G is a permutation of the vertices that preserves all adjacencies and preserves the signs of adjacencies. An automorphism must therefore send a charged vertex to a charged vertex of the same sign.

C_H.   The class graph of the charged signed graph H.

$d_H(x, y)$.   The number of edges on the shortest path from x to y in the connected graph H. We have $d_H(x, x) = 0$.

H_x.   The charged signed graph produced by deleting vertex x from H.

H_{xy}.   The charged signed graph produced by deleting vertices x and y from H. Note that $H_{xy} = (H_x)_y = (H_y)_x$.

P_H.   The parity check group of the charged signed graph H.

U_H.   The underlying graph of a charged signed graph H. This preserves charges (with their signs), but replaces any negative edges between distinct vertices by positive edges.

chordless cycle.   A cycle in a charged signed graph H is called chordless if it equals the subgraph of H induced by its vertices.

chordless path.   A path $v_1 \bullet v_2 \bullet \cdots \bullet v_r$ in a charged signed graph H is called chordless if the only edges in H between any of the v_i are those that appear already in the path. Thus, a path is chordless if the subgraph induced by its vertices is a path.

class graph.   The class graph of H is the graph whose vertices are the distinct classes $[x]$ where x runs over the vertices of H, and where $[x]$ is adjacent to $[y]$ if and only if x is adjacent to y. One may check that this definition of adjacency is well-defined.

common neighbour class.   The common neighbour class of x in a charged signed graph G is the set of vertices of G that share precisely the same neighbours as x. As ever, a charged vertex is adjacent to itself, from which one sees that a common neighbour class either has all elements charged or has all elements neutral. We always have that x is in its own common neighbour class.

conjugates.   The conjugates of a vertex x in a charged signed graph G are the elements of $[x]$ other than x itself. If $[x] = \{x, y\}$ has precisely two elements, then we write $y = \bar{x}$ for the unique conjugate of x.

long cycle.   A cycle of length k in T_{2k} ($k \geq 3$) is called a long cycle. In the standard drawing of T_{2k}, there are two obvious long cycles, one having all edges positive and one having all edges negative. There are usually 2^k long cycles in T_{2k}, but the case $k = 4$ is special (and left as an exercise).

long path.   A long path in $C_{2k}^{\bullet\bullet}$ is a chordless path of length $2k$. The vertices at the ends of a long path must be charged.

minimal noncyclotomic.   An integer symmetric matrix A is minimal noncyclotomic if it is (i) not cyclotomic (Chap. 6) and (ii) any proper induced submatrix (Chap. 6) is cyclotomic. Any such matrix must be connected (Chap. 6).

parity check group.   The parity check group of a charged signed graph H has elements that are ordered pairs (S, ε), where S is some subset of the edges of H, and $\varepsilon \in \{0, 1\}$ is the reduction modulo 2 of the number of positive edges in S. To add two elements of the parity check group, one computes the symmetric difference of the two subsets of edges, and adds the parities modulo 2. Any charged vertex x in H contributes an edge (which may be viewed as a signed loop) to the edges of H. It is sometimes convenient to write a trail or circuit in place of S to mean the set of edges that appear in that trail or circuit.

underlying graph.   The underlying graph U_H of a charged signed graph H has the same vertex set as H, and the same adjacencies, but edges between distinct vertices are all positive.

Chapter 10
The Transfinite Diameter and Conjugate Sets of Algebraic Integers

10.1 Introduction

Let E be a compact subset of the complex plane $\mathbb{C}$, symmetric about the real axis. Does it contain any conjugate sets of algebraic integers (Sect. 1.1.2)? If so, finitely many or infinitely many? The transfinite diameter $\tau(E)$ can inform us about this question. If $\tau(E) < 1$, then S contains only finitely many such sets. On the other hand, if $\tau(E) > 1$ then any open neighbourhood of E contains infinitely many. If $\tau(E) = 1$, however, the situation is complicated, and far from completely resolved.

10.2 Analytic Properties of the Transfinite Diameter

There is no number theory in this section! For our compact set $E \subset \mathbb{C}$ and a given positive integer n, we choose n points $z_1, \ldots, z_n$ in E for which the geometric mean G_n of the $\binom{n}{2}$ distances between them is as large as possible. Each of $z_1, \ldots, z_n$ depends on n, but to avoid double subscripts we have not labelled them as such. Thus

$$G_n = G_n(E) = \max_{z_1, \ldots, z_n \in E} \left(\prod_{1 \le j < k \le n} |z_j - z_k| \right)^{2/n(n-1)} . \tag{10.1}$$

Then the *transfinite diameter $\tau(E)$ of E* is defined as $\lim_{n \to \infty} G_n$. The concept is due to Fekete.

It is clear that the transfinite diameter is translation-invariant and scales linearly: for $\lambda, \mu \in \mathbb{C}$

$$\tau(\lambda E + \mu) = |\lambda| \tau(E) .$$

© Springer Nature Switzerland AG 2021
J. McKee and C. Smyth, *Around the Unit Circle*, Universitext,
https://doi.org/10.1007/978-3-030-80031-4_10

Also $\tau(E') \leq \tau(E)$ for any compact subset E' of E. We now give some other less-obvious properties of the transfinite diameter. That its defining limit exists follows from Proposition 10.1(a) below.

Proposition 10.1 *Let E be a compact subset of $\mathbb{C}$, and define G_n by (10.1).*

(a) The sequence $\{G_n\}$ is monotonically decreasing.

(b) For $n = 0, 1, 2, \ldots$ let $T_n(z) \in \mathbb{C}[z]$ be monic of degree n and be chosen so that its maximum modulus on E

$$m_n := \max_{z \in E} |T_n(z)|$$

is a small as possible. Then $m_n^{1/n}$ tends to a limit as $n \to \infty$, which is the transfinite diameter $\tau(E)$.

(c) For any monic polynomial $h(z)$ of degree d, the transfinite diameter of $h^{-1}(E)$ is $\tau(E)^{1/d}$.

(d) Define E_ε, the ε-thickening of E, to be the set of all $z \in \mathbb{C}$ distant at most ε from some element of E. Then the limit $\lim_{\varepsilon \searrow 0} \tau(E_\varepsilon)$ exists and equals $\tau(E)$.

(e) The boundary ∂E of E has the same transfinite diameter as E.

Version (b) of $\tau(E)$ is also called the **Chebyshev constant of** E. The polynomials $T_n(z)$ are called **Chebyshev polynomials of** E.

Proof (a) For $z_1, \ldots, z_n \in E$, define

$$V(z_1, \ldots, z_n) = \prod_{\substack{j, k=1 \\ j<k}}^{n} (z_j - z_k) \qquad (10.2)$$

and let V_n be the maximum of $|V(z_1, \ldots, z_n)|$ as $z_1, \ldots, z_n$ range over E. Because E is compact, this maximum clearly exists. Then

$$G_n = V_n^{\frac{2}{n(n-1)}} .$$

Now choose $n + 1$ new points $z_1, \ldots, z_n, z_{n+1} \in E$ such that $|V(z_1, \ldots, z_n, z_{n+1})| = V_{n+1}$. Then, since

$$V(z_1, \ldots, z_n, z_{n+1}) = (z_1 - z_2)(z_1 - z_3) \cdots (z_1 - z_{n+1}) V(z_2, \ldots, z_n, z_{n+1}) ,$$

and $|V(z_2, \ldots, z_n, z_{n+1})| \leq V_n$ we have

$$V_{n+1} \leq |z_1 - z_2| \cdot |z_1 - z_3| \cdots |z_1 - z_{n+1}| V_n .$$

Similarly, doing the same for all sets of n of these $n + 1$ points, we have

$$V_{n+1} \leq |z_2 - z_1| \cdot |z_2 - z_3| \cdots |z_2 - z_{n+1}| V_n \,,$$

$$\vdots \qquad \vdots \qquad \vdots \qquad \vdots \qquad \vdots$$

$$V_{n+1} \leq |z_{n+1} - z_1| \cdot |z_{n+1} - z_2| \cdots |z_{n+1} - z_n| V_n \,,$$

so that on multiplying all these inequalities we obtain

$$V_{n+1}^{n+1} \leq V_{n+1}^2 V_n^{n+1} \,.$$

Hence $V_{n+1}^{n-1} \leq V_n^{n+1}$, giving $G_{n+1} \leq G_n$.

(b) We first remark that the sequence $\tau_n := m_n^{1/n}$ $(n = 1, 2, \dots)$ is bounded since, for the particular polynomial $(z - z_0)^n$ with $z_0 \in E$ we have

$$\tau_n \leq (\max_{z \in E} |z - z_0|^n)^{1/n} = \max_{z \in E} |z - z_0| \leq \max_{z, z_0 \in E} |z - z_0| = D \,,$$

the (standard) diameter of E. Next, we show that this sequence converges. Define $a := \liminf_{n \to \infty} \tau_n$ and $b := \limsup_{n \to \infty} \tau_n$. We have $a \leq b$ and aim to show that $b \leq a$. Let $\varepsilon > 0$ and choose n such that $\tau_n < a + \varepsilon$. Then we have $|T_n(z)| < (a + \varepsilon)^n$ for $z \in E$. If k and ℓ are positive integers and $z_0 \in E$ is fixed, then, for all $z \in E$

$$|(z - z_0)^\ell T_n(z)^k| \leq D^\ell (a + \varepsilon)^{nk} \,.$$

Hence

$$\tau_{nk+\ell} \leq D^{\frac{\ell}{nk+\ell}} (a + \varepsilon)^{\frac{nk}{nk+\ell}} \,.$$

Now choosing a sequence $\tau_{n_\nu} \to b$ as $n_\nu \to \infty$ and putting $n_\nu = nk_\nu + \ell_\nu$, where $0 < \ell_\nu \leq n$, we see that $b \leq a + \varepsilon$. Since ε can be arbitrarily small, we have $b \leq a$, as was our aim. Hence $a = b$, so that the sequence $\{\tau_\nu\}$ converges.

We must now show that in fact $\{\tau_\nu\}$ converges to $\tau(E)$. This is clearly true if E is finite, so we can assume that E is infinite, and V_n is never 0. We claim that

$$m_n \leq \frac{V_{n+1}}{V_n} \leq (n + 1)m_n \,. \tag{10.3}$$

To prove these inequalities, note that

$$m_n \leq \max_{z \in E} |(z - z_1) \cdots (z - z_n)| = \frac{\max_{z \in E} |V(z, z_1, \dots, z_n)|}{|V(z_1, \dots, z_n)|} \leq \frac{V_{n+1}}{V_n}$$

when $z_1, \dots, z_n \in E$ are chosen so that $|V(z_1, \dots, z_n)| = V_n$. This proves the first inequality of (10.3).

For the second inequality, we write $V(z_1, \dots, z_{n+1})$ in its well-known Vandermonde determinantal form, and then add suitable multiples of the first n rows to the bottom row to obtain

$$V(z_1, \ldots, z_{n+1}) = \begin{vmatrix} 1 & 1 & \cdots & 1 \\ z_1 & z_2 & \cdots & z_{n+1} \\ \vdots & \vdots & \vdots & \vdots \\ z_1^{n-1} & z_2^{n-1} & \cdots & z_{n+1}^{n-1} \\ T_n(z_1) & T_n(z_2) & \cdots & T_n(z_{n+1}) \end{vmatrix}.$$

On expanding this determinant by its bottom row, we have that

$$|V(z_1, \ldots, z_{n+1})| \leq |T_n(z_1)| \cdot |V(z_2, \ldots, z_{n+1})| \\ + |T_n(z_2)| \cdot |V(z_1, z_3, \ldots, z_{n+1})| + \cdots + |T_n(z_{n+1})| \cdot |V(z_1, \ldots, z_n)|.$$

Now, choosing $z_1, \ldots, z_{n+1}$ so that $V(z_1, \ldots, z_{n+1}) = V_{n+1}$, we obtain the bound $V_{n+1} \leq (n+1)m_n V_n$, giving the second inequality of (10.3). Thus, we have

$$\tau_n^n \leq \frac{V_{n+1}}{V_n} \leq (n+1)\tau_n^n.$$

Multiplying these inequalities for $n = 2$ up to n', and then replacing n' by n, we obtain

$$(\tau_2^2 \cdots \tau_n^n)^{\frac{2}{n(n+1)}} \cdot V_2^{\frac{2}{n(n+1)}} \leq G_{n+1} \leq ((n+1)!)^{\frac{2}{n(n+1)}}(\tau_2^2 \cdots \tau_n^n)^{\frac{2}{n(n+1)}} \cdot V_2^{\frac{2}{n(n+1)}}.$$

Since both $V_2^{\frac{2}{n(n+1)}}$ and $((n+1)!)^{\frac{2}{n(n+1)}}$ tend to 1 as $n \to \infty$, it remains only to show that $(\tau_2^2 \cdots \tau_n^n)^{\frac{2}{n(n+1)}} \to \tau$ as $n \to \infty$. This follows from Exercise 10.5 below. Then $\tau(E) = \lim_{n \to \infty} G_n = \tau$.

(c) Let $T_n(z)$ and $T_n^*(z)$ denote the nth Chebyshev polynomial for E and $E^* := h^{-1}(E)$, respectively, and put

$$m_n^* := \max_{z \in E^*} |T_n^*(z)|.$$

Then it is easy to check that

$$m_{dn}^* = \max_{z \in E^*} |T_{dn}^*(z)| \leq \max_{z \in E^*} |T_n(h(z))| = \max_{z \in E} |T_n(z)| = m_n.$$

This gives

$$(m_{dn}^*)^{1/(dn)} \leq (m_n^{1/n})^{1/d}$$

which, on letting $n \to \infty$, shows that $\tau(E^*) \leq \tau(E)^{1/d}$.

We must now prove this inequality in the other direction. For an arbitrary fixed $w_0 \in E$, we define $z_1, \ldots, z_d$ by $h(z) - w_0 = \prod_{j=1}^{d}(z - z_j)$ and define $z_1^*, \ldots, z_n^*$ in E^* by $T_n^*(z) := \prod_{k=1}^{n}(z - z_k^*)$. Then from the identity

$$\left| \prod_{j=1}^{d} \prod_{k=1}^{n} (z_j - z_k^*) \right| = \left| \prod_{k=1}^{n} \prod_{j=1}^{d} (z_k^* - z_j) \right|$$

we have

$$\left| \prod_{j=1}^{d} T_n^*(z_j) \right| = \left| \prod_{k=1}^{n} (h(z_k^*) - w_0) \right| = |q_n(w_0)|,$$

where

$$q_n(w) := \prod_{k=1}^{n} (w - h(z_k^*)).$$

Since $h(z_k^*) \in E$, we can deduce that

$$m_n \le \max_{w \in E} |q_n(w)| \le \left(\max_{z \in E^*} |T_n^*(z)| \right)^d = (m_n^*)^d,$$

showing that

$$\left(m_n^{1/n} \right)^{1/d} \le (m_n^*)^{1/n}$$

and thence that $\tau(E)^{1/d} \le \tau(E^*)$.

(d) We know that $\tau(E) \le \tau(E_\varepsilon)$. Given $\delta > 0$, we will show that $\tau(E_\varepsilon) \le \tau(E) + 2\delta$ for all small enough $\varepsilon > 0$. Put $\nu = n(n-1)/2$ and suppose that

$$G_n(E_\varepsilon) = \left| \prod_{\substack{j,k=1 \\ j<k}}^{n} (z_j^* - z_k^*) \right|^{1/\nu},$$

where $z_j^* = z_j + \mu_j \varepsilon \in E_\varepsilon$, with $z_j \in E$ and $|\mu_j| \le 1$. Then

$$G_n(E_\varepsilon)^\nu = \left| \prod_{\substack{j,k=1 \\ j<k}}^{n} (z_j - z_k + (\mu_j - \mu_k)\varepsilon) \right|$$

$$\le \prod_{\substack{j,k=1 \\ j<k}}^{n} (|z_j - z_k| + 2\varepsilon)$$

$$= \prod_{\substack{j,k=1 \\ j<k}}^{n} |z_j - z_k| + \prod_{\substack{j,k=1 \\ j<k}}^{n} (|z_j - z_k| + 2\varepsilon) - \prod_{\substack{j,k=1 \\ j<k}}^{n} |z_j - z_k|$$

$$\leq \prod_{\substack{j,k=1 \\ j<k}}^{n} |z_j - z_k| + (D + 2\varepsilon)^{\nu} - D^{\nu},$$

where D is the diameter of E. For the last line, Exercise 10.7 below is used. So since the sequence $G_n(E_\varepsilon)$ is monotonically decreasing, we have

$$\tau(E_\varepsilon) \leq (G_n^\nu + (D + 2\varepsilon)^\nu - D^\nu)^{1/\nu}.$$

On choosing n sufficiently large so that $G_n \leq \tau(E) + \delta$, we then have

$$\tau(E_\varepsilon) \leq ((\tau(E) + \delta)^\nu + (D + 2\varepsilon)^\nu - D^\nu)^{1/\nu}$$

$$\leq (\tau(E) + \delta)\left(1 + \frac{(D + 2\varepsilon)^\nu - D^\nu}{\delta^\nu}\right)^{1/\nu},$$

using $\tau(E) + \delta \geq \delta$. Then we can choose $\varepsilon > 0$ small enough so that

$$\left(1 + \frac{(D + 2\varepsilon)^\nu - D^\nu}{\delta^\nu}\right)^{1/\nu} \leq 1 + \frac{\delta}{\tau(E) + \delta},$$

giving $\tau(E_\varepsilon) \leq \tau(E) + 2\delta$, as we wanted.

(e) As earlier, let $V_n(E)$ be the maximum of $|V(z_1, \ldots, z_n)|$, defined in (10.2), as $z_1, \ldots, z_n$ range over E. For any particular z_j, this maximum occurs for z_j on the boundary of E, by the maximum principle. Hence $V_n(E) = V_n(\partial E)$, and so $\tau(E) = \tau(\partial E)$. $\square$

Exercise 10.2 Use the polynomial $h(z) := z^2 - 2$ to show that the interval $E := [-2, 2]$ has $\tau(E) = 1$. Deduce that an interval of length L has transfinite diameter $L/4$.

Exercise 10.3 Show that the derived set of E (its set of limit points) has transfinite diameter $\tau(E)$.

Problem 10.4 Describe the compact subsets E of the complex plane having $\tau(E) = 0$.

Exercise 10.5 Let $\{a_n\}$ be a sequence of positive numbers with finite limit $a > 0$. Show that then also $(a_1^1 a_2^2 \cdots a_n^n)^{\frac{2}{n(n+1)}} \to a$ as $n \to \infty$.

Exercise 10.6 (Tsuji [Tsu75, p. 72]) Suppose that a sequence of nonnegative numbers $a_1, a_2, \ldots,$ satisfies $a_{\ell+n} \leq a_\ell a_n$ for all $\ell, n \geq 1$. Show that the sequence $\{a_n^{1/n}\}$ converges.

Exercise 10.7 Let δ and $a_1, \ldots, a_n$ be positive numbers, with $A := \max_{j=1}^n a_j$. Show that

$$\prod_{j=1}^{n}(a_j + \delta) - \prod_{j=1}^{n} a_j \le (A + \delta)^n - A^n .$$

[The expansion of the left side in powers of δ has all positive terms, so an upper bound is obtained by replacing each a_j with A, giving $(A + \delta)^n - A^n$.]

Lemma 10.8 *Suppose that $E \in \mathbb{C}$ is compact and symmetric about the real axis. Then the polynomial $T_n(z)$ can be chosen to have real coefficients.*

Proof Let $\max_{z \in E} |T_n(z)| = m_n$. Then for any $z \in E$

$$\left|\tfrac{1}{2}(T_n + \overline{T_n})(z)\right| \le \tfrac{1}{2}\left(|T_n(z)| + |\overline{T_n}(z)|\right) = \tfrac{1}{2}\left(|T_n(z)| + |T_n(\overline{z})|\right) \le m_n ,$$

the final inequality using the fact that $\overline{z} \in E$. Hence, $\tfrac{1}{2}(T_n + \overline{T_n})(z) \in \mathbb{R}[z]$ is also a Chebyshev polynomial of degree n for E. $\square$

The next result will be needed for the proof of Langevin's theorem in Chap. 11.

Proposition 10.9 *Let S be a proper closed subset of the unit circle $\mathbb{T}$. Then $\tau(S) < 1$.*

Proof By rotating S about the origin—this does not affect its transfinite diameter—we can assume that it does not contain a neighbourhood of -1. Thus, it is contained in some closed arc, A say, of $\mathbb{T}$ subtending an angle $\theta < 2\pi$. Such an arc is known [Hay66, Lan86] to have transfinite diameter $\sin(\theta/4) < 1$. Since S is a subset of a rotation of such an arc, it too has $\tau(S) < 1$. $\square$

10.3   Application to Conjugate Sets of Algebraic Integers

Before we apply the transfinite diameter to conjugate sets of algebraic integers, we need some results involving integer vectors and integer polynomials.

Lemma 10.10 *Given a real $n \times n$ matrix B of positive determinant, there exists a nonzero vector $\mathbf{x} \in \mathbb{Z}^n$ such that the components of $B\mathbf{x} \in \mathbb{R}^n$ all have modulus at most $(\det B)^{1/n}$.*

Proof Let $t := (\det B)^{1/n}$. Then $B^{-1}[0, t]^n$ has volume 1 in $\mathbb{R}^n$, and is closed. By partitioning this set and translating the resulting pieces by different integer vectors, we can map it to $[0, 1)^n$, necessarily with some overlap. Thus, there exist two different vectors $\mathbf{q}_1, \mathbf{q}_2 \in [0, t]^n$ say, such that $B^{-1}\mathbf{q}_1 \equiv B^{-1}\mathbf{q}_2 \pmod 1$. So we have that $B^{-1}(\mathbf{q}_1 - \mathbf{q}_2) \in \mathbb{Z}^n$ and is nonzero. $\square$

Theorem 10.11 *Let E be a compact subset of $\mathbb{C}$, symmetric about the real axis, and with transfinite diameter $\tau(E) < 1$. Then there is a polynomial $P(z)$ with integer coefficients for which $|P(z)| < 1$ for all $z \in E$.*

Proof From Lemma 10.8, we can take an nth Chebyshev polynomial $T_n(z)$ for E to have real coefficients; it is monic of degree n. Also $T_0(z) \equiv 1$. Let $m_n := \max_{z \in E} |T_n(z)|$. We know that $m_n^{1/n} \to \tau(E) < 1$ as $n \to \infty$.

Given any polynomial $P(z) = p_0 + p_1 z + \cdots + p_n z^n \in \mathbb{C}[z]$, we want to express it as a sum $\lambda_0 T_0(z) + \lambda_1 T_1(z) + \cdots + \lambda_n T_n(z)$. To do this, first, write

$$z^k = T_k(z) + \alpha_{k-1}^{(k)} T_{k-1}(z) + \cdots + \alpha_0^{(k)} T_0(z) \quad (1 \le k \le n).$$

Then

$$P(z) = \sum_{k=0}^{n} p_k z^k$$

$$= \sum_{k=0}^{n} p_k \left(T_k(z) + \sum_{j=0}^{k-1} \alpha_j^{(k)} T_j(z) \right)$$

$$= \sum_{j=0}^{n} T_j(z) \left(p_j + \sum_{k=j+1}^{n} \alpha_j^{(k)} p_k \right).$$

So

$$\lambda_j = p_j + \sum_{k=j+1}^{n} \alpha_j^{(k)} p_k \quad (0 \le j \le n),$$

and hence also

$$m_j \lambda_j = m_j p_j + \sum_{k=j+1}^{n} m_j \alpha_j^{(k)} p_k \quad (0 \le j \le n).$$

This can be written in vector form as

$$\begin{pmatrix} m_0 \lambda_0 \\ \vdots \\ m_n \lambda_n \end{pmatrix} = B \begin{pmatrix} p_0 \\ \vdots \\ p_n \end{pmatrix},$$

where B is an upper-triangular $(n+1) \times (n+1)$ matrix of determinant $m_0 \cdots m_n$. Hence by Lemma 10.10, we can choose $p_0, p_1, \ldots, p_n$ to be integers, not all 0, such that

$$|m_j \lambda_j| \le (m_0 m_1 \cdots m_n)^{1/(n+1)} \quad (j = 0, \ldots, n).$$

Hence for $z \in E$

$$|P(z)| \leq |m_0\lambda_0| + \cdots + |m_n\lambda_n| \leq (n+1)(m_1m_2\cdots m_n)^{1/(n+1)}$$
$$= (n+1)\left((\tau_1^1\tau_2^2\cdots\tau_n^n)^{2/(n(n+1))}\right)^{n/2}.$$

Now since $\tau_n \to \tau(E) < 1$ as $n \to \infty$, we have, for any t with $\tau(E) < t < 1$, that from Exercise 10.5, $|P(z)| < (n+1)t^{n/2}$ for n sufficiently large, and hence that $|P(z)| < 1$ for n sufficiently large. $\qquad\square$

Exercise 10.12 For a compact subset E of $\mathbb{C}$, symmetric about the real axis, use the above proof to show that there is a sequence $P_j(z)$ of integer polynomials of degree n_j with maximum modulus m_j on E such that $\limsup m_j^{1/n_j} \leq \sqrt{\tau(E)}$.

We can now apply this result to conjugate sets.

Theorem 10.13 *Let E be a compact subset of the complex plane, symmetric about the real axis, and of transfinite diameter less than 1. Then E contains only finitely many conjugate sets of algebraic integers.*

Proof From Theorem 10.11, we know that there is a polynomial $P(z) \in \mathbb{Z}[z]$ with $|P(z)| < 1$ for all z in E. So for any conjugate set $\alpha_1, \ldots, \alpha_n$ of algebraic integers, we have $|\prod_{j=1}^n P(\alpha_j)| < 1$. But this is an integer, being a symmetric function of the conjugates α_j, so it must be zero. So each α_j is one of the (finite number of) zeros of P. $\qquad\square$

Thus, the monic irreducible factors of $P(z)$ that have all their zeros in E give a complete list of the minimal polynomials of conjugate sets of algebraic integers lying in E.

On the other hand, if we know that a compact set E contains infinitely many conjugate sets of algebraic integers then, by this result, its transfinite diameter must be at least 1. For instance, the circle $\mathbb{T} := \{z : |z| = 1\}$ contains the conjugate sets of all roots of unity, so $\tau(\mathbb{T}) \geq 1$. Then, applying Proposition 10.1(c) with $h(z) = z^2$, we have $h^{-1}(\mathbb{T}) = \mathbb{T}$, giving $\tau(\mathbb{T}) = 1$.

Similarly, taking a monic integer polynomial $P(z)$ and defining the lemniscate $L_P := \{z : |P(z)| = 1\}$, we see by the same argument that $\tau(L_P) = 1$.

Proposition 10.14 *For P a monic integer polynomial, the lemniscate $L_P = \{z : |P(z)| = 1\}$ contains infinitely many conjugate sets of algebraic integers.*

Proof The polynomials $P(z)^k - 1$ have all their zeros on L_P. Note that the polynomials
$$\frac{P(z)^k - 1}{P(z) - 1}$$

have no factors in common when k runs over all the primes. To see this, note that if $P(\alpha)^k = P(\alpha)^{k'} = 1$ then $1 = P(\alpha)^{\gcd(k,k')} = P(\alpha)$. Thus the irreducible factors of these polynomials give rise to infinitely many conjugate sets of algebraic integers on L_P. $\qquad\square$

Thus, one way of proving that a set E in the complex plane contains infinitely many conjugate sets of algebraic integers would be to prove that it contains such a lemniscate. We now do this for certain E having $\tau(E) > 1$.

Proposition 10.15 *Let $b(z)$ be a real monic polynomial of degree n, and $R > 1$. The filled-in lemniscate $|b(z)| \leq R$ contains a lemniscate $|P(z)| = 1$ for some monic $P(z) \in \mathbb{Z}[z]$.*

Proof First note that since (by scaling the unit circle) the circle $|z| = R$ has transfinite diameter R, we see from Proposition 10.1(c) that the lemniscate $|b(z)| = R$ has transfinite diameter $R^{1/n} > 1$.

We first need to find a lemniscate $|b_1(z)| = R_1$ say in $|b(z)| \leq R$ where $1 < R_1 < R$ and $b_1(z)$ is monic with rational coefficients. To do this, take $b(z)$ and perturb its coefficients (except the leading one) so that they are all rational with positive integer denominator N, where N is chosen large enough that the perturbations of the coefficients can be made small enough so that $||b(z)| - |b_1(z)|| < R - R_1$. Then for $|b_1(z)| = R_1$, we have

$$|b(z)| \leq ||b(z)| - |b_1(z)|| + |b_1(z)| < R,$$

so that the lemniscate $|b_1(z)| = R_2$ lies in $|b(z)| \leq R$.

We now write

$$b_1(z) = z^k + \frac{\gamma_1 z^{k-1} + \cdots + \gamma_k}{N},$$

put $\nu := (\mu k)! N^{\mu k}$ and consider $b_1(z)^\nu$. Here the integer μ is to be chosen later. One can readily check that the coefficients of $z^{k\nu}, z^{k\nu-1}, \ldots, z^{k\nu-k\mu}$ are all integers. Then we perturb the other coefficients of $b_1(z)^\nu$ to obtain an integer polynomial, as follows. There exist coefficients $\lambda_i^{(j)} \in [0, 1)$ so that

$$P_\mu(z) := b_1(z)^\nu + b_1(z)^{\nu-\mu-1} \left(\lambda_1^{(1)} z^{k-1} + \cdots + \lambda_k^{(1)} \right)$$
$$+ b_1(z)^{\nu-\mu-2} \left(\lambda_1^{(2)} z^{k-1} + \cdots + \lambda_k^{(2)} \right)$$
$$+ \quad \vdots \qquad\qquad \vdots \qquad\qquad \vdots$$
$$+ b_1(z)^0 \left(\lambda_1^{(\nu-\mu)} z^{k-1} + \cdots + \lambda_k^{(\nu-\mu)} \right)$$
$$= b_1(z)^\nu + \Delta_\mu(z),$$

say, is monic with integer coefficients. Now on $|z| = R_1$,

$$\left| \frac{\Delta_\mu(z)}{b_1(z)^\nu} \right| < \left(\frac{1}{R_1^{\mu+1}} + \cdots + \frac{1}{R_1^\nu} \right) B,$$

where $B := \left(\max_{|b_1(z)|=R_1} (1 + |z| + \cdots + |z|^{k-1}) \right)$

$$< \frac{B}{R_1^\mu (R_1 - 1)}$$

$$< \frac{1}{2} \quad \text{on taking } \mu \geq \mu_0 \text{ say.}$$

So, by Rouché's theorem, all zeros of $P_\mu(z)$ lie in $|b_1(z)| < R_1$. Also, on $|b_1(z)| = R_1$

$$|P_\mu(z)| \geq |b_1(z)|^\nu - |\Delta_\mu(z)| = R_1^\nu \left(1 - \frac{|\Delta_\mu(z)|}{R_1^\nu} \right) > \frac{R_1^\nu}{2} > \frac{R_1^\mu}{2} .$$

So $|P_\mu(z)| > 1$ on $|b_1(z)| = R_1$ for μ sufficiently large, and then the lemniscate $|P_\mu(z)| = 1$ lies in $|b_1(z)| \leq R_1$ and thus also in $|b(z)| \leq R$. We can therefore take $P = P_\mu$. $\qquad\square$

10.4   Integer Transfinite Diameters

We know from Proposition 10.1(b) that the transfinite diameter $\tau(E)$ of a compact set E can be defined using the Chebyshev polynomials for E. Recall that these are monic polynomials, and can be assumed to have real coefficients if $E \subset \mathbb{R}$. Polynomials with integer coefficients can be used to define two different variants of $\tau(E)$, at least for E real.

10.4.1   The Integer Transfinite Diameter

In this variant, the Chebyshev polynomials must have integer coefficients, but are not restricted to be monic. Otherwise, the definition is as in Proposition 10.1(b). We call such a polynomial of degree n an ***integer Chebyshev polynomial***, denoted by $T_{n,\mathbb{Z}}(z)$. For its maximum m_n on E, we see from Exercise 10.6 that $\lim_{n\to\infty} m_n^{1/n}$ exists. This limit is the ***integer transfinite diameter of*** E, denoted by $\tau_{\mathbb{Z}}(E)$. Clearly, $\tau_{\mathbb{Z}}(E) \geq \tau(E)$, and we see from Exercise 10.12 that $\tau_{\mathbb{Z}}(E) \leq \sqrt{\tau(E)}$.

The integer Chebyshev polynomials for the interval $[0, 1]$ have been much studied, since they have an interesting connection with the Prime Number Theorem. We now describe this, following [HS97]. The idea is due to Gelfond and Shnirelman from 1936. See Pritsker [Pri05a], as well as H. Montgomery's lecture [Mon94, Chap. 10] on the topic for a survey, as well as historical remarks and many references.

Let $d_n := \text{lcm}(1, 2, \dots, n)$. Now one form of the Prime Number Theorem is

$$\lim_{n\to\infty} \frac{\log d_n}{n} = 1 .$$

Then for any positive integer k (and with $T_{n,\mathbb{Z}}$, the integer Chebyshev polynomial for $[0, 1]$), we have that $\int_0^1 T_{n,\mathbb{Z}}^{2k}(x)\, dx$ is a positive integer multiple of d_{2kn+1}^{-1}. As this integral is at most m_n^{2k}, we have

$$\frac{\log(d_{2kn+1})}{2kn} \geq -\frac{2k\log(m_n)}{2kn} = -\frac{\log(m_n)}{n},$$

so that

$$\liminf_{k\to\infty} \frac{\log(d_k)}{k} \geq \log\left(\frac{1}{m_n^{1/n}}\right).$$

Now letting $n \to \infty$, we have

$$\liminf_{k\to\infty} \frac{\log(d_k)}{k} \geq \log\left(\frac{1}{\tau_{\mathbb{Z}}([0,1])}\right). \tag{10.4}$$

Lemma 10.16 *Suppose that $P(z)$ is an integer polynomial of degree n, and $Q(z) = az^k + \ldots$ is an integer polynomial of degree k having all its zeros in E. Suppose too that P and Q are coprime. Then*

$$m_P^{1/n} \geq |a|^{-1/k},$$

where m_P is the maximum modulus of $P(z)$ on E.

Proof Let $\beta_1, \ldots, \beta_k \in E$ be the zeros of $Q(z)$. Then because P and Q are coprime, their z-resultant is a nonzero integer, and thus of modulus at least 1. But from Appendix A, this resultant also has modulus

$$|a^n P(\beta_1) \cdots P(\beta_k)| \geq 1.$$

Hence $m_P^k \geq |a|^{-n}$, giving the result.   □

Following [FRS97], we say than an irreducible integer polynomial $Q(t) = at^k + \cdots$ is **critical for** E if $|a|^{-1/k} > \tau_{\mathbb{Z}}(E)$. Then for any polynomial $P(z)$ as in the above lemma for which $m_P^{1/n}$ is sufficiently close to $\tau_{\mathbb{Z}}(E)$, the lemma implies that Q must divide P. By this method, factors of P, for instance, for $P = T_{n,\mathbb{Z}}$, can be found.

Most of the work on $\tau_{\mathbb{Z}}(E)$ has concentrated on the case of E a real interval. For E an interval of length at least 4, it is known that $\tau_{\mathbb{Z}}(E) = \tau(E) = |E|/4$ [Gol69, p. 298]. However, $\tau_{\mathbb{Z}}(E)$ is not known exactly for any interval E of length less than 4. There are not even any conjectured values! For $E = [0, 1]$, quite tight bounds are known: Pritsker [Pri05b] showed that

$$0.4213 \leq \tau_{\mathbb{Z}}([0,1]) \leq 0.4232,$$

improving the bounds of earlier authors. Thus, (10.4) gives

$$\liminf_{k\to\infty} \frac{\log(d_k)}{k} \geq \log\left(\frac{1}{0.4232}\right) \geq 0.8599.$$

   Ten critical polynomials for the interval $[0, 1]$ are known ([FRS97]). For such polynomials $R(t)$ say, positive constants b_R are known such that the polynomials $T_{n,\mathbb{Z}}(t)$ are divisible by $R^{\lfloor b_R n \rfloor}$. For four of these polynomials, upper bounds b_R^+ for the largest possible value of b_R are also known. Borwein and Erdélyi [BE96] pointed out that critical polynomials $R(t)$ can often be produced from the minimal polynomials $Q(x)$ of small totally positive algebraic integers, via the transformation

$$R(t) := Q\left(\frac{t}{1-t}\right)(1-t)^{\deg Q}. \tag{10.5}$$

We apply this idea to some of the polynomials Q_j in Table 14.1 of Chap. 14. These polynomials have small Mahler measure. Let $R_0(t) := t - 1$, R_j ($j = 1, 2, 3, 4, 6,$ $8, 9$) defined from Q_j by (10.5) and $R_{10}(t) := 6t^2 - 6t + 1$. Then for N large, we put

$$R_N(t) := \prod_{j=0,1,2,3,4,6,8,9,10} R_j(t)^{\lfloor e_j N \rfloor},$$

where

$$[e_0, e_1, e_2, e_3, e_4, e_6, e_8, e_9, e_{10}] = [0.31784899, 0.31784899, 0.11621266,$$
$$0.03824029, 0.01501115, 0.00624421,$$
$$0.00575228, 0.00321130, 0.00119514].$$

Then [Fla95] $m_{R_N}^{1/\deg R_N} = 0.42353115$, so that any irreducible integer polynomial $at^d + \cdots$ with all zeros in $[0, 1]$ and $a^{-1/d} > 0.42353115$ is critical [FRS97]. This gives the ten critical polynomials of Table 10.1. In it, R_5 and R_6 are redefined as the two irreducible factors of (the original) R_6, and R_7 and R_8 are redefined as the two irreducible factors of R_8. The upper bounds for $b_{R_j}^+$ ($j = 0, 1, 2, 3$), as well as some lower bounds $b_{R_j}^-$, come from [Pri05b]. The corresponding critical values $a^{-1/k}$ for $R_j(t) = at^k + \cdots$ are denoted a_j.

Problem 10.17 (*open problem*) Does the interval $[0, 1]$ have infinitely many critical polynomials?

   Related to this question is a result of Pritsker [Pri05b, Theorem 1.8]. For closed intervals I of length less than 4, his result states that any infinite sequence $\{T_{n_j,\mathbb{Z}}(I)\}$ of integer Chebyshev polynomials has infinitely many distinct factors with integer coefficients occurring among them. Hare and Smyth [HS06, Prop. 2.2] showed that if an interval of length less than 4 has infinitely many critical polynomials $Q_j(t) = a_{d_j,j}t^{d_j} + \cdots + a_{0,i}$ then $\tau_{\mathbb{Z}}(I) = \inf a_{d_j,j}^{-1/d_j}$.
   It is natural to think it likely, as in [BE96], that all factors of the integer Chebyshev polynomials $T_{n,\mathbb{Z}}([0, 1])$ would have all zeros in $[0, 1]$. In fact, it is not the case: Habsieger and Salvy [HS97] showed that $T_{78,\mathbb{Z}}([0, 1])$ has an irreducible factor of degree 10 with four nonreal zeros. (These zeros have real parts in $[0, 1]$, and imaginary parts $\pm 0.0227 \cdots$.)

Table 10.1 The ten known critical polynomials R_j for $[0, 1]$

j	a_j	$b_{R_j}^-$	$b_{R_j}^+$	$R_j(t)$
0	1	0.31	0.34	$t - 1$
1	1	0.31	0.34	t
2	0.5	0.11	0.14	$2t - 1$
3	0.4472	0.035	0.057	$5t^2 - 5t + 1$
4	0.4309	0.001065		$29t^4 - 58t^3 + 40t^2 - 11t + 1$
5	0.4252	0.000232		$13t^3 - 20t^2 + 9t - 1$
6	0.4252	0.000232		$13t^3 - 19t^2 + 8t - 1$
7	0.4249	0.000136		$941t^8 - 3764t^7 + 6349t^6 - 5873t^5$ $+3243t^4 - 1089t^3 + 216t^2 - 23t + 1$
8	0.4237	0.000026		$31t^4 - 63t^3 + 44t^2 - 12t + 1$
9	0.4237	0.000026		$31t^4 - 61t^3 + 41t^2 - 11t + 1$

10.4.2 The Monic Integer Transfinite Diameter

In this variant of transfinite diameter, the Chebyshev polynomials must again have integer coefficients, and also, like the standard Chebyshev polynomials, be monic. Thus, for a compact subset $E \in \mathbb{C}$, the *nth monic integer Chebyshev polynomial* $T_{n,\mathbb{Z}}^M(z)$ has the smallest maximum m_n say, on E among monic integer polynomials of degree n. The *monic integer transfinite diameter of* E, $\tau_{\mathbb{Z}}^M(E)$ say, is then $\lim_{n \to \infty} m_n^{1/n}$. This limit exists, by Exercise 10.6. Clearly

$$\tau(E) \le \tau_{\mathbb{Z}}(E) \le \tau_{\mathbb{Z}}^M(E).$$

The study of the monic integer transfinite diameter, initiated by Borwein, Pinner and Pritsker in 2003 [BPP03], concentrated on the case of E being a real interval.

We summarise without proof some basic properties of $\tau_{\mathbb{Z}}^M(E)$.

Proposition 10.18 ([BPP03]) *Let $E \subset \mathbb{C}$ be compact, and symmetric about the real axis. Then*

(a) *If $\tau(E) \ge 1$, then $\tau_{\mathbb{Z}}^M(E) = \tau(E)$.*
(b) *If $\tau(E) < 1$, then $\tau_{\mathbb{Z}}^M(E) < 1$.*
(c) *If $E \subset F \subset \mathbb{C}$, then $\tau_{\mathbb{Z}}^M(E) \le \tau_{\mathbb{Z}}^M(F)$.*
(d) *For any monic integer polynomial P of degree n, we have*

$$\tau_{\mathbb{Z}}^M(P^{-1}(E)) = \tau_{\mathbb{Z}}^M(E)^{1/n}.$$

(e) *In fact, $\tau_{\mathbb{Z}}^M(E) = \inf_n m_n^{1/n}$.*

Exercise 10.19 ([BPP03]) Let $k \geq 2$. With the help of Proposition 10.18, prove the following.

(a) Show that

$$\tau_{\mathbb{Z}}^M \left([0, \tfrac{1}{k}]\right) = \tau_{\mathbb{Z}}^M \left([1 - \tfrac{1}{k}, 1]\right) = \tau_{\mathbb{Z}}^M \left([-\tfrac{1}{k}, \tfrac{1}{k}]\right) = \tfrac{1}{k}.$$

(b) Show that

$$\tau_{\mathbb{Z}}^M \left([-\tfrac{1}{\sqrt{k}}, \tfrac{1}{\sqrt{k}}]\right) = \sqrt{\tau_{\mathbb{Z}}^M \left([0, \tfrac{1}{k}]\right)} = \tfrac{1}{\sqrt{k}}.$$

(c) Show that

$$\tau_{\mathbb{Z}}^M \left([0, 1]\right) = \sqrt{\tau_{\mathbb{Z}}^M \left([0, \tfrac{1}{4}]\right)} = \tfrac{1}{2}.$$

(d) Show that

$$\tau_{\mathbb{Z}}^M \left([-1, 1]\right) = \sqrt{\tau_{\mathbb{Z}}^M \left([0, 1]\right)} = \tfrac{1}{\sqrt{2}}.$$

Related to (a) (for $k = 2$) and (c) is the result that $\tau_{\mathbb{Z}}^M \left([0, x]\right) = 1/2$ for $1/2 \leq x \leq 1.26$, while $\tau_{\mathbb{Z}}^M \left([0, 1.328]\right) > 1/2$ ([HS06, Theorem 9.2]).

For intervals E of length less than 4, it turns out that, in contrast to $\tau_{\mathbb{Z}}(E)$, sometimes $\tau_{\mathbb{Z}}^M(E)$ can be evaluated exactly. This is done with the help of Lemma 10.16, applied to polynomials $Q(z) = az^k + \cdots$ as in the lemma, with the extra constraint that $a > 1$. Such polynomials, being nonmonic, cannot be factors of any $P(z)$ of degree n with $\max_{z \in E} |P(z)|^{1/n} = \tau_{\mathbb{Z}}^M(E)$. Thus, $m_n^{1/n} \geq \sup_Q a^{-1/k}$ and so $\tau_{\mathbb{Z}}^M(E) \geq \sup_Q a^{-1/k}$. If this supremum, s say, is actually attained, and furthermore there is a monic polynomial of degree n with maximum modulus m satisfying $m^{1/n} = s$, then, by Proposition 10.18(e), $\tau_{\mathbb{Z}}^M(E) = s$.

It can be the case, however, that for certain intervals I the supremum s is attained, but there is no monic polynomial of degree n with maximum modulus m satisfying $m^{1/n} = s$. This is shown for $I = [-0.864, 0.517]$ in [HS06, Counterexample 2.1], where $s = 7^{-1/3}$. Thus, here we know only that $\tau_{\mathbb{Z}}^M(I) \geq 7^{-1/3}$.

10.4.2.1 The Monic Integer Transfinite Diameter of Farey Intervals

Now we consider $\tau_{\mathbb{Z}}^M(I)$, where I is a Farey interval $[a_1/b_1, a_2/b_2]$ with $a_1 b_2 - a_2 b_1 = 1$, of length $1/(b_1 b_2)$. Borwein, Pinner and Pritsker [BPP03] made the conjecture that $\tau_{\mathbb{Z}}^M(I) = \max(b_1^{-1}, b_2^{-1})$, and verified this conjecture for all such intervals with b_1 and b_2 at most 22. Also, in [HS06] the conjecture is verified for an infinite family of such intervals.

On choosing an integer n such that $a_j^n \equiv 1 \pmod{b_j}$ $(j = 1, 2)$, the monic integer polynomial

$$P(t) = t^n - \left(\frac{a_1^n - 1}{b_1}\right)(b_2 t - a_2)^{n-1} - \left(\frac{a_2^n - 1}{b_2}\right)(b_1 t - a_1)^{n-1}$$

has $P(a_j/b_j) = 1/b_j^n$ $(j = 1, 2)$. This shows that the two-point set

$$\{a_j/b_j \ (j = 1, 2)\}$$

has

$$\tau_{\mathbb{Z}}^{\mathrm{M}}(\{a_j/b_j \ (j = 1, 2)\}) = \max(b_1^{-1}, b_2^{-1}),$$

giving a lower bound for $\tau_{\mathbb{Z}}^{\mathrm{M}}([a_1/b_1, a_2/b_2])$. However $P(t)$ may have a larger maximum modulus than $\max(b_1^{-1}, b_2^{-1})$ on the Farey interval. Thus, in order to prove the conjecture a replacement polynomial without this drawback must be found. This is what was done in [BPP03] for Farey intervals with $\max(b_1, b_2) \leq 22$.

To prove the conjecture, of course, it would be sufficient to show for every Farey interval $[a_1/b_1, a_2/b_2]$ the existence of a monic integer polynomial of some degree N whose maximum on the interval occurred at one of its endpoints, say a_1/b_1.

10.5 Notes

Parts (a) and (b) of Proposition 10.1 are due to Fekete [Fek23]. The account here is based on that in Goluzin [Gol69, Chapter VII,§1]. For (c), see Fekete [Fek30b]. For (d), see Fekete [Fek30a]. For (e), see Ransford [Ran95, p. 152]. Lemma 10.10 is due to Minkowski—see [Fek23, p. 247]. Theorem 10.11 is due to Fekete [Fek23]. Proposition 10.15 is due to Fekete and Szegő [FS55].

10.6 Glossary

$G_n = G_n(E)$.   Let E be a compact subset of the complex plane. For a given positive integer n, define

$$G_n = G_n(E) = \max_{z_1, \ldots, z_n \in E} \left(\prod_{1 \leq j < k \leq n} |z_j - z_k| \right)^{2/n(n-1)}.$$

This maximum exists by compactness. Thus, G_n is the maximum of the geometric mean of the distances between n points in E

$\tau(E)$.   The transfinite diameter of E.

$\tau_{\mathbb{Z}}(E)$.   The integer transfinite diameter of E.

E_ε.   The ε-thickening of a set E.

V, V_n.   For a compact set E, and $z_1, \ldots, z_n \in E$, define

$$V(z_1, \ldots, z_n) := \prod_{1 \le j < k \le n} (z_j - z_k).$$

Then V_n is the maximum of $|V(z_1, \ldots, z_n)|$ as $z_1, \ldots z_n$ range over E (this maximum exists by compactness). Note that $G_n = V_n^{2/n(n-1)}$.

ε-thickening.　This is the set of all $z \in \mathbb{C}$ for which there is a point $y \in E$ with $|z - y| < \varepsilon$.

Chebyshev constant of E.　The limit $\lim_{n \to \infty} m_n^{1/n}$, where m_n is the maximum modulus on E of the nth Chebyshev polynomial for E, $T_n(z)$.

Chebyshev polynomials for E.　The nth Chebyshev polynomial for E, $T_n(z)$, is the monic polynomial in $\mathbb{C}[z]$ for which the maximum modulus on E is as small as possible.

critical.　An irreducible integer polynomial $at^d + \cdots$ is critical for a compact set E if $|a|^{-1/d} > \tau_{\mathbb{Z}}(E)$.

integer Chebyshev polynomial for E.　The nth integer Chebyshev polynomial for E, $T_{n,\mathbb{Z}}(z)$, is the polynomial with integer coefficients that makes the maximum modulus on E as small as possible. Note that there is no requirement that $T_{n,\mathbb{Z}}(z)$ should be monic.

integer transfinite diameter.　Let m_n be the maximum of $T_{n,\mathbb{Z}}$ on E. Then the integer transfinite diameter of E is $\tau_{\mathbb{Z}}(E) = \lim_{n \to \infty} m_n^{1/n}$.

monic integer Chebyshev polynomial for E.　The nth monic integer Chebyshev polynomial for E, $T_{n,\mathbb{Z}}^{M}(z)$, is the monic polynomial with integer coefficients that makes the maximum modulus on E as small as possible.

monic integer transfinite diameter.　Let m_n be the maximum of $T_{n,\mathbb{Z}}^{M}(z)$ on E. Then the monic integer transfinite diameter of E is $\tau_{\mathbb{Z}}^{M}(E) = \lim_{n \to \infty} m_n^{1/n}$.

transfinite diameter.　Let $E \subseteq \mathbb{C}$ be a compact set. The transfinite diameter of E, written $\tau(E)$, is the limit as $n \to \infty$ of the sequence $\{G_n(E)\}$.

Chapter 11
Restricted Mahler Measure Results

This chapter contains inequalities of various kinds for the Mahler measure, for restricted classes of polynomials.

11.1 Monic Integer Irreducible Noncyclotomic Polynomials

We first consider monic integer irreducible polynomials that are noncyclotomic. From Kronecker's first theorem (Theorem 1.3), we know that such polynomials have Mahler measure greater than 1.

Proposition 11.1 *Let $P(z)$ be a monic integer irreducible noncyclotomic polynomial of length L and degree d. Then*

$$M(P) \geq 2^{1/(4L)}. \tag{11.1}$$

Recall that the length of a polynomial is the sum of the moduli of its coefficients.

Proof Let α_j ($j = 1, \ldots, d$) be the zeros of P, and p be a prime. Then

$$|P(\alpha_j)| \leq L \max(1, |\alpha_j|)^d,$$

so that, on applying Dobrowolski's lemma A.23 we have

$$p^d \leq \prod_{i,j=1}^{d} |\alpha_i^p - \alpha_j| = \prod_{i=1}^{d} |P(\alpha_i^p)| \leq L^d M(P)^{pd}.$$

By Bertrand's postulate (see Chap. 3), we can fix p to be a prime in the range $2L < p < 4L$. Then

© Springer Nature Switzerland AG 2021
J. McKee and C. Smyth, *Around the Unit Circle*, Universitext,
https://doi.org/10.1007/978-3-030-80031-4_11

$$M(P) \geq \left(\frac{P}{L}\right)^{1/p} \geq 2^{1/(4L)} .$$

□

Next, we consider monic integer irreducible noncyclotomic polynomials that are the sum of k monomials; we state the following without proof.

Proposition 11.2 (Dobrowolski [Dob06, Prop. 1]) *Let $P(z)$ be a monic integer irreducible noncyclotomic polynomial of degree d that is the sum of k monomials. Then putting $k' = k/2$, we have*

$$M(P) \geq 1 + \frac{0.17}{2^{k'}k'!} . \tag{11.2}$$

For results of this kind where P is not assumed irreducible, see the chapter notes. Our next stated result concerns a very select class of polynomials.

Proposition 11.3 (Borwein, Dobrowolski, Hare and Mossinghoff [BHM04], [BDM07]) *Let $P(z)$ be an integer irreducible noncyclotomic polynomial of degree d, all of whose coefficients are odd. Then $M(P) \geq 5^{1/4} = 1.49538 \cdots$. Furthermore, if $P(z)$ is nonreciprocal then $M(P) \geq M(z^2 - z - 1) = (1 + \sqrt{5})/2$.*

The smallest known Mahler measure $M(P)$ of this kind with P reciprocal is

$$M(z^6 + z^5 - z^4 - z^3 - z^2 + z + 1) = 1.556030 \cdots , \qquad ([BHM04]).$$

11.2   Complex Polynomials That are Sums of a Bounded Number of Monomials

The main result of this section concerns general polynomials with complex coefficients. Let $P(z) = c_0 z^{m_0} + \cdots + c_n z^{m_n}$, where the c_i are complex numbers, the m_i are strictly increasing positive integers and $m_0 = 0$. Our aim is to obtain a lower bound for the Mahler measure $M(P)$ of P that is independent of the exponents m_i.

Theorem 11.4 *For a polynomial $P(z) = c_0 z^{m_0} + \cdots + c_n z^{m_n} \in \mathbb{C}[z]$ where $0 = m_0 < m_1 < \cdots < m_n$, and P not identically zero, we have*

$$M(P) \geq \max_{0 \leq j \leq n} \frac{|c_j|}{\binom{n}{j}} .$$

Proof We can suppose that all the coefficients c_i are nonzero, and use induction on n. For $n = 0$ (i.e., P consists of one monomial), the result is trivial. For $n = 1$, $P(z) = c_0 + c_1 z^{m_1}$, so that, by Lemma 2.9 and Jensen's theorem, $M(P) = \max(|c_0|, |c_1|)$.

We can now assume that $n \geq 2$, and that the result holds for all polynomials that are sums of at most n monomials. We define

$$Q(z) = z^{m_n} P(z^{-1}) = c_0 z^{m_n - m_0} + c_1 z^{m_n - m_1} + \cdots + c_n.$$

Then, by Lemma 2.9, $M(P) = M(Q)$. Further, by Proposition 1.17,

$$M(P') \le m_n M(P) \quad \text{and} \quad M(Q') \le m_n M(Q).$$

But P' and Q' are the sum of at most n monomials, so we can apply the induction hypothesis to each of them. Hence we obtain, for $j = 1, \ldots, n$ that

$$|c_j| m_j \le \binom{n-1}{j-1} M(P') \le m_n \binom{n-1}{j-1} M(P).$$

This also holds, trivially at $j = 0$. Similarly, for $j = 0, 1, \ldots, n-1$ we have

$$|c_j|(m_n - m_j) \le \binom{n-1}{n-1-j} M(P') \le m_n \binom{n-1}{j} M(P).$$

This holds trivially at $j = n$, too. Adding these inequalities we obtain for $j = 0, 1, \ldots, n$ that

$$\begin{aligned}
|c_j| m_n &= |c_j| m_j + |c_j|(m_n - m_j) \\
&\le m_n \binom{n-1}{j-1} M(P) + m_n \binom{n-1}{j} M(P) \\
&= m_n \binom{n}{j} M(P).
\end{aligned}$$

Hence $M(P) \ge |c_j|/\binom{n}{j}$ for $j = 0, 1, \ldots, n$. $\qquad\square$

11.3 Some Sets of Algebraic Numbers with the Bogomolov Property

A set $\mathcal{A}$ of algebraic numbers is said to have the **Bogomolov Property** if there is a (Bogomolov) constant $c > 0$ such that for all $\alpha \in \mathcal{A}$ either α is a root of unity or $h(\alpha) \ge c$. In terms of Mahler measure, this states that either $M(\alpha) = 1$ or $M(\alpha) \ge e^{c \deg(\alpha)}$.

11.3.1 Totally p-Adic Fields

Let $\mathbb{Q}^{tr}$ denote the field of totally real algebraic numbers. These are the real algebraic numbers all of whose conjugates are also real. Similarly, let $\mathbb{Q}^{tp}$ denote the field of totally p-adic algebraic numbers. These are the algebraic numbers whose minimal polynomials split into linear factors over $\mathbb{Q}_p$. The field $\mathbb{Q}^{tr}$ has the Bogomolov property. It is discussed in Chap. 14. Here, we show that the fields $\mathbb{Q}^{tp}$ also have the Bogomolov property. This has been known for some time—see Bombieri and Zannier [BZ01]. They proved this by finding upper bounds for $\liminf_{\alpha \in \mathbb{Q}^{tp}} h(\alpha)$. This result does not give a lower bound for the Bogomolov constant c. However, recently Pottmeyer has produced a delightfully simple argument to explicitly bound the Bogomolov constant c from below, for each prime p.

Theorem 11.5 *Let $\alpha \in \mathbb{Q}^{tp}$, with $\alpha^p \neq \alpha$. Then*

$$h(\alpha) > \begin{cases} \frac{\log(p/2)}{p+1} & \text{for } p > 2; \\ \frac{\log 2}{4} & \text{for } p = 2. \end{cases}$$

Proof First take $p > 2$. For the conjugates $\alpha = \alpha_1, \ldots, \alpha_d$ of α embedded in $\mathbb{Q}_p$, let r of them have $|\alpha|_p < 1$ and s of them have $|\alpha|_p > 1$. Because $h(\alpha^{-1}) = h(\alpha)$, we can assume that $r \geq s$. As α is not 0 or a root of unity, $r \geq 1$. Then from Corollary 1.46,

$$h(\alpha) = h(\alpha^{-1}) \geq \frac{r \log p}{d} .$$

Now $d - r - s$ of the α_i have $|\alpha_i|_p = 1$. For such α_i, we have from Fermat's little theorem that $|(\alpha_i^{p-1} - 1)^{-1}|_p \geq p$, so that

$$h((\alpha_i^{p-1} - 1)^{-1}) \geq \frac{(d - s - r) \log p}{d} \geq \frac{(d - 2r) \log p}{d} .$$

In the other direction, $h((\alpha_i^{p-1} - 1)^{-1}) < (p - 1)h(\alpha) + \log 2$ from Exercise 1.42, giving

$$h(\alpha) > \frac{\log(p/2)}{p - 1} - \frac{2r \log p}{(p - 1)d} .$$

Hence

$$h(\alpha) > \max \left(\frac{r \log p}{d}, \frac{\log(p/2)}{p - 1} - \frac{2r \log p}{(p - 1)d} \right) \geq \frac{\log(p/2)}{p + 1} ,$$

on choosing the least favourable value for r. This lower bound is positive as p is odd.

For the proof in the case $p = 2$, we refer to Pottmeyer [Pot18, Theorem 1.2]. The only different information used in that proof is that $a^2 \equiv 1 \pmod 8$ for odd integers a, rather than Fermat. □

Exercise 11.6 ([Pot18, Theorem 1.1]) Modify the above proof to show that for p an odd prime and α as in Theorem 11.5

$$h(\alpha) > \begin{cases} \frac{\log(p/2)}{p} & \text{if } \alpha \text{ is an algebraic integer;} \\ \frac{\log(p/2)}{p - 1} & \text{if } \alpha \text{ is an algebraic unit.} \end{cases}$$

Exercise 11.7 Show that for every prime p, the polynomial $z^p - z + p$ is irreducible, and splits completely over $\mathbb{Q}_p$. Deduce that the Bogomolov constant c for $\mathbb{Q}^{tp}$ cannot be greater than $(\log p)/p$.

11.3.2 Abelian Extensions of $\mathbb{Q}$

In this section, we show that the union $\mathbb{Q}^{ab} := \bigcup_{m=1}^{\infty} \mathbb{Q}(\omega_m)$ of all cyclotomic fields has the Bogomolov property. As usual, ω_m is a primitive mth root of 1. We shall prove the following.

Theorem 11.8 Let α be a nonzero algebraic number, not a root of unity, with $\mathbb{Q}(\alpha)$ an abelian extension of $\mathbb{Q}$. Then

$$h(\alpha) \geq \frac{\log(5/2)}{10} = 0.09163 \cdots .$$

The proof first requires a preliminary lemma which allows us to work entirely with algebraic integers. After another preliminary lemma, the theorem is then an immediate corollary of Proposition 11.11 that follows it.

The definition of valuations as used here is the alternative approach given at the end of Sect. A.3.

Lemma 11.9 Let α be a nonzero algebraic number, and p be a prime number. Then there is an algebraic integer $\beta \in \mathbb{Q}(\alpha)$ such that $\alpha\beta$ is also an algebraic integer and

$$\max(|\beta|_p, |\alpha\beta|_p) = 1 . \tag{11.3}$$

Proof Let $\alpha = \alpha_1, \ldots, \alpha_d$ be the conjugates of α, and $\mathcal{Q}$ be the set $\{p\} \cup \mathcal{Q}'$, where $\mathcal{Q}'$ is the set of those primes q for which $|\alpha_j|_q > 1$ for some j. Then the Strong Approximation Theorem [CF10, Ch. II, Sect. 15, p. 67] shows the existence of some $\beta \in \mathbb{Q}(\alpha)$ such that, for β_j the conjugate of β corresponding to α_j,

$$|\beta - 1|_p < 1 \qquad \qquad \text{if } \ |\alpha|_p \leq 1 , \tag{11.4}$$

$$|\beta_j - \alpha_j^{-1}|_q < |\alpha_j|_q^{-1} \quad \text{if } \ |\alpha_j|_q > 1 \ (j = 1, \ldots, d) \ (q \in \mathcal{Q}), \tag{11.5}$$

with $|\beta_j|_q \leq 1$ for all other pairs (j, q), with q any prime. Then from (11.5), we have $|\beta_j|_p = |\alpha_j|_p^{-1}$ for $|\alpha_j|_p > 1$, while $|\beta|_p = 1$ for $|\alpha|_p \leq 1$ from (11.4). Hence, β is an algebraic integer, and (11.3) holds. $\qquad \square$

Lemma 11.10 Let $\gamma \in \mathbb{Z}[\omega_m]$ for some m. Then for each prime p, there is an automorphism $\sigma \in \mathrm{Gal}(\mathbb{Q}(\omega_m)/\mathbb{Q})$ such that

$$p \mid (\gamma^p - \sigma\gamma) \qquad\qquad\qquad if\ p \nmid m, \qquad (11.6)$$
$$p \mid (\gamma^p - \sigma\gamma^p) \qquad\qquad\qquad if\ p \mid m. \qquad (11.7)$$

Moreover, in the case $p \mid m$, if $\sigma\gamma^p = \gamma^p$ then there is a root of unity $\omega \in \mathbb{Q}(\omega_m)$ such that $\omega\gamma$ is contained in $\mathbb{Q}(\omega_{m/p})$.

Proof We have $\gamma = f(\omega_m)$ for some $f \in \mathbb{Z}[z]$.

First consider the case $p \nmid m$. Define σ by $\sigma\omega_m := \omega_m^p$. Then

$$\gamma^p \equiv f(\omega_m^p) \equiv f(\sigma\omega_m) \equiv \sigma\gamma \quad (\mathrm{mod}\ p).$$

For the case $p \mid m$, note that the Galois group $G := \mathrm{Gal}(\mathbb{Q}(\omega_m)/\mathbb{Q}(\omega_{m/p}))$ is cyclic of order p or $p - 1$ according to whether $p^2 \mid m$ or not. We can identify G with the subgroup of $\mathrm{Gal}(\mathbb{Q}(\omega_m)/\mathbb{Q})$ that fixes $\mathbb{Q}(\omega_{m/p})$ pointwise. Choose a generator σ of G, taken to be in $\mathrm{Gal}(\mathbb{Q}(\omega_m)/\mathbb{Q})$. Then $\sigma\omega_m = \omega_p\omega_m$ for some primitive pth root of unity ω_p, so that

$$\gamma^p \equiv f(\omega_m^p) \equiv f(\sigma\omega_m^p) \equiv \sigma\gamma^p \quad (\mathrm{mod}\ p).$$

Finally, suppose that $\sigma\gamma^p = \gamma^p$. Then $\sigma\gamma = \omega_p^k\gamma$ for some integer k. Then

$$\sigma\left(\frac{\gamma}{\omega_m^k}\right) = \frac{\omega_p^k\gamma}{\sigma(\omega_m^k)} = \frac{\omega_p^k\gamma}{(\omega_p\omega_m)^k} = \frac{\gamma}{\omega_m^k}.$$

Thus, for $\omega := \omega_m^{-k}$, we have that $\omega\gamma$ belongs to the fixed field $\mathbb{Q}(\omega_{m/p})$ of σ. □

Proposition 11.11 *Let p be an odd prime, and α be a nonzero element of $\mathbb{Q}(\omega_m)$, not a root of unity. Then*

$$h(\alpha) \geq \begin{cases} \frac{\log(p/2)}{p+1} & if\ p \nmid m; \\ \frac{\log(p/2)}{2p} & if\ p \mid m. \end{cases}$$

Proof For our given p and α, let β be chosen as in Lemma 11.9, and σ be as in Lemma 11.10. Note that then

$$|\beta|_p = \max(1, |\alpha|_p)^{-1}. \qquad (11.8)$$

Assume first that $p \nmid m$. Applying Lemma 11.10 to the integers $\alpha\beta$ and β, we have

$$|(\alpha\beta)^p - \sigma(\alpha\beta)|_p \leq \frac{1}{p} \quad \text{and} \quad |\beta^p - \sigma\beta|_p \leq \frac{1}{p}.$$

Then

$$|\alpha^p - \sigma\alpha|_p = |\beta|_p^{-p} |(\alpha\beta)^p - \sigma(\alpha\beta) + (\sigma\beta - \beta^p)\sigma\alpha|_p$$

$$\leq |\beta|_p^{-p} \max(|(\alpha\beta)^p - \sigma(\alpha\beta)|_p, |\sigma\beta - \beta^p|_p|\sigma\alpha|_p)$$

$$\leq \tfrac{1}{p} \max(1, |\alpha|_p)^p \max(1, |\sigma\alpha|_p), \tag{11.9}$$

the last line using (11.8).

We now apply the Product Rule A.15 to $\gamma := \alpha^p - \sigma\alpha$, known to be nonzero by Lemma A.26 of Appendix A. We obtain, using all primes q and the conjugates α_i ($i = 1, \ldots, d$) of α, and also (11.9), that

$$0 = \sum_i \log |(\alpha^p - \sigma\alpha)_i|_p + \sum_{q \neq p} \sum_i \log |(\alpha^p - \sigma\alpha)_i|_q + \sum_i \log |(\alpha^p - \sigma\alpha)_i|$$

$$\leq \sum_i (p \log_+ |\alpha_i|_p + \log |(\sigma\alpha)_i|_p - \log p) + \sum_{q \neq p} \sum_i (p \log |\alpha_i|_q + \log |(\sigma\alpha)_i|_q)$$

$$+ \sum_i (p \log_+ |\alpha_i| + \log_+ |(\sigma\alpha)_i| + \log 2)$$

$$\leq \sum_q \left(\sum_i (p+1) \log_+ |\alpha_i|_q \right) - d \log p + \sum_i (p+1) \log_+ |\alpha_i| + d \log 2$$

$$= d \left((p+1)h(\alpha) - \log(p/2) \right).$$

Here, we have also used the simple inequality

$$x + y \leq 2 \max(1, x) \max(1, y) \quad \text{for } x, y > 0,$$

as well as Corollary 1.46. Hence

$$h(\alpha) \geq \frac{\log(p/2)}{p+1}.$$

The case $p \mid m$ is similar, although an inductive argument is needed. From Lemma 11.10, we have

$$|(\alpha\beta)^p - \sigma(\alpha\beta)^p|_p \leq \tfrac{1}{p} \quad \text{and} \quad |\beta^p - \sigma\beta^p|_p \leq \tfrac{1}{p}.$$

Then

$$|\alpha^p - \sigma\alpha^p|_p = |\beta|_p^{-p} |(\alpha\beta)^p - \sigma(\alpha\beta)^p + (\sigma\beta^p - \beta^p)\sigma\alpha^p|_p$$

$$\leq |\beta|_p^{-p} \max(|(\alpha\beta)^p - \sigma(\alpha\beta)^p|_p, |\sigma\beta^p - \beta^p|_p|\sigma\alpha^p|_p)$$

$$\leq \tfrac{1}{p} \max(1, |\alpha|_p)^p \max(1, |\sigma\alpha|_p)^p, \tag{11.10}$$

using (11.8) again. Furthermore, we can assume that $\alpha^p \neq \sigma\alpha^p$. Otherwise, by Lemma 11.10 there is a root of unity ω in $\mathbb{Q}(\omega_m)$ such that $\omega\alpha$ is contained in

$\mathbb{Q}(\omega_{m/p})$. Since $h(\alpha) = h(\omega\alpha)$ by Exercise 1.42, we either have recursively that $h(\alpha) \geq \log(p/2)/(2p)$ or, from the first case, $h(\alpha) \geq \log(p/2)/(p+1)$, depending on whether or not $p \mid m/p$. This time, the product rule gives us

$$0 \leq \sum_i (p\log_+ |\alpha_i|_p + p\log|(\sigma\alpha)_i|_p - \log p) + \sum_{q \neq p}\sum_i p(\log|\alpha_i|_q + \log|(\sigma\alpha)_i|_q)$$

$$+ \sum_i (p\log_+ |\alpha_i| + p\log_+ |(\sigma\alpha)_i| + \log 2)$$

$$\leq \sum_q \left(\sum_i 2p\log_+ |\alpha_i|_q\right) - d\log p + \sum_i 2p\log_+ |\alpha_i| + d\log 2$$

$$\leq d\left((2p)h(\alpha) - \log(p/2)\right),$$

and hence

$$h(\alpha) \geq \frac{\log(p/2)}{2p}.$$

□

Then Theorem 11.8 follows by choosing $p = 5$ in the proposition.

11.3.3 Langevin's Theorem

Theorem 11.12 (Langevin [Lan86]) *Let V be a closed subset of $\mathbb{C}$* **not** *containing the whole of the unit circle $\mathbb{T}$. Then there is a constant $C_V > 1$ such that, for all conjugate sets of algebraic integers $\{\alpha = \alpha_1, \ldots, \alpha_d\}$ lying in V we have*

$$M(\alpha) = 1 \quad \text{if } \alpha \text{ is cyclotomic, or } \alpha = 0;$$
$$M(\alpha) \geq C_V^d \quad \text{otherwise}.$$

The inequality (14.17), a result of Schinzel [Sch73], can be interpreted as the particular case $V = \mathbb{R}$ of this result.

Proof First note that since any conjugate set of algebraic integers is symmetric about the real axis, if it belongs to V then it also belongs to its complex conjugate $\overline{V}$ and hence to $V \cap \overline{V}$. Thus, we can assume that V is symmetric about the real axis. Further, $V \cap \mathbb{T}$ is closed, and is not the whole of $\mathbb{T}$. Thus by Proposition 10.9, its transfinite diameter is less than 1. Hence by Theorem 10.11, there is a nonzero polynomial $P(z) \in \mathbb{Z}[z]$ such that $|P(z)| < 1$ for all $z \in V \cap \mathbb{T}$. We can clearly assume that $z \nmid P(z)$. Then, for the reciprocal polynomial $P^*(z)$ of $P(z)$, the self-reciprocal polynomial $R(z) := P(z)P^*(z)$ is also of modulus less than 1 on $V \cap \mathbb{T}$.

Now put $V_1 := \{z \in V \cup V^{-1} : |z| \leq 1\}$. We next claim that there is an integer a and a real number $m < 1$ such that for all $z \in V_1$ we have $|z^a R(z)| \leq m$.

To prove the claim, let $W := V_1 \cap \{z \in \mathbb{C} : |R(z)| \geq 1\}$, a closed set. Since $|R(z)| < 1$ for $z \in V_1 \cap \mathbb{T}$, the set $W \cap \mathbb{T}$ is empty, and so the maximum modulus of $z \in W$ is r say, where $r < 1$. Let M be the maximum of $|R(z)|$ for $z \in W$. Then for any $m < 1$, we have for a sufficiently large that

$$|z^a R(z)| \leq r^a M < m < 1 \qquad \text{for all } z \in W . \tag{11.11}$$

For z in the closure of $V_1 \setminus W = \{z \in V_1 : |R(z)| < 1\}$, the maximum of $|R(z)|$ is attained for z on its boundary. If this maximum is 1, then $z \in W$ and so (11.11) holds for all $z \in V_1$. Otherwise the maximum is say $m' < 1$, in which case (11.11) holds for all $z \in V_1$, albeit with m replaced by $\max(m, m')$. This proves the claim.

Let α be a nonzero algebraic integer with conjugates $\alpha_1 = \alpha, \alpha_1, \ldots, \alpha_d$ all lying in V, and minimal polynomial $P_\alpha(z)$. We now separate two cases. If $P_\alpha(z) \mid R(z)$ then either $M(\alpha) = 1$ (when α is a root of unity) or else $M(\alpha) > 1$ and

$$M(\alpha)^{1/\deg \alpha} \geq \min_{\substack{\text{all such } \alpha}} M(\alpha)^{1/\deg \alpha} = C_V' \tag{11.12}$$

say, where $C_V' > 1$.

On the other hand, if $P_\alpha(z) \nmid R(z)$, then since $R(z) = z^{\deg R} R(1/z)$,

$$
\begin{aligned}
1 &\leq \left| \prod_{i=1}^{d} \alpha_i^a R(\alpha_i) \right| \\
&= \left| \prod_{\substack{i=1 \\ |\alpha_i| \leq 1}}^{d} \alpha_i^a R(\alpha_i) \right| \cdot \left| \prod_{\substack{i=1 \\ |\alpha_i| > 1}}^{d} \alpha_i^a R(\alpha_i) \right| \\
&= \left| \prod_{\substack{i=1 \\ |\alpha_i| \leq 1}}^{d} \alpha_i^a R(\alpha_i) \right| \cdot \left| \prod_{\substack{i=1 \\ |\alpha_i| > 1}}^{d} \alpha_i^{-a} R(\alpha_i^{-1}) \right| \cdot \prod_{\substack{i=1 \\ |\alpha_i| > 1}}^{d} |\alpha_i|^{2a + \deg R} \qquad (11.13) \\
&\leq m^d M(\alpha)^{2a + \deg R} ,
\end{aligned}
$$

where for the final inequality we have used the fact that $\alpha_i \in W$ if $|\alpha_i| \leq 1$ and $\alpha_i^{-1} \in W$ if $|\alpha_i| > 1$. Hence in this case

$$M(\alpha)^{1/d} \geq \left(\frac{1}{m} \right)^{2a + \deg R} = C_V'' ,$$

say. Combining the two cases, we see that the result holds with $C_V := \min(C_V', C_V'')$.
$\qquad \square$

We remark that if it happens that $C_V = C_V' = M(\beta)^{1/\deg \beta}$ where β is an algebraic integer lying with its conjugates in V, then the constant C_V is best possible. In Sect. 14.8, we give some examples of this.

11.4   The Height of Zhang and Zagier and Generalisations

For an algebraic number α, we define its ***Zhang–Zagier height*** to be $h(\alpha) + h(1 - \alpha)$, where h is, as usual, the Weil height. Zhang [Zha92] proved that there is a positive constant c such that there are only finitely many α with $h(\alpha) + h(1 - \alpha) = 0$, and that $h(\alpha) + h(1 - \alpha) \geq c$ for all other α. Zagier [Zag93] found the best value of c, in the following result.

Theorem 11.13 *For all algebraic numbers $\alpha \neq 0$, 1 or $\frac{1}{2}(1 \pm \sqrt{-3})$ we have*

$$h(\alpha) + h(1 - \alpha) \geq \tfrac{1}{2} \log \left(\tfrac{1}{2}(1 + \sqrt{5}) \right), \tag{11.14}$$

with equality if and only if one of α and $1 - \alpha$ is a primitive 10th root of unity.

The proof is based on the following lemma from [Zag93].

Lemma 11.14 *For p prime and $z \in \overline{\mathbb{Q}}_p$ we have*

$$\log_+ |z|_p + \log_+ |1 - z|_p \geq \tfrac{\sqrt{5}-1}{2\sqrt{5}} \log |z^2 - z|_p + \tfrac{1}{2\sqrt{5}} \log |z^2 - z + 1|_p, \tag{11.15}$$

while for $z \in \mathbb{C}$ the corresponding inequality is

$$\log_+ |z| + \log_+ |1 - z| \geq \tfrac{\sqrt{5}-1}{2\sqrt{5}} \log |z^2 - z| + \tfrac{1}{2\sqrt{5}} \log |z^2 - z + 1|$$
$$+ \tfrac{1}{2} \log \left(\tfrac{1}{2}(1 + \sqrt{5}) \right). \tag{11.16}$$

The second inequality is an equality if and only if z or $1 - z$ is a primitive 10th root of unity (i.e., $e^{\pm \pi i/5}$ or $e^{\pm 3\pi i/5}$.)

Exercise 11.15 (Zagier [Zag93]) Prove Lemma 11.14. (For (11.15), separate the cases $|z|_p > 1$ and $|z|_p \leq 1$. For (11.16), show using the Maximum Modulus Principle (Theorem C.3) that the maximum of the difference of the right and left sides of (11.16) occurs on one of the circles $|z| = 1$ and $|1 - z| = 1$, and that by symmetry it is enough to consider only $z = e^{i\theta}$ for $0 \leq \theta \leq \pi$. Separate into two cases $0 \leq \theta \leq \pi/3$ and $\pi/3 \leq \theta \leq \pi$.)

Proof (of Theorem 11.13) Let α be an algebraic number of degree d. Recall that we obtain d p-adic valuations $|\alpha|_p$ for each prime p by embedding the α_i into $\overline{\mathbb{Q}}_p$, and d valuations $|\alpha|$ by embedding the α_i into $\mathbb{C}$. Then, for $\alpha \neq 0$, the Product Rule (Proposition A.15) tells us that for all but finitely many primes p these valuations are all 1, and that the product of all the valuations of α is 1.

We now put $z := \alpha$, where α is any algebraic number such that none of z, $1 - z$ or $z^2 - z + 1$ is 0. This means that the product rule can be applied to these terms, and that $\alpha = 0$, 1 and $\frac{1}{2}(1 \pm \sqrt{-3})$ are excluded. We then sum (11.15) over all p where not all terms in (11.15) are zero, and over the d p-adic valuations for such p.

We add to this sum the sum of (11.16) over the d complex valuations. Finally, on dividing by d and using the definition (1.15) of Weil height, we obtain (11.14). □

Zagier also showed that there was a gap beyond c, that is a number $c' > c$ so that if $h(\alpha) + h(1 - \alpha)$ is greater than c then it is greater than c'. Doche [Doc01a] showed that one could take $c' = 0.2482474$. He also showed [Doc01b] that the smallest limit point of the spectrum of positive values of $h(\alpha) + h(1 - \alpha)$ lies in the interval $[0.2482474, 0.2544368]$. Thus, there are only a finite number of values less than 0.2482474 in this spectrum. This means that the second smallest value in the spectrum exists and is isolated; however, it is currently unknown.

Problem 11.16 (*open problem*) Find the second smallest positive value of $h(\alpha) +$ $h(1 - \alpha)$ among all algebraic numbers α.

Theorem 11.13 can be reformulated as stating that the Mahler measure of any irreducible nonconstant polynomial in $\mathbb{Z}[z^2 - z]$, apart from the four polynomials $\pm(z^2 - z)$ and $\pm(z^2 - z + 1)$, has Mahler measure at least $\left(\frac{1}{2}(1 + \sqrt{5})\right)^{d/2}$. Thus, the theorem is a restricted Mahler measure result.

Rhin and Smyth [RS97] generalised Zagier's result by replacing polynomials in $z^2 - z$ by polynomials in $T(z)$, where $T(z) \in \mathbb{Z}[z]$.

Theorem 11.17 ([RS97, Theorem 2]) *Let $T(z) \in \mathbb{Z}[z]$ be a fixed polynomial, of degree $d \geq 2$, and be divisible by z but not equal to z^d or $-z^d$. Let $\mathcal{A}_T$, denote the set of all algebraic numbers α whose minimal polynomial is of the form $P(T(z))$, where $P(z) \in \mathbb{Z}[z]$ is of degree at least 2. Then for all $\alpha \in \mathcal{A}_T$*

$$\overline{M}(\alpha) \geq c_T := 1 + \frac{1}{d - d_0 + 4dL(T)}.$$

Here, z^{d_0} is the highest power of z dividing $T(z)$, and $L(T)$ is the length of T (the sum of the absolute values of its coefficients.).

11.5 The Weil Height of α When $\mathbb{Q}(\alpha)/\mathbb{Q}$ is Galois

Suppose that α is an algebraic number of degree d such that $h(\alpha) > 0$ (so α is not 0 or a root of unity). Further, suppose $\mathbb{Q}(\alpha)$ is a Galois extension of $\mathbb{Q}$, so that all conjugates α_j $(j = 1, \ldots, d)$ of α lie in $\mathbb{Q}(\alpha)$. Here we state, without proof, recent work of Amoroso and Masser concerning such numbers.

Theorem 11.18 (Amoroso and Masser [AM16]) *If $\mathbb{Q}(\alpha)$ is Galois with $h(\alpha) > 0$, then for any $\varepsilon > 0$ there is a constant $C(\varepsilon) > 0$ such that $h(\alpha) > C(\varepsilon)/d^\varepsilon$.*

Example 11.19 Let $n \geq 2$ and β be a zero of $z^n - z - 1$, known to be irreducible, by Exercise 12.4, and with Galois group the full symmetric group $\mathfrak{S}_n$ on n symbols.

Define $\alpha := \beta_1^1 \beta_2^2 \cdots \beta_{n-1}^{n-1}$, where $\beta_1 = \beta, \beta_2, \ldots, \beta_n$ are the conjugates of β. The following table gives $h(\alpha)$ for $n \leq 9$.

n	$d = n!$	$M(\alpha)^{1/d}$	$h(\alpha) = \log(M(\alpha)^{1/d})$
2	2	1.2720196495	0.2406059129
3	6	1.1509639252	0.1405997869
4	24	1.2428334720	0.2173938309
5	120	1.2292495215	0.2064038385
6	720	1.2846087150	0.2504541700
7	5040	1.2833028970	0.2494371427
8	40320	1.3243452986	0.2809182234
9	362880	1.3307248410	0.2857237870

This example suggests that perhaps more is true: that there may be a positive constant c such that $h(\alpha) \geq c$ provided that $h(\alpha) > 0$. In fact, for α of the special kind as in the example, even more *is* true.

Theorem 11.20 (Amoroso [Amo18]) *Let β be an algebraic unit of degree n, with conjugates $\beta_1 = \beta, \beta_2, \ldots, \beta_n$, and Galois group $\mathfrak{S}_n$. Let $a_1, \ldots, a_n$ be distinct integers, and define $\alpha := \beta_1^{a_1} \beta_2^{a_2} \cdots \beta_n^{a_n}$. Then $\mathbb{Q}(\alpha) = \mathbb{Q}(\beta_1, \ldots, \beta_n)$ and as $n \to \infty$*

$$h(\alpha) \geq (1 + o(1))\sqrt{\frac{n}{8\pi}} \log M(\beta).$$

In Example 11.19 above, we have $M(\beta) \geq M(z^3 - z - 1) = 1.3247 \cdots$ (indeed, by Proposition 2.17, $M(\beta) \to M(y + z + 1) = 1.38135 \cdots$ as $n \to \infty$), so that $h(\alpha) \to \infty$ as $n \to \infty$. Amoroso conjectures that this holds more generally.

Conjecture 11.21 (Amoroso [Amo18]) *Let $\alpha \in \overline{\mathbb{Q}}$ be the generator of a Galois extension of $\mathbb{Q}$ of degree $n!$, with Galois group $\mathfrak{S}_n$. Then $h(\alpha) \to \infty$ as $n \to \infty$.*

11.6 Notes

Proposition 11.1 is due to Mignotte [Mig78, Prop.5].

Proposition 11.2 still holds, albeit with a smaller lower bound, when the restriction that P be irreducible is removed. Dobrowolski, Lawton and Schinzel [DLS83] first gave such a bound. This was later improved by Dobrowolski [Dob91] to

$$M(P) \geq 1 + \frac{1}{13911} e^{-2.27k^k}.$$

Theorem 11.4 is due to Akhtari and Vaaler [AV19], improving a slightly weaker lower bound of the same kind by Dobrowolski and Smyth [DS17].

The proof of Theorem 11.5 is from Pottmeyer [Pot18] . Of course, examples of specific algebraic numbers α with $h(\alpha)$ small give upper bounds for the greatest lower bound for the constant c. One instance of this is some recent work by Emerald Stacy on computing minimal heights for totally p-adic cubic polynomials [Sta20].

In the unit case of Exercise 11.6, Dubickas and Mossinghoff [DM05, eqn. (14)] had earlier given a slightly larger lower bound than that given here. (When $p = 2$, corresponding lower bounds are $\frac{2}{5} \log 2$ (integer case) and $\log 2$ (unit case), again due to Pottmeyer [Pot18, Theorem 1.2].)

Theorem 11.8 and its proof are due to Amoroso and Dvornicich [AD00]. In fact, they also proved the result with a better lower bound:

$$h(\alpha) \geq \frac{\log 5}{12} = 0.1931 \cdots .$$

The proof of this requires detailed considerations at the prime 2.

Dubickas and Smyth [DS01b] applied Theorem 11.12 to the annulus $V(R, \gamma) := \{z \in \mathbb{C} \mid R^{-\gamma} < |z| < R\}$, where $R > 1$ and $\gamma > 0$, proving that the best constant $C_{\mathbb{C}\backslash V(R,\gamma)} > 1$ is $R^{\gamma/(1+\gamma)}$.

Zagier's work was motivated by a far-reaching result of Zhang [Zha92] (see also [Wal00, p. 103]) for curves on a linear torus. He proved that, apart from curves of the type $x^i y^j = \omega$, where $i, j \in \mathbb{Z}$ and ω is a root of unity, for all other curves there is a constant $c > 0$ such that the curve has only finitely many algebraic points (x, y) with $h(x) + h(y) \leq c$. Zagier's result was for the curve $x + y = 1$.

Theorem 11.13 is, as we have seen, proved by the use of an auxiliary function defined as the difference between the right and left sides of (11.16), and which is bounded on $\mathbb{C}$. It is thus akin to the problems and methods described in Chap. 14. In extending Theorem 11.13 as described above, Doche also used an auxiliary function method.

Noticing that Zagier's theorem 11.13 has the same lower bound (after taking logs) as Schinzel's result (14.17) for totally real α, Samuels [Sam06] showed that the same lower bound holds for a more general height function, including those of both Zagier and Schinzel. His proof is based on [BZ97].

Dresden [Dre98] and later van Ittersum [vI17] generalised Theorem 11.13 in a different way. They regarded the set $\{\alpha, 1 - \alpha\}$ as the group of Möbius transformations $\{x, 1 - x\}$ acting on α by substitution, and extended the result to other such groups. Thus, for the group $\{x, 1/(1 - x), 1 - 1/x\}$ Dresden proved that $h(\alpha) + h(1/(1 - \alpha)) + h(1 - 1/\alpha) \geq \log|\beta|$, where β is a zero of largest absolute value of $(z^2 - z + 1)^3 - (z^2 - z)^2$, and that equality occurs when $\alpha = \beta$. He also states a result of a similar type for a 4-element group. This idea was extended by van Ittersum to other finite substitution groups.

The proof of Theorem 11.17 uses a general result of Beukers and Zagier [BZ97] on heights of points on projective hypersurfaces.

In 1999, Amoroso and David [AD99] answered Lehmer's question positively for algebraic integers α with $\mathbb{Q}(\alpha)$ Galois. This result has been superseded by Theorem 11.18 which, in terms of Mahler measure, says that $M(\alpha) > \exp(C(\varepsilon)d^{1-\varepsilon})$.

11.7 Glossary

$\log_+ x$.   This is the maximum of $\log x$ and 0, where x is a positive real number.

$\mathfrak{S}_n$.   The symmetric group on n symbols.

Bogomolov Property.   Let $\mathcal{A}$ be a set of algebraic numbers. If there is a constant $c > 0$ such that every $\alpha \in \mathcal{A}$ that is not a root of unity satisfies $h(\alpha) \geq c$, then $\mathcal{A}$ is said to satisfy the Bogomolov property (with Bogomolov constant c).

Weil height.   Written as $h(\alpha)$, the Weil height of an algebraic number α is defined to be the logarithm of its Mahler measure, divided by its degree: $h(\alpha) = \log M(\alpha)/\deg(\alpha)$.

Chapter 12
The Mahler Measure of Nonreciprocal Polynomials

12.1 Mahler Measure of Nonreciprocal Polynomials

In this section, we show that Lehmer's Conjecture is true for integer nonreciprocal polynomials. In fact, we prove slightly more, as follows:

Theorem 12.1 *Suppose that $P(z) \neq z$ is a nonreciprocal irreducible polynomial with integer coefficients. Then*

$$M(P(z)) \geq M(z^3 - z - 1) = 1.32471 \cdots . \qquad (12.1)$$

Furthermore, for monic polynomials, equality occurs precisely for the polynomials $P(z^n) = z^{3n} - z^n - 1$ and $-z^{3n}P(1/z) = z^{3n} + z^{2n} - 1$ for $n = 1, 2, 3, \ldots$, as well as for those polynomials with z replaced by $-z$.

Finally, if $M(P(z)) > M(z^3 - z - 1)$ then $M(P(z)) > \sqrt{(93 + \sqrt{2249})/80} = 1.32487 \cdots$, so that $M(z^3 - z - 1)$ is an isolated point in the spectrum of such Mahler measures.

The idea of the proof, pioneered by Salem and Siegel for the study of Pisot numbers, is to look at the quotient $P(z)/Q(z)$, where $Q(z)$ is the reciprocal polynomial of $P(z)$. For P reciprocal, this quotient is just ± 1, but for P nonreciprocal, it has an integer power series expansion with infinitely many nonzero terms. Moreover, it can be written as a quotient $f(z)/g(z)$ of Hardy H^1 functions f, g. It is the study of these functions in the next section which forms the basis of the proof of the theorem.

It would be interesting to know more about the spectrum of values of $M(P)$ for P nonreciprocal. All of the known small points in this spectrum come from trinomials, or their irreducible factors:

© Springer Nature Switzerland AG 2021
J. McKee and C. Smyth, *Around the Unit Circle*, Universitext,
https://doi.org/10.1007/978-3-030-80031-4_12

$1.324717959\cdots = M(z^3 - z - 1) = M(z^5 - z^4 - 1)/(z^2 - z + 1));$

$1.349716105\cdots = M(z^5 - z^4 + z^2 - z + 1) = M((z^7 + z^2 + 1)/(z^2 + z + 1));$

$1.359914149\cdots = M(z^6 - z^5 + z^3 - z^2 + 1) = M((z^8 + z + 1)/(z^2 + z + 1));$

$1.364199545\cdots = M(z^5 - z^2 + 1);$

$1.367854634\cdots = M(z^9 - z^8 + z^6 - z^5 + z^3 - z + 1)$

$$= M((z^{11} + z^4 + 1)/(z^2 + z + 1)).$$

The smallest known limit point of nonreciprocal measures is

$$\lim_{n \to \infty} M(z^n + z + 1) = M(x + z + 1) = 1.38135 \cdots \quad \text{([Boy 78])}.$$

The theorem readily generalises to polynomials in several variables, as follows:

Corollary 12.2 *Suppose that $k \geq 2$, $\mathbf{z}_k = (z_1, \ldots, z_k)$ and that $P(\mathbf{z}_k) \in \mathbb{Z}[\mathbf{z}_k]$ is nonreciprocal. Then*

$$M(P(\mathbf{z}_k)) \geq M(z^3 - z - 1) = 1.3247\cdots . \tag{12.2}$$

Proof We know from Theorem 2.5 and its proof that $M(P(\mathbf{z}_k))$ is a limit of Mahler measures $M(P(z^{r_1}, \ldots, z^{r_k}))$ for an infinite sequence of integer k-tuples $(r_1, \ldots, r_k)$. Since $P(\mathbf{z}_k)$ is nonreciprocal, the polynomials $P(z^{r_1}, \ldots, z^{r_k})$ are nonreciprocal too, by Exercise 2.3. Thus, $M(P(\mathbf{z}_k)) \geq M(z^3 - z - 1)$, by the theorem. $\qquad \square$

Corollary 12.3 *An integer polynomial $P(z)$ has at most*

$$\left\lfloor \frac{\log M(P(z))}{\log M(z^3 - z - 1)} \right\rfloor$$

irreducible nonreciprocal factors, counted with multiplicity.

Exercise 12.4 (Irreducibility of $z^n - z - 1$) Let $n \geq 2$. Show that $z^n - z - 1$ has no self-reciprocal factors. Use Gonçalves' Inequality (Proposition 1.12) and Corollary 12.3 to deduce that $z^n - z - 1$ is irreducible.

12.1.1 The Set $\mathcal{H}$ of Rational Hardy Functions

The space H^1 consists of those functions $f(z)$ of a complex variable z that are analytic for $|z| < 1$ and for which $\int_0^1 |f(re^{2\pi it})|\, dt$ is bounded for $0 < r < 1$. We are interested in the subset $\mathcal{H}$ of H^1 consisting of those f in H^1 that are rational functions and have modulus at most 1 on $|z| = 1$ (there might be removable singularities on the unit circle, which we remove). By the Maximum Modulus Principle (Theorem C.3), for nonconstant $f \in \mathcal{H}$, $|f(z)| \leq 1$ for all z in the unit disc $|z| \leq 1$, with $|f(z)| < 1$ when $|z| < 1$.

Proposition 12.5 *The set $\mathcal{H}$ has the following properties:*

(a) *We have $a \in \mathcal{H}$ for all $a \in \mathbb{C}$ with $|a| \leq 1$;*
(b) *We have $(z - c)/(1 - \overline{c}z) \in \mathcal{H}$ for all $c \in \mathbb{C}$ with $|c| \leq 1$. In particular $(c = 0)$ $z \in \mathcal{H}$. These functions have modulus 1 on the whole of $|z| = 1$;*
(c) *The set $\mathcal{H}$ is a semigroup: if $f(z)$ and $g(z) \in \mathcal{H}$, then $f(z)g(z) \in \mathcal{H}$;*
(d) *The set $\mathcal{H}$ is a closed under composition of functions: so if $f(z)$ and $g(z) \in \mathcal{H}$, then $g(f(z)) \in \mathcal{H}$;*
(e) *The set $\mathcal{H}$ is convex: if $f_1(z)$ and $f_2(z) \in \mathcal{H}$ and $\lambda \in [0, 1]$, then $\lambda f_1(z) + (1 - \lambda) f_2(z) \in \mathcal{H}$. More generally, if $f_1, f_2, \ldots, f_k \in \mathcal{H}$ and $\lambda_1, \ldots, \lambda_k$ are real, nonnegative and sum to 1, then $\lambda_1 f_1(z) + \cdots + \lambda_k f_k(z) \in \mathcal{H}$;*
(f) *If $f(z) \in \mathcal{H}$ and $f(0) = 0$, then $f(z)/z \in \mathcal{H}$.*
(g) *If $f \in \mathcal{H}$ and $f(z) = g(z^k)$, then $g \in \mathcal{H}$;*
(h) *Every $f(z) \in \mathcal{H}$ defined on the disc $|z| \leq 1$ attains its maximum modulus on the boundary $|z| = 1$.*

Exercise 12.6 Prove Proposition 12.5.

Since $f(z) \in \mathcal{H}$ is analytic, we can expand it as

$$f(z) = c_0 + c_1 z + c_2 z^2 + \cdots + c_n z^n + \cdots, \tag{12.3}$$

where $c_n = f^{(n)}(0)/n! \in \mathbb{C}$.

Proposition 12.7 *For $f(z)$ as in (12.3), any positive integer k the function*

$$f_k(z) = c_0 + c_k z + c_{2k} z^2 + \cdots + c_{nk} z^n + \cdots \tag{12.4}$$

belongs to $\mathcal{H}$.

Proof Using Exercise 5.5 from Chapter 5, we have (with $\omega = \omega_k$) that

$$f_k(z^k) = \frac{1}{k} \left(f(z) + f(\omega z) + f(\omega^2 z) + \cdots + f(\omega^{k-1} z) \right).$$

Hence, by Proposition 12.5(a), (b), (c), (d), (e) and (g), we see that $f_k(z) \in \mathcal{H}$. $\square$

More generally, one can show the following:

Exercise 12.8 For $f(z)$ as in (12.3), any positive integer k and any integer ℓ with $0 \leq \ell \leq k - 1$ show that the function

$$f_{k,\ell}(z) = c_\ell + c_{\ell+k} z + c_{\ell+2k} z^2 + \cdots + c_{\ell+nk} z^n + \cdots \tag{12.5}$$

belongs to $\mathcal{H}$.

Proposition 12.9 *Suppose that $f(z) \in \mathcal{H}$ has modulus 1 on $|z| = 1$. Then f is an extreme point of $\mathcal{H}$.*

Proof Suppose that $f(z) = \lambda f_1(z) + (1 - \lambda) f_2(z)$ for some distinct f_1, $f_2 \in \mathcal{H}$ and $0 \leq \lambda \leq 1$. Since $f_1 - f_2$ has its maximum modulus on $|z| \leq 1$ on its boundary $|z| = 1$, it cannot be 0 everywhere there. Hence, there is some z_0 on $|z| = 1$ with $f_1(z_0) \neq f_2(z_0)$. But if $f(z_0)$ lay on the open chord joining $f_1(z_0)$ and $f_2(z_0)$, it would have modulus less than 1. Hence, $f = f_1$ or f_2. □

Problem 12.10 (open problem)

What are the extreme points of $\mathcal{H}$? Do they all have modulus 1 on $|z| = 1$?

Next, we study the coefficient sequence $\{c_n\}$ of power series coefficients of $f(z) \in \mathcal{H}$, as in (12.3).

Proposition 12.11 *For nonconstant $f(z) = \sum_{n=0}^{\infty} c_n z^n \in \mathcal{H}$:*

(a) *We have $\sum_{n=0}^{\infty} |c_n|^2 \leq 1$, with equality if and only if $|f(z)| = 1$ on $|z| = 1$.*
(b) *For $k = 1, 2, \ldots$, we have $|c_k| \leq 1 - |c_0|^2$;*
(c) *For $k = 1, 2, \ldots$, we have*

$$\left| c_{2k} + \frac{\overline{c_0} c_k^2}{1 - |c_0|^2} \right| \leq 1 - |c_0|^2 - \frac{|c_k|^2}{1 - |c_0|^2};$$

(d) *When $f(z)$ has real coefficients, we have, for $k = 1, 2, \ldots$ that*

$$-(1 - c_0^2) + \frac{c_k^2}{1 + c_0} \leq c_{2k} \leq 1 - c_0^2 - \frac{c_k^2}{1 - c_0}.$$

Proof For part (a), we have

$$1 \geq \int_0^1 |f(e^{2\pi i t})|^2 \, dt = \sum_{n=0}^{\infty} |c_n|^2,$$

since this sum is the constant term of the Fourier expansion of $|f(e^{2\pi i t})|^2 = f(e^{2\pi i t}) \overline{f(e^{2\pi i t})}$ (Parseval's Identity: Theorem C.2).

For the remaining parts, we can assume that $k = 1$. Then the results for general k follow from applying the $k = 1$ result to the function $f_k(z)$ from (12.4).

For (b), we put $g(z) = (z - c_0)/(1 - \overline{c_0} z)$ as in Proposition 12.5(b). Applying parts (d) and (f) of that Proposition, we have $g(f(z))/z \in \mathcal{H}$. On expanding in series, we have

$$g(f(z))/z = \frac{c_1 + c_2 z + \cdots}{1 - |c_0|^2 - \overline{c_0} c_1 z + \cdots}.$$

Its constant term $c_1/(1 - |c_0|^2)$ is at most 1 in modulus, giving the result.

For (c), we note that

$$g(f(z))/z = \frac{1}{1 - |c_0|^2}\left(c_1 + \left(c_2 + \frac{\overline{c_0}c_1^2}{1 - |c_0|^2}\right)z + \cdots\right).$$

Applying (b) to this function we obtain the inequality

$$\left|c_2 + \frac{\overline{c_0}c_1^2}{1 - |c_0|^2}\right| \leq 1 - |c_0|^2 - \frac{|c_1|^2}{1 - |c_0|^2},$$

which gives the result for general k. Part (d) then follows straight away. □

Exercise 12.12 Show that in Proposition 12.11(d) (with $k = 1$), the first inequality is an equality for

$$f(z) = \frac{(1 - c_0)(z^2 + c_0) + c_1 z}{(1 - c_0)(c_0 z^2 + 1) + c_1 z},$$

while the second inequality is an equality for

$$f(z) = \frac{(1 + c_0)(z^2 - c_0) - c_1 z}{(1 + c_0)(c_0 z^2 - 1) + c_1 z}.$$

Exercise 12.13 For $f(z) = \sum_{n=0}^{\infty} c_n z^n \in \mathcal{H}$ having real coefficients with $c_0 > 0$ and c_2 given, show that

$$c_1^2 \leq \begin{cases} (1 - c_0)(1 - c_0^2 - c_2) & \text{if } -c_0(1 - c_0^2) \leq c_2 \leq 1 - c_0^2; \\ (1 + c_0)(1 - c_0^2 + c_2) & \text{if } -(1 - c_0^2) \leq c_2 \leq -c_0(1 - c_0^2). \end{cases}$$

Show, also, that equality in the first case occurs when $f(z) = f_+(\pm z)$, where

$$f_+(z) := \frac{\sqrt{1 - c_0}(z^2 + c_0) + z\sqrt{1 - c_0^2 - c_2}}{\sqrt{1 - c_0}(c_0 z^2 + 1) + z\sqrt{1 - c_0^2 - c_2}},$$

while equality in the second case occurs when $f(z) = f_-(\pm z)$, where

$$f_-(z) := \frac{\sqrt{1 + c_0}(z^2 - c_0) - z\sqrt{1 - c_0^2 + c_2}}{\sqrt{1 + c_0}(c_0 z^2 - 1) + z\sqrt{1 - c_0^2 + c_2}}.$$

Problem 12.14 (open problem)

Given nonconstant $f(z) = \sum_{n=0}^{\infty} c_n z^n \in \mathcal{H}$ with real coefficients and c_0 and c_k known, find sharp upper and lower bounds for c_ℓ as a function of c_0 and c_k. Here $\ell > k > 0$.

Proposition 12.11(d) solves this problem for $\ell = 2k$. If one could solve it for other values of ℓ, it may be possible to strengthen Theorem 12.1 to say more about the spectrum of values of $M(P)$ for irreducible nonreciprocal integer polynomials P.

Exercise 12.15 Suppose that the irreducible polynomial $P(z) \in \mathbb{Z}[z]$ has a zero of modulus 1. Prove that then $P(z)$ is self-reciprocal.

12.2   Proof of Theorem 12.1

12.2.1   Start of the Proof

We work with a monic irreducible nonreciprocal polynomial $P(z) \in \mathbb{Z}[z]$, with $P(z) \neq z$ and α a zero of $P(z)$. Denote by α_j $(j = 1, \ldots, d)$ the conjugates of α, with $\alpha = \alpha_1$ and d being the degree of $P(z)$. Let $\theta_0 := M(z^3 - z - 1) = 1.3247\cdots$, the real zero of $z^3 - z - 1$, and the smallest Pisot number [Sie44]. We need to show that either $M(P) = \theta_0$ or $M(P) > \theta_0 + 10^{-4}$. We also need to find all P for which the equality $M(P) = \theta_0$ holds. We put $Q(z) = z^d P(1/z)$. As P is nonreciprocal, $P(z) \neq \pm Q(z)$. For the proof, it is convenient to assume that $M(P) \leq 4/3$, as we clearly are entitled to do.

We first note that $P(0) = \pm 1$ for otherwise $M(P) \geq 2$. Secondly, by Exercise 12.15, $P(z)$ has no zeros of modulus 1. We work with the nonconstant rational function $P(0)\frac{P(z)}{Q(z)}$, and its power series

$$P(0)\frac{P(z)}{Q(z)} = 1 + a_k z^k + a_\ell z^\ell + \cdots,$$

where k and ℓ are the first two indices for which the corresponding coefficients are nonzero. Since $Q(0) = 1$, these coefficients are rational integers, and so certainly $|a_k| \geq 1$ and $|a_\ell| \geq 1$. Now

$$P(0)\frac{P(z)}{Q(z)} = \frac{f(z)}{g(z)},$$

say, where

$$f(z) = \pm \prod_{|\alpha_j|<1} \left(\frac{z - \alpha_j}{1 - \overline{\alpha_j}z}\right) = c + c_1 z + c_2 z^2 + \cdots \in \mathcal{H}$$

and

$$g(z) = \pm \prod_{|\alpha_j|>1} \left(\frac{1 - \overline{\alpha_j}z}{z - \alpha_j}\right) = d + d_1 z + d_2 z^2 + \cdots \in \mathcal{H},$$

say, choosing the signs so that $c > 0, d > 0$. Then from

$$1 + a_k z^k + a_\ell z^\ell + \cdots = \frac{c + c_1 z + c_2 z^2 + \cdots}{d + d_1 z + d_2 z^2 + \cdots}$$

we have

$$d = c = \prod_{|\alpha_j| > 1} |\alpha_j|^{-1} = M(P)^{-1}$$

$$d_i = c_i \quad (i = 1, 2, \ldots, k - 1) \tag{12.6}$$

$$a_k c + d_k = c_k \tag{12.7}$$

$$a_k d_i + d_{k+i} = c_{k+i} \quad (i = 1, 2, \ldots, \ell - k - 1) \tag{12.8}$$

$$a_\ell c + a_k d_{\ell-k} + d_\ell = c_\ell, \tag{12.9}$$

using $c = d$ to obtain (12.7) and (12.9). Now since $|a_k| \geq 1$, we have from (12.7) that $\max(|c_k|, |d_k|) \geq c/2$. Hence, by Proposition 12.11(b), we see that $c/2 \leq 1 - c^2$. This gives

$$M(P) \geq \frac{1 + \sqrt{17}}{4} = 1.2807 \cdots . \tag{12.10}$$

In fact, the inequality must be strict, since $M(P)$ is an algebraic integer (Proposition 1.9). Furthermore, we cannot have $|a_k| \geq 2$; for then by a similar argument we would have $\max(|c_k|, |d_k|) \geq c$ and $c \leq 1 - c^2$, contradicting $c = M(P)^{-1} \geq 3/4$. So $a_k = \pm 1$. It is also clear that

$$|c_k| + |d_k| = c, \tag{12.11}$$

for otherwise we would again have $\max(|c_k|, |d_k|) \geq c$. Further, $a_\ell = \pm 1$; for if $|a_\ell| \geq 2$ then, by (12.9),

$$\max(|d_{\ell-k}|, |d_\ell|, |c_\ell|) \geq \tfrac{2}{3} c,$$

and so $\tfrac{2}{3} c \leq 1 - c^2$. This gives $M(P) \geq (1 + \sqrt{10})/3$, contradicting $M(P) \leq 4/3$.

The argument now divides into two cases.

12.2.2   The Case $\ell < 2k$

Firstly, we may assume for this case that $a_k = +1$; for otherwise, by interchanging the roles of P and Q, allowed because $M(P) = M(Q)$, we may replace the series $1 + a_k z^k + a_\ell z^\ell + \cdots$ by its formal reciprocal, and so change the sign of a_k.

We now apply Parseval's Identity (Theorem C.2) to

$$(1 + \gamma z^{\ell-k} - z^k + \beta z^\ell) f(z),$$

and to

$$(-1 - \gamma z^{\ell-k} - z^k + \beta z^\ell) g(z) \,,$$

where β, γ are real numbers. We obtain, by omitting all but the first four terms on the left hand side, that

$$c^2 + (c_{\ell-k} + \gamma c)^2 + (c_k + \gamma c_{2k-\ell} - c)^2 + (c_\ell + \gamma c_k - c_{\ell-k} + \beta c)^2 \le 2 + \gamma^2 + \beta^2$$

and

$$c^2 + (-c_{\ell-k} - \gamma c)^2 + (-d_k - \gamma c_{2k-\ell} - c)^2 + (-d_\ell - \gamma d_k - c_{\ell-k} + \beta c)^2 \le 2 + \gamma^2 + \beta^2 \,.$$

Now, averaging and using the inequality

$$\frac{1}{2}(a^2 + b^2) \ge \left(\frac{1}{2}(a+b) \right)^2 \tag{12.12}$$

we obtain

$$c^2 + (c_{\ell-k} + \gamma c)^2 + \left(\tfrac{1}{2}(c_k - d_k) - c \right)^2$$
$$+ \left(\tfrac{1}{2}(c_\ell - d_\ell) + \tfrac{\gamma}{2}(c_k - d_k) - c_{\ell-k} + \beta c \right)^2 \le 2 + \gamma^2 + \beta^2 \,.$$

From (12.7), (12.9), (12.6) ($\ell - k < k$ here, so $c_{\ell-k} = d_{\ell-k}$) and the fact that $a_k = 1$, this simplifies to

$$\tfrac{5}{4}c^2 + (c_{\ell-k} + \gamma c)^2 + \left(\tfrac{1}{2}(a_\ell c + c_{\ell-k}) + \tfrac{\gamma}{2}c - c_{\ell-k} + \beta c \right)^2 \le 2 + \gamma^2 + \beta^2 \,.$$

Multiplying both terms in brackets by a_ℓ $(= \pm 1)$, putting $x := -a_\ell c_{\ell-k}$, and replacing β, γ by $a_\ell \beta, -a_\ell \gamma$, respectively, we get

$$\tfrac{5}{4}c^2 + (x + \gamma c)^2 + \left(\tfrac{1}{2}c + \tfrac{1}{2}x - \tfrac{\gamma}{2}c + \beta c \right)^2 \le 2 + \gamma^2 + \beta^2 \,.$$

We complete the square for β, and choose its value so that the square is 0. This gives

$$\tfrac{5}{4}c^2 + (x + \gamma c)^2 + \frac{(c + x - \gamma c)^2}{4(1 - c^2)} \le 2 + \gamma^2 \,.$$

We then do the same for γ, obtaining

$$\tfrac{5}{4}c^2 + \frac{4(1 - c^2)x^2 + (x + c)^2 - c^2(2x + c)^2}{4c^4 - 9c^2 + 4} \le 2 \,.$$

Now complete the square for x. Equation (12.10) implies that the coefficient of x^2 is positive. While we cannot choose x, we know that the completed square is nonnegative. This gives the inequality

$$\tfrac{5}{4}c^2 + \frac{c^2((5-8c^2)(1-c^2)-(1-2c^2)^2)}{(5-8c^2)(4c^4-9c^2+4)} \le 2$$

which, on simplification, yields

$$40c^4 - 93c^2 + 40 \ge 0,$$

or, with $M(P) = 1/c$,

$$M(P)^4 - 2.325M(P)^2 + 1 \ge 0. \tag{12.13}$$

But for θ_0, the real zero of $z^3 - z - 1$, we have

$$\begin{aligned}
\theta_0^4 - 2.325\theta_0^2 + 1 &= \theta_0(\theta_0 + 1) - 2.325\theta_0^2 + 1 \\
&= \theta_0 + 1 - 1.325\theta_0^2 \\
&= \theta_0^2(\theta_0 - 1.325) \\
&< 0
\end{aligned}$$

as $\theta_0 = 1.3247\cdots$. Hence, $M(P) > \theta_0$. And from (12.13), along with Proposition 1.9, we get

$$M(P) > \sqrt{\frac{93 + \sqrt{2249}}{80}} = 1.3248\cdots = \theta_0 + 0.0001576\cdots.$$

Thus, the proof for $\ell < 2k$ is complete.

Problem 12.16 (open problem)
    What is the least value (or the infimum of values) of $M(P)$ in the case $\ell < 2k$?

12.2.3 The Case $\ell \ge 2k$: Proof that $M(P) \ge \theta_0$

First of all, we claim that we may assume that $\ell > 2k$. For if $a_{2k} = \pm 1$, then by interchanging $P(z)$ and $Q(z)$, the resulting new value for a_{2k} is either 0 or 2 (we are no longer assuming that $a_k = +1$, though we still, of course, have $a_k = \pm 1$). But we have shown that $a_\ell = 2$ contradicts our assumptions, so we may take $a_{2k} = 0$, and hence have $\ell > 2k$.

   Now (12.8) is valid for $i = k$, giving

$$a_k d_k + d_{2k} = c_{2k}. \tag{12.14}$$

We now apply Proposition 12.11(d) to $f(z)$ and $g(z)$, and obtain, with $d_0 = c_0 = c$,

$$-(1-c^2) + \frac{c_k^2}{1+c} \le c_{2k} \le 1 - c^2 - \frac{c_k^2}{1-c}$$

$$-(1-c^2) + \frac{d_k^2}{1-c} \le -d_{2k} \le 1 - c^2 - \frac{d_k^2}{1+c}.$$

$$(12.15)$$

Adding these inequalities, and using (12.14), we have

$$-2(1-c^2) + \frac{c_k^2}{1+c} + \frac{d_k^2}{1-c} \le a_k d_k \le 2(1-c^2) - \left(\frac{c_k^2}{1-c} + \frac{d_k^2}{1+c} \right).$$
$$(12.16)$$

From (12.7) and (12.11), we know that

$$a_k d_k = -|d_k|.$$ $$(12.17)$$

Thus, from (12.11) and the first inequality of (12.16), with $x := |c_k|$, we have

$$c - x \le 2(1-c^2) - \frac{x^2}{1+c} - \frac{(c-x)^2}{1-c}.$$ $$(12.18)$$

Clearing the denominators gives

$$c - c^3 - (1-c^2)x \le 2(1-c^2)^2 - x^2(1-c) - (c-x)^2(1+c)$$

which can be rewritten as

$$2\left(x - (1-c^2)\right)^2 + (3-5c)(1+c)(x - (1-c^2)) - (1+c)(1-c^2-c^3) \le 0.$$

Dropping the first term, and using the fact that $5c - 3 > 0$ because $c \ge 3/4$, this gives

$$1 - c^2 - x \le \frac{1 - c^2 - c^3}{5c - 3}.$$

So, with Proposition 12.11(b), we obtain

$$1 - c^2 - \frac{1 - c^2 - c^3}{5c - 3} \le |c_k| \le 1 - c^2.$$ $$(12.19)$$

This clearly shows that $1 - c^2 - c^3 \ge 0$, so that $M(P)^3 - M(P) - 1 \ge 0$ and hence $M(P) \ge \theta_0$.

12.2.4 The Case $\ell \geq 2k$: Existence of a δ, Part 1

We next need to show that if $M(P) > \theta_0$, then $M(P) > \theta_0 + \delta$, where we claim that for the case $\ell \geq 2k$ we can take $\delta = 0.0002 > \sqrt{(93 + \sqrt{2249})/80} - \theta_0$.

Firstly, we will put (12.19) into a more convenient form, and derive from it similar inequalities for d_k and d_{2k}. Let $C = \theta_0^{-1}$. Then

$$1 - c^2 - c^3 = (C - c)(C^{-1} + (1 + C)c + c^2). \tag{12.20}$$

If $C - c \geq 10^{-3}$, then $1/c - 1/C \geq 10^{-3}/C^2 > 0.0002$. Since we are aiming to prove that if $c \neq C$ then $c^{-1} - C^{-1} > 0.0002$, we can, therefore, assume that

$$C - c < 10^{-3}. \tag{12.21}$$

We shall also make standard use of the inequality

$$\tfrac{3}{4} \leq c \leq C < \tfrac{4}{5} < 1. \tag{12.22}$$

Lemma 12.17 *The following inequalities hold:*

$$
\begin{aligned}
1 - c^2 - 6(C - c) &\leq & |c_k| &\leq & 1 - c^2; & (12.23)\\
c^2 + c - 1 &\leq & |d_k| &\leq & c^2 + c - 1 + 6(C - c); & (12.24)\\
& & |d_{2k}| &\leq & 28(C - c). & (12.25)
\end{aligned}
$$

Proof From (12.20) and (12.22)

$$1 - c^2 - c^3 \leq (C - c)(\tfrac{4}{3} + \tfrac{4}{5} + \tfrac{32}{25}) \leq 4(C - c). \tag{12.26}$$

So

$$\frac{1 - c^2 - c^3}{5c - 3} \leq \frac{4(C - c)}{5 \cdot \tfrac{3}{4} - 3} \leq 6(C - c)$$

which, with (12.19), gives (12.23). Equation (12.24) follows from (12.23), using (12.11).

To obtain (12.25), we substitute (12.23) into (12.15), yielding

$$-\left(1 - c^2 - \frac{(1 - c^2 - 6(C - c))^2}{1 + c}\right) \leq c_{2k} \leq 1 - c^2 - \frac{(1 - c^2 - 6(C - c))^2}{1 - c}.$$

Subtracting $36(C - c)^2/(1 + c)$ from the left and adding $36(C - c)^2/(1 - c)$ to the right gives, after simplification,

$$-c(1 - c^2) - 12(C - c)(1 - c) \leq c_{2k} \leq -c(1 - c^2) + 12(C - c)(1 + c).$$

We now use (12.14), (12.17) and (12.24) to get, from this equation

$$c^3 + c^2 - 1 - 12(C - c)(1 - c) \le d_{2k} \le c^3 + c^2 - 1 + 12(C - c)(1 + c) + 6(C - c)$$
$$= c^3 + c^2 - 1 + (C - c)(18 + 12c).$$

Now, from (12.26) and the fact that $c^3 + c^2 - 1 \le 0$, we get

$$-(16 - 12c)(C - c) \le d_{2k} \le (18 + 12c)(C - c),$$

which gives (12.25).   □

We now let $\alpha_0 := -\theta_0^{-1} = -C$. Then α_0 has minimal polynomial $P_0(z) = z^3 - z^2 + 1$, and $Q_0(z) := z^3 P_0(1/z) = 1 - z + z^3$. Let

$$\frac{P_0(z)}{Q_0(z)} = 1 + \sum_{i=1}^{\infty} A_i z^i ;$$

we note that α_0 has been chosen so that $P_0(0) = 1$, $A_1 = 1$ and $A_2 = 0$.

We next compare the power series related to α and α_0. Consider

$$\frac{P(z)}{Q(z)} \cdot \frac{Q_0(a_k z^k)}{P_0(a_k z^k)} = (1 + a_k z^k + a_\ell z^\ell + \cdots)(1 + A_1(a_k z^k) + A_2(a_k z^k)^2 + \cdots)^{-1}.$$
(12.27)

If the power series for $P(z)/Q(z)$ and $P_0(a_k z^k)/Q_0(a_k z^k)$ are not identical, we have that both sides of (12.27) are equal to $1 + b_p z^p + \cdots$, where p is the first index where these series do not agree, so that b_p is a nonzero integer. We note for future reference that $p > 2k$, since both $P(z)/Q(z)$ and $P_0(a_k z^k)/Q_0(a_k z^k)$ have constant term 1, coefficient of z^k equal to a_k, and all other coefficients 0 for $i \le 2k$. Again, (12.27) equals

$$\left(\frac{c + c_1 z + c_2 z^2 + \cdots}{c + d_1 z + d_2 z^2 + \cdots} \right) \cdot \frac{1 - a_k z^k + (a_k z^k)^3}{1 - (a_k z^k)^2 + (a_k z^k)^3}$$
$$= \frac{\sum_{r=0}^{\infty} (c_r - a_k c_{r-k} + a_k c_{r-3k}) z^r}{\sum_{r=0}^{\infty} (d_r - d_{r-2k} + a_k d_{r-3k}) z^r} = \frac{\sum_{r=0}^{\infty} \gamma_r z^r}{\sum_{r=0}^{\infty} w_r z^r}$$
(12.28)

say, where $a_k^2 = 1$, $c_0 = d_0 = c$, and the terms with negative indices are taken to be 0. So

$$c_p - a_k c_{p-k} + a_k c_{p-3k} = c b_p + d_p - d_{p-2k} + a_k d_{p-3k}$$
(12.29)
$$\gamma_p = c b_p + w_p.$$

Hence

$$|\gamma_p - w_p| \ge c \ge \tfrac{3}{4}.$$
(12.30)

12.2.5 The Case $\ell \geq 2k$: Existence of a δ, Part 2

We now find upper bounds for γ_p and w_p in terms of $(C - c)$, and so obtain a lower bound for $(C - c)$. Now

$$P_0(z) = 1 - z^2 + z^3 = (z + C)(z^2 - (1 + C)z + C^{-1})$$
$$Q_0(z) = 1 - z + z^3 = (1 + Cz)(1 - (1 + C)z + C^{-1}z^2).$$

We look at

$$\sum_{i=0}^{\infty} f_i z^i := \left(\sum_{i=0}^{\infty} c_i z^i \right) (1 + a_k C z^k)$$

and

$$\sum_{i=0}^{\infty} g_i z^i := \left(\sum_{i=0}^{\infty} d_i z^i \right) (C^{-1} - a_k(1 + C)z^k + z^{2k}).$$

Then from (12.28)

$$\sum_{i=0}^{\infty} f_i z^i (1 - a_k(1 + C)z^k + C^{-1}z^{2k}) = \sum_{i=0}^{\infty} \gamma_i z^i \qquad (12.31)$$

and

$$\sum_{i=0}^{\infty} g_i z^i (C + a_k z^k) := \sum_{i=0}^{\infty} w_i z^i, \qquad (12.32)$$

and it is algebraically easier to get bounds for the f_i, g_i than for the γ_i, w_i directly.

Lemma 12.18 *The following (in)equalities hold:*

$$|f_0 - C| = (C - c) \qquad (12.33)$$
$$|f_k - a_k| \leq 7(C - c) \qquad (12.34)$$
$$|f_i| \leq 4\sqrt{C - c} \quad (i \neq 0, k) \qquad (12.35)$$
$$|\gamma_{3k} - C^{-1}a_k| \leq 12\sqrt{C - c} \qquad (12.36)$$
$$|\gamma_i| \leq 17\sqrt{C - c} \quad (i \neq 0, k, 2k, 3k) \qquad (12.37)$$

(We can obtain similar inequalities for γ_0, γ_k, γ_{2k} and γ_{3k}, but these are not needed.)

Proof Now $f_0 = c$, so (12.33) is immediate. Also, $f_k = c_k + a_k c C$ and

$$|f_k - a_k| = |c_k + a_k cC - a_k(1 - c^2 + c^2)|$$
$$\leq |c_k - a_k(1 - c^2)| + c|a_k C - a_k c|$$
$$\leq (6 + c)(C - c)$$

by (12.23) (as $c_k = a_k|c_k|$ by (12.7) and (12.11))

$$\leq 7(C - c),$$

which is (12.34).

For (12.35), we have, by Parseval's Identity,

$$\sum_{i=0}^{\infty} f_i^2 = 1 + C^2. \tag{12.38}$$

So

$$\sum_{\substack{i=0 \\ i \neq 0,k}}^{\infty} f_i^2 = 1 - f_k^2 + C^2 - f_0^2$$

$$= (a_k - f_k)(a_k + f_k) + (C - f_0)(C + f_0)$$
$$\leq 7(C - c)(2 + 7(C - c)) + (C - c)(2C + (C - c))$$

by (12.34) (and using $c < C$)

$$= (14 + 2C)(C - c) + 50(C - c)^2$$
$$\leq 16(C - c)$$

by (12.21) and (12.22). This gives (12.35).

Now from (12.31)

$$\gamma_i = f_i - f_{i-k}a_k(1 + C) + f_{i-2k}C^{-1}. \tag{12.39}$$

So

$$|\gamma_{3k} - a_k C^{-1}| = |f_{3k} - f_{2k}(1 + C)a_k + (f_k - a_k)C^{-1}|$$
$$\leq |f_{3k}| + |f_{2k}|(1 + C) + |f_k - a_k|C^{-1}$$
$$\leq 4(2 + C)\sqrt{C - c} + 7(C - c)C^{-1}$$

by (12.34) and (12.35)

$$\leq \left(8 + \tfrac{16}{5} + \tfrac{28}{3} \cdot 10^{-3/2}\right)\sqrt{C - c}$$

by (12.21) and (12.22)

$$\leq 12\sqrt{C-c},$$

which is (12.36).

Finally, for $i \neq 0, k, 2k, 3k$, we have from (12.39) and (12.35) that

$$
\begin{aligned}
|\gamma_i| &\leq 4\sqrt{C-c}\left(1 + |a_k(1+C)| + C^{-1}\right) \\
&= 4\sqrt{C-c}\left(2 + C + C^{-1}\right) \\
&\leq 17\sqrt{C-c}
\end{aligned}
$$

by (12.22).   □

We now go through the same process for the g_i and the w_i.

Lemma 12.19 *The following (in)equalities hold:*

$$|g_0 - 1| = (C - c)C^{-1} \tag{12.40}$$

$$|g_k - (-a_k(1+C))| \leq 14(C - c) \tag{12.41}$$

$$|g_{2k} - C^{-1}| \leq 54(C - c) \tag{12.42}$$

$$|g_i| \leq 15\sqrt{C-c} \quad (i \neq 0, k, 2k) \tag{12.43}$$

$$|w_{3k} - a_k C^{-1}| \leq 14\sqrt{C-c} \tag{12.44}$$

$$|w_i| \leq 27\sqrt{C-c} \quad (i \neq 0, k, 2k, 3k) \tag{12.45}$$

Again, we could obtain similar inequalities for w_0, w_k, w_{2k}, w_{3k}.

Proof Now $g_0 = cC^{-1}$, $g_k = d_k C^{-1} - ca_k(1+C)$, and

$$g_{2k} = d_{2k}C^{-1} - d_k a_k(1+C) + c. \tag{12.46}$$

Thus, (12.40) is immediate. Further

$$
\begin{aligned}
g_k - &(-a_k(C^2 + C - 1)C^{-1} - C(1+C)a_k)| \\
&\leq C^{-1}|d_k - (-a_k(C^2 + C - 1))| + (C - c)(1 + C) \\
&\leq C^{-1}|d_k - (-a_k(c^2 + c - 1))| + C^{-1}|(C^2 + C - 1) - (c^2 + c - 1)| \\
&\quad + (C - c)(1 + C) \\
&\leq 6C^{-1}(C - c) + C^{-1}(C - c)(C + c + 1) + (C - c)(1 + c)
\end{aligned}
$$

by (12.24) and (12.17)

$$
\begin{aligned}
&= (C - c)(7C^{-1} + 1 + cC^{-1} + 1 + c) \\
&\leq 14(C - c)
\end{aligned}
$$

by (12.22). Now

$$(C^2 + C - 1)C^{-1} + C(1 + C) = C + 1 + (-1 + C^2 + C^3)C^{-1} = C + 1.$$

So we obtain (12.41).

For g_{2k}, we have from (12.46) and (12.17) that

$$
\begin{aligned}
|g_{2k} - ((C^2 + C - 1)(1 + C) + C)| \\
\le |d_{2k}|C^{-1} + (1 + C)\big||d_k| - (C^2 + C - 1)\big| + (C - c) \\
\le |d_{2k}|C^{-1} + (1 + C)\big||d_k| - (c^2 + c - 1)\big| \\
+ (1 + C)((C^2 + C - 1) - (c^2 + c - 1)) + (C - c) \\
\le (C - c)(28C^{-1} + 6(1 + C) + (1 + C)(2C + 1) + 1)
\end{aligned}
$$

by (12.24) and (12.25)

$$
\begin{aligned}
&\le (C - c)(28C^{-1} + 8 + 9C + 2C^2) \\
&\le (C - c)\big(\tfrac{112}{3} + 8 + \tfrac{36}{5} + \tfrac{32}{25}\big) \quad \text{by (12.22)} \\
&\le 54(C - c).
\end{aligned}
$$

To prove (12.43), we use Parseval again:

$$\sum_{i=0}^{\infty} g_i^2 = 1 + (1 + C)^2 + C^{-2}. \tag{12.47}$$

So we have

$$\sum_{\substack{i=0 \\ i \ne 0, k, 2k}}^{\infty} g_i^2 = C^{-2} - g_{2k}^2 + (1 + C)^2 - g_k^2 + 1 - g_0^2$$

$$
\begin{aligned}
&= (C^{-1} - g_{2k})(C^{-1} + g_{2k}) + (1 + C - g_k)(1 + C + g_k) \\
&\qquad\qquad + (1 - g_0)(1 + g_0) \\
&\le 54(C - c)(2C^{-1} + 54(C - c)) + 14(C - c)(2(1 + C) + 14(C - c)) \\
&\qquad\qquad + C^{-1}(C - c)(2 + C^{-1}(C - c))
\end{aligned}
$$

by (12.40), (12.41) and (12.42)

$$
\begin{aligned}
&= 2(54C^{-1} + 14(1 + C) + C^{-1})(C - c) + (54^2 + 14^2 + C^{-2})(C - c)^2 \\
&\le 2\big(\tfrac{220}{3} + 14 + \tfrac{56}{5}\big)(C - c) + 3114(C - c)^2 \quad \text{by (12.22)} \\
&\le 201(C - c) \quad \text{by (12.21).}
\end{aligned}
$$

Hence, (12.43) holds.

For the inequalities (12.44) and (12.45), we use

$$w_i = Cg_i + a_k g_{i-k} \tag{12.48}$$

from (12.32). Then

$$
\begin{aligned}
|w_{3k} - a_k C^{-1}| &= |Cg_{3k} + a_k(g_{2k} - C^{-1})| \\
&\le 15C\sqrt{C-c} + 54(C-c) \quad \text{by (12.43) and (12.44)} \\
&\le 14\sqrt{C-c} \quad \text{by (12.21) and (12.22)}.
\end{aligned}
$$

This proves (12.44). Finally, for (12.45), we have from (12.48) for $i \ne 0, k, 2k, 3k$ that

$$
\begin{aligned}
|w_i| &\le 15(1+C)\sqrt{C-c} \quad \text{by (12.44)} \\
&\le 27\sqrt{C-c} \quad \text{by (12.22)}.
\end{aligned}
$$

$\square$

12.2.6   The Case $\ell \ge 2k$: Completion of the Proof

On comparing equations (12.36) and (12.37) with (12.44) and (12.45), respectively, we obtain for all $i > 2k$

$$|w_i - \gamma_i| \le \max(26\sqrt{C-c}, 44\sqrt{C-c}) = 44\sqrt{C-c}. \tag{12.49}$$

So, as $p > 2k$, we have a contradiction to (12.30) if $44\sqrt{C-c} < 3/4$. Now (12.30) holds if $C \ne c$, so in this case

$$C - c \ge \left(\frac{3}{4 \cdot 44}\right)^2 > 2 \cdot 10^{-4}.$$

Hence, if $C \ne c$,

$$M(P) - \theta_0 = c^{-1} - C^{-1} = \frac{(C-c)}{cC} > (C-c) > 2 \cdot 10^{-4}.$$

This establishes the existence of a δ in the case $l \ge 2k$, which together with the sharper result for the case $\ell < 2k$ gives the last line of Theorem 12.1.

It remains only to find for which polynomials P we have $M(P) = \theta_0$. To do this, we put $c = C$ in (12.49), giving

$$\gamma_i = w_i \quad (i > 2k).\tag{12.50}$$

However, from the definition of p, we know that $\gamma_i = w_i$ for $i < p$, and hence for $i \le 2k$; thus $\gamma_i = w_i$ for all i. Hence, from (12.27) and (12.28),

$$\frac{P(z)}{Q(z)} = \frac{P_0(a_k z^k)}{Q_0(a_k z^k)}.$$

So for our zero α of $P(z)$, the number $a_k \alpha^k$ is a conjugate of $\alpha_0 = -\theta_0^{-1}$, and therefore, $\alpha_0 = a_k \alpha_i^k$ for some conjugate α_i of α. Hence, $\alpha_i^k = a_k \alpha_0 = \pm\theta_0^{-1}$, giving $\alpha_i = (\pm\theta_0)^{-1/k}$. However, earlier in the proof we made some assumptions which could have necessitated the interchanging of $P(z)$ and $Q(z)$; this is equivalent to replacing α by α^{-1}. Thus, we also have the possibility $\alpha_i = (\pm\theta_0)^{+1/k}$.

12.3 Notes

Theorem 12.1 is from [Smy71, Smy72]. It generalises Siegel's result [Sie44] that $z^3 - z - 1$ is the minimal polynomial of the smallest Pisot number. It also strengthens an earlier result of Breusch [Bre51], who proved, for $P(z)$ as in the Theorem, that $M(P) \ge M(z^3 - z^2 - 1/4) = 1.17965$. Dixon and Dubickas [DD04] improved the lower bound $1.32487\cdots$ when $M(P(z)) > M(z^3 - z - 1)$ in Theorem 12.1 to $1.32497\cdots$. The idea of using the inequality (12.12) in the proof of the $\ell < 2k$ case of the Theorem followed a suggestion of Schinzel (personal communication).

Schinzel [Sch73] and then Bazylewicz [Baz77] generalised Theorem 12.1 to polynomials over *Kroneckerian* fields. These are either totally real number fields or a totally complex quadratic extensions of such a field. If the field does not contain a primitive cube root of unity ω_3, then the best constant is again θ_0, while if it does contain ω_3, then the best constant is the maximum modulus of the roots of θ of $\theta^2 - \omega_3\theta - 1 = 0$.

An extension of Theorem 12.1 to algebraic numbers was proved by Notari [Not78]. See also Schinzel [Sch82, Sch00].

Dubickas [Dub00] proved an interesting conditional strengthening of Theorem 12.1: if α is a nonzero algebraic integer such that at most $1/1000$th of its conjugates lie outside the unit circle, then $M(\alpha) \ge 1.3259$.

12.4 Glossary

$\mathcal{H}$. The set of rational Hardy functions (see the start of Sect. 12.1.1). If $f \in \mathcal{H}$, then f is a rational function that is analytic in the open disc $|z| < 1$ and has modulus at most 1 on the circle $|z| = 1$ (there might be removable singularities on the unit circle, which we remove).

Chapter 13
Minimal Noncyclotomic Integer Symmetric Matrices

For Mahler measures of integer symmetric matrices, we can settle the analogue of Lehmer's Conjecture. Indeed, we shall see that if the Mahler measure is greater than 1, then it is at least Lehmer's number, and that this bound is attained. The analysis splits into: (i) minimal noncyclotomic examples that can be produced by adding a vertex to a subgraph of one of the sporadic cyclotomic examples; (ii) minimal noncyclotomic examples that cannot be so produced.

13.1 Supersporadic Matrices and Other Sporadic Examples

There are finitely many sporadic maximal cyclotomic charged signed graphs (S_7, S_8, S_8', S_{14}, S_{16}), so finitely many subgraphs of them, and finitely many ways to add a vertex to any one of these subgraphs, so there are finitely many minimal noncyclotomic charged signed graphs with the property that deleting some single vertex leaves a connected subgraph equivalent to a subgraph of one of the sporadic cyclotomic charged signed graphs. These we will call *supersporadic*. Finding all of them requires some ingenuity if the search is to be manageable, but at least the search is finite. In this section, we give the details.

We start by recording the small examples of minimal noncyclotomic charged signed graphs: those with 6 or fewer vertices. Not all of these are supersporadic (see Exercise 13.1). Even working up to equivalence, there are quite a lot of them: 99 to be precise, of which 91 are supersporadic. The search can be done simply by connected growing of charged signed graphs. From a list of all cyclotomic charged signed graphs on n vertices, one considers all possible ways of adding a vertex to the graph; those that are (connected and) cyclotomic go towards the next list of cyclotomic graphs; those that are not cyclotomic are tested for minimality (does deleting *any*

© Springer Nature Switzerland AG 2021

J. McKee and C. Smyth, *Around the Unit Circle*, Universitext,

https://doi.org/10.1007/978-3-030-80031-4_13

vertex leave a cyclotomic subgraph). In this way, the examples below were found: 8 on 3 vertices, 35 on 4 vertices, 40 on 5 vertices and 16 on 6 vertices. The diagrams are labelled with a number that indicates the number of vertices, and a distinguishing letter.

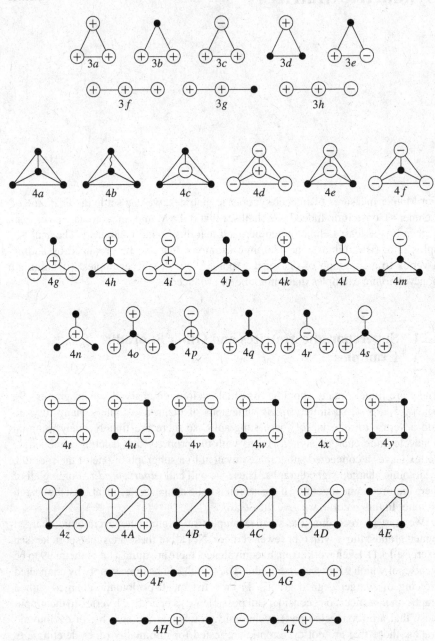

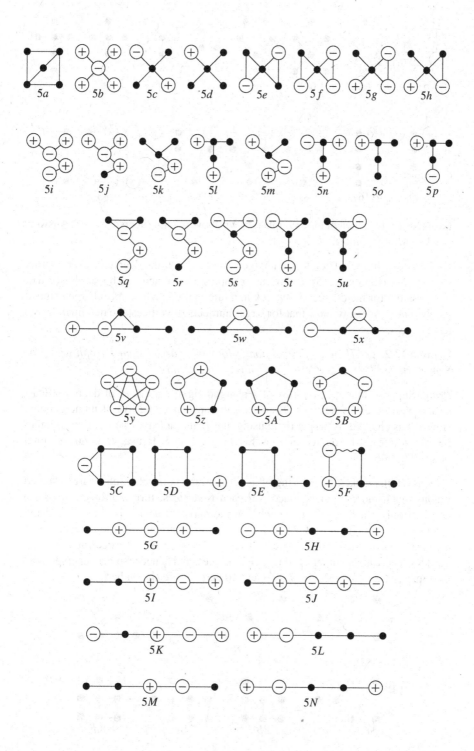

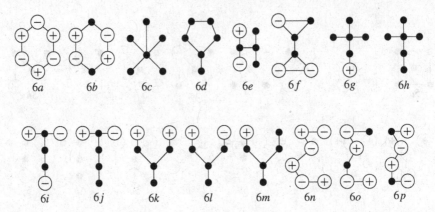

Exercise 13.1 Which of the examples shown above are not supersporadic? [Answer: $3a, 3f, 4d, 4f, 4n, 5y, 6k$ and $6l$.]

We know that in a cyclotomic charged signed graph the degree of every vertex is at most 4 (Lemma 6.15). From the examples $5b$, $5y$ and $6c$ above, we see that it is possible for the degree of a vertex in a minimal noncyclotomic charged signed graph to be 5. We now show that for larger numbers of vertices, the maximal degree reduces to 4.

Lemma 13.2 *Let H be a minimal noncyclotomic charged signed graph with 7 or more vertices. Then every vertex of H has degree at most 4.*

Proof Suppose that x is a vertex of a charged signed graph H of degree at least 5, and with at least 7 vertices. Then the subgraph on x and five of its neighbours (four if x is charged) is proper (here using that H has at least 7 vertices), but cannot be cyclotomic as the degree of x in this subgraph is 5. Hence, H is not minimal noncyclotomic.   □

This lemma makes the growing of larger minimal noncyclotomic charged signed graphs much more efficient: once we reach 6 vertices, further growing needs to consider only attaching vertices such that the maximal degree is at most 4. Using echelon growing (Exercise 6.22) one finds that there are 3 minimal noncyclotomic charged signed graphs on 7 vertices, 4 on 8 vertices, 5 on 9 vertices and 6 on 10 vertices. There are no others on 11, ..., 17 vertices, and hence no other supersporadic examples. The examples with 7, 8, 9 and 10 vertices are shown here.

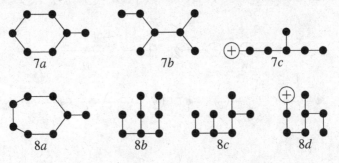

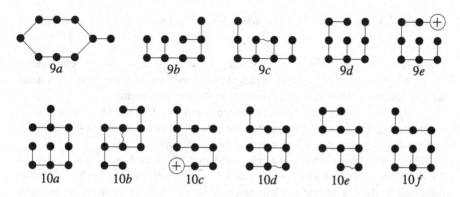

9a   9b   9c   9d   9e

10a   10b   10c   10d   10e   10f

Exercise 13.3 Verify that all the examples $7a$–$10f$ are supersporadic.

13.2 Minimal Noncyclotomic Charged Signed Graphs: Any that Are Not Supersporadic

Having dealt with the supersporadic minimal noncyclotomic charged signed graphs, any other minimal noncyclotomic charged signed graph H has a rather strong property: if we delete any vertex in such a way as to leave a connected subgraph H', then H' is equivalent to a subgraph of a member of at least one of the infinite families T_{2k}, C_{2k}^{++}, and C_{2k}^{+-}, and H' is *not* equivalent to a subgraph of any of S_7, S_8, S_8', S_{14} and S_{16}. In this section, we show that there are finitely many such H, and that they are amongst the examples listed already, namely, $3a$, $3f$, $4d$, $4f$, $4n$, $5y$, $6k$ and $6l$.

Our strategy will be to take two vertices x and y that are as far apart as possible, and consider deleting each in turn to give subgraphs H_x and H_y. Each of H_x and H_y is equivalent to a subgraph of some T_{2k}, C_{2k}^{++}, or C_{2k}^{+-}. We shall use our strong knowledge of the structure of these infinite families to show that we can embed the underlying graph of the whole of H into the underlying graph of one of these family members; then we shall check the 4-circuit rule to deduce that H is in fact cyclotomic. This 4-circuit check will be trivial provided x and y are far enough apart, since the 4-circuit rule holds in both H_x and H_y.

13.2.1 The Uncharged Case

We tackle first the case where H is minimal noncyclotomic, uncharged, and with the property that for any connected subgraph H_v obtained by deleting a single vertex v, we have that H_v is equivalent to a subgraph of T_{2k} for some k (depending on v).

The aim of this subsection is to prove the following lemma.

Lemma 13.4 *There is no minimal noncyclotomic signed graph that is not super-sporadic.*

Proof Let H be a minimal noncyclotomic signed graph that is not supersporadic. We may suppose that H contains at least 17 vertices: smaller cases have been dealt with already by a finite search, as outlined in the previous section.

Let $d_H(x, y)$ denote the distance between x and y in H. The subscript is used as we shall sometimes be considering this distance for vertices in some subgraph of H, and the distance can grow when we move to a subgraph.

Take $x, y \in H$ such that $d_H(x, y)$ is maximal. We must have $d_H(x, y) \geq 2$, as $4a$ and $4b$ are already noncyclotomic; thus $x \not\sim y$. Let H_x and H_y be the subgraphs produced by deleting x and y, respectively, from H. By hypothesis, these are both cyclotomic, and each is equivalent to a subgraph of some T_{2k} (*a priori* k might differ for x and y). Let H_{xy} be the subgraph produced by deleting both x and y from H. By maximality of $d_H(x, y)$, both H_x and H_y are connected, but H_{xy} might not be.

The class graph of H_x is either a path or a cycle, with each class containing either 1 or 2 vertices (here we need at least 9 vertices in H_x, to avoid the unusual features of T_8 and its subgraphs).

First, consider the case where C_{H_x} is a cycle $[y] : [v_1] : \cdots : [v_r]$, with $r \geq 7$. Then, $d_{H_x}(y, v_3) = 3$ and $d_{H_x}(y, v_4) = 4$ (this is where we use that H has at least 17 vertices). We must have $d_H(y, v_3) = 3$, as there can be no shorter path in H via x (since $x \not\sim y$). Thus, $d_H(x, y) \geq 3$. Then $d_H(y, v_4) = 4$, as there can be no shorter path in H via x (now we know that $d_H(x, y) \geq 3$). Hence, $d_H(x, y) \geq 4$.

If $[y] = \{y, \overline{y}\}$, then $C_{H_{xy}}$ is still a cycle, $[\overline{y}] : [v_1] : \cdots : [v_r]$, with the classes $[v_i]$ in H_{xy} being the same as in H_x. To what can x be adjacent? Since $d_H(x, y) > 3, x \not\sim \overline{y}$ (else there would be the path $x \bullet \overline{y} \bullet v_1 \bullet y$), $x \not\sim v_1, x \not\sim v_2, x \not\sim v_r$, and $x \not\sim v_{r-1}$; nor is x adjacent to the conjugate of any of v_1, v_2, v_r and v_{r-1} (should any of these be present in H_{xy}). Given that C_{H_y} is a path or a cycle, we conclude that in C_H, the vertex x joins one of the classes $[v_i]$, with $4 \leq i \leq r - 3$. Then $U_H \hookrightarrow U_{T_{2(r+1)}}$, and with x and y so far apart, the 4-cycle rule in H follows from H_x and H_y.

If $[y] = \{y\}$, then $C_{H_{xy}}$ is the path $[v_1] \bullet \cdots \bullet [v_r]$. (Since $r > 3$, the classes $[v_1]$ and $[v_r]$ in C_{H_x} do not merge in $C_{H_{xy}}$.) Again our bound on $d_H(x, y)$ forces $x \not\sim v_1$, $x \not\sim v_2, x \not\sim v_r$, and $x \not\sim v_{r-1}$; nor is x adjacent to the conjugate of any of v_1, v_2, v_r and v_{r-1} (should any of these be present in H_{xy}). In C_H the vertex x joins one of the classes $[v_i]$, with $4 \leq i \leq r - 3$. Then $U_H \hookrightarrow U_{T_{2(r+1)}}$, and with x and y so far apart the 4-cycle rule in H follows from H_x and H_y.

Next consider that case where C_{H_x} is a path $[v_1] \bullet \cdots \bullet [v_r] \bullet [y] \bullet [w_1] \bullet \cdots \bullet [w_s]$, for some $0 \leq r \leq s, r + s \geq 7$. We may suppose that C_{H_y} is also a path, or we are in the previous case with the roles of x and y swapped. Again we see that x is not adjacent to any of v_{r-1}, v_r, w_1 and w_2, or their conjugates, should any of these vertices be present. With $d_H(x, y)$ maximal and C_{H_y} not a cycle, we see that the only possibilities are: (i) $r = 0$ and either $x \in [w_s]$ (in H) or $[x] = \{x\}$ with $x \sim w_s$ (and $x \sim \overline{w}_s$ if present); (ii) $s \leq r + 1$, $[y] = \{y\}$, $[x] = \{x\}$, $x \sim v_1$ and $x \sim v_s$. In either case, U_H embeds in some $U_{T_{2k}}$, and the 4-cycle rule holds in H ($d_H(x, y) > 2$).

We conclude in all cases that H is in fact cyclotomic, using Lemma 9.15, contradicting the hypothesis. So no such H exists: there is no minimal noncyclotomic signed graph that is not supersporadic.   □

13.2.2   The Charged Case

Lemma 13.5 *The only minimal noncyclotomic charged signed graphs that are not supersporadic are* $3a$, $3f$, $4d$, $4f$, $4n$, $5y$, $6k$ *and* $6l$.

Proof Let H be a minimal noncyclotomic charged signed graph that is not supersporadic. Having disposed of the uncharged case, we now assume that H contains at least one charged vertex. Taking x and y as before (i.e., with $d_H(x, y)$ maximal), we may assume that H_x contains a charged vertex, although H_y might or might not. The absence of long cycles makes the charged case simpler. Again, having performed our previous computations, we can suppose that H has at least 17 vertices, and that $d_H(x, y) \geq 4$ (more than we need in this charged case).

Given that C_{H_x} and C_{H_y} are paths, and that $d_H(x, y)$ is maximal, one soon checks that the only possibility for C_H is a path of the shape $[x] \bullet [v_1] \bullet \cdots \bullet [v_r] \bullet [y]$. Then $U_H \hookrightarrow C_{2k}^{\bullet\bullet}$ for suitable k, and with x and y so far apart the 4-circuit rule for H follows from H_x and H_y. We conclude by Lemma 9.19 that H is cyclotomic.   □

13.3   Completing the Classification

The hard work in finding all the minimal noncyclotomic charged signed graphs has been done. We can now fill in the final few steps to complete the classification of all minimal noncyclotomic integer symmetric matrices, and settle the analogue of Lehmer's Conjecture in this setting.

Theorem 13.6 *Up to equivalence, the minimal noncyclotomic integer symmetric matrices are the* 117 *sporadic examples* $3a$–$10f$ *of Sect. 13.1, the* 8 *sporadic examples*

$$\begin{pmatrix} 0 & 2 & 2 \\ 2 & 0 & 2 \\ 2 & 2 & 0 \end{pmatrix}, \begin{pmatrix} 0 & 2 & 2 \\ 2 & 0 & 1 \\ 2 & 1 & 0 \end{pmatrix}, \begin{pmatrix} 0 & 2 & 1 \\ 2 & 0 & 1 \\ 1 & 1 & 0 \end{pmatrix}, \begin{pmatrix} 0 & 2 & 0 \\ 2 & 0 & 2 \\ 0 & 2 & 0 \end{pmatrix},$$

$$\begin{pmatrix} 0 & 2 & 1 \\ 2 & 0 & 1 \\ 1 & 1 & 1 \end{pmatrix}, \begin{pmatrix} 0 & 2 & 1 \\ 2 & 0 & 1 \\ 1 & 1 & -1 \end{pmatrix}, \begin{pmatrix} 0 & 2 & 0 \\ 2 & 0 & 1 \\ 0 & 1 & 1 \end{pmatrix}, \begin{pmatrix} 0 & 2 & 0 \\ 2 & 0 & 1 \\ 0 & 1 & 0 \end{pmatrix},$$

and the four infinite families

$$(a)\,,\ a \geq 3,\quad \begin{pmatrix} 2 & b \\ b & c \end{pmatrix},\quad |b| \geq 1, |c| \leq 2,$$

$$\begin{pmatrix} 1 & d \\ d & e \end{pmatrix},\quad |d| \geq 2, |e| \leq 1,\quad \begin{pmatrix} 0 & f \\ f & 0 \end{pmatrix},\quad f \geq 3.$$

Proof Let $A = (a_{ij})$ be a minimal noncyclotomic integer symmetric matrix.

Suppose that $|a_{ij}| \geq 3$ for some i and j. If $i = j$, then since (a_{ii}) is already noncyclotomic this must be the whole of A. Changing sign if necessary, we have a member of our first infinite family. If $i \neq j$, then by symmetry we may suppose that $i < j$, and since the submatrix

$$\begin{pmatrix} a_{ii} & a_{ij} \\ a_{ij} & a_{jj} \end{pmatrix}$$

is noncyclotomic, this must be the whole of A. By minimality of A, we must have $|a_{ii}| \leq 2$ and $|a_{jj}| \leq 2$. By permuting and switching, we can suppose that $|a_{jj}| \leq a_{ii} \leq 2$, and if $a_{ii} = a_{jj} = 0$ then $a_{ij} > 0$. We see that A is a member of one of the other infinite families in the statement of the theorem.

Now we can suppose that all entries of A have modulus at most 2, and consider the case where some entry has modulus 2, and changing signs if necessary we can suppose that $a_{ij} = 2, i \leq j$. If $i = j$, then since (2) is cyclotomic A must have at least two rows, and by connectedness the ith row must contain some nonzero off-diagonal entry. Permuting if necessary, A contains a submatrix

$$\begin{pmatrix} 2 & a \\ a & b \end{pmatrix},$$

where $a \neq 0$ and $|b| \leq 2$. This matrix is not cyclotomic, so must be (equivalent to) the whole of A, and we have a member of our second infinite family. If $i < j$, then A contains a submatrix of the shape

$$\begin{pmatrix} a & 2 \\ 2 & b \end{pmatrix}.$$

If a and b are not both 0, then this submatrix is not cyclotomic, so equals A, and we have that A is equivalent to a member of either our second or third infinite family. If $a = b = 0$, then this submatrix is cyclotomic, so is not the whole of A. One checks that up to equivalence every connected 3×3 matrix containing this submatrix is noncyclotomic, and the only minimal ones are equivalent to one of the eight sporadic examples (Exercise 13.7).

We are now reduced to the case where all entries of A have modulus at most 1: A is the adjacency matrix of a charged signed graph. We have seen that up to equivalence there are 117 such charged signed graphs, listed in Sect. 13.1. □

Exercise 13.7 Verify the claim concerning connected 3×3 matrices that contain

$$\begin{pmatrix} 0 & 2 \\ 2 & 0 \end{pmatrix}$$

as a submatrix. (Write the larger matrix as

$$\begin{pmatrix} 0 & 2 & a \\ 2 & 0 & b \\ a & b & c \end{pmatrix},$$

where a and b are not both 0. Verify first that $|c| < 2$, else there is a noncyclotomic 2×2 submatrix.)

Corollary 13.8 *Let A be an integer symmetric matrix. Then either $M(A) = 1$, or $M(A)$ is at least Lehmer's number.*

Proof If $M(A) \neq 1$, then A is not cyclotomic, so contains some minimal noncyclotomic submatrix B and $M(A) \geq M(B)$. One checks that every member of any of the infinite families of Theorem 13.6 has Mahler measure at least 1.722, and that each of the 117 sporadic examples has Mahler measure at least Lehmer's number, with equality for the charged signed graphs $5u$, $5N$ and $10f$. □

13.4 Notes

Theorem 13.6 is due to McKee and Smyth, and appeared in [MS12a], with a different proof that used the concept of graphs that have a *profile*. In the language of this chapter, a graph has a profile if the common neighbour classes all have size one or two, and the class graph is a path or a cycle.

Corollary 13.8 shows that the smallest Mahler measure of noncyclotomic integer symmetric matrix is Lehmer's number. By growing minimal noncyclotomic examples, one can find integer symmetric matrices that have some other small Mahler measures. This exercise was done in [MS12a], finding all connected integer symmetric matrices that have Mahler measure below 1.3. The matrices in question are all adjacency matrices of charged signed graphs. The following table records the results from [MS12a]. Some of the charged signed graphs appeared above; the new ones are $5O$, $6q$, ..., $6v$, $7d$, ..., $7h$, $8e$, ..., $8h$, $9f$, ..., $9j$, $10g$, ..., $10i$, $11a$, ..., $11d$, $12b$ (the name comes from [MS12a] where there is also a $12a$ that has larger Mahler measure), $13a$ and $14a$, which are drawn below.

Mahler measure	Charged signed graph
1.17628	$5u$, $5N$, $10f$
1.18837	$9e$
1.20003	$8d$
1.21639	$5p$, $5M$, $7c$, $10e$
1.21972	$9g$
1.23039	$5L$, $6v$, $11c$
1.23632	$8e$
1.24073	$6m$
1.25364	$10c$
1.25622	$9h$
1.26123	$5F$, $5K$, $6t$, $6u$, $7d$, $10d$, $12b$
1.26730	$8h$, $9i$, $10i$
1.28064	$4I$, $5o$, $5x$, $6q$, $6r$, $6s$, $8f$, $9d$, $10g$, $11a$, $11b$, $13a$
1.28929	$11d$
1.29349	$5J$, $5O$, $7e$, $7f$, $7g$, $8g$, $9j$, $10h$, $14a$
1.29568	$9f$

Mossinghoff's online table [Mos11] gives thousands of Mahler measures below 1.3. There are 236 known below 1.25, of which only 8 come from integer symmetric matrices. Although the theory for this special class of Mahler measures is more advanced than for the general case, we see that only a tiny proportion of small Mahler measures arise this way. To make a significant advance, we need to break symmetry, but we lose the tight control that we have for symmetric matrices.

The related question of classifying integer symmetric matrices that have small **spectral radius** (the largest modulus of any eigenvalue) was also considered by McKee and Smyth in [MS12a], including a list of all the noncyclotomic examples whose spectral radius is at most 2.019.

The connected charged signed graphs on five, six or seven vertices (other than the minimal noncyclotomic ones) whose Mahler measure lies in the interval (1, 1.3):

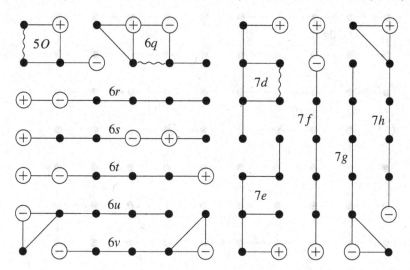

The connected charged signed graphs on eight or nine vertices (other than the minimal noncyclotomic ones) whose Mahler measure lies in the interval (1, 1.3):

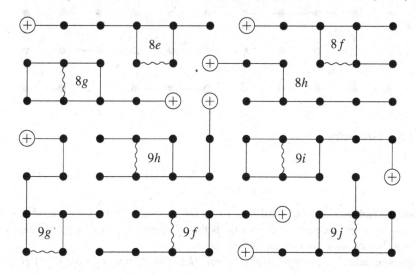

The connected charged signed graphs on ten or more vertices (other than the minimal noncyclotomic ones) whose Mahler measure lies in the interval (1, 1.3):

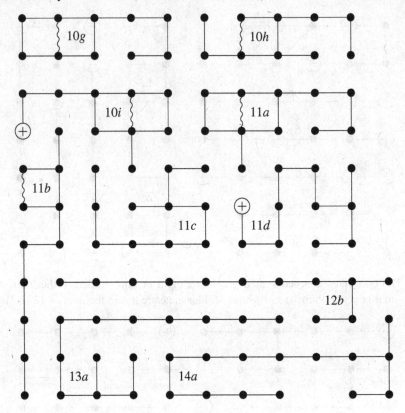

13.5 Glossary

Much of the terminology and notation was introduced in Chap. 9, and we refer the reader to the Glossary at the end of that chapter.

spectral radius.   The spectral radius of a square matrix is the largest modulus of any of its eigenvalues. If the matrix is the adjacency matrix of a graph, then its spectral radius is, in fact, an eigenvalue.

supersporadic.   A charged signed graph H is called supersporadic if: (i) H is a minimal noncyclotomic charged signed graph, and (ii) there exists some vertex of H whose deletion leaves a connected subgraph that it is equivalent to a subgraph of at least one of S_7, S_8, S_8', S_{14} and S_{16}.

Chapter 14
The Method of Explicit Auxiliary Functions

14.1 Conjugate Sets of Algebraic Numbers

Recall from Chap. 1 that a conjugate set of algebraic numbers is a finite subset of the field $\overline{\mathbb{Q}}^{\times}$ of algebraic numbers whose elements are the zeros of an irreducible polynomial with integer coefficients. Certainly, we can assume that the content of such a polynomial (the gcd of its coefficients) is 1 and that its leading coefficient is positive. This is the **minimal polynomial** of the set, or indeed of any member of it. A fundamental question about conjugate sets is where in the complex plane they can lie. The method of explicit auxiliary functions can help us answer this question. Most of the applications of this method have been to conjugate sets of algebraic integers; however, the method also can be applied to conjugate sets of algebraic numbers.

But two different conjugate sets are more than simply disjoint: they cannot be too close together, in the following sense. Suppose that $P(x)$ and $Q(x)$ are the minimal polynomials (coefficients in $\mathbb{Z}$, as above) of two different conjugate sets. Then the **resultant**, defined in Sect. A.2 of Appendix A, is a rational integer. Moreover, from (A.2) this integer is not zero. Then we can invoke the Fundamental Theorem of Arithmetic: *Every nonzero integer has modulus at least* 1. (Masser [Mas16, p. 7] calls this the Fundamental Theorem of Transcendence.)

Now fix a monic minimal polynomial P and let Q be any other minimal polynomial. The Eq. (A.2) now reads

$$\operatorname{res}_x(P, Q) = \prod_{\alpha : P(\alpha) = 0} Q(\alpha). \tag{14.1}$$

Taking the absolute value, invoking the Fundamental Theorem, and taking logs gives

$$\sum_{\alpha} \log |Q(\alpha)| \geq 0, \tag{14.2}$$

where the sum is over the conjugate set $\{\alpha : P(\alpha) = 0\}$.

© Springer Nature Switzerland AG 2021
J. McKee and C. Smyth, *Around the Unit Circle*, Universitext,
https://doi.org/10.1007/978-3-030-80031-4_14

14.2 The Optimisation Problem

In this section, we describe the method of explicit auxiliary functions, and later in the chapter, we apply it to several problems concerning the distribution of the conjugates of algebraic integers and, more generally, algebraic numbers α on the complex plane.

Suppose that we are given a closed subset S of the complex plane, and also a continuous map $f : S \to \mathbb{R}_{\geq 0} \cup \{\infty\}$. The problem that the method of explicit auxiliary functions tries to solve is the following:

Problem 14.1 Describe the spectrum of values of $f(P) := \frac{1}{d} \sum_{i=1}^{d} f(\alpha_i)$, where

$$P(z) = \prod_{i=1}^{d} (z - \alpha_i) \in \mathbb{Z}[z] \tag{14.3}$$

is irreducible, with all the α_i lying in S, as $\alpha_1, \ldots, \alpha_d$ range over all such conjugate sets of algebraic integers.

Exercise 14.2 Show that the spectrum of values of $f(P)$ for $\alpha_1, \ldots, \alpha_d \in S$ is the same as the spectrum of such values for $\alpha_1, \ldots, \alpha_d \in S \cap \overline{S}$.

Thus, we may as well assume that $S = \overline{S}$, i.e., that S is symmetric about the real axis. In fact, the method typically (when it works!) shows that the smallest elements of the spectrum are discrete, and finds some of them explicitly. Sometimes one can prove by other methods that the spectrum is dense in $\mathbb{R}$ from some point on, i.e., in an interval (c, ∞) for some c.

The idea of the method is the following: given any list of minimal polynomials $P_1, \ldots, P_k, \ldots$ say, we know that for any minimal polynomial P as in (14.3) not on the list, we have (by (14.2)) for all k that

$$\sum_{\alpha} \log |P_k(\alpha)| \geq 0, \tag{14.4}$$

where the sum is over the zeros of P.

Thus, to find the smallest $f(P)$ where P is not equal to any P_k, we need to solve the optimisation problem

$$\text{Minimise } f(P) \text{ subject to (14.4).} \tag{14.5}$$

We can regard an algebraic number α of degree d with conjugates α_i $(i = 1, \ldots, d)$ as an atomic measure with atoms of weight $1/d$ at each α_i. Thus, it makes sense to expand the optimisation problem to all probability measures with support in S. This gives us the measure problem

Find the infimum of $\int_S f(z)\mu(z)$ $\qquad$ (14.6)

subject to $\int_S \log|P_k(z)|\mu(z) \geq 0$ $\qquad$ for all k, $\qquad$ (14.7)

$$\int_S \mu(z) \geq 1 \qquad \text{and } \mu(z) \geq 0 \text{ on } S.$$

If there is a minimal measure μ whose support is a finite number of points α_i ($i = 1, \ldots, d$) with weights $1/d$ and, furthermore, the α_i form a conjugate set, then we will have solved the original problem, the minimum being $m = \frac{1}{d}\sum_{i=1}^d f(\alpha_i)$. In general, although the minimal measure will be atomic with finite support, the nonzero atomic weights will not always be equal. However, we will then know that the spectrum of values of $f(P)$ in the interval $[0, m)$ consists exactly of those $f(P_k)$ that lie in this interval.

The modified problem is a doubly infinite linear programming (LP) problem, as it generally has uncountably many variables—the weights at all the points of S—and countably many constraints (14.4). While it may be possible in the future to deal with infinitely many constraints, it is currently only practicable to have finitely many, say for $k = 1, \ldots, K$. Then the problem becomes a *semi-infinite* LP problem. A big issue is then: which polynomials P_k to choose for the constraints (14.7)? Well, if we want to find all α with $\frac{1}{d}\sum_{i=1}^d f(\alpha_i) \leq m$, then certainly we will need to include all P_α for all the α we know of with $m' := \frac{1}{d}\sum_{i=1}^d f(\alpha_i) < m$. Otherwise the measure μ with weights $1/d$ at each of the conjugates of α that optimised the primal problem would have infimum at most m', which is less than m. But we need that infimum to be *at least* m so that we know all the values of $\frac{1}{d}\sum_{i=1}^d f(\alpha_i) \leq m$ in $(0, m]$.

14.2.1 Dualising the Problem

Recall that for a standard LP problem (finitely many variables, finitely many constraints), the dual problem to the (primal) problem

Minimise $\qquad \sum_{i=1}^I b_i x_i \qquad$ with $x_i \geq 0$ $\quad (i = 1, \ldots, I)$ $\qquad$ (14.8)

subject to $\qquad \sum_{i=1}^I a_{ik} x_i \geq c_k \quad (k = 1, \ldots, K)$.

is the (dual) problem

$$\text{Maximise} \qquad \sum_{k=1}^{K} c_k y_k \qquad \text{with } y_k \geq 0 \quad (k = 1, \ldots, K) \qquad (14.9)$$

$$\text{subject to} \qquad \sum_{k=1}^{K} a_{ik} y_k \leq b_i \quad (i = 1, \ldots, I).$$

These problems can be simultaneously solved by the simplex method. They have the same optimum value, and the solutions to both problems can be read off from a single optimal simplex tableau. Analogously, we can find the dual of (14.6) by assigning the variable y_0 to the constraint $\int_S \mu(z) \geq 1$ and, for $k = 1, \ldots, K$, the variable y_k to the constraint $\int_S \log |P_k(z)| \mu(z) \geq 0$. Then its dual is easily seen to be

$$\text{Maximise} \qquad y_0 \qquad \text{with all } y_k \geq 0 \quad (k = 1, \ldots, K)$$

$$\text{subject to} \qquad y_0 + \sum_{k=1}^{K} y_k \log |P_k(z)| \leq f(z) \qquad \text{for all } z \in S.$$

Thus, the dual problem to (14.6) is to evaluate

$$\sup_{y_1 \geq 0, \, \ldots, \, y_K \geq 0} \min_{z \in S} \left(f(z) - \sum_{k=1}^{K} y_k \log |P_k(z)| \right). \qquad (14.10)$$

14.2.2   Method Outline

Here is an outline of the main steps of one solution method.

1. By some kind of search procedure, find as many monic integer irreducible polynomials P as possible with $f(P)$ small. Let us call them $P_1, P_2, \ldots, P_k$.
2. Choose a collection $P_1, P_2, \ldots, P_\ell$ of monic integer irreducible polynomials with $\ell \geq k$, so that $P_1, P_2, \ldots, P_k$ are included, and define the auxiliary function

$$A(z, \mathbf{y}) := f(z) - \sum_{j=1}^{\ell} y_j \log |P_j(z)|,$$

where $\mathbf{y} = (y_1, \ldots, y_\ell)$ and the y_j are real and nonnegative.
3. Choose $\mathbf{y}$ such that $\min_{z \in S} A(z, \mathbf{y})$ is as large as possible, say at least m. (See below for one approach to this step.) The P_j should have been chosen earlier to make m large.
4. Then we have, for every monic $P(z)$ not equal to any P_j, with $\alpha_1, \ldots, \alpha_d$ the zeros of P, that

$$md \leq \sum_{i=1}^{d} A(\alpha_i, \mathbf{y}) = \sum_{i=1}^{d} f(\alpha_i) - d \sum_{j=1}^{\ell} y_j \log \left| \prod_{i=1}^{d} P_j(\alpha_i) \right| .$$

Now

$$\prod_{i=1}^{d} P_j(\alpha_i) = \mathrm{res}_z(P, P_j)$$

which, since $\gcd(P, P_j) = 1$, is a nonzero integer; see (14.1). Hence, it is at least 1 in modulus, and so $\log | \mathrm{res}_z(P, P_j)| \geq 0$ for all j. Thus

$$f(P) \geq m + \frac{1}{d} \sum_{j=1}^{\ell} c_j \log | \mathrm{res}_z(P, P_j)| \geq m .$$

5. We now have $f(P) \geq m$ for every $P(z)$ not equal to any P_j. Some of the $f(P_j)$ could, however, be less than m. Indeed, supposing that

$$f(P_1) \leq f(P_2) \leq \cdots \leq f(P_r) < m ,$$

then we have that the set of all $f(P)$ in the interval $(-\infty, m)$ is

$$\{f(P_1), f(P_2), \cdots, f(P_r)\} .$$

14.2.2.1 A Method for Step 3

In all applications so far, we can assume that the set $S \subset \mathbb{C}$, where our conjugate sets lie, is actually one-dimensional. This is because, even when it is two-dimensional, we always have $A(z, \mathbf{y})$ being harmonic on S, so that its minimum is attained on its boundary. Thus, we can replace S by its boundary in the optimisation problem. To avoid technicalities, however, we shall assume for the purposes of discussion that S is simply a real interval $[a, b]$, say, and so replace z by the real variable x. One way of proceeding is by the following 'substeps'.

1. Choose a large finite set $X \subset [a, b]$. One possibility is a fine mesh of values, spanning X. Another possibility is to take such a mesh between each pair of consecutive real zeros of $\prod_{j=1}^{\ell} P_j(x)$, and take X to be the union of these meshes. This averts the possibility that a mesh point could be at, or very close to, a zero of some P_j.
2. Solve the linear programming (LP) problem

$$\text{Maximise}_{y_1 \geq 0, \dots, y_\ell \geq 0} \min_{x \in X} A(x, \mathbf{y}) .$$

Denote by $\mathbf{y}(X)$ a vector $\mathbf{y}$ where the maximum occurs, and let $\overline{m}(\mathbf{y}(X))$ be that maximum.

3. Now enlarge X by adding to it the locations of the local minima in $[a, b]$ of the function $A(x, \mathbf{y}(X))$. Denote by $\underline{m}(\mathbf{y}(X))$ the global minimum of $A(x, \mathbf{y}(X))$ on $[a, b]$.

4. Repeat substeps 2 and 3 above, with the enlarged set X, until the difference $\overline{m}(\mathbf{y}(X)) - \underline{m}(\mathbf{y}(X))$ is very small. Then we can take $m := \underline{m}(\mathbf{y}(X))$.

14.2.2.2 Seeking New Constraint Polynomials

Denote by $\mathbf{y}^* := \mathbf{y}(X)$ the value of $\mathbf{y}$ obtained above in the final LP step. Corresponding to $\mathbf{y}^*$, which gives the maximum $\overline{m}(\mathbf{y}^*)$ for the current set X is a nonnegative vector at which the maximum of the dual problem occurs. (This is the dual of the dual, so is essentially the original problem (14.6), since we have been working with the dual (14.10).) Each component, $v(x)$ say, of this vector corresponds to a value of x, so that we can consider the vector as an atomic measure $\mu^*(x)$ ($x \in [a, b]$). Suppose we have an integer polynomial $P(x)$ that we are considering adding to our list $P_1, \ldots, P_\ell$ of constraint polynomials. Now if $\int_a^b \log |P(x)| \mu^*(x) \geq 0$, then μ^* will still be an optimal measure. However, if $\int_a^b \log |P(x)| \mu^*(x) < 0$, then μ^* will no longer be optimal, so that adding P to our list of polynomials may give a better value of $\overline{m}$, and so also a larger value of m.

14.2.2.3 Semi-infinite Programming Literature

The algorithm described above is tailored specifically to the type of problems discussed here. For more general semi-infinite programming methods by specialists in the field, one could refer, for instance, to the books [GG83, GL98], the more recent survey [LS07] and the updates [GL17, GL18].

In the next sections, we look at some specific applications of this auxiliary function method.

14.3 The Schur–Siegel–Smyth Trace Problem

Let α be a totally positive algebraic integer of degree d, so that its conjugates $\alpha = \alpha_1, \ldots, \alpha_d$ are all positive. The trace problem addresses the question of how small the mean trace $\overline{\mathrm{tr}}(\alpha) := \frac{1}{d} \sum_{i=1}^{d} \alpha_i$ can be. Specifically, what is $\liminf \overline{\mathrm{tr}}(\alpha)$, where the infimum is taken over all such α? Thus, defining $\mathcal{T}$ to be the set of all such mean traces, we are interested in finding the smallest limit point of $\mathcal{T}$.

In 1918 Schur [Sch18] and in 1945 Siegel [Sie45] used discriminant information to study this problem. However, Smyth [Smy84] and later authors used resultants for semi-infinite programming, as described in this chapter, and obtained improved bounds. For an excellent survey of the trace problem, see Aguirre and Peral [AP08].

We start with an exercise that uses an auxiliary function in a typical, yet simple, way.

Exercise 14.3 Let $\varphi = \frac{1}{2}(1 + \sqrt{5})$. Show that for $x > 0$

$$x - 1.54 - 0.65 \log(x) - 0.95 \log|x - 1| - 0.14 \log|x^2 - 3x + 1| > 0.$$

Deduce that a totally positive algebraic integer $\alpha \neq 1$ has $\overline{\mathrm{tr}}(\alpha) \geq \frac{3}{2}$, with equality if and only if $\alpha = \varphi^2$ or φ^{-2}.

This is a very simple example of the type shown in Fig. 14.1.

14.3.1 Totally Positive Algebraic Integers with Small Mean Trace

If ω is a primitive nth root of unity, and $n \neq 2$, then $\alpha := \omega + \omega^{-1} + 2$ is a totally positive algebraic integer of degree $d := \varphi(n)$ (where φ is Euler's φ-function). Its mean trace is $2 + 2\mu(n)/\varphi(n)$, where μ is the Möbius function (see Exercise 14.4 below). This shows that 2 is a limit point of $\mathcal{T}$. But is it the smallest one?

A related question is: given a rational number $q \in (0, 2)$, are there only finitely many totally positive algebraic integers α with $\overline{\mathrm{tr}}(\alpha) \leq q$? For instance, is this true for $q = 9/5$? Schur [Sch18] proved that this was true for $q = \sqrt{e}$, while Siegel [Sie45] proved it for $q = 1.7336 \cdots$. Using the method of auxiliary functions, this was improved in [Smy84] to $q < 1.77193$.

Exercise 14.4 Let ω be a primitive nth root of unity, where $n > 2$.

(a) Show that $\overline{\mathrm{tr}}(\omega + \omega^{-1} + 2) = 2 + 2\mu(n)/\varphi(n)$. (See also Exercise 5.28(a).)
(b) Show that $\mathcal{T} \cap [0, 2)$ contains all numbers of the form $2 - 2/\varphi(P)$, where P is the product of an odd number of distinct primes.
(c) Find all known values of P for which $\varphi(P)$ is a power of 2 and $\mu(P) = -1$. Which powers of 2 are thereby produced?

In the next exercise, we find totally positive algebraic integers of mean trace less than 2 whose mean traces are of the form $2 - 1/2^k$ for all $k \in \mathbb{N}$.

Exercise 14.5 Define $\beta^{(0)} = 1$, and $\beta^{(n)} > 1$ for $n \geq 1$ by

$$\beta^{(n)} + 1/\beta^{(n)} = 2 + \beta^{(n-1)}. \tag{14.11}$$

It is known that $\beta^{(n)}$ has degree 2^n; see Proposition A.28 in Appendix A.

(a) Show that $\overline{\mathrm{tr}}(\beta^{(n)}) = 2 - 1/2^n$.
(b) Show that for $n \geq 2$ the numbers $\beta^{(n)}$ are greater than 4, and so not of the form
     $\omega + \omega^{-1} + 2$ as in Exercise 14.4.

For more information on $\beta^{(n)}$ and its conjugates, see Appendix A.5.1. Figure 14.1 shows a plot using the logs of 97 polynomials (taken from [Fla16]), showing that $\overline{\mathrm{tr}}(\alpha) \geq 1.7928125$, apart from for $\alpha = \beta^{(n)}$ $(n = 0, 1, 2, 3)$, $(2\cos(2\pi/7))^2$ and $(2\cos(2\pi/60))^2$. At the time of writing, the record is $\overline{\mathrm{tr}}(\alpha) \geq 1.793145$ [WWW21].

On the other hand, it is shown in [Smy99] that the semi-infinite programming method using resultants could not prove, for *every* q less than 2, that there were only finitely many totally positive algebraic integers of mean trace less than q. Specifically, the method could not work if $q > 2 - 10^{-41}$. Serre [AP08, Appendix], greatly improved this condition to $q > 1.898302 \cdots$.

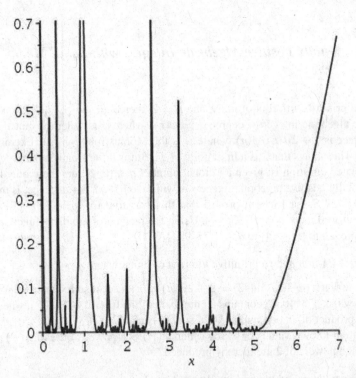

Fig. 14.1 A plot of a function $x - 1.79281259997 - \sum_{i=1}^{97} c_i \log |P_i(x)|$ for specific constants $c_i > 0$ and polynomials $P_i(x)$ (see [Fla16]). Careful evaluation of the function at its local minima shows that it is nonnegative for all $x > 0$

14.4 The Mean Trace of α Less Its Least Conjugate

In this section, we compare the mean trace $\overline{\mathrm{tr}}(\alpha)$ of a totally real algebraic integer to its smallest conjugate (assumed to be α itself), obtaining a lower bound for their difference.

Theorem 14.6 *Suppose that $\alpha \in [0, 1)$ is a totally real algebraic integer with all its conjugates lying in the interval $[\alpha, \infty)$. Then*

$$\overline{\mathrm{tr}}(\alpha) \geq 1.68 + \alpha \qquad (14.12)$$

unless α has minimal polynomial given in the following table.

$P_\alpha(x)$	$\overline{\mathrm{tr}}(\alpha) - \alpha$
x	0
$x^2 - 3x + 1$	1.1180340
$x^2 - 5x + 5$	1.1180340
$x^2 - 4x + 2$	1.4142136
$x^3 - 5x^2 + 6x - 1$	1.4686044
$x^3 - 8x^2 + 19x - 13$	1.4686044
$x^4 - 7x^3 + 13x^2 - 7x + 1$	1.5222229
$x^3 - 6x^2 + 9x - 3$	1.5320889
$x^3 - 9x^2 + 24x - 19$	1.5320889
$x^4 - 7x^3 + 14x^2 - 8x + 1$	1.5770909
$x^3 - 7x^2 + 14x - 7$	1.5803129
$x^4 - 9x^3 + 27x^2 - 31x + 11$	1.6056743
$x^5 - 9x^4 + 26x^3 - 29x^2 + 11x - 1$	1.6688217
$x^6 - 11x^5 + 42x^4 - 68x^3 + 46x^2 - 12x + 1$	1.6702846
$x^6 - 13x^5 + 64x^4 - 151x^3 + 177x^2 - 96x + 19$	1.6703513
$x^3 - 6x^2 + 8x - 2$	1.6751309

It is far from clear how much further this constant can be increased. But certainly, as the examples $\alpha = 2 - 2\cos(2\pi/n)$ $(n = 1, 2, \cdots)$ show, it must be less than 2.

For the proof of Theorem 14.6, we need the following lemma, and its corollary.

Lemma 14.7 *Suppose that we have a positive constant C, a positive integer ℓ, polynomials $P_j(x) \in \mathbb{Z}[x]$ $(j = 0, 1, \ldots, \ell - 1)$, constants*

$$0 = b_0 < b_1 < b_2 < \cdots < b_{\ell-1} < b_\ell = 1$$

and constants $c_j > 0$ $(j = 0, 1, \ldots, \ell - 1)$ such that for $j = 0, 1, \ldots, \ell - 1$

$$x - c_j \log |P_j(x)| \geq C + b_{j+1} \qquad \text{for all } x \geq b_j. \qquad (14.13)$$

Then for each $b \in [0, 1)$ there is some $j \in \{0, 1, \ldots, \ell - 1\}$ such that, for $c' := c_j$ and $P(x) := P_j(x)$ we have

$$x - c' \log |P(x)| \geq C + b \qquad \text{for all } x \geq b. \tag{14.14}$$

The proof is clear, by choosing $j \in \{0, 1, \ldots, \ell - 1\}$ such that $b \in [b_j, b_{j+1})$. Then in (14.13), we can replace c_j by c', both b_j and b_{j+1} by b and P_j by P.

Corollary 14.8 *Suppose that the hypotheses of the lemma hold, for some values of all the parameters. Then every totally real algebraic integer $\alpha \in [0, 1)$ with conjugates in $[\alpha, \infty)$ satisfies*

$$\overline{\text{tr}}(\alpha) \geq C + \alpha, \tag{14.15}$$

with a finite number of explicit exceptions, namely the real zeros α of $\prod_{j=0}^{\ell-1} P_j(x)$ that have all their conjugates in $[\alpha, \infty)$ but do not satisfy (14.15).

Proof Take $b := \alpha$ and, for $x := \alpha_i$, average (14.14) over the conjugates α_i $(i = 1, \ldots, d)$ of α. This gives

$$\overline{\text{tr}}(\alpha) \geq C + \alpha + \frac{c'}{d} \log \left(\prod_i |P(\alpha_i)| \right).$$

Thus, (14.15) holds provided that α is not a zero of P. Since P could be any of the P_j, depending on α, the result follows. $\square$

The proof of Theorem 14.6 then follows from the corollary, on taking $C = 1.68$, $\ell = 36$ and choosing the b_j, c_j and P_j for $j = 0, \ldots, 35$ to satisfy the conditions of Lemma 14.7. For this data, we refer to Flammang [Fla19].

14.5   An Upper Bound Trace Problem

This application of the explicit auxiliary function method considers finding an *upper* bound for the mean trace of totally positive algebraic integers whose conjugates are confined to a fixed interval. We restrict ourselves to an example. This result is applied in Chap. 5.

Proposition 14.9 *Suppose that α is a totally real algebraic integer, lying with all its conjugates in the interval $(0, 5.04)$. Then the mean trace of α is less than 3.25 unless α is conjugate to $\alpha' := \frac{1}{4} \left(13 + \sqrt{5} + \sqrt{38 - 14\sqrt{5}} \right)$ or to $3 + 2\cos(2\pi/n)$ for $n = 1, 6, 10, 14$ or 15.*

The proof comes from consideration of the auxiliary function

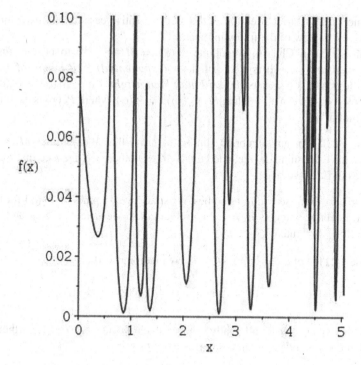

Fig. 14.2 A plot of $f(x)$, used in the proof of Proposition 14.9

$$f(x) = 3.250 - x$$
$$- 0.110 \log |x - 3| - 0.530 \log |x - 4| - 0.620 \log |x - 5|$$
$$- 0.018 \log |x^2 - 6x + 6| - 0.028 \log |x^2 - 6x + 7| - 0.194 \log |x^2 - 7x + 11|$$
$$- 0.130 \log |x^3 - 10x^2 + 31x - 29| - 0.045 \log |x^4 - 13x^3 + 58x^2 - 98x + 41|$$
$$- 0.040 \log |x^4 - 13x^3 + 59x^2 - 107x + 61|,$$

which is positive on $(0, 5.04)$, as illustrated in Fig. 14.2. The polynomials defining this function are the minimal polynomial of α' along with the minimal polynomials of $3 + 2\cos(2\pi/n)$ for $n = 1, 4, 6, 8, 10, 12, 14$ and 15. Then suppose α is as in Proposition 14.9, but is not a zero of any of these polynomials.

14.6 Mahler Measure of Totally Real Algebraic Integers

For the remaining sections of this chapter we use the explicit auxiliary function method to obtain lower bounds for the Mahler measures of restricted sets of algebraic numbers. We start with the case of totally real algebraic numbers (α and its conjugates all real), and in this case we split the work into two subsections, treating algebraic integers in this section and then seeing what can be done more generally with algebraic numbers in the following section. In fact, for the totally real case, we can immediately restrict attention to the totally positive case (α and its conjugates all positive), in view of the following fact.

As in (1.15) in Chap. 1, we define $\overline{M}(\alpha) := M(\alpha)^{1/\deg\alpha}$ to be the **absolute Mahler measure** of α. It is also called the **(exponential) Weil height of** α, with $h(\alpha) := \log(\overline{M}(\alpha))$ being its **(logarithmic) Weil height**. Since $\overline{M}(\alpha) = \overline{M}(\alpha')$ for conjugate α, α', we can also define $\overline{M}(P_\alpha(x)) := \overline{M}(\alpha)$, where $P_\alpha(x)$ is the minimal polynomial of α.

Exercise 14.10 For any algebraic number α, prove that $M(\alpha)$ equals $M(\alpha^2)^{1/2}$ if α and α^2 have the same degree, and equals $M(\alpha^2)$ otherwise. Deduce that $h(\alpha^2) = 2h(\alpha)$ (as in Exercise 1.42).

In our next exercise, we find the best possible lower bound for $\overline{M}(\alpha)$ for totally positive algebraic integers $\alpha \neq 1$. In Problem 14.1, we take $S = \mathbb{R}_{\geq 0}$ and define $f(x) := \log_+ x = \max(0, \log x)$.

Exercise 14.11 Let $\varphi := \frac{1}{2}(1 + \sqrt{5})$. Show that for $x > 0$

$$\log_+ x - \tfrac{1}{2}(1 - \tfrac{1}{\sqrt{5}}) \log x - \tfrac{1}{\sqrt{5}} \log |x - 1| \geq \log \varphi. \tag{14.16}$$

Deduce that $\overline{M}(\alpha) \geq \varphi$ for all totally positive algebraic integers $\alpha \neq 1$. Furthermore, show that equality occurs if and only if $\alpha = \varphi^2$ or φ^{-2}.

Note that if $\alpha \neq \pm 1$ is totally real (rather than totally positive), then

$$\overline{M}(\alpha) \geq \varphi^{1/2}, \tag{14.17}$$

using Exercise 14.10.

This auxiliary function can be expanded, leading to theorems that give a spectrum of the smallest values of $\overline{M}(\alpha)$.

An example of a good explicit bound without using too many polynomials is the following:

Theorem 14.12 *For α a totally positive algebraic integer, the only $\overline{M}(\alpha)$ in the interval $(0, 1, 720566]$ are those for $\alpha = \beta^{(n)}$ ($n = 0, 1, 2, 3, 4$), $(2\cos(2\pi/7))^2$ and $(2\cos(2\pi/60))^2$.*

Proof We have

$$\log_+ x - \sum_{j=1}^{10} c_j \log |Q_j(x)| \geq \log(1.720566), \tag{14.18}$$

where the c_j and Q_j are given in Table 14.1.

Now define $\beta^{(0)} := 1$ and $\beta^{(n)}$ by $\beta^{(n)} + 1/\beta^{(n)} = 2 + \beta^{(n-1)}$ for $n \geq 1$, as in (14.11). Then the minimal polynomials of $\beta^{(0)}$, $\beta^{(1)}$, $\beta^{(2)}$, $\beta^{(3)}$ and $\beta^{(4)}$ are Q_2, Q_3, Q_4, Q_8, Q_{10} respectively. All have $\overline{M}(\beta^{(n)}) < 1.720566$. Also

Table 14.1 The c_j and Q_j for Eq. (14.18)

j	c_j	$Q_j(x)$
1	0.20333607	x
2	0.32506177	$x - 1$
3	0.04536243	$x^2 - 3x + 1$
4	0.01688943	$x^4 - 7x^3 + 13x^2 - 7x + 1$
5	0.00118560	$x^4 - 8x^3 + 15x^2 - 8x + 1$
6	0.00292360	$x^6 - 11x^5 + 41x^4 - 63x^3 + 41x^2 - 11x + 1$
7	0.00087933	$x^6 - 12x^5 + 44x^4 - 67x^3 + 44x^2 - 12x + 1$
8	0.00569441	$x^8 - 15x^7 + 83x^6 - 220x^5 + 303x^4 - 220x^3 + 83x^2 - 15x + 1$
9	0.00069160	$x^8 - 15x^7 + 84x^6 - 225x^5 + 311x^4 - 225x^3 + 84x^2 - 15x + 1$
10	0.00195846	$x^{16} - 31x^{15} + 413x^{14} - 3141x^{13} + 15261x^{12} - 50187x^{11} + 115410x^{10} - 189036x^9 + 222621x^8 - 189036x^7 + 115410x^6 - 50187x^5 + 15261x^4 - 3141x^3 + 413x^2 - 31x + 1$

$$Q_6(x) = (x^3 - 6x^2 + 5x - 1)(x^3 - 5x^2 + 6x - 1),$$

which is the product of the minimal polynomials of $\alpha_7 := (2\cos(2\pi/7))^2$ and α_7^{-1}. We have $\overline{M}(\alpha_7) = \overline{M}(\alpha_7^{-1}) = 1.715534$. Furthermore

$$Q_9(x) = (x^4 - 7x^3 + 14x^2 - 8x + 1)(x^4 - 8x^3 + 14x^2 - 7x + 1),$$

which is the product of the minimal polynomials of $\alpha_{60} := (2\cos(2\pi/60))^2$ and α_{60}^{-1}. We have $\overline{M}(\alpha_{60}) = \overline{M}(\alpha_{60}^{-1}) = 1.719388$. The polynomials $Q_5(x)$ and $Q_7(x)$, while contributing to raising the minimum of the auxiliary function, both have absolute Mahler measure greater than 1.720566. □

The numbers $\beta^{(n)}$, as well as α_7 and α_{60}, can be described in terms of iterates H^n of the function $H(x) = x - 1/x$, as in (A.13). Specifically, $H^n(\sqrt{\beta^{(n)}}) = 1$, while $H^3(\alpha_7) = -\alpha_7$ and $H^4(\alpha_{60}) = \alpha_{60}$ (Sect. A.5.1).

Corollary 14.13 *Suppose that α is totally positive and $\mathrm{Gal}(\mathbb{Q}(\alpha)/\mathbb{Q})$ is abelian. Then the only $\overline{M}(\alpha)$ in the interval $(0, 1.720566]$ are for α equal to 1, $(2\cos(2\pi/5))^2$ $(2\cos(2\pi/7))^2$ and $(2\cos(2\pi/60))^2$.*

Of course, other possibilities for α can give one one these values, namely the reciprocals and the conjugates of the stated α.

Proof This follows from $\beta^{(0)} = 1$, $\beta^{(1)} = (3 + \sqrt{5})/2 = (2\cos(2\pi/5))^{-2}$, while that for $n = 2, 3, 4$ the Galois group $\mathrm{Gal}(\mathbb{Q}(\beta^{(n)})/\mathbb{Q})$ is nonabelian. □

14.7 Mahler Measure of Totally Real Algebraic Numbers

As with integers, questions about the Mahler measure of totally real algebraic numbers can be turned into questions about totally positive ones, adapting the previous exercise. Recall that we are taking the minimal polynomial of an algebraic number to have integer coefficients, content 1, with leading coefficient positive.

The following theorem bounds from below the Mahler measure of totally positive algebraic numbers, and then its corollary is immediate from Exercise 14.10.

Theorem 14.14 *Suppose that α is a totally positive algebraic number of degree d whose minimal polynomial has leading coefficient a. Then for $\varphi = (1 + \sqrt{5})/2$ as earlier, we have*

$$\overline{M}(\alpha) \geq \varphi \cdot a^{\frac{5-\sqrt{5}}{10d}}. \tag{14.19}$$

Corollary 14.15 *If in the theorem we specify only that α is totally real, then*

$$\overline{M}(\alpha) \geq \varphi^{1/2} \cdot a^{\frac{5-\sqrt{5}}{20d}}. \tag{14.20}$$

Thus, the lower bounds (14.16) and (14.17) apply to real algebraic non-integers too.

Proof *(of Theorem 14.14)* We modify our standard auxiliary function argument, to take account of α not being an algebraic integer. From (14.16), we sum over the conjugates α_j of α to get

$$\sum_j \log_+ \alpha_j - \tfrac{1}{2}(1 - \tfrac{1}{\sqrt{5}}) \sum_j \log \alpha_j - \tfrac{1}{\sqrt{5}} \sum_j \log |\alpha_j - 1| \geq d \log \varphi.$$

We now add in a multiple of $\log a$, and use the fact that $d \log \overline{M}(\alpha) = \log a + \sum_j \log_+ \alpha_j$. Also, we know that

$$\log a + \sum_j \log \alpha_j = \log(a \prod_j \alpha_j) \geq 0$$

and similarly that $\log a + \sum_j \log |\alpha_j - 1| \geq 0$. Thus, we obtain

$$d \log \overline{M}(\alpha) \geq d \log \varphi + \tfrac{1}{2}(1 - \tfrac{1}{\sqrt{5}}) \log a.$$

$\square$

More generally, by the same method it is easy to prove that, given any inequality

$$\log_+ x - \sum_{j=1}^{10} c_j \log |Q_j(x)| \geq c \text{ for } x > 0, \tag{14.21}$$

where the polynomials Q_j have integer coefficients, we can prove the following.

Theorem 14.16 *Let α be a totally positive algebraic number of degree d that is not a zero of any Q_j in (14.21), and has a as the leading coefficient of its minimal polynomial. Then*

$$\overline{M}(\alpha) \geq e^c \cdot a^{(1-\sum_j c_j \deg Q_j)/d}. \tag{14.22}$$

In particular, we obtain the following from (14.18).

Corollary 14.17 *Let α be any totally positive non-integer algebraic number whose minimal polynomial has degree d and leading coefficient a (since it is a non-integer we have $a > 1$). Then*

$$\overline{M}(\alpha) \geq 1.720566 \cdot a^{0.15104806/d}.$$

For α totally real, the corresponding result is

$$\overline{M}(\alpha) \geq 1.311703 \cdot a^{0.07552403/d}.$$

14.8 Langevin's Theorem for Sectors

In this section, we consider the Mahler measure of conjugate sets of algebraic integers lying in sectors $S_\theta := \{z \in \mathbb{C} : |\arg(z)| \leq \theta\}$, where $0 \leq \theta < 180°$. Because such sets are closed and do not include the whole unit circle, we can apply Langevin's Theorem 11.12 from Chap. 1. Thus, we know that there is a constant, c_θ say, with $c_\theta > 1$, such that for any nonzero noncyclotomic algebraic integer α lying with its conjugates in S_θ we have $M(\alpha) \geq c_\theta^{\deg \alpha}$. For some particular values of θ, we can, in fact, find the largest possible value of c_θ. This is possible for $\theta = 90°$, for example, as we shall see in Theorem 14.18 below.

Now $c_0 = \frac{1}{2}(1 + \sqrt{5}) = 1.618 \cdots$, from Exercise 14.11. It is clear from the example polynomials $z^{2k+1} - 2$ that $c_\theta \to 1$ as $\theta \to 180°$. In [RS95] Rhin and Smyth, on the basis of computational evidence, conjectured that c_θ is a decreasing 'staircase' function, with countably many horizontal 'stairs', with left discontinuities separating them. In that paper, 9 intervals over which c_θ is constant were identified—partial stairs. Later, Rhin and Wu [RW08] and Flammang and Rhin [FR15] extended this to 14 such intervals, including one complete stair. One of the partial stairs found includes the angle $\theta = 90°$, which gives Theorem 14.18 below. The complete stair is described in Theorem 14.19 below.

The idea for the choice of auxiliary function for these results comes essentially from the proof of Langevin's Theorem. For a particular sector S_θ, let

$$A(z) := \log \max(1, |z|) - c_0 \log |z| - \sum_{j=1}^{J} c_j \log |Q_j(z)|,$$

where the constants c_j $(j = 0, \ldots, J)$ are positive, and the $Q_j(z)$ are integer polynomials. They are chosen so that $A(z) \geq m > 0$ on the sector $|\arg(z)| \leq \theta$, where m is as large as can be found.

We can simplify this method, using the particular geometry of sectors. Because $1/z$ is in S_θ when z is, we have

$$A\left(\frac{1}{z}\right) := \log \max(1, |z|) - (1 - c_0 - \sum_{j=1}^{J} c_j d_j) \log |z| - \sum_{j=1}^{J} c_j \log |Q_j^*(z)|,$$

where $Q_j^*(z) = z^{d_j} Q_j(1/z)$ is the reciprocal polynomial of $Q_j(z)$, and $d_j := \deg Q_j$. Again, $A(1/z) \geq m$ on the sector $|\arg(z)| \leq \theta$. Averaging, we get

$$A_1(z) := \frac{1}{2}\left(A(z) + A\left(\frac{1}{z}\right)\right)$$

$$= \log \max(1, |z|) - \frac{1}{2}(1 - \sum_{j=1}^{J} c_j d_j) \log |z| - \frac{1}{2} \sum_{j=1}^{J} c_j \log |Q_j(z) Q_j^*(z)| \geq m,$$

$$(14.23)$$

again for $z \in S_\theta$. The minimum of $A_1(z)$ may in fact be strictly greater than m. The polynomials $Q_j(z) Q_j^*(z)$ are self-reciprocal. Of course, if Q_j was already self-reciprocal, then the term $-\frac{1}{2} c_j \log |Q_j(z) Q_j^*(z)|$ is simply $-c_j \log |Q_j(z)|$. Thus, we can rewrite $A_1(z)$ as

$$A_1(z) = \log \max(1, |z|) - \frac{1}{2}(1 - \sum_{j=1}^{J} c_j d_j) \log |z| - \frac{1}{2} \sum_{j=1}^{J} c_j \log |R_j(z)|,$$

where R_j is self-reciprocal and of degree $2d_j$.

Note that the coefficient c_0 of $\log |z|$ in $A(z)$ has disappeared, and that the coefficient of $\log |z|$ in A_1 is determined by the other c_j.

From the fact that $A_1(z) = A_1(1/z)$, it is enough to find the minimum of $A_1(z)$ on

$$W_\theta := \{z = re^{i\theta} : |\arg z| \leq \theta, 0 \leq r \leq 1\},$$

where, on redefining the c_j, A_1 takes the simpler form

$$A_1(z) = -(\tfrac{1}{2} - \sum_{j=1}^{J} c_j d_j) \log |z| - \sum_{j=1}^{J} c_j \log |R_j(z)| . \qquad (14.24)$$

Thus, A_1 is harmonic on W_θ, and so its minimum occurs on its boundary. (Note however that A_1 is not harmonic on the whole of S_θ.) Thus, for optimising the coefficients c_j to make m as large as possible, from the fact that $A_1(\overline{z}) = A_1(z)$ for $z \in W_\theta$ it is enough find a good lower bound m for $A_1(z)$ on the radius $\{z : \arg z = \theta, 0 \leq r \leq 1\}$ and on the arc $\{z = e^{it} : 0 \leq t \leq \theta\}$—the **half-boundary** of W_θ.

Let α be a noncyclotomic algebraic integer lying with its conjugates in S_θ. Then by the method outlined in Sect. 14.2.2, we obtain

$$\log M(\alpha) \geq m \deg(\alpha) ,$$

provided α is not a zero of any of the R_j. Hence, if $M(\alpha) > 1$ then

$$M(\alpha)^{1/d} \geq \min(e^m, \min_{\substack{j=1,\cdots,J \\ M(R_j)>1}} M(R_j)^{1/(2d_j)}) ,$$

where $2d_j := \deg R_j$.

Theorem 14.18 *For any nonzero noncyclotomic algebraic integer α lying with its conjugates in the right half-plane $\mathrm{Re}(z) \geq 0$, its Mahler measure satisfies*

$$M(\alpha)^{1/\deg \alpha} \geq M(\alpha')^{1/6} = 1.12933793 ,$$

where α' has minimal polynomial $z^6 - 2z^5 + 4z^4 - 5z^3 + 4z^2 - 2z + 1$. Since α' and its conjugates lie in $\mathrm{Re}(z) \geq 0$, this lower bound is best possible.

Proof We in fact prove a slightly stronger version of the theorem, by claiming that the conclusion holds when α and its conjugates lie in the sector $|\arg z| \leq 91.40°$. We use the auxiliary function

$$A_1(z) = -0.12168 \log |z| - 0.06679 \log |z^2 - 2z + 1| - 0.13137 \log |z^2 - z + 1|$$
$$-0.09200 \log |z^2 + 1| - 0.00808 \log |z^4 - z^3 + z^2 - z + 1|$$
$$-0.02400 \log |z^6 - 2z^5 + 4z^4 - 5z^3 + 4z^2 - 2z + 1| .$$

Computation shows that for $\theta = 91.40°$ the auxiliary function $A_1(z)$ has a maximum m greater than $\log(1.12935)$ on the half-boundary of W_θ, and therefore, on the whole of W_θ. Because all polynomials in A_1 except z and $z^6 - 2z^5 + 4z^4 - 5z^3 + 4z^2 - 2z + 1$ are cyclotomic, and $M(\alpha')^{1/6} = 1.12933793 < 1.12935$, the theorem follows. $\qquad\square$

The technicalities of finding a good auxiliary function $A(z)$ are similar to the earlier problems described in this chapter, although they are more complicated. A significant issue is the difficulty of finding good polynomials to use in the auxiliary function. These should include all monic irreducible integer polynomials having small Mahler measure, and all conjugates in or close to the sector $\{z \in \mathbb{C} : |\arg(z)| \leq \theta\}$ being studied. Flammang, Rhin and Wu have developed special algorithms [Wu03, RW08, FR15] to search for such polynomials effectively. Some of these involve the use of the LLL algorithm.

Flammang and Rhin [FR15] extended the endpoint of the arc over which the lower bound in Theorem 14.18 holds from $91.40°$ to $92.56°$, by using an auxiliary function containing 10 polynomials.

We now give the details of the only known complete 'stair' of the conjectured 'staircase' graph for c_θ.

Theorem 14.19 *Let*

$$P_1(z) := z^6 - 5z^5 + 13z^4 - 17z^3 + 13z^2 - 5z + 1,$$
$$P_2(z) := z^8 - 7z^7 + 26z^6 - 53z^5 + 67z^4 - 53z^3 + 26z^2 - 7z + 1,$$

and let $\theta_1 = 47.941432°$ and $\theta_2 = 50.830864°$ be the maximum phase (argument) of a zero of P_1 and P_2, respectively. Then

$$c_\theta \geq 1.303116 \qquad \text{for } 0 \leq \theta < \theta_1; \qquad (14.25)$$
$$c_\theta = M(P_1)^{1/6} = 1.303055 \qquad \text{for } \theta_1 \leq \theta < \theta_2; \qquad (14.26)$$
$$c_\theta = M(P_2)^{1/8} = 1.300734 \qquad \text{for } \theta_2 \leq \theta \leq 51.388°. \qquad (14.27)$$

Proof Define

$A_1(z) :=$

$\quad - 0.12632313 \log |z| - 0.16414061 \log |z^2 - 2z + 1|$

$\quad - 0.00410301 \log |2z^2 - 3z + 2|$

$\quad - 0.14506049 \log |z^2 - z + 1| - 0.00471334 \log |z^4 - 3z^3 + 5z^2 - 3z + 1|$

$\quad - 0.00694054 \log |z^6 - 5z^5 + 13z^4 - 17z^3 + 13z^2 - 5z + 1|$

$\quad - 0.00114188 \log |z^6 - 4z^5 + 10z^4 - 13z^3 + 10z^2 - 4z + 1|$

$\quad - 0.00355721 \log |z^8 - 7z^7 + 26z^6 - 53z^5 + 67z^4 - 53z^3 + 26z^2 - 7z + 1|$

$\quad - 0.00052416 \log |z^{12} - 11z^{11} + 62z^{10} - 212z^9 + 487z^8 - 788z^7$
$\qquad\qquad + 923z^6 - 788z^5 + 487z^4 - 212z^3 + 62z^2 - 11z + 1|$

$\quad - 0.00155417 \log |z^{12} - 11z^{11} + 60z^{10} - 199z^9 + 448z^8 - 717z^7$
$\qquad\qquad + 837z^6 - 717z^5 + 448z^4 - 199z^3 + 60z^2 - 11z + 1|.$

Let m_θ be the minimum of $A_1(z)$ for $z \in W_\theta$, evaluated by finding the minimum on its half-boundary, as described above. Obviously m_θ is a nonincreasing function of θ. Now all the polynomials P composing A_1, except for P_1 and P_2, either have $M(P)^{1/\deg P} > M(P_1)^{1/6} = 1.303055$, or have $M(P) = 1$, or have some zero whose phase is greater than $51.388°$. Now computation gives $\exp(m_{\theta_1}) = 1.303116$, which shows that (14.25) holds. Also, $\exp(m_{\theta_2}) > M(P_1)^{1/6} = 1.303055$, which shows that (14.26) holds. Finally, $\exp(m_{51.388°}) > 1.300734$, proving (14.27).   □

14.8.1 Further Remarks

The first 'stair' has height $M(z^2 - 3z + 1)^{1/2} = \frac{1}{2}(1 + \sqrt{5}) = 1.618034$. It extends at least from $\theta = 0$ to $\theta = 17.50°$ [FR15]. Although its exact extent is not known, it does not extend beyond $18.863480°$, which is the maximum phase of the zeros of

$$P(z) := z^8 - 12z^7 + 58z^6 - 143z^5 + 193z^4 - 143z^3 + 58z^2 - 12z + 1$$
$$= z^4((z + 1/z - 3)^4 + z + 1/z - 2),$$

for which $M(P)^{1/8} = 1.610559$. This value is not known to be a stair height, however. The next known stair, of height 1.539222, starts at $26.408740°$. There are presumably other, as yet undiscovered, stairs between them.

In [RS95, Table 4] there is a list of polynomials P which, for a given θ, have the smallest known values of $\overline{M}(P) = M(P)^{1/\deg P}$ for P with all their zeros in the sector S_θ. Some further such polynomials appear in [FR15, Table 3].

14.9 Notes

The first result similar to Theorem 14.6 (with 1.6 instead of 1.68) was given by Flammang, Rhin and Smyth [FRS97]. They used Lemma 14.7 with $\ell = 10$. Aguirre, Bilbao and Peral [ABP06] improved this constant to 1.66. Theorem 14.6, with constant 1.68, is due to Flammang [Fla19]. She used Lemma 14.7, combined with an improved method for choosing the polynomials P_j.

The result that $\overline{M}(\alpha) \geq \varphi$ for all totally positive algebraic integers $\alpha \neq 1$ in Exercise 14.11 is originally due to Schinzel [Sch73]. Höhn and Skoruppa [HS93] gave the short proof of the result, essentially as in the exercise.

The first application of the auxiliary function method in number theory was in [Smy81c]. This was an extension of Schinzel's result, similar to Theorem 14.12, which is due to Flammang [Fla96].

Proposition 14.9 comes from Robinson and Wurtz [RW13].

In fact, the largest published lower bound in Theorem 14.12 and Corollary 14.13 is 1.722069 [Fla15]. It is known from [Smy80] that this constant cannot be larger than

$\ell := 1.7273\cdots$, because the set $\{\overline{M}(\alpha) \mid \alpha \text{ totally positive}\}$ is dense in the interval (ℓ, ∞).

Theorem 14.18 is due to Rhin and Smyth [RS95], and Theorem 14.19 is a result of Flammang and Rhin [FR15], slightly modified.

14.10   Glossary

$\operatorname{res}_x(P, Q)$.   The resultant of P and Q, as defined in Sect. A.2 of Appendix A.

S_θ.   The sector of the complex plane for which the argument is bounded in modulus by θ, where $0 \le \theta < 180°$.

W_θ.   The portion of S_θ where the modulus is bounded by 1, i.e., the sector of the closed unit disc for which the argument is bounded in modulus by θ.

$\mathcal{T}$.   The set of all mean traces of totally positive algebraic integers.

absolute Mahler measure.   The absolute Mahler measure of an algebraic number α that has degree d is $\overline{M}(\alpha) = M(\alpha)^{1/\deg\alpha}$, as defined in Chap. 1.

auxiliary function.   In an effort to obtain lower bounds for the average of a function f over the zeros of a polynomial, we consider auxiliary functions of the shape $f(z) - \sum_{j=1}^{\ell} y_j \log |P_j(z)|$, where the y_j are positive real numbers and the P_j are a well-chosen set of polynomials. If an auxiliary function of this shape can be bounded below on a set S, then one can infer a lower bound for the average of f over the zeros of any polynomial that is prime to all the P_j and has all its zeros in S, using resultant information.

half-boundary.   The portion of the boundary of W_θ for which the argument is nonnegative (and including 0) is called the half-boundary of W_θ.

mean trace.   Let α be an algebraic number. The trace of α is the sum of its conjugates; the mean trace is the arithmetic mean of its conjugates, namely the trace divided by the degree.

minimal polynomial.   Let α be an algebraic number. Then α is a zero of a unique monic irreducible polynomial $f_\alpha(x) \in \mathbb{Q}[x]$. Clear denominators of f_α to get a polynomial $m_\alpha(x) \in \mathbb{Z}[x]$ that has positive leading coefficient and content 1. We call $m_\alpha(x)$ the minimal polynomial of α, or of its conjugate set. Thus our minimal polynomial is defined over $\mathbb{Z}$ rather than $\mathbb{Q}$, so might not be monic.

(Schur–Siegel–Smyth) trace problem.   What is the smallest limit point of $\mathcal{T}$? What is the spectrum of small elements of $\mathcal{T}$? Given $\rho < 2$, determine all the totally positive algebraic integers that have mean trace at most ρ.

semi-infinite LP problem.   A linear programming (LP) problem which has finitely many constraints but infinitely many variables.

stair.   Given θ with $0 \le \theta < 180°$, take c_θ to be maximal such that $\overline{M}(\alpha) \ge c_\theta$ for any α that lies with all its conjugates in the sector S_θ (see Theorem 11.12). It is conjectured that the function $\theta \to c_\theta$ is a decreasing step function, or staircase function, with left discontinuities separating the stairs. A stair is a maximal interval $[\theta_1, \theta_2)$ on which c_θ is constant. The value of c_θ on a stair is the height of that stair. There is currently only one stair for which both endpoints are known.

Chapter 15
The Trace Problem for Integer Symmetric Matrices

Recall the Schur-Siegel-Smyth trace problem (Sect. 14.3): what is the smallest limit point for the set of mean traces of totally positive algebraic integers? We can formulate an analogous problem for the mean traces of positive definite integer symmetric matrices. As with Lehmer's Conjecture, we find that the restricted combinatorial problem is more tractable, and indeed, we can solve it. We shall also develop a structure theory for minimal-trace examples.

15.1 The Mean Trace of a Positive Definite Matrix

An $n \times n$ integer symmetric matrix is said to be *positive definite* if all its eigenvalues are strictly positive. Each eigenvalue is then a totally positive algebraic integer, although the characteristic polynomial of the matrix might not be irreducible, so the degree of the algebraic integer might be smaller than n.

If A is an $n \times n$ positive definite matrix, then by analogy with the mean trace of algebraic integers, we define the *mean trace* $\overline{\mathrm{tr}}(A)$ of A to be $\mathrm{trace}(A)/n$. In the event that the characteristic polynomial is irreducible, the mean trace of A equals the mean trace of any of its eigenvalues. Since the product of the eigenvalues of A is a strictly positive rational integer, it is at least 1, and the inequality for arithmetic and geometric means implies that the trace is at least n, and the mean trace is at least 1. Moreover, there is equality here if and only if A is the identity matrix. Now for $n > 1$, the identity matrix is not connected. Although we shall not require our matrices to have irreducible characteristic polynomials, we shall restrict attention to those that are connected: otherwise there is little of interest to say about the spectrum of possible mean traces.

Exercise 15.1 Allowing disconnected matrices, show that the set of all mean traces of positive definite integer symmetric matrices is $\mathbb{Q} \cap [1, \infty)$.

© Springer Nature Switzerland AG 2021
J. McKee and C. Smyth, *Around the Unit Circle*, Universitext,
https://doi.org/10.1007/978-3-030-80031-4_15

For *connected* positive definite $n \times n$ integer symmetric matrices, however, we can settle the analogue of the Schur-Siegel-Smyth trace problem. The mean trace will be shown to be at least $2 - 1/n$, and this bound is attained for each n, from which, one easily deduces that the smallest limit point is 2.

Since we are considering positive definite matrices, the appropriate notion of equivalence is strong equivalence (Sect. 6.3): we can conjugate by signed permutation matrices without changing the eigenvalues.

15.2   The Trace Problem for Integer Symmetric Matrices

Theorem 15.2 *Let A be an integer symmetric $n \times n$ matrix. If A is positive definite and connected, then its trace is at least $2n - 1$. Moreover, this bound is best-possible for every $n \geq 1$.*

Proof We get the last point out of the way first. Let $\mathbf{e}_1, \mathbf{e}_2, \ldots, \mathbf{e}_n$ be an orthonormal basis for $\mathbb{R}^n$. Then

$$\mathbf{f}_1 = \mathbf{e}_1, \ \mathbf{f}_2 = \mathbf{e}_1 + \mathbf{e}_2, \ \mathbf{f}_3 = \mathbf{e}_2 + \mathbf{e}_3, \ \ldots, \ \mathbf{f}_n = \mathbf{e}_{n-1} + \mathbf{e}_n$$

is a linearly independent list of vectors, and hence the corresponding Gram matrix $A = (\mathbf{f}_i \cdot \mathbf{f}_j)$ is positive definite (use the argument in the proof of Lemma 16.34). Moreover, A is connected (the underlying graph is a path on n vertices). We have

$$A = \begin{pmatrix} 1 & 1 & 0 & 0 & \cdots & 0 & 0 & 0 \\ 1 & 2 & 1 & 0 & \cdots & 0 & 0 & 0 \\ 0 & 1 & 2 & 1 & \cdots & 0 & 0 & 0 \\ 0 & 0 & 1 & 2 & \cdots & 0 & 0 & 0 \\ \vdots & \vdots & \vdots & \vdots & \ddots & \vdots & \vdots & \vdots \\ 0 & 0 & 0 & 0 & \cdots & 2 & 1 & 0 \\ 0 & 0 & 0 & 0 & \cdots & 1 & 2 & 1 \\ 0 & 0 & 0 & 0 & \cdots & 0 & 1 & 2 \end{pmatrix},$$

which has trace $2n - 1$, attaining the claimed lower bound.

Now suppose that the bound can be beaten: we shall derive a contradiction. Take A to be a counterexample to the theorem for which

- n is minimal, and   (15.1)
- the trace $t \leq 2n - 2$ is minimal for this value of n.   (15.2)

Let $\mathbf{e}_1, \ldots, \mathbf{e}_n$ be a basis for $\mathbb{R}^n$. The matrix $A = (a_{ij})$ defines a symmetric bilinear form $\langle \cdot, \cdot \rangle$ on $\mathbb{R}^n \times \mathbb{R}^n$ via $\langle \mathbf{e}_i, \mathbf{e}_j \rangle = a_{ij}$. Since A is positive definite, this form is positive definite, and in particular, $a_{ii} > 0$ for all i.

Given trace(A) $\leq 2n - 2$, at least two of the (positive rational integral) diagonal entries must equal 1. In particular, $n \geq 2$, and after relabelling our basis, if necessary, we may assume that $a_{11} = \langle \mathbf{e}_1, \mathbf{e}_1 \rangle = 1$. Since A is connected, some off-diagonal entry in the first row is nonzero, and after a relabelling we may suppose that $a_{12} \neq 0$.

Define $\mathbf{e}_2' = \mathbf{e}_2 - a_{12}\mathbf{e}_1$, noting that $\mathbf{e}_1, \mathbf{e}_2', \mathbf{e}_3, \ldots \mathbf{e}_n$ is another basis for $\mathbb{R}^n$. We let $A' = (a_{ij}')$ be the matrix for our bilinear form with respect to this new basis. Since the form is positive definite, so is A'. Moreover, A' is an integer symmetric matrix. The matrices A and A' agree except perhaps for entries in the second row and column.

The only new diagonal entry is a_{22}', and

$$
\begin{aligned}
a_{22}' &= \mathbf{e}_2' \cdot \mathbf{e}_2' \\
&= (\mathbf{e}_2 - a_{12}\mathbf{e}_1) \cdot (\mathbf{e}_2 - a_{12}\mathbf{e}_1) \\
&= a_{22} - 2a_{12}^2 + a_{12}^2 \\
&= a_{22} - a_{12}^2,
\end{aligned}
$$

which is strictly smaller than a_{22}, so A' has smaller trace than A.

By our minimality condition (15.2), we must have that A' is not connected.

Let G and G' be the underlying graphs of A and A', respectively. We know that G is connected, and the only possible changes in the edges when we move from G to G' are those edges involving vertex 2. We shall show that every vertex in G' is in the same component (in G') as either 1 or 2, and hence that G' (being disconnected) has exactly two components. Let K_1, K_2 be the components of G' containing 1, 2, respectively. A priori we might have $K_1 = K_2$, but once we have shown that every vertex is in either K_1 or K_2, then since G' is not connected, these components must be distinct.

Take any $j \in \{3, 4, \ldots, n\}$. There is a path in G from j to 2. If all the edges of this path lie in G', then $j \in K_2$. Otherwise, the path must finish with an edge from some vertex i to the vertex 2 that is not present in G' (no other edges in the path involve the vertex 2). Then from

$$
a_{2i}' = \mathbf{e}_i \cdot (\mathbf{e}_2 - a_{12}\mathbf{e}_1) = a_{2i} - a_{12}a_{1i}
$$

we see that $a_{1i} \neq 0$ (else $a_{2i}' = a_{2i} \neq 0$). Then we can follow our path along edges in G' from j as far as i, and from there to 1, to see that $j \in K_1$. Hence, as claimed, G' has exactly two components (and the components K_1 and K_2 must be different).

The adjacency matrix of each component is positive definite (since the form is positive definite), and of course, each component is connected. If the two components of A' have r_1 and r_2 rows, respectively ($r_1 + r_2 = n$), and traces t_1 and t_2, respectively, then $t_1 + t_2 < t < 2n - 1$, so $t_1 + t_2 \leq 2(r_1 + r_2) - 3$. Thus, either $t_1 < 2r_1 - 1$ or $t_2 < 2r_2 - 1$, and one of the two components would give a counterexample to the theorem, contradicting our minimality assumption (15.1). $\square$

Exercise 15.3 Proofs by contradiction do not usually lend themselves to illustration by examples, as it turns out once the contradiction has been established that there are no examples to use. The *idea* in the above proof, however, can be applied to actual examples.

Let $A = (a_{ij})$ be a positive definite connected $n \times n$ matrix ($n \geq 2$) that has trace $2n - 1$. Show that A has at least one diagonal entry that is equal to 1; by relabelling suppose that $a_{11} = 1$, and that $a_{12} \neq 0$.

Apply the change of basis in the proof to produce a new matrix A'. Use the known lower bound on the trace to deduce that A' is not connected, and argue that A' has exactly two components, each of minimal trace for their size.

Illustrate with the example given in the proof for which the trace is $2n - 1$.

(The next section pursues this idea in reverse to construct *all* minimal-trace examples.)

Later (Chap. 21), we shall explore the question of which polynomials arise as *minimal* polynomials of integer symmetric matrices. To this end, the bound in the following Corollary will be a powerful tool.

Corollary 15.4 *Let P be an irreducible monic polynomial with integer coefficients. If all the zeros of P are strictly positive, and P has degree d, and P is the minimal polynomial of some integer symmetric matrix, then the trace of P is at least $2d - 1$.*

Proof Let t be the trace of P. Since P is irreducible, if it is the minimal polynomial of some $n \times n$ integer symmetric matrix A, then we must have $\chi_A = P^r$ for some r. Also, using irreducibility of P, if A is not connected then P is the minimal polynomial of each component; moving to one of the components we may assume that A is connected. Now A has trace rt, and is positive definite. From Theorem 15.2, $rt \geq 2n - 1 = 2dr - 1 \geq r(2d - 1)$, so that we, indeed, have $t \geq 2d - 1$.   □

We also note that Theorem 15.2 immediately implies that the smallest limit point of the set of mean traces of connected positive definite integer symmetric matrices is 2.

Corollary 15.5 *The smallest limit point of the set of mean traces of connected positive definite integer symmetric matrices is 2.*

Proof Since the bound on the trace given in Theorem 15.2 is attained, we have examples of $n \times n$ connected positive definite integer symmetric matrices that have mean trace $2 - 1/n$ for each n, so that 2 is certainly a limit point.

We now show that 2 is the smallest limit point. It will be enough to show that for any $\varepsilon > 0$, only finitely many examples have mean trace below $2 - \varepsilon$. By the Theorem, having mean trace below $2 - \varepsilon$ implies that n is at most $1/\varepsilon$, so is bounded. Since the diagonal entries are positive, and the trace is bounded, and the number of diagonal entries is bounded, it follows that the set of possibilities for the diagonal of our matrix is finite. Each 2×2 principal submatrix is positive definite, and this bounds the off-diagonal entries in terms of the diagonal ones. Hence, only finitely many connected positive definite integer symmetric matrices have mean trace below $2 - \varepsilon$.   □

15.3 Constructing Examples that Have Minimal Trace

Suppose that A_1 and A_2 are positive definite connected integer symmetric matrices, say with A_1 being $n_1 \times n_1$ and A_2 being $n_2 \times n_2$. By Theorem 15.2, they have traces at least $2n_1 - 1$ and $2n_2 - 1$, respectively. Suppose that, in fact, both matrices attain these minimal bounds. Then A_1 has at least one diagonal entry equal to 1; working up to strong equivalence, we may suppose that this is in the $(1, 1)$ position. (We could make a similar statement concerning A_2, but in fact, we wish to break the symmetry in our construction in order to produce all minimal-trace examples.)

Take a Gram basis $\mathbf{e}_1, \ldots, \mathbf{e}_{n_1}, \mathbf{f}_1, \ldots, \mathbf{f}_{n_2}$ for the matrix

$$\begin{pmatrix} A_1 & O_{n_1,n_2} \\ O_{n_2,n_1} & A_2 \end{pmatrix},$$

where $O_{r,s}$ is the $r \times s$ zero matrix. By switching, we can assume that all entries in the first row and column are nonnegative, and similarly for all entries in the $(n_1 + 1)$th row and column. Reversing the construction of the proof of Theorem 15.2, change the basis by replacing $\mathbf{f}_1$ by $\mathbf{f}_1 + \mathbf{e}_1$. This adds the first row to the $(n_1 + 1)$th, similarly for the columns, and adds 1 to the $(n_1 + 1, n_1 + 1)$ entry. The new matrix is connected, and has trace $(2n_1 - 1) + (2n_2 - 1) + 1 = 2(n_1 + n_2) - 1$, so is still of minimal trace. We write

$$A_1 * A_2$$

for the new matrix.

For example, we have

$$\begin{pmatrix} 1 & 1 \\ 1 & 2 \end{pmatrix} * \begin{pmatrix} 2 & 1 & 0 \\ 1 & 2 & 1 \\ 0 & 1 & 1 \end{pmatrix} = \begin{pmatrix} 1 & 1 & 1 & 0 & 0 \\ 1 & 2 & 1 & 0 & 0 \\ 1 & 1 & 3 & 1 & 0 \\ 0 & 0 & 1 & 2 & 1 \\ 0 & 0 & 0 & 1 & 1 \end{pmatrix}. \tag{15.3}$$

To illustrate such compositions, we could draw weighted graphs for which A_1 and A_2 are the adjacency matrices. But the diagonal entries in our examples will usually lie in the set $\{1, 2, 3\}$, and so it will be convenient in our pictures to draw weighted graphs for which $A_1 - 2I$ and $A_2 - 2I$ are the adjacency matrices. Then most of our examples can be drawn as charged signed graphs: negatively charged vertices now correspond to entries 1 on the diagonal of our matrix; neutral vertices correspond to 2; positive charges correspond to 3.

To indicate the vertices that correspond to the first rows of the two matrices being composed (since these distinguished vertices play a key role in the composition), we underline those vertices. The above example (15.3) is illustrated thus:

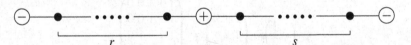

(Recall that here $\ominus$ corresponds to a diagonal entry of 1, $\bullet$ to 2, and $\oplus$ to 3.)

Of course, this composition depends on the choice of distinguished vertices. If these vertices are, respectively, v_1 and v_2, then it is more precise to include this information and write

$$(A_1, v_1) * (A_2, v_2)$$

for the operation.

Exercise 15.6 Show that the above operation $*$ is neither commutative nor associative (even when the various compositions all make sense).

Exercise 15.7 Use the above idea to give another proof that all the eigenvalues of the path P_r^-

$$\ominus \!\!-\!\!\bullet \!\!-\!\! \cdots \!\!-\!\! \bullet$$

(one negative vertex, $r - 1$ neutral vertices) are strictly greater than -2, for all $r \geq 1$.

(Take G_1 to be a single negatively charged vertex, and G_2 to be P_r^-. Let A_i be the adjacency matrix of G_i with twice the identity matrix added (so that all eigenvalues increase by 2). Compute $A_1 * A_2$ and make an inductive argument. The distinguished vertex of G_2 should be the negatively charged vertex.)

Exercise 15.8 It is possible for a minimal-trace example to have diagonal entries greater than 2. Let $P_{r,s}^{-+-}$ be the path shown.

$$\ominus \!\!-\!\!\bullet \!\!-\!\! \cdots \!\!-\!\! \bullet \!\!-\!\! \oplus \!\!-\!\! \bullet \!\!-\!\! \cdots \!\!-\!\! \bullet \!\!-\!\! \ominus$$

$$\underbrace{\hspace{3em}}_{r} \qquad\qquad \underbrace{\hspace{3em}}_{s}$$

Using the construction above, prove that for all $r \geq 0$, $s \geq 0$, all eigenvalues of $P_{r,s}^{-+-}$ are strictly greater than -2, and that by adding twice the identity matrix to the adjacency matrix, these provide further minimal-trace examples.

In fact, it is possible to have a diagonal entry as large as n (no larger, as each of the other $n - 1$ diagonal entries is at least 1). Indeed, let G_1 be the weighted graph that has a single negatively charged vertex, and for $n \geq 2$ let G_n be the weighted graph that has one vertex of weight $n - 2$ and $n - 1$ vertices of weight -1, with $n - 1$ edges joining each of the vertices of weight -1 to the vertex of weight $n - 1$: G_5 is shown here.

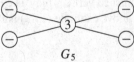

$$G_5$$

Let A_n be the adjacency matrix of G_n plus twice the identity matrix, so that A_n has trace $2n - 1$ and has n as one of its diagonal entries. Then note that $A_n = A_1 * A_{n-1}$ for $n \geq 2$, and since A_1 is a connected positive definite integer symmetric matrix of trace $2 \times 1 - 1 = 1$, one sees inductively that A_n is a connected positive definite integer symmetric matrix of trace $2n - 1$, and one of the diagonal entries equals n.

Repeated application of the process used in the proof of Theorem 15.2 splits a minimal-trace example into smaller and smaller examples, until reduced to the 1×1 matrix (1). Reversing this process, one can build *all* minimal-trace examples of each size, up to strong equivalence. To grow all $n \times n$ examples, run through each partition of n as $r + s$ with $1 \leq r, s \leq n$ (consider (r, s) and (s, r) as distinct unless $r = s$). For each such partition, consider all $r \times r$ and $s \times s$ minimal-trace examples. For each pair of examples, represented by A and B, run through all essentially distinct choices for the distinguished vertices. Working to strong equivalence, make the distinguished vertices correspond to the first rows and columns of A and B, and make all entries in those rows and columns nonnegative. Then compute $A * B$.

For example, the only 2×2 case up to strong equivalence is

$$\ominus \quad * \quad \ominus \quad = \quad \ominus\!\!-\!\!\bullet \quad .$$

And knowing the 1×1 and 2×2 cases, we can quickly find all the 3×3 examples:

$$\ominus \quad * \quad \ominus\!\!-\!\!\bullet \quad = \quad \ominus\!\!-\!\!\bullet\!\!-\!\!\bullet \quad ,$$

$$\ominus \quad * \quad \ominus\!\!-\!\!\bullet \quad = \quad \ominus\!\!-\!\!\oplus\!\!-\!\!\ominus \quad ,$$

$$\ominus\!\!-\!\!\bullet \quad * \quad \ominus \quad = \quad \ominus\!\!\diagdown\!\!\bullet \quad .$$

Exercise 15.9 Find all eight 4×4 examples, up to strong equivalence.

We can glean further information from this process about the structure of minimal-trace examples.

Lemma 15.10 *Let A be an $n \times n$ connected positive definite matrix, and suppose that trace $A = 2n - 1$ (so that A has minimal trace for an $n \times n$ connected positive definite matrix). Then all off-diagonal entries of A are either -1, 0, or 1.*

Proof At least one of the diagonal entries in A is 1, and we may suppose that this is the $(1, 1)$ entry. If $n = 1$, then there is nothing to prove. If $n \geq 2$, then vertex 1 is connected to at least one other vertex, which we may suppose is vertex 2: $a_{21} = a_{12} \neq 0$. As in the proof of Theorem 15.2, we make a change of basis that transforms A to a disconnected matrix A', trace $2n - 1 - a_{12}^2$. As in that proof, one argues that A' has exactly two components, say on r and s vertices, respectively

$(r + s = n)$, and these have traces at least $2r - 1$ and $2s - 1$, respectively. Then $2n - 1 - a_{12}^2 = 2r - 1 + 2s - 1 = 2n - 2$, so $a_{12} = \pm 1$. By an inductive argument, each off-diagonal of each component of A' (and hence each off-diagonal entry of A') has modulus at most 1. Moreover, vertices 1 and 2 are in different components of A', so that if $a'_{2i} \neq 0$, then $a'_{1i} = a_{1i} = 0$. Therefore, for $i \neq 2$,

$$a_{2i} = a'_{2i} + a_{12}a_{1i} = \begin{cases} a_{12}a_{1i} & \text{if } a'_{2i} = 0, \\ a'_{2i} & \text{if } a'_{2i} = \pm 1. \end{cases}$$

We see inductively that all off-diagonal entries of A have modulus at most 1. □

15.4 Notes

Apart from Sect. 15.3, the results of this chapter first appeared in work by McKee and Yatsyna [MY14]. In [Mit20], Mitchell extended these results to positive semidefinite hermitian matrices, allowing the off-diagonal entries to be arbitrary complex numbers that have modulus at most 1, merely keeping integers on the diagonal. For §15.3, see McKee and Smyth [MS20b].

An integer matrix A is said to be symmetrizable if there is a diagonal matrix D with all diagonal entries real and strictly positive such that $D^{-1}AD$ is symmetric. In particular, all eigenvalues of A are real. The set of characteristic polynomials of symmetrizable matrices is strictly larger than that of symmetric matrices. Nevertheless, one can show that the trace bound for positive definite connected symmetric matrices also holds for connected positive definite symmetrizable integer matrices. Moreover, if the bound of Theorem 15.2 holds, then, in fact, the matrix is symmetric. For symmetrizable matrices that are not symmetric, the analogous lower bound for the trace is $2n$, and this bound is achieved for each n. For all of this, see Chap. 19 (and [MS20b]).

15.5 Glossary

mean trace. The mean trace $\overline{\mathrm{tr}}(A)$ of an $n \times n$ matrix A is $\mathrm{trace}(A)/n$. One generally uses this only for positive definite matrices.

positive definite. An integer symmetric matrix is positive definite if all its eigenvalues are strictly positive.

Chapter 16
Small-Span Integer Symmetric Matrices

Let $P(x)$ be a monic polynomial with integer coefficients and all zeros real. The span of P is the difference between the largest and smallest zeros. How small can the span of such a polynomial be? Some open problems will be discussed in the first section of the chapter, before moving to a case that has been settled: which integer symmetric matrices have characteristic polynomials whose span is below 4?

16.1 Small-Span Polynomials

Let $P(x) \in \mathbb{Z}[x]$ be a monic polynomial for which all the zeros are real: order them $\alpha_1 \leq \cdots \leq \alpha_d$, where d is the degree of P. There is no requirement that P be irreducible. The **span** of P is defined to be $\alpha_d - \alpha_1$. It is the length of the shortest real interval that contains all the zeros, namely, $[\alpha_1, \alpha_d]$. If, in fact, P is irreducible, then $\alpha_d - \alpha_1$ may be referred to as the span of any of the α_i: the span of a totally real algebraic integer is defined to be the span of its minimal polynomial. As in Chap. 1, when P is irreducible, we refer to $\{\alpha_1, \ldots, \alpha_d\}$ as a **conjugate set**.

If $\omega_n = e^{2\pi i/n}$ is a root of unity, then $\theta_n = \omega_n + 1/\omega_n = 2\cos(2\pi/n)$ is a totally real algebraic integer whose Galois conjugates all lie in the interval $[-2, 2]$, and moreover, θ_n has span strictly less than 4 (since -2 and 2 are not Galois conjugates). Indeed by Kronecker's Second Theorem, Theorem 1.4, these θ_n (and their conjugates) are the only totally real algebraic integers whose Galois conjugates lie in the interval $[-2, 2]$. Schur [Sch18] gave a different argument to show that for $0 \leq a < 2$, the interval $[-a, a]$ contains only finitely many conjugate sets of algebraic integers. This is implied by Kronecker's result, but Pólya noted that Schur's argument could be applied to any interval of length strictly less than 4, not necessarily centred on the origin (a footnote in [Sch18]). Robinson [Rob62] (or see [McK08]) showed that by contrast any interval of length strictly greater than 4 contains infinitely many conjugate sets of algebraic integers. Translates of the θ_n show that there are infinitely

© Springer Nature Switzerland AG 2021
J. McKee and C. Smyth, *Around the Unit Circle*, Universitext,
https://doi.org/10.1007/978-3-030-80031-4_16

many conjugate sets of algebraic integers in any interval of the shape $[a, a + 4]$, where $a \in \mathbb{Z}$, but other than this case, it is still open as to what happens for any other interval of length exactly 4.

Problem 16.1 (*open problem*)

Is there an interval I of length 4 that does not have integer endpoints such that I contains infinitely many conjugate sets of algebraic integers?

Problem 16.2 (*open problem*)

Is there an interval I of length 4 that does not have integer endpoints such that I contains only finitely many conjugate sets of algebraic integers?

Problem 16.3 (*open problem*)

For intervals I that have length 4, which contain infinitely many conjugate sets of algebraic integers, and which contain only finitely many?

In view of the special significance of the number 4 in the above results and open problems, it is natural to say that a totally real algebraic integer has **small span** (or is a **small-span** algebraic integer) if all its Galois conjugates lie in an interval of length strictly less than 4. We extend this definition to totally real monic polynomials that have integer coefficients, whether or not they are irreducible: if $P(x) \in \mathbb{Z}[x]$ is a monic polynomial with all zeros real, then we say that P has **small span** (or that P is a **small-span** polynomial) if the span of P is strictly less than 4.

The minimal polynomials of the θ_n provide an infinite supply of small-span polynomials. Robinson [Rob64] found all the totally real small-span algebraic integers of degree up to 8, although he could only prove his list was complete up to degree 6. At the time of writing, the complete list is known up to degree 15 ([CFS10] up to degree 14, [FRW11] for degree 15), and examples are known other than the θ_n for degrees 16, 17 [CFS10] and 18 [OMRSÉ13]. Of course, if θ has small span, then so does $-\theta$, and so does $\theta + n$ for any integer n. In compiling lists, one views all such trivially-related examples as being equivalent, and needs to give only one representative of each equivalence class.

Exercise 16.4 Suppose that θ is a small-span algebraic integer. Show that for some integer n, either $\theta + n$ or $-\theta + n$ has all its Galois conjugates in the interval $[-2, 2.5)$.

Apart from the classes of the θ_n, it is not known whether or not there is an infinite supply of equivalence classes of small-span algebraic integers. By Exercise 16.4, we can search for representatives that have all Galois conjugates in the interval $[-2, 2.5)$.

Problem 16.5 (*open problem*)

Show that there are infinitely many totally real small-span algebraic integers other than the $\theta_n = 2\cos(2\pi/n)$ that have all their Galois conjugates in the interval $[-2, 2.5)$, or prove that there are only finitely many such algebraic integers.

In the next section, we shall consider the analogous problems for integer symmetric matrices. As with Lehmer's Conjecture, the combinatorial constraints will allow the analogous problems to be solved: we shall give a complete description of all small-span integer symmetric matrices, once we have defined what we mean by the span of a matrix.

We say that two totally real, monic polynomials $P(x)$, $Q(x) \in \mathbb{Z}[x]$ of the same degree d are **span-equivalent** if $Q(x) = \varepsilon^d P(\varepsilon x + n)$ for some $\varepsilon = \pm 1$ and some $n \in \mathbb{Z}$. If two polynomials are span-equivalent, then they have the same span. From Exercise 16.4, we see that every small-span irreducible monic integer polynomial is span-equivalent to one that has all its zeros in the interval $[-2, 2.5)$.

Exercise 16.6 It is possible for distinct span-equivalent polynomials $P(x)$ and $Q(x)$ to have all their zeros in the interval $[-2, 2.5)$.

For example, verify for each of the following two pairs of polynomials that they are span-equivalent and have all their zeros in the interval $[-2, 2.5)$:

(i) $x^2 + x - 1$ and $x^2 - x - 1$ (more generally, if $P(x)$ has all its zeros in the interval $[-2, 2]$, then so does $P(-x)$);
(ii) $x^3 - x^2 - 3x + 1$ and $x^3 - 2x^2 - 2x + 2$.

Our focus is on monic integer polynomials, but the definition of span naturally applies to all integer polynomials, whether or not monic. In this more general setting, we make the following observation.

Proposition 16.7 *If $P(x) \in \mathbb{Z}[x]$ has all zeros real and has rational span, then the largest and smallest zeros are themselves rational. In the monic case, this implies that the largest and smallest zeros are in $\mathbb{Z}$. An irreducible integer polynomial has rational span if and only if the degree is 1 (and the span is 0).*

Proof Suppose that α and β are the largest and smallest zeros of $P(x)$, that $\alpha - \beta = r > 0$, with $r \in \mathbb{Q}$, and that at least one of α or β is irrational. Apply an automorphism of the normal closure of $\mathbb{Q}(\alpha, \beta)$ that maps α to $\alpha' \leq \alpha$ and β to $\beta' \geq \beta$, where at least one of these inequalities is strict. Then

$$r = \alpha' - \beta' < \alpha - \beta = r ,$$

a contradiction. □

Proposition 16.8 *Let $P(x)$ be a monic polynomial with integer coefficients, all zeros real, with none repeated. All such P with span $s = 0, 1, 2, 3$ or 4 are, up to translation by an integer, as follows:*

- $s = 0$: $P(x) = x - 2$;
- $s = 1$: $P(x) = (x - 2)(x - 1)$;
- $s = 2$: $P(x) = (x - 2)(x - 1)x$ or $(x - 2)x$;
- $s = 3$: $P(x) = (x - 2)(x + 1)Q(x)$, where $Q(x)$ is a factor of $(x - 1)x(x^2 - x - 1)$;
- $s = 4$: $P(x) = (x - 2)(x + 2) \prod_{j \in J} P_j(x)$, where J is any finite subset of $\{3, 4, 5, \ldots\}$, and $P_j(x)$ is the minimal polynomial of $2 \cos(2\pi/j)$.

Proof By Proposition 16.7, we know that the largest and smallest zeros of $P(x)$ are in $\mathbb{Z}$. Thus, by translation by an integer, we can assume that the largest zero is 2. The case $s = 0$ is trivial. For $s = 1, 2, 3$ define $H_s(x)$ by $H_1(x) := (x - 2)(x - 1)$, $H_2(x) := (x - 2)(x - 1)x$, and $H_3(x) := (x - 2)(x - 1)x(x^2 - x - 1)$. Then it is easily checked that $|H_s(x)| < 1$ on $[2 - s, s]$ for $s = 1, 2, 3$. We claim that any irreducible factor, $F(x)$ say, of $P(x)$ must divide $H_s(x)$. This is because, by (A.2), the resultant $\mathrm{res}_x(F, H_s)$ has modulus $|\prod_{\alpha:F(\alpha)=0} P(\alpha)|$, which is less than 1. Hence, by Proposition A.20, F and H_s must have a common zero. Because F is assumed irreducible, F divides H_s. Thus, we have shown that the possible $F(x)$ are those factors of $H_s(x)$ that have $(x - 2)(x - 2 + s)$ as a factor, so as to have span s.

For $s = 4$, we know that $P(x)$ has $(x - 2)(x + 2)$ as a factor, and that all other factors must have all their zeros in $(-2, 2)$. Such factors are given by Kronecker's Second Theorem, Theorem 1.4.   □

Exercise 16.9 Extend the results of Proposition 16.8 to the case where $P(x)$ is allowed to have repeated zeros.

16.2   Small-Span Integer Symmetric Matrices

Let A be an integer symmetric matrix, and let χ_A be its characteristic polynomial. Then χ_A is a monic polynomial that has integer coefficients and all zeros real, so we may define the *span* of A to be the span of χ_A. We say that A has *small span* (or that A is a *small-span* matrix) if the span of A is strictly less than 4. If A has small span, then so do $-A$, $A + nI$ and $-A + nI$, for any $n \in \mathbb{Z}$.

Recall the notion of strong equivalence from Chap. 6: integer symmetric matrices A and B are strongly equivalent if there is a signed permutation matrix P such that $P^\mathsf{T} A P = P^{-1} A P = B$. Strongly equivalent matrices have the same characteristic polynomial, and hence the same span. For the purposes of classifying integer matrices that have small span, we define A to be *span-equivalent* to B if B is strongly equivalent to one of $\pm A + nI$ for some $n \in \mathbb{Z}$. Span-equivalent matrices have the same span, and to classify small-span matrices, we seek to find a representative for each equivalence class. We write

$$[A]_{\mathrm{sp}}$$

for the set of integer symmetric matrices that are span-equivalent to A. Thus

$$[A]_{\mathrm{sp}} = \bigcup_{n \in \mathbb{Z}} \left([A + nI]_{\mathrm{str}} \cup [-A + nI]_{\mathrm{str}}\right).$$

Recall (see Glossary for Chap. 6) that $[B]_{\mathrm{str}}$ denotes the set of integer symmetric matrices that are strongly equivalent to B.

Exercise 16.10 Suppose that A is an integer symmetric matrix with characteristic polynomial $P(x)$, and that $Q(x)$ is span-equivalent to $P(x)$ (as defined in the previous section). Show that there is some $B \in [A]_{sp}$ such that B has characteristic polynomial $Q(x)$.

After Exercises 16.4 and 16.10, we seek to find integer symmetric matrices that have all their eigenvalues in the interval $[-2, 2.5)$, and in addition are small-span. We call a small-span matrix **reduced** if all its eigenvalues are in the interval $[-2, 2.5)$. Note that a small-span matrix might be span-equivalent to more than one reduced matrix.

Exercise 16.11

(i) To which reduced small-span matrices is the 1×1 matrix (0) span-equivalent?
(ii) Show that a small-span matrix is span-equivalent to only finitely many reduced matrices.

Exercise 16.12 (*computational exercise*)

Write a program to compute all reduced matrices in $[A]_{sp}$, where A is a small-span matrix.

The following simple lemma will be repeatedly helpful.

Lemma 16.13 *If A is a reduced small-span integer symmetric matrix, then so is any induced submatrix of A.*

Proof Interlacing shows that any induced submatrix of A has span at most that of A, and so must have small span. Moreover, if all eigenvalues of A lie in the interval $[-2, 2.5)$, then by interlacing the same is true for any induced submatrix.   □

Our definition of small-span requires span strictly less than four, but if the span of A equals 4, then after Proposition 16.7, we see that A is span-equivalent to a cyclotomic integer symmetric matrix that has eigenvalues at both -2 and 2.

The cyclotomic matrices of Chap. 6 have all their eigenvalues in the interval $[-2, 2]$, and are, therefore, reduced small-span matrices unless both -2 and 2 are eigenvalues. One thing we wish to do in the context of classifying small-span matrices is to identify which cyclotomic matrices do not have both -2 and 2 as eigenvalues. For examples that are not cyclotomic, however, we seek integer symmetric matrices A such that the following three properties hold:

- A has small span;
- all eigenvalues of A lie in the interval $[-2, 2.5)$;
- at least one eigenvalue of A lies in the interval $(2, 2.5)$.   (16.1)

We aim to find span-equivalent representatives of such matrices, and will see that there are only finitely many satisfying (16.1). This contrasts with the situation for polynomials generally, where it is still an open question as to whether or not there are finitely many span equivalence classes of small-span polynomials other than the ones that correspond to roots of unity (Problem 16.5).

16.3   Bounds on Entries and Degrees

Lemma 16.14 *Let A be a reduced small-span matrix. Then each diagonal entry of A has modulus at most* 2.

Proof If A has a diagonal entry a, then (a) is an induced submatrix of A that has eigenvalue a. Since A is reduced, so is (a) (Lemma 16.13), which implies that $|a| \leq 2$.
□

Lemma 16.15 *Let A be a reduced small-span matrix. Then each off-diagonal entry of A has modulus at most* 1.

Proof Let b be an off-diagonal entry of A. By Lemma 16.13, A contains an induced submatrix of the shape $\begin{pmatrix} a & b \\ b & c \end{pmatrix}$ that has span strictly less than 4. This implies that $\sqrt{(a-c)^2 + 4b^2} < 4$, which, in turn, implies that $|b| \leq 1$.
□

This last inequality also shows that A cannot have both 2 and -2 on the diagonal.

Exercise 16.16 Lemma 16.15 shows one contrast between small-span matrices and cyclotomic matrices: the latter could have off-diagonal entries ± 2. Verify that all cyclotomic matrices that have off-diagonal entries either 2 or -2 have span 4, so are not small-span.

Exercise 16.17 (*computational exercise*) Adapt the connected growing code from the exercises of Chap. 6 to give connected growing of reduced small-span matrices, using the bounds on entries from Lemmas 16.14 and 16.15.

Lemma 16.18 *Let A be a connected matrix that contains a diagonal entry with modulus* 2. *Then A is span-equivalent to one of*

$$(2), \quad \begin{pmatrix} 2 & 1 \\ 1 & -1 \end{pmatrix}, \quad \begin{pmatrix} 2 & 1 \\ 1 & 0 \end{pmatrix}, \quad \begin{pmatrix} 2 & 1 & 0 \\ 1 & 0 & 1 \\ 0 & 1 & 0 \end{pmatrix}, \quad \begin{pmatrix} 2 & 1 & 0 & 0 \\ 1 & 0 & 1 & 0 \\ 0 & 1 & 0 & 1 \\ 0 & 0 & 1 & 0 \end{pmatrix}.$$

Proof Starting with the matrix (2), grow connectedly keeping only those matrices that have all eigenvalues in the interval $[-2, 2.5)$. The process terminates with a finite set of matrices, each of which is span-equivalent to one of those listed. The full computation is left as an exercise.
□

Exercise 16.19 (*computational exercise*)
Use the program written for Exercise 16.17 to perform the connected growing required for the proof of Lemma 16.18.

Exercise 16.20 Show that apart from $\begin{pmatrix} 2 & 1 \\ 1 & -1 \end{pmatrix}$, each of the matrices in Lemma 16.18 is span-equivalent to the reduced adjacency matrix of a charged signed graph.

In view of Lemmas 16.15 and 16.14, and Exercise 16.20, we may now restrict our search for reduced small-span matrices to those that have all entries coming from the set $\{-1, 0, 1\}$, i.e., we can restrict to adjacency matrices of charged signed graphs. Our search is further helped, as it was in the classification of cyclotomic integer symmetric matrices, by bounding the number of nonzero entries in each row.

Lemma 16.21 *Let A be a connected reduced small-span integer symmetric matrix. Then each row of A has at most four nonzero entries.*

Proof If any entry of A has modulus greater than 1, then after Lemmas 16.14, 16.15 and 16.18, we are done by computing all the reduced matrices span-equivalent to those of Lemma 16.18. Otherwise, we are done (by interlacing) by computing all examples on up to 6 rows (by connected growing), and observing that none has five nonzero entries in any row.   □

Exercise 16.22 (*computational exercise*)
Perform the computation indicated in the proof of Lemma 16.21.

16.4 Growing Small Examples

Our main classification theorem (Theorem 16.39) has the striking consequence that if A is a connected reduced small-span integer symmetric matrix and has at least 13 rows, then, in fact, A is cyclotomic. The number 13 is best-possible here, and in this section, we detail the finite search to find representatives up to span equivalence of every connected small-span integer symmetric matrix that has up to 12 rows.

This is 'just' an exercise in connected growing, adapting the code used for finding small cyclotomic charged signed graphs. That we can restrict to charged signed graphs is a consequence of Lemma 16.18 and Exercise 16.20, and we also exploit Lemma 16.21 to speed the computations.

Of course, the characteristic polynomials of these small-span matrices need not be irreducible. There is some discussion in [McK10] as to which irreducible characteristic polynomials arise for these small examples. All the irreducible small-span polynomials of degree up to 5 appear as characteristic polynomials, but some of degree 6 and higher are missing. We shall return to this in Chap. 21.

By interlacing, it is enough to list (representatives of span equivalence classes of) small-span matrices that are maximal with respect to being connected small-span integer symmetric matrices. There is a slight subtlety here, in that, a reduced small-span matrix might not be maximal, yet perhaps all connected small-span matrices strictly containing it are not reduced. For example, (-2) is a reduced small-span matrix, yet any larger connected small-span matrix has at least one eigenvalue below -2. One way around this issue is simply to catalogue connected integer symmetric matrices that have all their eigenvalues in the interval $[-2, 2.5)$. This was the approach taken in [McK10], which worked up to strong equivalence. Here, however, our focus is on classifying the maximal small-span matrices up to *span* equivalence, and so we wish to filter the output of the search to produce a list of reduced representatives of the equivalence classes (one representative per class). For example, the matrix (-2) would be included amongst the matrices maximal subject to having all eigenvalues in the interval $[-2, 2.5)$, yet it is not maximal as a small-span matrix (Exercise 16.23).

Exercise 16.23 Suppose that $A = \begin{pmatrix} -2 & 1 \\ 1 & a \end{pmatrix}$ is a small-span matrix. Find all possible values of a, and in particular, show that such a matrix A exists. Show that A is not reduced.

With these remarks in mind, we can adapt our growing code from Chap. 6, and produce the results shown below. The code adapts trivially to find first of all those matrices that are maximal with respect to having all eigenvalues in the interval $[-2, 2.5)$, up to strong equivalence, and one then needs to test which of these are maximal as small-span matrices. Finally, one needs to check if any of the resulting matrices are span-equivalent. We list only those that are not cyclotomic. In the next section, we shall identify all cyclotomic examples, and later, we shall see that these together with the 163 noncyclotomic examples computed in this section exhaust all the possibilities for the classes of matrices that are maximal subject to being small-span.

16.4.1 Two Rows

The only connected small-span example, up to span equivalence, is

$$\begin{pmatrix} 2 & 1 \\ 1 & -1 \end{pmatrix}.$$

The span is 3.60556 (in this and all the examples below, the span is rounded to five decimal places). This example is unique amongst all the connected small-span examples, in that, there is no representative of the class that is the adjacency matrix of a charged signed graph.

Exercise 16.24 Verify the above claim: show that there is no charged signed graph whose adjacency matrix is span equivalent to $\begin{pmatrix} 2 & 1 \\ 1 & -1 \end{pmatrix}$.

16.4.2 Three Rows

There is one example, up to span equivalence. It is the adjacency matrix of a charged signed graph, and the graph is drawn rather than giving the matrix. The span is shown under the graph.

3.88945

16.4.3 Four Rows

There are six examples up to span equivalence.

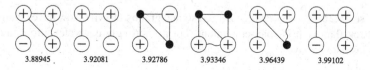

3.88945   3.92081   3.92786   3.93346   3.96439   3.99102

Exercise 16.25 The small-span matrix

$$\begin{pmatrix} 2 & 1 & 0 & 0 \\ 1 & 0 & 1 & 0 \\ 0 & 1 & 0 & 1 \\ 0 & 0 & 1 & 0 \end{pmatrix}$$

of Lemma 16.18 is span-equivalent to the adjacency matrix of one of the above examples: which?

16.4.4 Five Rows

There are seventeen examples up to span equivalence. We see some examples that have the same span but are not span-equivalent. Rounding to five decimal places was adequate to distinguish between different spans for these examples: if the spans agree to five decimal places, then they are actually equal.

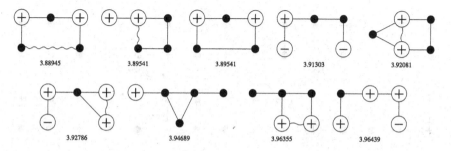

3.88945   3.89541   3.89541   3.91303   3.92081

3.92786   3.94689   3.96355   3.96439

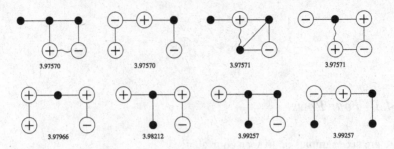

3.97570   3.97570   3.97571   3.97571

3.97966   3.98212   3.99257   3.99257

Exercise 16.26 Verify that the charged signed graphs shown below have small span, and check that each is span-equivalent (but not strongly equivalent) to one of the examples listed above.

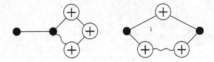

16.4.5 Six Rows

There are forty examples up to span equivalence.

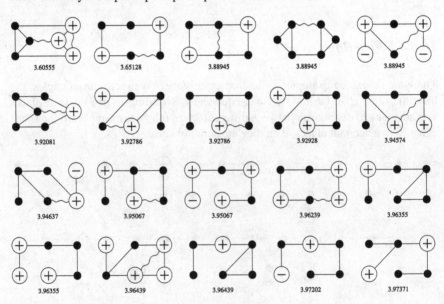

3.60555   3.65128   3.88945   3.88945   3.88945

3.92081   3.92786   3.92786   3.92928   3.94574

3.94637   3.95067   3.95067   3.96239   3.96355

3.96355   3.96439   3.96439   3.97202   3.97371

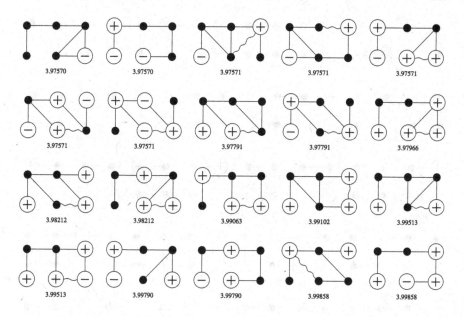

Exercise 16.27 Verify that the charged signed graphs shown below have small span, and check that each is span-equivalent (but not strongly equivalent) to one of the examples listed above.

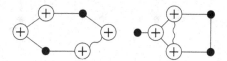

16.4.6 Seven Rows

There are twenty-seven examples up to span equivalence.

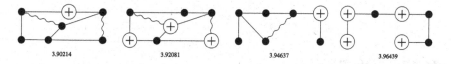

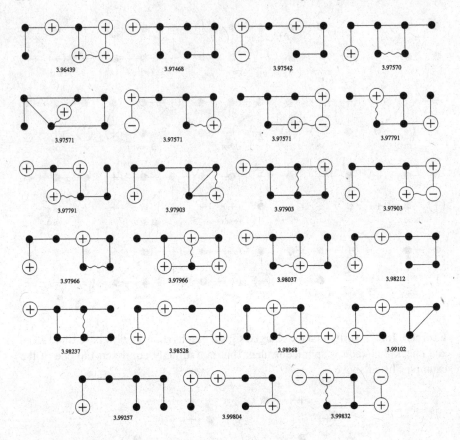

Exercise 16.28 Verify that the charged signed graph shown below has small span, and check that it is span-equivalent (but not strongly equivalent) to one of the examples listed above.

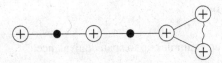

16.4.7   Eight Rows

There are thirty-eight examples up to span equivalence.

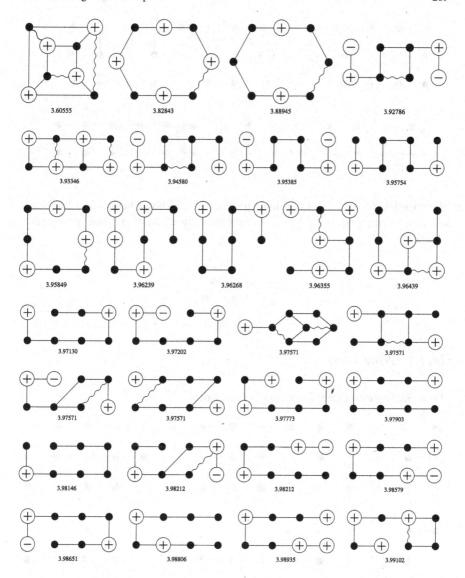

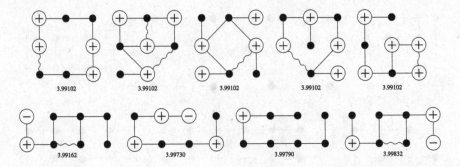

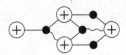

Exercise 16.29 Verify that the charged signed graph shown below has small span, and check that it is span-equivalent (but not strongly equivalent) to one of the examples listed above.

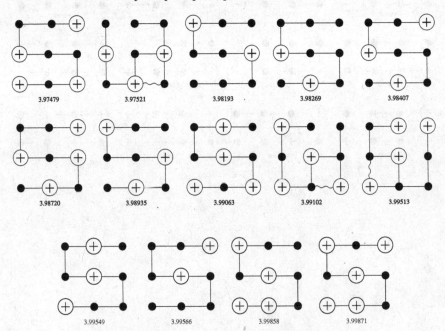

16.4.8   Nine Rows

There are fourteen examples up to span equivalence.

Exercise 16.30 Verify that the charged signed graph shown below has small span, and check that it is span-equivalent (but not strongly equivalent) to one of the examples listed above.

16.4.9 Ten Rows

There are fifteen examples up to span equivalence.

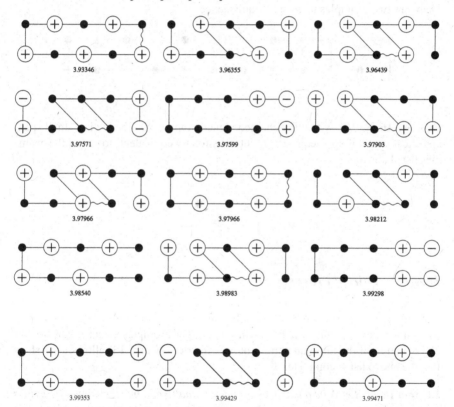

16.4.10   Eleven Rows

There are two examples up to span equivalence.

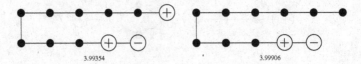

3.99354   3.99906

16.4.11   Twelve Rows

There are two examples up to span equivalence.

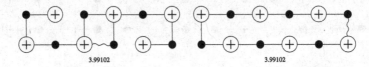

3.99102   3.99102

Exercise 16.31 Verify that the charged signed graph shown below has small span, and check that it is span-equivalent (but not strongly equivalent) to one of the examples listed above.

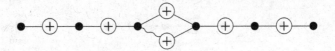

16.4.12   Thirteen Rows

Not only are there no maximal examples, there are no examples at all. This does not immediately preclude the possibility that larger examples might occur, but our classification of the minimal noncyclotomic matrices (Chap. 13) allows us to show that the above list is complete.

Lemma 16.32 *Let A be a maximal connected small-span integer symmetric matrix that is not span-equivalent to a cyclotomic matrix. Then A is span-equivalent to one of the* 163 *sporadic maximal examples listed above in this section.*

Proof If A were not span-equivalent to one of the above examples, then by our computations it would have at least 14 rows. If any such A exists, we can take one that has the smallest number of rows.

Let B be *any* connected induced submatrix of A that can be produced by deleting one row and column. Then B has at least 13 rows, and our minimality condition on A tells us that B must be span-equivalent to a cyclotomic matrix. Hence, there is a matrix in $[A]_{sp}$ that is a minimal noncyclotomic matrix. By Theorem 13.6, there is no minimal noncyclotomic matrix as large as this, and we are done.   □

Exercise 16.33 Any small-span noncyclotomic connected matrix must contain an induced submatrix that is equivalent to one of the minimal noncyclotomic matrices. For each of the examples above, identify a suitable minimal noncyclotomic matrix.

16.5 Cyclotomic Small-Span Matrices

We now turn to span-equivalent classes $[A]_{sp}$ that contain at least one cyclotomic element. We shall use our classification of cyclotomic integer symmetric matrices (Theorem 6.31), and determine for all the possible examples, which ones have small span, and hence determine the maximal classes. Note that a cyclotomic matrix is small-span if and only if not both of -2 and 2 are eigenvalues.

16.5.1 Examples with an Entry of Modulus Greater Than 1

Up to strong equivalence, the only connected cyclotomic integer symmetric matrices that have an entry with modulus greater than 1 are (2) and $\begin{pmatrix} 0 & 2 \\ 2 & 0 \end{pmatrix}$. The former has small span, is span-equivalent to (0), and is not maximal. The latter is not small-span, and the submatrix (0) is not maximal.

16.5.2 Subgraphs of the Sporadic Examples

Suppose that A is a small-span matrix that is the adjacency matrix of a subgraph of one of the sporadic maximal cyclotomic charged signed graphs, S_7, S_8, S_8', S_{14}, or S_{16}. Note that the characteristic polynomials of S_{14} and S_{16} are, respectively, $(x - 2)^7(x + 2)^7$ and $(x - 2)^8(x + 2)^8$. By interlacing, we need to remove 7 or 8 vertices, respectively, for there to be any chance of a subgraph being small-span (removing fewer vertices would leave both -2 and 2 as eigenvalues). We see that we need only search through cyclotomic charged signed graphs on up to eight vertices to find all maximal cyclotomic classes of small-span matrices.

An alternative route to the same conclusion is via Gram vectors. We saw in Exercise 7.9 that one could find Gram vector representations of S_{14} and S_{16} in 7-dimensional and 8-dimensional space. It follows that, any set of eight of the vectors for S_{14}, or nine of the vectors for S_{16} would be linearly dependent. One can then apply the following general lemma to conclude that -2 would be an eigenvalue; then by bipartiteness (recall that for signed graphs, this means that negating all the edges gives a signed graph in the same strong equivalence class) 2 would also be an eigenvalue; hence the induced subgraph would not have small span.

Lemma 16.34 *Let* $\mathbf{v}_1, ..., \mathbf{v}_r$ *be a set of Gram vectors for a charged sign graph G. Then* $\mathbf{v}_1, ..., \mathbf{v}_r$ *are linearly dependent if and only if* -2 *is an eigenvalue for G.*

Proof *(Compare with Exercise 6.30.)*

Let A be the adjacency matrix of G, with rows ordered to match the Gram vectors. Thus, if B is the (possibly not square) matrix whose columns are the $\mathbf{v}_i$, then $A + 2I = B^\top B$. If the Gram vectors are linearly dependent, then there is a nonzero vector $\mathbf{c}$ such that $B\mathbf{c} = \mathbf{0}$. Then also $(A + 2I)\mathbf{c} = B^\top B\mathbf{c} = \mathbf{0}$. Hence, 0 is an eigenvalue of $A + 2I$, and -2 is an eigenvalue of A.

Conversely, suppose that -2 is an eigenvalue for G. To sidestep the issue that B might not be square, we take an alternative set of Gram vectors $\mathbf{w}_1, ..., \mathbf{w}_r \in \mathbb{R}^r$, and let C be the matrix whose columns are the $\mathbf{w}_i$. Then $A + 2I = C^T C$, and C is square, so we can compute $0 = \det(A + 2I) = \det(C^T C) = \det(C^T)\det(C) = \det(C)^2$, and so $\det(C) = 0$. Hence, there is some nonzero vector $\mathbf{c}$, such that $C\mathbf{c} = \mathbf{0}$, and this reveals a linear dependency amongst the $\mathbf{w}_i$, say $\sum_i \lambda_i \mathbf{w}_i = \mathbf{0}$. Then, since $\mathbf{w}_i \cdot \mathbf{w}_j = \mathbf{v}_i \cdot \mathbf{v}_j$ for all i and j, we have

$$0 = \left(\sum_i \lambda_i \mathbf{w}_i\right) \cdot \left(\sum_i \lambda_i \mathbf{w}_i\right) = \left(\sum_i \lambda_i \mathbf{v}_i\right) \cdot \left(\sum_i \lambda_i \mathbf{v}_i\right),$$

and hence the same linear dependency holds for the $\mathbf{v}_i$.   □

We need, therefore, to search for connected small-span cyclotomic charged signed graphs on up to eight vertices. To this end, we adapt the code used for growing all connected cyclotomic charged signed graphs, keeping only those that are small-span. A set of inequivalent representatives (initially using equivalence as for cyclotomics, and then inspecting the output to see if any are span-equivalent) is produced, and these are tested for maximality: i.e., maximality as connected small-span charged signed graphs, not as connected cyclotomic charged signed graphs.

Some of the examples found are also subgraphs of the toroidal or cylindrical tessellations, which will be covered later. We record here those that appear only as subgraphs of the sporadic maximal connected cyclotomic charged signed graphs.

The graphs S_7, S_8 and S_8' each yield one example, which is shown here.

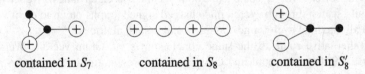

contained in S_7 contained in S_8 contained in S_8'

The example contained in S_7 is distinguished amongst all the maximal small-span cyclotomic examples in that the largest eigenvalue equals 2 (but -2 is not an eigenvalue, so this cyclotomic example does, indeed, have small span).

The sporadic signed graph S_{14} contains no maximal small-span subgraphs that are not also contained in one of the tessellations. By contrast, S_{16} provides a rich yield of eleven maximal small-span examples, shown here.

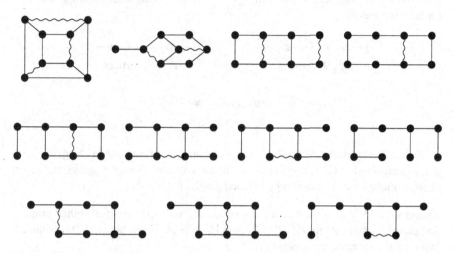

Together with the 163 maximal small-span noncyclotomic examples, we, therefore, have a total of 177 sporadic maximal small-span examples. After Lemma 16.32 and the computations reported in this section, we know that all other connected small-span integer symmetric matrices are span-equivalent to a subgraph of one of the toroidal or cylindrical tessellations, T_{2k}, C_{2k}^{++}, or C_{2k}^{+-}. It will be convenient to start with the cylindrical case.

16.5.3 Subgraphs of Cylindrical Tessellations

We consider the possibilities for connected charged signed graphs G maximal subject to:

- G is small-span;
- G is contained in some $C_{2k}^{\bullet\bullet}$;
- G contains at least one charged vertex.

(If G had no charged vertices, it would be contained in some T_{2k}, and this possibility will be considered in the next section.)

First, suppose that G contains two charged vertices of opposite sign, and hence contains a long path in C_{2k}^{+-}. Starting at the end with the negative charge, our path is represented by Gram vectors $\mathbf{e}_1$, $\varepsilon_1 \mathbf{e}_1 + \mathbf{e}_2$, $\varepsilon_2 \mathbf{e}_2 + \mathbf{e}_3$, ..., $\varepsilon_{k-2} \mathbf{e}_{k-2} + \mathbf{e}_{k-1}$, $\varepsilon_{k-1} \mathbf{e}_{k-1} + \sqrt{2} \mathbf{e}_k$, where each $\varepsilon_i = \pm 1$. These Gram vectors span $\mathbb{R}^k$. Adding any conjugate vertex would give a charged signed graph H with linearly dependent Gram vectors, hence by Lemma 16.34, it would have -2 as an eigenvalue. Similarly, we see that a set of Gram vectors for $-H$ would also be linearly dependent, so that, -2 would be an eigenvalue of $-H$, and H would have both -2 and 2 as eigenvalues, so would not be small-span. Hence, our path is all of G, and after switching, we see that G is span-equivalent to the path $P_k^{\pm}$ shown here, along with paths P_k^+ and P_k^- for later reference:

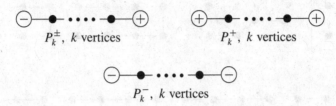

$P_k^{\pm}$, k vertices P_k^+, k vertices

P_k^-, k vertices

We see that the paths $P_k^{\pm}$ produce maximal examples of small-span charged signed graphs contained in C_{2k}^{+-}, and inspection of the sporadic examples shows that these are maximal as small-span charged signed graphs for $k \geq 6$.

Exercise 16.35 Amongst the sporadic maximal small-span charged signed graphs, find examples that contain $P_2^{\pm}$, $P_3^{\pm}$, $P_4^{\pm}$ and $P_5^{\pm}$. Check that no larger $P_k^{\pm}$ is contained in any of the sporadic examples.

Next suppose that G contains a path with vertices of the same charge at both ends: something in the same switching class as one of P_k^+ or P_k^-, shown above. Note that P_k^- has -2 as an eigenvalue (either write down an eigenvector with components of alternating signs, or note that the Gram vectors $e_1 + e_2, e_2 + e_3, \ldots, e_{r-2} + e_{r-1}, e_{r-1}$ are linearly dependent). The critical question, then, is how many conjugate vertices can be added to P_k^+ before -2 becomes an eigenvalue.

Our Gram vectors for P_k^+ are $\sqrt{2}e_0 + e_1$, $e_1 + e_2$, $e_2 + e_3$, $\ldots$, $e_{k-2} + e_{k-1}$, $e_{k-1} + \sqrt{2}e_k$, where $e_0, \ldots, e_k$ form an orthonormal basis of $\mathbb{R}^{k+1}$. Adding any conjugate vertex, charged or uncharged, produces a set of Gram vectors that form a basis for $\mathbb{R}^{k+1}$; adding a second conjugate vertex would introduce a dependency amongst the Gram vectors, so that -2 would become an eigenvalue. We conclude that the maximal small-span examples (in this case, and maximal as subgraphs of C_{2k}^{++}) are of the shape P_k^+ with a single conjugate vertex added. If there are $\ell \geq 0$ vertices on P_k^+ that lie to the left of the conjugate pair and $r \geq \ell$ to the right (up to equivalence we can suppose that $r \geq \ell$, and then since $k \geq 2$ we have $r \geq 1$), then we get the small-span example $Z_{\ell,r}$:

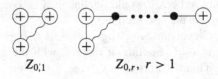

$Z_{0,1}$ $Z_{0,r}$, $r > 1$

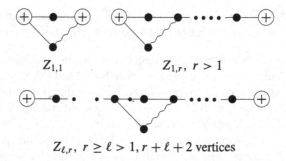

$$Z_{1,1} \qquad\qquad Z_{1,r},\ r > 1$$

$$Z_{\ell,r},\ r \geq \ell > 1, r + \ell + 2 \text{ vertices}$$

Inspection of the sporadic maximal examples shows that $Z_{\ell,r}$ is maximal unless $(\ell, r) \in \{(0, 1),\ (0, 2),\ (0, 3)\ (1, 1),\ (1, 2),\ (1, 3),\ (2, 2),\ (2, 3),\ (3, 3),\ (3, 4)\}$.

Exercise 16.36 Check that each of $Z_{0,1}$, $Z_{0,2}$, $Z_{0,3}$, $Z_{1,1}$, $Z_{1,2}$, $Z_{1,3}$, $Z_{2,2}$, $Z_{2,3}$, $Z_{3,3}$, and $Z_{3,4}$ is contained in at least one of the maximal noncyclotomic small-span charged signed graphs, and that no other $Z_{\ell,r}$ ($r \geq \ell \geq 0, r \geq 1$) is contained in any of those sporadic examples. (Remember that 'contained in' means that it is equivalent to a subgraph.)

16.5.4 Subgraphs of Toroidal Tessellations

Now suppose that G is a maximal small-span subgraph of some T_{2k}. At this point, we mean maximal subject to being both small-span and a subgraph of T_{2k}; later we shall check if G is actually maximal as a small-span charged signed graph. First, consider the case where G contains a long cycle (length k).

If $k = 2r + 1$ is odd, then all such cycles are equivalent, and we may suppose that our cycle is the graph O_{2r+1} having $2r + 1$ positive edges (we also draw here O_{2r}^- for later use)

$$O_{2r+1},\ 2r + 1 \text{ vertices} \qquad\qquad O_{2r}^-,\ 2r \text{ vertices}$$

Now 2 is an eigenvalue for O_{2r+1} (take the vector of ones for an eigenvector), but not -2 (the graph is not bipartite). A Gram representation of O_{2r+1} is given by $\mathbf{e}_1 + \mathbf{e}_2$, $\mathbf{e}_2 + \mathbf{e}_3, \ldots, \mathbf{e}_{2r} + \mathbf{e}_{2r+1}, \mathbf{e}_{2r+1} + \mathbf{e}_1$, where $\mathbf{e}_1, \ldots, \mathbf{e}_{2r+1}$ form an orthonormal basis of $\mathbb{R}^{2r+1}$. Given that $k = 2r + 1$ is odd, these Gram vectors themselves form a basis of $\mathbb{R}^{2r+1}$. Adding any conjugate vertex to the graph then gives a vector dependent on those already present, and hence -2 would become an eigenvalue, making the signed graph not small-span. Hence, as a subgraph of T_{2k}, this example is maximal. Looking through the noncyclotomic maximal examples, we see that O_3 and O_5 are not maximal as small-span charged signed graphs, but O_{2r+1} is maximal for $r \geq 3$.

Exercise 16.37 Verify that there are sporadic noncyclotomic small-span charged signed graphs containing O_3 and O_5, but none containing O_{2r+1} for $r \geq 3$.

If $k = 2r$ is even, then our cycle of length k contained in G is bipartite. If the cycle is equivalent to one with all edges positive, then both 2 and -2 are eigenvalues and we do not have small span. Up to equivalence, we are reduced to considering O_{2r}^- as a subgraph of T_{4r}, a cycle with $2r - 1$ positive edges and one negative edge, drawn above. We now have a Gram representation $\mathbf{e}_1 + \mathbf{e}_2, \mathbf{e}_2 + \mathbf{e}_3, \ldots, \mathbf{e}_{2r-1} + \mathbf{e}_{2r}, \mathbf{e}_{2r} - \mathbf{e}_1$. Given $k = 2r$ is even, these vectors span $\mathbb{R}^{2r}$. If we add any conjugate vertex, we would have -2 as an eigenvalue, and since T_{4r} is bipartite, we would also have 2 as an eigenvalue and would not be small-span. Thus, the O_{2r}^- are maximal as small-span subgraphs of T_{4r}. Again, we must look through the sporadic noncyclotomic examples, and discover that O_4^- and O_6^- are not maximal as small-span charged signed graphs, but O_{2r}^- is maximal for $r \geq 4$.

Exercise 16.38 Verify that there are sporadic noncyclotomic small-span charged signed graphs containing O_4^- and O_6^-, but none containing O_{2r}^- for $r \geq 4$.

Now we suppose that G does not contain a long cycle. We take a path P that is part of a long cycle, and is of maximal length as such. By switching, we may suppose that all edges are positive, and P has Gram representation $\mathbf{e}_1 + \mathbf{e}_2, \mathbf{e}_2 + \mathbf{e}_3$, $\ldots, \mathbf{e}_{r-1} + \mathbf{e}_r$ for some $r \leq k$ (here P has $r - 1$ vertices). If we add any conjugate vertex (Gram vector $\mathbf{e}_i - \mathbf{e}_{i+1}$), then we get a basis for $\mathbb{R}^r$, and as before our signed graph is already maximal (given that the path cannot be lengthened within G, adding any further vertex would be to add another conjugate vertex, giving a dependency, hence -2 would be an eigenvalue, and G is necessarily bipartite, given that it is does not contain a long cycle). Thus, P with a single conjugate vertex added is maximal as a small-span subgraph of G that does not contain a long cycle. All such signed graphs are contained in larger charged signed small-span cyclotomic graphs, simply by adding positively charged vertices to each end of the path to produce some $Z_{\ell,r}$ as described in the previous section.

Thus, from the T_{2k}, we find precisely the following maximal small-span examples (maximal as small-span charged signed graphs), working up to span equivalence: O_{2r+1} for $r \geq 3$, and O_{2r}^- for $r \geq 4$.

16.6   The Classification Theorem

The hard work has been done. We can draw together Lemma 16.32 and the work of Sect. 16.5 to give a complete classification of all connected small-span integer symmetric matrices.

Theorem 16.39 *Let A be a connected, small-span, integer symmetric matrix. Then either $[A]_{\mathrm{sp}} = \left[\begin{pmatrix} 2 & 1 \\ 1 & -1 \end{pmatrix} \right]_{\mathrm{sp}}$ or A is span-equivalent to the adjacency matrix of a*

charged signed graph G. Every such charged signed graph G is contained in a
maximal one, and the maximal examples, up to span equivalence, are:

- *the* $163 + 14 = 177$ *sporadic examples presented in Sects. 16.4 and 16.5.2;*
- *one of the infinite cyclotomic families of examples listed in Sects. 16.5.3 and 16.5.4,*
  namely, $P_k^\pm$ for $k \geq 6$, $Z_{\ell,r}$ for $r \geq \ell \geq 0$ and (ℓ, r) not one of $(0, 0)$, $(0, 1)$, $(0, 2)$,
  $(0, 3)$, $(1, 1)$, $(1, 2)$, $(1, 3)$, $(2, 2)$, $(2, 3)$, $(3, 3)$, $(3, 4)$, O_{2r+1} for $r \geq 3$, or O_{2r}^- for
  $r \geq 4$.

This settles the integer symmetric matrix analogues of all the open problems
from the introductory section of this chapter. For intervals $I = [a, a + 4]$ of length
4, there are infinitely many connected integer symmetric matrices whose eigenvalues
all lie in I if and only if $a \in \mathbb{Z}$. If A is a connected integer symmetric matrix, all of
whose eigenvalues lie in I, then either A is small-span (in which case we appeal to
the classification—there are only finitely many possibilities up to span equivalence
that are not span-equivalent to cyclotomic examples; and infinitely many connected
cyclotomic examples would include arbitrarily long paths, and hence eigenvalues
arbitrarily close to ± 2, so that the endpoints of I would have to be in $\mathbb{Z}$) or A
has span exactly 4. In the latter case, Proposition 16.7 shows that the endpoints of
I, which must now be algebraic integers, are in $\mathbb{Z}$ (and A is span-equivalent to a
cyclotomic matrix; there are infinitely many possibilities).

16.7 Notes

The classification of small-span integer symmetric matrices appeared first in McKee's
work [McK10], although the focus there was on which irreducible polynomials
appeared as characteristic and minimal polynomials. We shall return to this topic
in Chap. 21. The classification in [McK10] was done up to strong equivalence rather
than span equivalence, so that some examples which we regard here as equivalent
were, there, considered inequivalent (for example, (-2), (-1), (0), (1) and (2)
were considered inequivalent in [McK10]). The number of cyclotomic families was
roughly double that here, and the number of maximal noncyclotomic examples was
increased to 197. Some sporadic cyclotomic examples were overlooked, and we set
the record straight here. For an extension by Greaves to hermitian matrices over
quadratic integer rings, see [Gre15].

The integer symmetric matrices having all eigenvalues in the interval $(-2, 2)$ were
described in [MS07]. Not all connected examples are contained in maximal ones.
These, of course, are particular examples of small-span matrices, but a small-span
cyclotomic matrix might have one (but not both) of -2 or 2 as an eigenvalue.

16.8   Glossary

$[A]_{sp}$.   The set of matrices that are span-equivalent to A.

conjugate set.   If α is an algebraic number, then the conjugate set of α is the set of all its Galois conjugates.

reduced.   A small-span matrix A is called reduced if all its eigenvalues are in the interval $[-2, 2.5)$.

small-span.   A polynomial, matrix, charged signed graph, or algebraic integer is said to be small-span, if it has small span.

small span.   An algebraic integer α has small span, if all its Galois conjugates are real and lie in an interval of length strictly less than 4. A monic polynomial $f \in \mathbb{Z}[x]$ is said to have small span, if it is totally real and has span strictly less than 4. An integer symmetric matrix is said to have small span, if its characteristic polynomial has small span. A charged signed graph is said to have small span, if its characteristic polynomial has small span.

span.   If $P(x) \in \mathbb{Z}[x]$ is a monic polynomial that has only real zeros $\alpha_1 \leq \alpha_2 \leq \cdots \leq \alpha_d$, then the span of P is defined to be $\alpha_d - \alpha_1$. If A is an integer symmetric matrix, then its span is defined to be the span of its characteristic polynomial.

span-equivalent.   Two totally real, monic polynomials $P(x), Q(x) \in \mathbb{Z}[x]$ of the same degree d are said to be span-equivalent if $Q(x) = \varepsilon^d P(\varepsilon x + n)$ for some $\varepsilon = \pm 1$ and some $n \in \mathbb{Z}$. Span-equivalent polynomials certainly have the same span. Two integer symmetric matrices A and B are said to be span-equivalent, if B is strongly equivalent (Chap. 6) to one of $\pm A + nI$ for some $n \in \mathbb{Z}$.

Chapter 17
Symmetrizable Matrices I: Introduction

17.1 Introduction

In several earlier chapters, we have been working with integer symmetric matrices. There are many alternative, perhaps, more general classes of matrices which one might consider. Some are more easily exploitable than others. In the next three chapters, we consider one such larger class of matrices: *symmetrizable* integer matrices. These matrices are a real diagonal change of basis away from being symmetric, so that all their eigenvalues are real, and we may consider whether the results of earlier chapters can be extended to this setting. As ever, we use the language of graphs to describe our matrices, where we feel that this is helpful to understand the theory. In this asymmetric setting, we need to work with digraphs. For conventions on weighted digraphs and their drawings, see Appendix B.2.

Our work will start with some fundamental results concerning the structure of symmetrizable integer matrices. The theory of symmetrizable matrices was developed largely in the context of Lie theory: generalised Cartan matrices are important special cases of symmetrizable matrices. There is a beautiful combinatorial classification of these matrices as those that are sign symmetric and satisfy a certain cycle condition. The exact analogue of Cauchy interlacing will be seen to hold for symmetrizable matrices.

We shall explore connections with equitable partitions of signed graphs. It will transpire that symmetrizable integer matrices are precisely those that occur as quotient matrices for these equitable partitions.

In Chap. 18, we extend the classification of cyclotomic integer symmetric matrices (Chap. 6) to symmetrizable matrices. The maximal connected examples in the symmetrizable case that have nonnegative integer entries and zeros on the diagonal will be seen to correspond precisely to the complete list of affine Dynkin diagrams.

We then illustrate all the maximal examples of cyclotomic symmetrizable matrices as quotients of signed graphs that have spectral radius at most 2. We do this also for the maximal cyclotomic integer symmetric matrices of Chap. 6 that are not themselves signed graphs, i.e., those that have charged vertices.

© Springer Nature Switzerland AG 2021
J. McKee and C. Smyth, *Around the Unit Circle*, Universitext,
https://doi.org/10.1007/978-3-030-80031-4_17

In Chap. 19, we extend the matrix analogue of the Schur-Siegel-Smyth trace problem (Chap. 15) to symmetrizable matrices. Although there are now many more possible characteristic polynomials, we shall establish the same trace bound as for the symmetric case. Moreover, we shall see that this bound can be attained only by symmetric matrices, so that for symmetrizable but asymmetric matrices, a stronger bound holds.

17.2   Definitions and Immediate Consequences

An $n \times n$ real matrix B is said to be *symmetrizable* if there is a real diagonal matrix $D = \text{diag}(\delta_1, \ldots, \delta_n)$ with each $\delta_i > 0$ such that

$$S = D^{-1} B D \qquad (17.1)$$

is symmetric. We call S the *symmetrization* of B: we shall see below that it is unique if it exists.

Thus, a symmetrizable matrix can be transformed to a symmetric matrix by a diagonal change of basis (and with positive scaling of each basis vector). In particular, all the eigenvalues of a symmetrizable matrix are real, since they equal those of the symmetrization. Note that if B is symmetrizable, then the corresponding D is certainly not uniquely determined, as we may scale all the entries by any positive real number, but any such scaling preserves $D^{-1} B D$.

It feels artificial to require that all the diagonal entries of D in (17.1) are positive, but we lose no generality in doing so. If $D^{-1} B D = S$ is symmetric, where now D is an arbitrary invertible diagonal real matrix, then define the diagonal matrix E that has as its diagonal entries the signs of those of D:

$$E = \text{diag}\big(\text{sgn}(\delta_1), \ldots, \text{sgn}(\delta_n)\big).$$

Then DE has all diagonal entries strictly positive, $E^{-1} = E$, and $(DE)^{-1} B (DE) = ESE$ is symmetric. One convenience of restricting to positive diagonal entries as a canonical choice is that this makes the symmetrization matrix unique.

Lemma 17.1 *If $B = (b_{ij})$ is symmetrizable, then it has a unique symmetrization.*

Proof Let $S = D^{-1} B D = (s_{ij})$ be any symmetrization of B, with

$$D = \text{diag}(\delta_1, \ldots, \delta_n)$$

and each $\delta_i > 0$. We have

$$s_{ij} = \delta_i^{-1} b_{ij} \delta_j. \qquad (17.2)$$

One immediate consequence of (17.2) is that the signs of the entries of B and S agree:

$$\text{sgn}(s_{ij}) = \text{sgn}(b_{ij}) \text{ for all } i \text{ and } j. \tag{17.3}$$

Using (17.2) and the symmetry of S, we have

$$s_{ij}^2 = s_{ij}s_{ji} = \delta_i^{-1}b_{ij}\delta_j\delta_j^{-1}b_{ji}\delta_i = b_{ij}b_{ji}. \tag{17.4}$$

Hence, $|s_{ij}|$ is determined by B. Together with (17.3), this shows that S is uniquely determined by B. $\qquad\square$

There are other, equivalent, definitions of symmetrizable matrices, some of which appear in the literature. These are contained in the next Lemma.

Lemma 17.2 *The following conditions on an $n \times n$ real matrix B are equivalent:*

(i) *B is symmetrizable;*
(ii) *there exists a diagonal matrix D_1 with all diagonal entries positive such that $D_1 B$ is symmetric;*
(iii) *one can write $B = D_2 S_1$, where D_2 is diagonal with all diagonal entries positive and S_1 is symmetric;*
(iv) *B^T is symmetrizable;*
(v) *there exists a diagonal matrix D_2 with all diagonal entries positive such that $B D_2$ is symmetric;*
(vi) *one can write $B = S_2 D_1$, where D_1 is diagonal with all diagonal entries positive and S_2 is symmetric.*

Proof Suppose that (i) holds, with $S = D^{-1}BD$ symmetric. Then we discover that $D^{-2}B = D^{-1}SD^{-1}$ is symmetric, so (ii) holds (with $D_1 = D^{-2}$).

Now suppose that (ii) holds, with $S_1 := D_1 B$ symmetric. Then $B = D_1^{-1}S_1$, so (iii) holds, and $D_2 = D_1^{-1}$.

Now suppose that (iii) holds. Then with $D = \sqrt{D_2}$ we have $B^\mathsf{T} = S_1 D_2$, so that

$$DB^\mathsf{T}D^{-1} = DS_1 D_2 D^{-1} = DS_1 D,$$

which is symmetric. Hence, B^T is symmetrizable.

Now suppose that (iv) holds, with $DB^\mathsf{T}D^{-1} = S$ say, which is symmetric. Then $D^{-1}BD = S$ also, and so $BD^2 = DSD = S_2$ say, which is symmetric.

Now suppose that (v) holds, with $BD_2 = S_2$ symmetric. Then $B = S_2 D_1^{-1}$, so (vi) holds with $D_2 = D_1^{-1}$.

Finally, suppose that (vi) holds. Then

$$D^{-1}BD = D^{-1}BD_2 D^{-1} = D^{-1}S_2 D^{-1} = D^{-1}(DSD)D^{-1} = S$$

is symmetric, and (i) holds. $\qquad\square$

Note that B and B^T have the same symmetrization S, as in (17.1), but the diagonal matrix used is D^{-1} instead of D. Also, $S_1 = D^{-1}SD^{-1}$ and $S_2 = DSD$, as well as $D_1^{-1} = D_2 = D^2$.

When the symmetrizable matrix B has integer entries, we shall see below in Lemma 17.4 that D_1 or D_2 in Lemma 17.2 can be chosen to have integer entries, and that the diagonal entries of D in (17.1) can be chosen to be square roots of positive integers.

As in the symmetric case, we shall say that the two symmetrizable integer $n \times n$ matrices A and B are *equivalent* if A can be transformed to $\pm B$ by conjugating by an element of $O_n(\mathbb{Z})$.

Lemma 17.3 *If an integer matrix B is symmetrizable, then so is any matrix equivalent to B.*

Proof Suppose that $D^{-1}BD = S$, where D is diagonal with positive diagonal entries, and let P be any signed permutation matrix of the same size. Then we compute that $P^{-1}DP = P^{\mathsf{T}}DP$ is also diagonal with positive diagonal entries:

$$(P^{-1}DP)_{ij} = \sum_{k,\ell}(P^{\mathsf{T}})_{ik}D_{k\ell}P_{\ell j}$$

$$= \sum_{k} P_{ki}D_{kk}P_{kj}$$

$$= \begin{cases} 0 & i \neq j, \\ D_{i^*i^*} & i = j, \text{ where } P_{i^*i} = \pm 1. \end{cases}$$

Since $(P^{-1}D^{-1}P)(P^{-1}BP)(P^{-1}DP) = P^{\mathsf{T}}SP$ is symmetric, we see that $P^{-1}BP$ is symmetrizable. Moreover $D^{-1}(-B)D = -S$ is symmetric, so that the negative of a symmetrizable matrix is symmetrizable. Hence, any matrix equivalent to B is symmetrizable. □

17.3   The Structure of Symmetrizable Matrices

A real $n \times n$ matrix $B = (b_{ij})$ is called **sign symmetric** if

$$\operatorname{sgn}(b_{ij}) = \operatorname{sgn}(b_{ji}) \tag{17.5}$$

holds for all $i, j \in \{1, \ldots, n\}$.

Let B be a symmetrizable matrix, with symmetrization S. One immediate consequence of (17.3), together with symmetry of S, is that any symmetrizable matrix is sign symmetric.

We now show that the entries of D_1 or D_2 in Lemma 17.2 may be chosen to be integers if the symmetrizable matrix B has integer entries.

Lemma 17.4 *If B is an integer $n \times n$ symmetrizable matrix, then we can choose D in (17.1) to have all entries being square roots of positive integers. Furthermore, we can choose D_1 and D_2 in Lemma 17.2 to have all entries in $\mathbb{Z}$.*

Proof We deal first with D_1, where $S_1 = D_1 B$ and S_1 is symmetric. Writing $B = (b_{ij})$, $D_1 = \text{diag}(d_1, \dots, d_n)$ and $S_1 = (s'_{ij})$, we have

$$d_i b_{ij} = s'_{ij} = s'_{ji} = d_j b_{ji}. \tag{17.6}$$

So

$$d_i = (b_{ji}/b_{ij})d_j \text{ when } b_{ij} \neq 0. \tag{17.7}$$

Thus, for all indices i, k in the same connected component of B, we see by considering a chain of such identities that d_i/d_k is rational. Thus, on fixing some i in this component, for an appropriate positive integer N we can scale by N/d_i all the d_k in this component to make them integers, while at the same time scaling the rows and columns of S_1 corresponding to this component by d_i/N, preserving the relation $S_1 = D_1 B$, and with S_1 remaining symmetric. Doing this for all connected components of B makes D_1 a positive integer diagonal matrix. Also, S_1 remains symmetric, and its entries $s'_{ij} = d_i^{-1} b_{ij}$ are rational.

Next, doing the same for B^T, we have $S_2 = D_2 B^\mathsf{T}$, with S_2 symmetric, and where D_2 has positive integer entries. But then also $S_2 = S_2^\mathsf{T} = B D_2 = B D^2$, where the diagonal entries of D are square roots of positive integers. Hence, $D^{-1} B D = D^{-1} S_2 D^{-1}$ is symmetric. $\square$

The proof does not need that the matrix entries are integers: all that is required is that $b_{ij}/b_{ji} \in \mathbb{Q}$ whenever $b_{ji} \neq 0$, so that the Lemma applies to a larger class of matrices than those stated, but we shall need it only for the integer symmetrizable case. The proof works with the connected components of B. We wish to dig more deeply into the structure of B, and break these components up in such a way that the d_i are constant on each piece.

Let B be a symmetrizable $n \times n$ matrix. Define B^*, a symmetric $n \times n$ matrix, by

$$(B^*)_{ij} = \begin{cases} b_{ij} & \text{if } b_{ij} = b_{ji}, \\ 0 & \text{if } b_{ij} \neq b_{ji}. \end{cases}$$

(Thus, we set to zero any entries of B that were revealing asymmetry of B. If B is symmetric, then $B^* = B$.) Viewing B and B^* as weighted digraphs on the same set of vertices, define the **symmetric components** of B to be the subsets of the vertices of B that correspond to the components of B^*.

Lemma 17.5 *Suppose that B is a symmetrizable matrix, with $\text{diag}(d_1, \dots, d_n)B$ symmetric. Then*

(i) the d_i are constant on the symmetric components of B;
(ii) let C_1, C_2 be any two distinct symmetric components of B; then for $i \in C_1$, $j \in C_2$ with $b_{ij} \neq 0$, the ratio b_{ji}/b_{ij} is independent of i and j.

Proof The first part is immediate from (17.7). Then using (17.7) again, along with (i), the second part follows. $\square$

17.4   The Balancing Condition and Its Consequences

Let B be an $n \times n$ symmetrizable matrix, with $D_1 B$ symmetric, as in Lemma 17.2
(ii). Write $B = (b_{ij})$ and $D_1 = \text{diag}(d_1, \ldots, d_n)$. Then, as in (17.6), symmetry of
$D_1 B$ gives

$$d_i b_{ij} = d_j b_{ji} \text{ for all } i, \ j \in \{1, \ldots, n\}. \tag{17.8}$$

We call (17.8) the **balancing condition**.

Lemma 17.6 *An $n \times n$ matrix B is symmetrizable if and only if there exist positive
real numbers $d_1, \ldots, d_n$ such that the balancing condition (17.8) holds. An $n \times n$
integer matrix B is symmetrizable if and only if there exist positive integers $d_1, \ldots,
d_n$ such that the balancing condition (17.8) holds.*

Proof The balancing condition (17.8) is equivalent to the existence of a diagonal
matrix $D_1 = \text{diag}(d_1, \ldots, d_n)$ such that $D_1 B$ is symmetric. Now apply Lemma 17.2.
For the integer case, use also Lemma 17.4. $\square$

Note that (17.7), obtained using Lemma 17.2(iii), is the balancing condition
for B^{T}. The fact that symmetrizable matrices satisfy the balancing condition gives
another way to see that symmetrizable matrices are sign symmetric.

Suppose that B is an $n \times n$ symmetrizable matrix. Take any $i_1, i_2, \ldots, i_t \in
\{1, \ldots, n\}$. Multiplying (17.8) for $(i, j) = (i_1, i_2), (i_2, i_3), \ldots, (i_{t-1}, i_t), (i_t, i_1)$, we
have

$$d_{i_1} d_{i_2} \cdots d_{i_{t-1}} d_{i_t} b_{i_1 i_2} b_{i_2 i_3} \cdots b_{i_{t-1} i_t} b_{i_t i_1} = d_{i_2} d_{i_3} \cdots d_{i_t} d_{i_1} b_{i_2 i_1} b_{i_3 i_2} \cdots b_{i_t i_{t-1}} b_{i_1 i_t}.$$

Dividing by the (nonzero) product of all the d_{i_j}, we get

$$b_{i_1 i_2} b_{i_2 i_3} \cdots b_{i_{t-1} i_t} b_{i_t i_1} = b_{i_2 i_1} b_{i_3 i_2} \cdots b_{i_t i_{t-1}} b_{i_1 i_t} \tag{17.9}$$

for all sequences $i_1, i_2, \ldots, i_t$ of elements of $\{1, \ldots, n\}$.

An $n \times n$ real matrix B is said to satisfy the **cycle condition** if (17.9) holds for
all sequences $i_1, i_2, \ldots, i_t$ of elements of $\{1, \ldots, n\}$.

Thus, any symmetrizable matrix is sign symmetric and satisfies the cycle condition. It is a beautiful fact that these two conditions together are sufficient for a matrix
to be symmetrizable.

Proposition 17.7 *An $n \times n$ real matrix B is symmetrizable if and only if it is sign
symmetric and satisfies the cycle condition.*

Proof We have seen that any symmetrizable matrix is sign symmetric and satisfies
the cycle condition.

Now suppose that $B = (b_{ij})$ satisfies both these conditions. For simplicity, suppose that B is connected (else treat each component separately). Set $d_1 = 1$. For
each neighbour i of vertex 1, sign symmetry gives both b_{1i} and b_{i1} nonzero, so that

we can define $d_i := d_1 b_{1i}/b_{i1} = b_{1i}/b_{i1}$. Then the balancing condition holds when $j = 1$ (if any b_{1j} in (17.8) is zero, then both sides are zero). Next for neighbours k of neighbours i of 1, define d_k by $d_k := d_i b_{ik}/b_{ki}$. By the cycle condition, any vertex k for which d_k has been defined more than once will have received the same value each time. The balancing condition now holds for $j = 1$ and for j any neighbour of 1. Continuing in this way, we grow our labelling to all the vertices (consistently, thanks to (17.9)), and produce positive numbers d_i such that (17.8) holds. By Lemma 17.6, B is symmetrizable. □

The above proof also yields a method to test the cycle condition in practice. On each component, attempt to compute all the d_i by the above process. If no conflicts are found (and having labelled all vertices, push the process one step more to check any edges not yet processed), then the cycle condition holds.

Exercise 17.8 *One of the two matrices*

$$\begin{pmatrix} 0 & 1 & -2 \\ 3 & 0 & 1 \\ -3 & 2 & 0 \end{pmatrix} \quad and \quad \begin{pmatrix} 0 & 1 & -2 \\ 3 & 0 & 2 \\ -3 & 1 & 0 \end{pmatrix}.$$

is a symmetrizable matrix B, and the other is not. Follow the process of the proof of Proposition 17.7 to determine which is symmetrizable, and to find positive integers d_1, d_2 and d_3 such that $\mathrm{diag}(d_1, d_2, d_3)B$ is symmetric.

17.5 The Symmetrization Map

We have seen that if B is symmetrizable, then its symmetrization S is uniquely determined. The arguments in Sect. 17.3 show that D is unique up to scalings of the entries that correspond to the components of B: different scalings may be applied to different components. We saw in the proof of Lemma 17.1 that the entries in S can be computed without knowledge of D.

Lemma 17.9 *For $B = (b_{ij})$ an $n \times n$ real matrix that is sign symmetric, define the real symmetric $n \times n$ matrix $\varphi(B)$ by*

$$\varphi(B) := \left(\mathrm{sgn}(b_{ij}) \sqrt{b_{ij} b_{ji}} \right). \tag{17.10}$$

If B is symmetrizable, then $\varphi(B)$ is its symmetrization.

Proof This follows from the proof of Lemma 17.1. □

We call the map φ in Lemma 17.9 the **symmetrization map**. Generally we have a fixed n in mind, but we may view the symmetrization map as being defined on the set of all sign symmetric matrices of any size.

We define

$$\sqrt{\mathbb{N}_0} = \{a \in \mathbb{R} \mid a^2 \in \mathbb{N}_0\} = \{0, 1, -1, \sqrt{2}, -\sqrt{2}, \sqrt{3}, -\sqrt{3}, \dots\}.$$

Lemma 17.10 *If B is a symmetrizable integer matrix, then its symmetrization has all entries in $\sqrt{\mathbb{N}_0}$.*

Proof Clear from (17.10). $\square$

The cycle condition for symmetrizable integer matrices implies a corresponding cycle condition for their symmetrizations. Let $S = (s_{ij})$ be a real symmetric $n \times n$ matrix. We say that S satisfies the **rational cycle condition** if for every $t \geq 2$ and every sequence $i_1, \dots, i_t$ of elements of $\{1, \dots, n\}$ there holds

$$s_{i_1 i_2} s_{i_2 i_3} \cdots s_{i_{t-1} i_t} s_{i_t i_1} \in \mathbb{Q}. \tag{17.11}$$

Lemma 17.11 *If B is a symmetrizable integer matrix, then $\varphi(B)$ satisfies the rational cycle condition.*

Proof Let $S = (s_{ij}) = \varphi(B)$, where $B = (b_{ij})$ is an $n \times n$ symmetrizable integer matrix. Take any $t \geq 2$ and any $i_1, \dots, i_t \in \{1, \dots, n\}$. From (17.10), we have

$$s_{i_1 i_2} s_{i_2 i_3} \cdots s_{i_{t-1} i_t} s_{i_t i_1} = \pm\sqrt{b_{i_1 i_2} b_{i_2 i_1} \cdots b_{i_t i_1} b_{i_1 i_t}}.$$

By (17.9) this is rational, indeed in $\mathbb{Z}$, given that the b_{ij} are all in $\mathbb{Z}$. $\square$

Lemma 17.12 *If $S = (s_{ij})$ satisfies the rational cycle condition (17.11), then its diagonal entries are all rational.*

Proof Taking $t = 2$ and $i_1 = i_2 = i$ in the rational cycle condition gives $s_{ii}^2 \in \mathbb{Q}$. Taking $t = 3$ and $i_1 = i_2 = i_3 = i$ gives $s_{ii}^3 \in \mathbb{Q}$. Hence, $s_{ii} \in \mathbb{Q}$. $\square$

Exercise 17.13 *Show that the two matrices in Exercise 17.8 have the same symmetrization, say S. Verify that S satisfies the rational cycle condition.*

The previous exercise shows that just because $\varphi(B) = S$ satisfies the rational cycle condition, one cannot conclude that B satisfies the cycle condition. The next Lemma is rather more positive: it shows that if S satisfies the rational cycle condition, then there is at least one symmetrizable matrix B such that $\varphi(B) = S$.

Lemma 17.14 *Let $S = (s_{ij})$ be a symmetric matrix with entries in $\sqrt{\mathbb{N}_0}$ that satisfies the rational cycle condition. Then there exists some symmetrizable integer matrix B such that $\varphi(B) = S$.*

Proof We may suppose that S is connected: if not, then tackle each component in turn and glue things together.

For each i and j, define positive integers d_{ij} and nonnegative integers e_{ij} by $s_{ij}^2 = d_{ij} e_{ij}^2$, with d_{ij} square-free (if $s_{ij} = 0$, then $d_{ij} = 1$ and $e_{ij} = 0$). Also put

$$A^\dagger = \left(\mathrm{sgn}(s_{ij})e_{ij}\right),$$

so that $e_{ii} = s_{ii}$.

Since S is symmetric, so is $A^\dagger$.

Choose a spanning tree T for the underlying graph G of S (or of $A^\dagger$: the underlying graphs are the same), and choose some vertex v as the root.

For each prime p dividing any of the d_{ij}, define a colouring (depending on p) of the vertices of G as follows: Colour the root vertex v red. For each other vertex of G, there is a unique path in T to the root; if an odd number of edges (i, j) on this path have $p \mid d_{ij}$, then colour the vertex blue; if an even number, then red.

We claim that for nonzero s_{ij}, we have that $p \mid d_{ij}$ if and only if the vertices i and j have different colours. This is clear if the edge between i and j is in the tree T. If not, consider the closed walk in G defined as follows: start at i, use the edge from i to j, then follow the unique path in T to v, then follow the unique path in T to i. By the rational cycle condition, this closed walk uses an even number of edges xy with p dividing d_{xy}. If $p \mid d_{ij}$, then the edge from i to j is one of this even number, and there must be an odd number in the rest of the closed walk; but that part of the closed walk is in T, and hence i and j must have opposite colours. If $p \nmid d_{ij}$, then there is an even number of edges xy in the rest of the closed walk for which $p \mid d_{xy}$, and since all these edges are in T, we deduce that i and j have the same colour.

Now for each (i, j) with $i < j$ and $p \mid d_{ij}$ do the following:

- if vertex i is red and vertex j is blue, multiply $A^\dagger_{ij}$ by p;
- if vertex i is blue and vertex j is red, multiply $A^\dagger_{ji}$ by p.

Repeat this for each prime p dividing any of the d_{ij}, and let $A = (a_{ij})$ be the final matrix produced from $A^\dagger$ having done all the required multiplications of elements.

If $i_1, i_2, \ldots, i_t, i_1$ is any closed walk in G, then for each prime p the number of changes from red to blue (using the colouring for p) must equal the number of changes from blue to red (since the walk is closed). Hence, each side of (17.9) is divisible by the same power of p. Since this holds for all p, the matrix A satisfies the cycle condition. Sign symmetry is trivial from the initial construction of $A^\dagger$, so A is symmetrizable.   □

Note that the colouring of the vertices in this construction (given $A_\dagger$ and p) is independent of the spanning tree chosen, and of the root, except that the colours red and blue might be swapped. There could easily be symmetrizable integer matrices in the fibre of φ over S that cannot be generated by the above procedure, but only finitely many: for we must have $b_{ij}b_{ji} = s^2_{ij}$, giving a finite set of possibilities for each b_{ij}. The force of the Lemma is that provided S satisfies the rational cycle condition, there is at least one choice of these b_{ij} that makes B symmetrizable.

Exercise 17.15 *Carry out the construction of the matrix B in the above proof for the symmetric matrix*

$$S = \begin{pmatrix} 1 & \sqrt{6} & -\sqrt{2} & 1 \\ \sqrt{6} & 2 & \sqrt{3} & 0 \\ -\sqrt{2} & \sqrt{3} & -3 & \sqrt{8} \\ 1 & 0 & \sqrt{8} & 4 \end{pmatrix}.$$

Exercise 17.16 *For the matrix S of the previous exercise, find all integer matrices $B = (b_{ij})$ for which $b_{ij}b_{ji} = s_{ij}^2$ and $\mathrm{sgn}(b_{ij}) = \mathrm{sgn}(s_{ij})$ for all relevant i and j. Check which of these matrices B also satisfy the cycle condition. Hence compute the complete list of symmetrizable integer matrices in the fibre of φ over S. Check that it includes the matrix computed in the previous exercise. Compare the complexity of these two approaches for finding a symmetrizable matrix B that has a given symmetrization.*

17.6 Interlacing

It was noted earlier (near the start of Sect. 17.2) that the eigenvalues of a symmetrizable matrix are all real. We record here that the analogue of Cauchy's Interlacing Theorem (Theorem B.1) holds for symmetrizable matrices.

Theorem 17.17 *Let B be a real $n \times n$ symmetrizable matrix. Then the eigenvalues of every principal $(n-1) \times (n-1)$ submatrix of B are real and interlace with the eigenvalues of B. Furthermore, these submatrices are also symmetrizable.*

Proof Take D as in (17.1), so that $S = D^{-1}BD$ is symmetric and has the same eigenvalues as B (all real).

If B_i is obtained by deleting row i and column i from B, and similarly D_i is obtained from D and S_i from S, then we have $S_i = D_i^{-1}B_iD_i$, showing that B_i is symmetrizable. The symmetric matrix S_i has the same eigenvalues as B_i. By the Interlacing Theorem (Theorem B.1), the eigenvalues of S_i interlace with those of S, and hence the eigenvalues of B_i interlace with those of B.   □

17.7 Equitable Partitions of Signed Graphs

Let G be a signed graph, with vertex set $V = \{1, \ldots, n\}$. A *partition* of V is simply a way of writing it as a disjoint union of nonempty subsets: i.e., $V = V_1 \cup \cdots \cup V_r$, where $V_i \cap V_j = \emptyset$ if $i \neq j$ and $V_i \neq \emptyset$ for any i. An arbitrary partition may not be very informative. By contrast, an *equitable partition* reflects a strong symmetry within the signed graph.

A partition $V = V_1 \cup \cdots \cup V_r$ is **equitable** if there are constants b_{ij} ($i, j \in \{1, \ldots, r\}$) such that for any i and j in $\{1, \ldots, r\}$ (perhaps $i = j$), and any vertex $x \in V_i$, the sum of the weights of the edges from x to vertices in V_j is b_{ij}. (If G is a graph, then b_{ij} is simply the number of neighbours that $x \in V_i$ has in V_j.)

The matrix $B = (b_{ij})$ is called the **_quotient matrix_** of the equitable partition. We briefly describe any such matrix as 'a quotient matrix'.

Example 17.18 Let G be the signed graph shown below, with the six vertices split into three subsets V_1, V_2 and V_3 as indicated, where the vertices within each subset are stacked vertically to emphasise the partition structure.

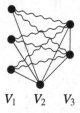

$$V_1 \quad V_2 \quad V_3$$

We see that every vertex in V_1 has no edges to other vertices in V_1, one positive edge to the (only) vertex in V_2 and two negative edges to vertices in V_3; the vertex in V_2 has three positive edges to vertices in V_1, none to vertices in V_2 and two positive edges to vertices in V_3; each vertex in V_3 has three negative edges to vertices in V_1, one positive edge to the vertex in V_2 and one to the other vertex in V_3. Hence, this partition is equitable, with quotient matrix

$$\begin{pmatrix} 0 & 1 & -2 \\ 3 & 0 & 2 \\ -3 & 1 & 1 \end{pmatrix}.$$

In this section, we show that quotient matrices are precisely symmetrizable integer matrices (Theorem 17.23). It is easy to see that quotient matrices are symmetrizable (Lemma 17.21), but the reverse implication that every symmetrizable integer matrix is in fact a quotient (Lemma 17.22) requires a little more work.

If the vertices are ordered to reflect an equitable partition of G, then the adjacency matrix A of G has block form

$$A = \begin{pmatrix} A_{11} & A_{12} & \cdots & A_{1r} \\ A_{21} & A_{22} & \cdots & A_{2r} \\ \vdots & \vdots & \ddots & \vdots \\ A_{r1} & A_{r2} & \cdots & A_{rr} \end{pmatrix} \tag{17.12}$$

where each submatrix A_{ij} has constant row sum b_{ij}.

Since A is symmetric, we have $A_{ji} = A_{ij}^\mathsf{T}$, and A_{ii} is always square. If $|V_i| = c_i$, then A_{ij} is a $c_i \times c_j$ matrix. Since $A_{ij} = A_{ji}^\mathsf{T}$, each A_{ij} also has constant column sum, namely, b_{ji}.

Example 17.19 Ordering the vertices in the previous example to reflect the equitable partition, we get the adjacency matrix

$$A = \left(\begin{array}{ccc|c|cc} 0 & 0 & 0 & 1 & -1 & -1 \\ 0 & 0 & 0 & 1 & -1 & -1 \\ 0 & 0 & 0 & 1 & -1 & -1 \\ \hline 1 & 1 & 1 & 0 & 1 & 1 \\ \hline -1 & -1 & -1 & 1 & 0 & 1 \\ -1 & -1 & -1 & 1 & 1 & 0 \end{array}\right).$$

The row sums in each block are constant, as are the column sums in each block.

The original matrix (a_{ij}) is symmetric, has zeros on the diagonal, and all off-diagonal entries are in the set $\{-1, 0, 1\}$. For the quotient matrix (b_{ij}), none of these properties need hold. Certainly, the b_{ij} are all integers, but the matrix need not be symmetric, and there are no general bounds on any of the entries.

We can view both A and its quotient B as weighted digraphs, and can draw the quotienting process by drawing first the original signed graph, with the vertices in each V_i stacked vertically, then a vertical arrow below the graph, pointing to the weighted digraph that represents the quotient matrix, with each vertex of the quotient drawn vertically below the corresponding V_i.

Here is a simple example:

In terms of matrices, $\begin{pmatrix} 0 & 1 & -1 \\ 1 & 0 & -1 \\ -1 & -1 & 0 \end{pmatrix}$ has quotient $\begin{pmatrix} 1 & -1 \\ -2 & 0 \end{pmatrix}$.

Exercise 17.20 *Let G be a cycle on n vertices. Find all possible equitable partitions of G. (Divisibility properties of n play a part in the answer.)*

If B is a quotient matrix for an equitable partition of a signed graph G with adjacency matrix A then (after permuting rows/columns so that the order of the vertices reflects the partition) A has block structure as in (17.12) with A_{ij} having constant row sum b_{ij}. If V_i contains c_i vertices, then the sum of all the entries in $A_{ij} = A_{ji}^{\mathsf{T}}$ can be computed either as $c_i b_{ij}$ or as $c_j b_{ji}$, so we must have:

$$c_i b_{ij} = c_j b_{ji}. \tag{17.13}$$

Comparing with (17.8) and Lemma 17.6, we see that if B is a quotient matrix, then it is symmetrizable. We record this as a Lemma.

Lemma 17.21 *Let B be the quotient matrix of an equitable partition of a signed graph. Then B is symmetrizable.*

Since quotient matrices are symmetrizable, they must satisfy the sign symmetry condition (17.5) and the cycle condition (17.9). We now show that the balancing condition (17.13) is sufficient to imply that B is a quotient matrix.

Lemma 17.22 *Let $B = (b_{ij})$ be an $r \times r$ integer matrix. If there are positive integers $c_1, \ldots, c_r$ such that (17.13) holds for all $1 \le i, j \le r$, then B is the quotient matrix of an equitable partition of a signed graph. If in addition the entries of B are all nonnegative, then B is the quotient matrix of an equitable partition of a graph.*

Proof Suppose that we have $B = (b_{ij})$ and $c_1, \ldots, c_r$ as in the hypothesis of the lemma. Define

$$M := \prod_{i=1}^{r}(1 + |b_{ii}|) \prod_{1 \le i, j \le r, \, i \ne j} \max(1, |b_{ij}|). \qquad (17.14)$$

For $1 \le i \le r$, put $c_i' = Mc_i$. We shall define a signed graph G on $n = c_1' + \cdots + c_r'$ vertices which has an equitable partition for which B is the quotient matrix. The construction will be such that if all the b_{ij} are nonnegative then the signed graph is actually a graph.

Take disjoint sets $V_1, \ldots, V_r$ with $|V_i| = c_i'$. The elements of these will be the vertices of G, and we shall now describe how to allocate signed edges so as to achieve our desired equitable partition.

For each i, split V_i (arbitrarily) into subsets of size $1 + |b_{ii}|$ (note from (17.14) that $1 + |b_{ii}|$ divides c_i'). Within each subset, put signed edges between every pair of vertices in the subset, all with the same sign $\mathrm{sgn}(b_{ii})$. These will be the only edges between vertices in the same V_i, so the sum of the weights of the edges between any one vertex in V_i and all other vertices in V_i is the constant b_{ii}.

For each $i < j$ for which $b_{ij} \ne 0$, split V_i into subsets $U_{i1}, \ldots, U_{ik}$ of size $|b_{ji}|$, where $k := c_i'/|b_{ji}| = Mc_i/|b_{ji}| = Mc_j/|b_{ij}| = c_j'/|b_{ij}|$, using (17.13) (which implies that $b_{ji} \ne 0$ too; and note from (17.14) that k is a positive integer). Hence, with the same k, we can split V_j into subsets $W_{j1}, \ldots, W_{jk}$ of size $|b_{ij}|$. For each $1 \le l \le k$, we put signed edges, all of sign $\mathrm{sgn}(b_{ij})$, between every vertex in U_{il} and every vertex in W_{jl}. Hence, the sum of the weights of the edges between any vertex in V_i and all vertices in V_j is the constant b_{ij}, and the sum of the weights of the edges between any vertex in V_j and all vertices in V_i is the constant b_{ji}.

We have, therefore, constructed a signed graph G that admits an equitable partition of its vertices as $V_1 \cup \cdots \cup V_r$ such that the quotient matrix is B, and if all entries of B are nonnegative then the construction produces a graph. $\qquad\qquad\square$

Combining this result with Lemmas 17.21 and 17.22 gives the following theorem.

Theorem 17.23 *A square integer matrix is symmetrizable if and only if it is the quotient matrix of an equitable partition of a signed graph.*

A square integer matrix with all entries nonnegative is symmetrizable if and only if it is the quotient matrix of an equitable partition of a graph.

Exercise 17.24 *The construction in the proof of Theorem 17.23 is not efficient. We saw earlier that* $\begin{pmatrix} 1 & -1 \\ -2 & 0 \end{pmatrix}$ *is the quotient of a signed graph on* 3 *vertices. Use the sizes of the classes in that 3-vertex signed graph to find c_i such that (17.13) holds. Follow the construction in the proof of Lemma 17.22 to give this same 2×2 matrix as the quotient of a signed graph on 12 vertices. Replace the definition of M by a suitable least common multiple, and repeat with the smaller value of M.*

17.8 Notes

Interlacing for symmetrizable matrices was shown by Kouachi [Kou16]. It is curious that Theorem 17.23 does not appear to have been noticed before the paper of McKee and Smyth [MS20b], given that it relates two important classes of matrices and one direction was known. Much of this chapter is adapted from [MS20a] and [MS20b].

17.9 Glossary

$\sqrt{\mathbb{N}_0}$ The set $\{0, 1, -1, \sqrt{2}, -\sqrt{2}, \sqrt{3}, -\sqrt{3}, \dots\}$.

balancing condition.   An $n \times n$ real matrix $B = (b_{ij})$ satisfies the balancing condition if there exist positive real numbers $d_1, \dots, d_n$ such that $d_i b_{ij} = d_j b_{ji}$ for all i and j. This condition is equivalent to saying that B is symmetrizable.

cycle condition.   An $n \times n$ real matrix $B = (b_{ij})$ satisfies the cycle condition if for every $t \geq 2$ and every sequence $i_1, \dots, i_t$ of elements of $\{1, \dots, n\}$ there holds $b_{i_1 i_2} b_{i_2 i_3} \cdots b_{i_{t-1} i_t} b_{i_t i_1} = b_{i_2 i_1} b_{i_3 i_2} \cdots b_{i_t i_{t-1}} b_{i_1 i_t}$. Any symmetrizable matrix satisfies this condition.

equitable partition.   An equitable partition of a signed graph G is a partition of its vertices as $V = V_1 \cup \cdots \cup V_r$ such that for any vertex $x \in V_i$ the sum of the weights of the edges from x to neighbours in V_j depends only on i and j, and not on the choice of $x \in V_i$. If the vertices are ordered so as to reflect such a partition, then the adjacency matrix has block form with the row sums of each block being constant within that block.

quotient matrix.   If G has an equitable partition, partitioning the vertices as $V = V_1 \cup \cdots \cup V_r$, then the corresponding quotient matrix $B = (b_{ij})$ is the $r \times r$ matrix for which b_{ij} is the sum of the weights of the edges joining any vertex $x \in V_i$ to its neighbours in V_j.

rational cycle condition.   A real symmetric matrix $S = (s_{ij})$ satisfies the rational cycle condition if for every $t \geq 2$ and every sequence $i_1, \dots, i_t$ of elements of $\{1, \dots, n\}$ there holds $s_{i_1 i_2} s_{i_2 i_3} \cdots s_{i_{t-1} i_t} s_{i_t i_1} \in \mathbb{Q}$. If B is symmetrizable, then its symmetrization satisfies the rational cycle condition.

sign symmetric.   An $n \times n$ matrix $B = (b_{ij})$ is sign symmetric if $\operatorname{sgn}(b_{ij}) = \operatorname{sgn}(b_{ji})$ for all relevant i and j. Any symmetrizable matrix is sign symmetric.

symmetric component.   If B is a symmetrizable matrix, then define the symmetric matrix B^* by $(B^*)_{ij} = b_{ij}$ if $b_{ij} = b_{ji}$, and $(B^*)_{ij} = 0$ if $b_{ij} \neq b_{ji}$. Viewing B and B^* as weighted digraphs on the same set of vertices, the symmetric components of B are (the induced subgraphs spanned by) the subsets of the vertices that correspond to the components of B^*. If $S = D^{-1}BD$ is the symmetrization of B, then although D is not uniquely determined, its diagonal entries must be constant on the symmetric components of B.

symmetrizable.   An $n \times n$ real matrix B is symmetrizable if there exists a real diagonal matrix $D = \text{diag}(d_1, \ldots, d_n)$ with positive entries on the diagonal such that $D^{-1}BD$ is symmetric.

symmetrization.   If B is a symmetrizable $n \times n$ matrix, with $D = \text{diag}(d_1, \ldots, d_n)$ having positive entries on the diagonal such that $S = D^{-1}BD$ is symmetric, then S is called the symmetrization of B. Although the diagonal matrix D is not unique, there is only one possible symmetrization S.

symmetrization map.   The map φ that sends real square sign symmetric matrices to real symmetric matrices via $\varphi\big((b_{ij})\big) = \big(\text{sgn}(b_{ij})\sqrt{b_{ij}b_{ji}}\big)$. If $B = (b_{ij})$ is symmetrizable, then $\varphi(B)$ is its symmetrization.

Chapter 18
Symmetrizable Matrices II: Cyclotomic Symmetrizable Integer Matrices

We now extend the results of Chap. 6 to the symmetrizable setting.

18.1 Cyclotomic Symmetrizable Integer Matrices

As one would by now expect, an integer symmetrizable matrix is said to be *cyclotomic* if all its eigenvalues are in the interval $[-2, 2]$. Naturally, the *transpose* of a weighted digraph is defined as the digraph whose adjacency matrix is the transpose of that of the original graph. We first note that the maximal connected cyclotomic integer symmetric matrices of Chap. 6 remain maximal amongst the larger set of symmetrizable matrices.

Lemma 18.1 *If A is a maximal connected cyclotomic integer symmetric matrix, then A is also maximal in the set of connected cyclotomic symmetrizable integer matrices.*

Proof From Theorem 6.31, we know that each row in A has degree 4 (i.e., the corresponding diagonal entry in A^2 is 4). If B were a connected symmetrizable integer matrix properly containing A as an induced submatrix, then some row of the symmetrization $\varphi(B)$, defined in (17.10), would have degree greater than 4 (else A would form a connected component of B). Thus, the matrix $\varphi(B)^2$ would have some diagonal entry greater than 4. By interlacing, the symmetric matrix $\varphi(B)^2$ would have some eigenvalue greater than 4. Hence, $\varphi(B)$ would have some eigenvalue outside $[-2, 2]$, and as this matrix has the same eigenvalues as B, we conclude that B would not be cyclotomic. □

Exercise 18.2 The proof of Lemma 18.1 uses the symmetrization of B. Complete the following outline of an alternative proof that works with B itself rather than its symmetrization.

(i) Show that if B is symmetrizable, then so is B^2.

(ii) Mimic the proof of Lemma 18.1 but using B wherever $\varphi(B)$ was used, and appealing to Theorem 17.17.

© Springer Nature Switzerland AG 2021
J. McKee and C. Smyth, *Around the Unit Circle*, Universitext,
https://doi.org/10.1007/978-3-030-80031-4_18

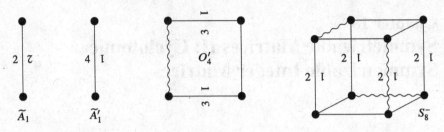

Fig. 18.1 $\tilde{A}_1$, $\tilde{A}'_1$, O'_4 and S_8^-

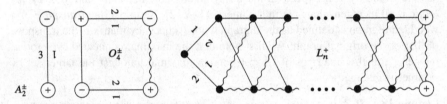

Fig. 18.2 L_n and L'_n, for $n = 2r + 2 \geq 4$ and even, where r is the number of vertices on the top row. The two weight pairs $2'$, $1'$ are $2, 1$ for L_n, but are swapped to $1, 2$ for L'_n. Note that L'_n is equivalent to its transpose, but L_n is not (Lemma 18.9 and Corollary 18.12)

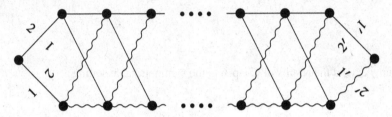

Fig. 18.3 $A_2^{\pm}$, $O_4^{\pm}$ and L_n^+, for $n = 2r + 1 \geq 3$, where r is the number of vertices on its top row, including the charged vertex

The main result in this section is the following theorem.

Theorem 18.3 *Let A be a maximal connected cyclotomic symmetrizable integer matrix. If A is not symmetric, then A is equivalent to one of the following weighted digraphs (see Figs. 18.1, 18.2, 18.3).*

Not charged: $\tilde{A}'_1$, O'_4, S_8^-, L_n *and* L'_n ($n \geq 4$ *and even*), L_n^{T} ($n \geq 6$, *even*);

Charged: $A_2^{\pm}$, $O_4^{\pm}$, L_n^+, $(L_n^+)^{\mathsf{T}}$ ($n \geq 3$ *and odd*).

All these digraphs (matrices) A have $A^2 = 4I$ *for the appropriate identity matrix I. Furthermore, every connected symmetrizable asymmetric matrix having all its eigenvalues in the interval* $[-2, 2]$ *is contained in one of these maximal ones.*

By combining our earlier work (Theorem 6.31) with these results, and separating the charged and noncharged cases, we obtain the following:

Corollary 18.4 *Let A be an* $n \times n$ *connected, sign symmetric matrix with nonnegative integer entries and maximal with respect to having all its eigenvalues in* $[-2, 2]$*. Suppose first that A is uncharged.*

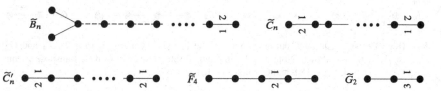

Fig. 18.4 The symmetrizable but asymmetric affine Dynkin diagrams $\widetilde{B}_n$ ($n \geq 3$), $\widetilde{C}_n$ ($n \geq 2$), $\widetilde{C}'_n$ ($n \geq 2$), $\widetilde{F}_4$, $\widetilde{G}_2$. Note that only $\widetilde{C}'_n$ is equivalent to its transpose (Lemma 18.9 and Corollary 18.12)

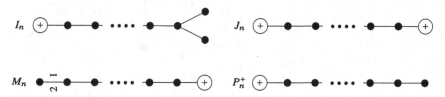

Fig. 18.5 The n-vertex digraphs I_n ($n \geq 3$), and J_n, M_n and P_n^+ for $n \geq 2$. Only M_n is asymmetric

- *If A is symmetric, then it is equivalent to one of: the 1×1 matrix (2), the matrix $\widetilde{A}_1$ (Fig. 18.1), or one of the simply laced Dynkin diagrams $\widetilde{A}_n$ ($n \geq 2$), $\widetilde{D}_n$ ($n \geq 4$), $\widetilde{E}_6$, $\widetilde{E}_7$ or $\widetilde{E}_8$ (Exercise 7.14).*
- *If A is asymmetric, then it is equivalent to one of the non-simply laced Dynkin diagrams $\widetilde{A}'_1$, $\widetilde{B}_n$, $\widetilde{B}_n^\top$, ($n \geq 2$), $\widetilde{C}_n$ ($n \geq 2$), $\widetilde{C}_n^\top$ ($n \geq 2$), $\widetilde{C}'_n$ ($n \geq 3$), $\widetilde{G}_2$, $\widetilde{G}_2^\top$, $\widetilde{F}_4$ or $\widetilde{F}_4^\top$. See Figs. 18.1 and 18.4.*

Now suppose that A is charged.

- *If A is symmetric then it is equivalent to one of I_n ($n \geq 3$) or J_n ($n \geq 2$) from Fig. 18.5.*
- *If A is asymmetric, then it is equivalent to the digraph M_n of Fig. 18.5 for some $n \geq 1$, or to its transpose $M_n^\top$.*

An important ingredient in the proofs is the generalisation by Greaves [Gre12b] of the classification of cyclotomic integer symmetric matrices (Theorem 6.31) to cyclotomic symmetric matrices whose entries are algebraic integers lying in the compositum of all real quadratic fields. It would take us too far afield to give a full account of this work (which is similar in spirit to the material in Chap. 6), and we shall simply quote the results that we need.

Corollary 18.4 implies the following:

Corollary 18.5 *The symmetrizable asymmetric affine Dynkin diagrams of Fig. 18.4 and $\widetilde{A}'_1$ of Fig. 18.1, along with their transposes, are the connected subgraphs of the diagrams of Theorem 18.3 that are maximal with respect to the property of having all eigenvalues in $[-2, 2]$ and no charges or negative edges.*

Exercise 18.6 Let A be an $n \times n$ symmetrizable matrix having all its eigenvalues at -2 and 2. Show that any induced subgraph of A having all its eigenvalues in the open interval $(-2, 2)$ has at most $\lfloor \frac{n}{2} \rfloor$ vertices.

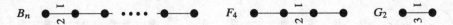

Fig. 18.6 The symmetrizable but asymmetric finite Dynkin diagrams B_n $(n \geq 2)$, F_4 and G_2; C_n $(n \geq 3) = B_n^{\mathsf{T}}$ not shown. Note that B_2, F_2 and G_2 are equivalent to their transposes, while B_n $(n \geq 3)$ is not (Lemma 18.9 and Corollary 18.12)

Fig. 18.7 The graphs O_4''
and $B_2^{\pm}$

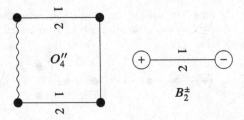

Exercise 18.7 Let A be a square integer connected symmetrizable asymmetric matrix having all its eigenvalues in the open interval $(-2, 2)$. Using Theorem 18.3 and Exercise 18.6, show that A is equivalent to an induced subgraph of one of the following graphs (all of which have all their eigenvalues in $(-2, 2)$; see Figs. 18.6 and 18.7): B_n $(n \geq 2)$, $C_n := B_n^{\mathsf{T}}$ $(n \geq 3)$, F_4, G_2, O_4'', or $B_2^{\pm}$.

Lemma 18.8 *Suppose that the weighted digraphs A and B are equivalent. Then so are A^{T} and B^{T}.*

Proof If $B = \pm P^{\mathsf{T}} A P$, then $B^{\mathsf{T}} = \pm P^{\mathsf{T}} A^{\mathsf{T}} P$. □

Lemma 18.9 *Each of the following asymmetric graphs is equivalent to its transpose: $\tilde{A}_1'$, O_4', S_8^- (Fig. 18.1), L_4, L_n', $(n \geq 4, even)$ (Fig. 18.2), $A_2^{\pm}$, $O_4^{\pm}$ (Fig. 18.3), $\tilde{C}_n'$ $(n \geq 3)$ (Fig. 18.4), B_2, F_4, G_2 (Fig. 18.6), O_4'', $B_2^{\pm}$ (Fig. 18.7).*

The proof for each of the digraphs listed is demonstrated by showing that each of these matrices A, the matrix A^{T} can be transformed back to the original matrix A by possibly first replacing A^{T} by $-A^{\mathsf{T}}$, then relabelling the vertices and finally applying a sign switching at certain vertices. The details are relegated to the following exercise.

Exercise 18.10 Verify that each of the graphs in Lemma 18.9 is equivalent to its transpose.

In the other direction, we need sometimes to show that two similar-looking weighted digraphs are *not* equivalent.

Lemma 18.11 *In a weighted digraph G, a 'weight modulus sequence' is a sequence of moduli of edge weights along an induced path in G (an induced subgraph that is a path). If G_1 and G_2 are weighted digraphs, and there is a weight modulus sequence in G_1 that does not appear in G_2, then G_1 and G_2 are not equivalent.*

Proof Any equivalence preserves the set of all weight modulus sequences. □

Corollary 18.12 *Each of the following weighted digraphs is not equivalent to its transpose:* M_n ($n \geq 2$) (Fig. 18.5), B_n ($n \geq 3$) (Fig. 18.6), $\widetilde{B}_n$ ($n \geq 3$), $\widetilde{C}_n$ ($n \geq 2$), $\widetilde{F}_4$, $\widetilde{G}_2$ (Fig. 18.4), L_n ($n \geq 6$, even) (Fig. 18.2), L_n^+ ($n \geq 3$, odd) (Fig. 18.3).

Furthermore, the graphs L_n *and* L_n' ($n \geq 4$, even) (Fig. 18.2) *are not equivalent.*

Proof To show that each of M_n, B_n, $\widetilde{B}_n$, $\widetilde{C}_n$, L_n^+ is not equivalent to its transpose, we apply Lemma 18.11 to the weight modulus sequence $1, 1, \ldots, 1, 2$ along a path from one end of the digraph to the other as drawn (right to left for M_n, B_n, L_n^T; left to right for $\widetilde{B}_n$, $\widetilde{C}_n$). The weighted digraphs M_2, L_3^+, L_5^+ are special cases. For M_2, the weight modulus sequence 2 appears in both M_2 and M_2^T, but in one case, the weight 2 is from a neutral vertex to a charged vertex, and in the other case it is not. The same argument applies to L_3^+. For L_5^+, note that the weight modulus sequence $1, 2$ appears in $(L_5^+)^\mathsf{T}$ on an induced path joining the three neutral vertices, whereas in L_5^+, a charged vertex is necessarily involved. For $\widetilde{F}_4$ take the sequence $1, 2, 1, 1$; for $\widetilde{G}_2$ the sequence $3, 1$; for L_n the sequence $2, 1, 1, \ldots, 1$. To show that L_n and L_n' are not equivalent, again take the weight modulus sequence $2, 1, 1, \ldots, 1$ in L_n from one end of the digraph to the other, and for $n = 4$, take the weight modulus sequence $2, 2$ in L_4'. $\qquad\square$

Lemma 18.13 *Let* $\begin{pmatrix} a & b \\ c & d \end{pmatrix}$ *be a sign symmetric matrix with both eigenvalues real and in the interval* $[-2, 2]$. *Then* $bc \leq 4$. *Moreover, if* $bc = 4$, *then* $a = d = 0$.

Proof The difference between the two eigenvalues is $\sqrt{(a-d)^2 + 4bc}$. This difference is at most 4, so $4bc \leq 16$ (here using that b and c have the same sign), so $bc \leq 4$. Moreover, if $bc = 4$, then we have $a = d$, and the eigenvalues are $a \pm 2$, whence $a = d = 0$. $\qquad\square$

Having established these preliminary lemmas, we are now ready to tackle the proofs of Theorem 18.3 and its corollaries.

Proof (*of Theorem* 18.3) Assume that A is symmetrizable, but not symmetric, is connected and has all its eigenvalues in $[-2, 2]$. Being symmetrizable, A is sign symmetric, so that a_{ij} and a_{ji} are either both zero, or else both nonzero and of the same sign. Hence, by asymmetry, either A or A^T is equivalent to an edge-labelled digraph with $a_{12} > a_{21}$, both positive integers. By repeated application of Theorem 17.17, we see that the induced subgraph on any two distinct vertices i and j has all its eigenvalues in $[-2, 2]$. By Lemma 18.13, the set $\{|a_{12}|, |a_{21}|\}$ is equal to either $\{4, 1\}$, $\{3, 1\}$ or $\{2, 1\}$. However, we note that if $a_{12}a_{21} = 4$, then by Lemma 18.13, the digraph A contains $\widetilde{A}_1'$ (Fig. 18.1) as an induced subgraph, and one checks that this is maximal, using the same argument as in Lemma 18.1. So we can assume that $\{|a_{12}|, |a_{21}|\}$ is equal to either $\{3, 1\}$ or $\{2, 1\}$.

Let $S = D^{-1}AD = (s_{ij})$ be the symmetrization of A, as in (17.1). For nonzero off diagonal entries of A that show symmetry, we note by Theorem 17.17 that if $a_{ij} = a_{ji} \neq 0$, then $a_{ij} = \pm 1$ or ± 2. But if $a_{ij} = \pm 2$, then A contains $\pm \widetilde{A}_1$ (Fig. 18.1) as a subgraph. But A_1 is maximal among integer symmetric matrices having all eigenvalues in $[-2, 2]$, and by Lemma 18.1, it is maximal amongst symmetrizable integer matrices having all eigenvalues in $[-2, 2]$. Thus, $A = \pm\widetilde{A}_1$, contradicting our assumption that A is not symmetric. Thus, $a_{ij} = \pm 1$, and hence by (17.4), we have $s_{ij} = \pm 1$.

The symmetrization of A, namely S, has the same eigenvalues as A, and all entries in $\sqrt{\mathbb{N}_0}$. Indeed, all nonzero entries of S are either ± 1, $\pm\sqrt{2}$, or $\pm\sqrt{3}$, and the eigenvalues of S all lie in the interval $[-2, 2]$. Moreover, the diagonal entries of S have modulus at most 1 (by interlacing the only other possibility would be ± 2, but then the now familiar argument about maximal cyclotomic submatrices kicks in).

Such matrices S were considered by Greaves [Gre12b, Theorems 3.2, 3.4 and 3.5]. Greaves gave the form of all maximal such S containing at least one $\pm\sqrt{2}$ or $\pm\sqrt{3}$ entry. Restricting to those which also satisfy our requirements for the diagonal entries, these Greaves graphs are shown in Figs. 18.8, 18.9 and 18.10. All satisfy $S^2 = 4I$.

Now we do not assume that our A is maximal, as we wish to establish the last sentence of the theorem, but we know that $S = D^{-1}AD$ is equivalent to a subgraph of one of the Greaves examples. A computation as in Lemma 17.3 shows that A is equivalent to a matrix whose symmetrization is a subgraph one of the Greaves graphs, so working up to equivalence, we may assume that S is actually a subgraph of one of them. Moreover, S must include at least one irrational edge, else A would be symmetric.

We now determine the A that correspond to such S, using $s_{ij}^2 = a_{ij}a_{ji}$ and $\mathrm{sgn}(s_{ij}) = \mathrm{sgn}(a_{ij})$ to constrain the possibilities. When $s_{ij} \in \{0, -1, 1\}$ we have $a_{ij} = a_{ji}$. When $s_{ij} = \sqrt{3}, -\sqrt{3}, \sqrt{2}$, or $-\sqrt{2}$, we have $\{a_{ij}, a_{ji}\} = \{1, 3\}, \{-1, -3\}$, $\{1, 2\}$, or $\{-1, -2\}$, respectively. Thus, an edge

in S, with $a \in \{1, 2\}$, corresponds to one of

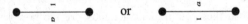

in A, and similarly for negative irrational edges. Moreover, by Lemma 17.5, we must make consistent choices for edges between the same pair of symmetric components of A. In nearly all cases, this immediately shows that A is a subgraph of one of the examples claimed: $\widetilde{A}_1'$, O_4', S_8^-, L_n, L_n', L_n^T, $O_4^\pm$, $A_2^\pm$, L_n^+, $(L_n^+)^\mathsf{T}$.

The most complicated case is when $S = L_{4,\mathrm{G}}$, or a subgraph of it, for then Lemma 17.5 places no restrictions on the choice of edges, so *a priori* there are 16 possibilities to check. For $S = L_{4,\mathrm{G}}$, we use the cycle condition (17.9) to limit the possibilities to

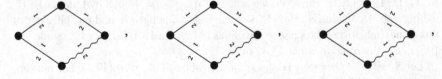

or the transposes of these, and working up to equivalence, one checks that the distinct possibilities for A are L_4 and L_4'.   □

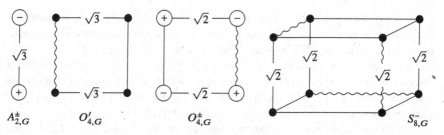

Fig. 18.8 The Greaves graphs $A_{2,G}^{\pm}$, $O_{4,G}'$, $O_{4,G}^{\pm}$ and $S_{8,G}^{-}$. Since they are symmetric, a single weight is given for each edge

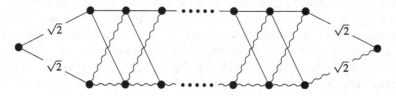

Fig. 18.9 $L_{n,G}$ for $n = 2r + 2 \geq 4$, where r is the number of vertices on the top row

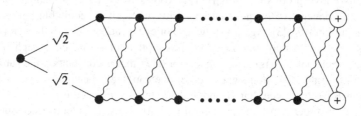

Fig. 18.10 $L_{n,G}^{+}$ for $n = 2r + 1 \geq 3$, where r is the number of vertices on the top row

Proof (*of Corollary* 18.4) Note that the hypothesis here merely requires A to be sign symmetric, rather than symmetrizable. But apart from $\widetilde{A}_n$, the digraphs we seek can contain no cycles (for the spectral radius of $\widetilde{A}_n$ is 2, and by Perron-Frobenius theory (Theorem B.6), any connected graph containing $\widetilde{A}_n$ as a (not necessarily induced) subgraph has strictly larger spectral radius). Hence, they satisfy (17.9), and so, by Lemma 17.7 must be symmetrizable.

Since the symmetric case is covered in Theorem 6.31, we can assume that our matrix (digraph) A is asymmetric but symmetrizable. From Theorem 18.3, we know that A is a subdigraph of one of the digraphs listed there. We must look for asymmetric, symmetrizable subdigraphs of these digraphs that are maximal with the property that they are equivalent to a digraph having no negative edges or any negative charges. Thus, we must remove as few vertices as possible to attain that aim. Trivially, there are no such subdigraphs from $\widetilde{A}_1'$ (other than $\widetilde{A}_1'$ itself), $A_2^{\pm}$ or $O_4^{\pm}$ (for this last, note that one must either remove both the negatively charged vertices or both the positively charged ones). From O_4', we remove one vertex to get $\widetilde{G}_2$ or its transpose, while from S_8^{-}, we remove three vertices to get a digraph equivalent to $\widetilde{F}_4$ or $\widetilde{C}_4$ (all 4-cycles in S_8^{-} must be broken).

To apply the same kind of argument to L_n, L_n^{T}, L_n', L_n^+ and $(L_n^+)^{\mathsf{T}}$, we note that these digraphs contain quadrilaterals with one or three negative edges. This number of negative edges may flip between 1 and 3 under equivalence, but is never even, and so never zero. Such quadrilaterals have either 2 or -2 as an eigenvalue, and by interlacing they must be excluded as subgraphs. We must, therefore, remove at least one vertex from every 'odd' quadrilateral. For L_n^+ and its transpose, we must also remove a vertex from the triangles containing two positively charged vertices. One way to do this is to remove the bottom row of r vertices from L_n, L_n^{T}, L_n', L_n^+ and $(L_n^+)^{\mathsf{T}}$. These are clearly maximal, as adding back any of these bottom vertices will produce an odd quadrilateral (or a forbidden triangle for L_n^+ or its transpose). So from L_{2r+2}, we obtain $\widetilde{C}_{r+1}^{\mathsf{T}}$, from L_{2r+2}^{T}, we obtain $\widetilde{C}_{r+1}$, from L_n', we obtain $\widetilde{C}_{r+1}'$, from L_{2r+1}^+, we obtain M_{r+1} and from $(L_{2r+1}^+)^{\mathsf{T}}$, we obtain M_{r+1}^{T}. A second way to obtain the kind of graphs we require is to remove from L_n and L_n' the leftmost vertex, as well as all bottom vertices, except the first. This gives the graphs $\widetilde{B}_r$, $\widetilde{B}_r^{\mathsf{T}}$. (Doing this for L_n^+ gives the symmetric graph I_r of Fig. 18.5.) We leave it as an exercise for the reader to check that these are the only graphs of the kind we are looking, for that, we can obtain from L_n, L_n', L_n^+ and their transposes.   □

An alternative proof of Corollary 18.4 comes from the identification of symmetrizable nonnegative integer matrices with quotients of equitable partitions of graphs (Theorem 17.23). Again we dispose of cycles and reduce to the symmetrizable case. Hence, we may suppose that G is a quotient of a connected graph H. Now each eigenvector of G lifts to an eigenvector of H that is constant on the subsets of the partition, and in the nonnegative case, one can apply Perron-Frobenius theory (Theorem B.6) to deduce that G and H have the same spectral radius, so that H also has all its eigenvalues in the interval $[-2, 2]$. Thus, H is an induced subgraph of one of the Smith graphs $\widetilde{A}_n$, $\widetilde{D}_n$, $\widetilde{E}_6$, $\widetilde{E}_7$, or $\widetilde{E}_8$. One can, therefore, identify all the possible G by considering all possible such H, and all possible ways of forming equitable partitions.

Exercise 18.14 Find all the equitable partitions of $\widetilde{A}_n$, $\widetilde{D}_n$, $\widetilde{E}_6$, $\widetilde{E}_7$, or $\widetilde{E}_8$ (see Exercise 17.20 for $\widetilde{A}_n$), and use these to reprove Corollary 18.4, as outlined above.

18.2   Quotients of Signed Graphs

Let A be (the adjacency matrix of) a signed graph, and let B be a quotient of an equitable partition of A. If the entries of B are nonnegative and A is, in fact, a graph, then the spectral radius of A equals that of B.

Exercise 18.15 Justify the previous comment. (Suppose that B is a quotient of an equitable partition of a graph that has adjacency matrix A. Let K be a component of B that has the same spectral radius λ as B. Suppose that $\mathbf{v}$ is an eigenvector of K with all entries positive (Theorem B.6), necessarily with eigenvalue λ. Note that $\mathbf{v}$ lifts to an eigenvector $\mathbf{w}$ of A with eigenvalue λ, with the components of $\mathbf{w}$ constant on the classes of the equitable partition. Moreover, this eigenvector has positive entries on every vertex in the component of A above K. Now use Theorem B.6 again.)

When there are negative edges in A (and negative weights in B), however, the situation is not so straightforward. For example, we can certainly produce the charged signed graph C_6^{++} (see Fig. 6.4 of Chap. 6) as a quotient of a signed graph (a special case of Theorem 17.23, as any symmetric matrix is symmetrizable) thus:

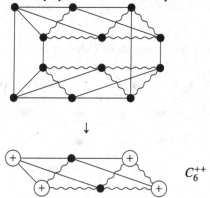

$$C_6^{++}$$

Here the drawing convention is as in Sect. 17.7, with vertices in the subsets of the partition stacked vertically above the image of that subset in the quotient. The quotient C_6^{++} has only -2 and 2 as eigenvalues, but the signed graph shown here, of which C_6^{++} is a quotient, has characteristic polynomial

$$x^2(x-2)^4(x+2)^2(x^2+2x-4)^2,$$

so has spectral radius $1+\sqrt{5} > 2$.

We shall see below, however, that C_6^{++} *can* be produced as the quotient of a signed graph that has spectral radius 2. Indeed, we now show that, in fact, all the examples of Theorem 18.3 can be produced as quotients of signed graphs that have all their eigenvalues in the intervals $[-2, 2]$, and the same is true for all the maximal charged signed graphs having all eigenvalues in $[-2, 2]$ that were classified in Chap. 6.

We start with the uncharged examples of Theorem 18.3. There are three sporadic cases, $\tilde{A}'_1$, O'_4 and S_8^- (Fig. 18.1). These can be produced as quotients of signed graphs as follows:

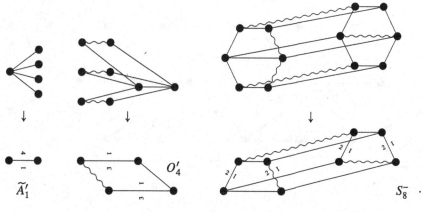

It can be checked that the signed graphs here are equivalent to subgraphs of S_{16} (Fig. 6.2), and hence have all their eigenvalues in $[-2, 2]$. Or one may compute their characteristic polynomials explicitly as

$$(x + 2)x^3(x - 2), \ (x + 2)^2(x + 1)^2(x - 1)^2(x - 2)^2, \ \text{and} \ (x + 2)^4(x - 2)^4(x^2 - 2)^2$$

respectively.

Now we move to the infinite families of Theorem 18.3. First, we have the family L_n:

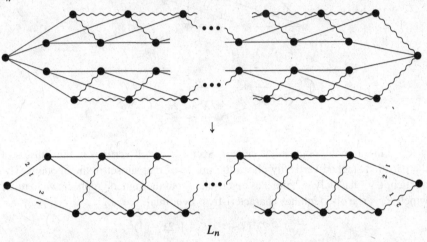

L_n

To see that the signed graph here is cyclotomic, apply Lemma 9.15.

For L'_n, we use a subgraph of the previous example (hence with all eigenvalues in $[-2, 2]$), but with the vertices at the right end stacked differently:

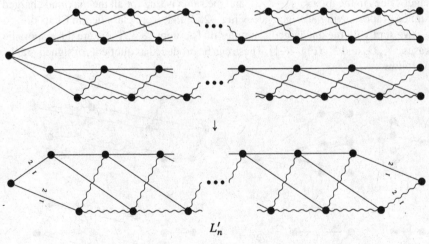

L'_n

Producing L_n^T as a quotient is slightly more straightforward than the previous two cases. Note the switch of a single vertex at the lower left end of the signed graph compared to the usual toroidal drawing. We see easily that the signed graph is equivalent to a subgraph of T_{n+4}, and hence has all its eigenvalues in $[-2, 2]$.

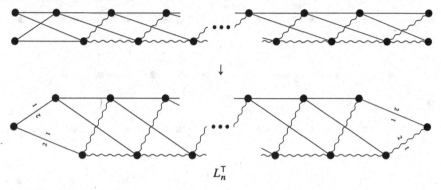

$$L_n^\mathsf{T}$$

Now we have the charged examples of Theorem 18.3. Again we start with the sporadic cases $O_4^\pm$ and $A_{12}^\pm$ (Fig. 18.3).

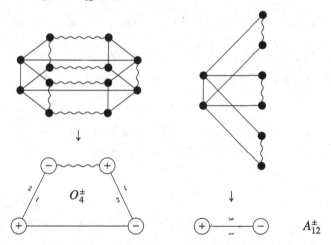

$$O_4^\pm \qquad A_{12}^\pm$$

Amusingly, we use the same signed graphs as for the uncharged sporadic examples, drawn differently, after some switching in the first case.

There are two charged infinite families in Theorem 18.3, namely, L_n^+ and $(L_n^+)^\mathsf{T}$. The first signed graph is equivalent to a subgraph of the second (with a different n), which, in turn, is equivalent to a subgraph of T_{2n}.

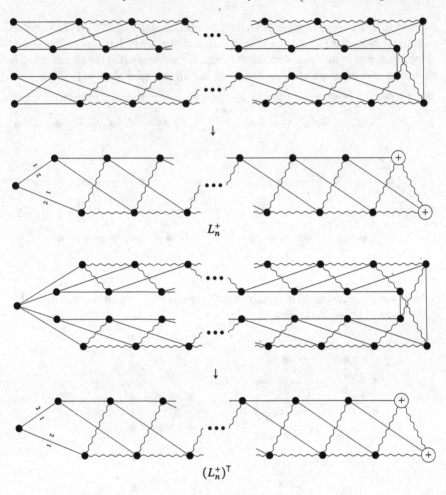

$$L_n^+$$

$$(L_n^+)^\mathsf{T}$$

Exercise 18.16 Exhibit the two infinite families B_n and $C_n = B_n^\mathsf{T}$ and the two sporadic examples F_4 and G_2 of Fig. 18.6 as quotients of connected graphs.

The connected charged signed graphs of Theorem 6.31 maximal subject to having all eigenvalues in $[-2, 2]$ must also be quotients of signed graphs, and ways of doing this are drawn below. The ones that have charges are the two infinite families C_{2k}^{++} and C_{2k}^{+-} (Fig. 6.4), and the three sporadic examples S_7, S_8 and S_8' (Exercises 6.17 and 6.20).

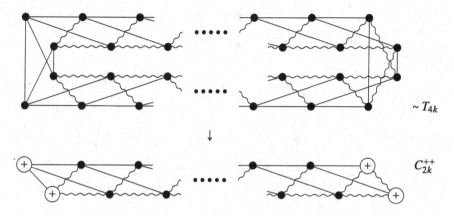

$\sim T_{4k}$

C_{2k}^{++}

Note the way that the edges have been placed near the ends of the toroidal tessellation so as to ensure that the 4-cycle rule holds. In particular, this shows how to recover C_6^{++} (seen at the start of this section) as a quotient of a signed graph that has spectral radius 2.

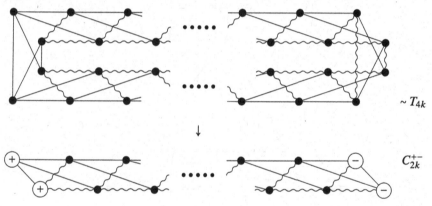

$\sim T_{4k}$

C_{2k}^{+-}

Now for the sporadic cases. The signed graphs S_{14} and S_{16} are from Figs. 6.1 and 6.2; the charged signed graphs S_7, S_8 and S_8' are from Exercises 6.17 and 6.20.

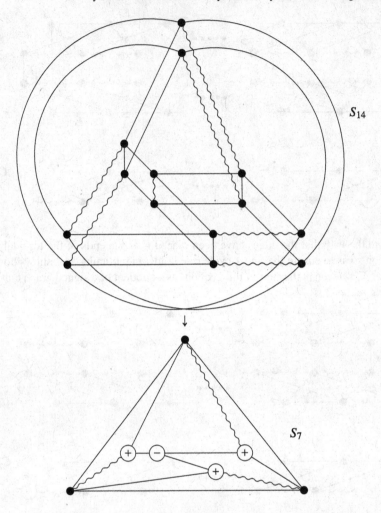

Note that the triangles in S_7 lift to hexagons in S_{14}.

The last two examples are both quotients of S_{16}, achieved by rather different equitable partitions.

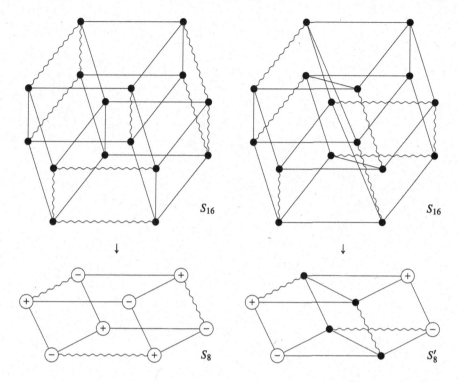

18.3 Notes

The material in Sect. 18.1 is adapted from [MS20a]. Corollary 18.4 was first shown by Sternberg [Ste04, Chap. 6].

18.4 Glossary

cyclotomic.   A symmetrizable integer matrix B is cyclotomic if all its eigenvalues are in the interval $[-2, 2]$.

transpose.   The transpose of a weighted digraph is defined as the digraph whose adjacency matrix is the transpose of the adjacency matrix of the original graph.

Chapter 19
Symmetrizable Matrices III: The Trace Problem

19.1 The Trace Problem for Symmetrizable Matrices

Symmetrizable integer matrices admit more possibilities for their characteristic polynomials than do integer symmetric matrices. For example, see the following exercise.

Exercise 19.1 Show that

- $x^2 - d$ is the characteristic polynomial of a symmetrizable integer matrix for any integer $d \geq 0$;
- $x^2 - d$ is the characteristic polynomial of an integer symmetric matrix if and only if d can be written as a sum of two squares of integers.

Our lower bound on the trace of a positive definite connected integer symmetric matrix (Theorem 15.2) warrants revisiting in the context of this larger class of characteristic polynomials: perhaps we can manage to make the trace smaller. It turns out that we cannot. Moreover if the trace of an $n \times n$ positive definite connected symmetrizable integer matrix is $2n - 1$, then in fact the matrix must be symmetric. In the asymmetric case, this lower bound on the trace can be improved to $2n$.

19.1.1 Definitions, Notation and Statement of the Results

We remarked in Sect. 17.2, symmetrizable matrices have all eigenvalues real. As in the symmetric case, a symmetrizable integer matrix is called *positive definite* if all its eigenvalues are strictly positive.

For $n \times n$ matrices that have integer entries, we define a chain of sets of matrices

$$\mathcal{M}_0 \supseteq \mathcal{M}_1 \supseteq \mathcal{M}_2 \supseteq \mathcal{M}_2^+$$

by imposing increasingly stringent conditions on the matrices.

- The set $\mathcal{M}_0$ comprises those integer matrices $A = (a_{ij})$ that satisfy the sign symmetry condition (17.5).

© Springer Nature Switzerland AG 2021
J. McKee and C. Smyth, *Around the Unit Circle*, Universitext,
https://doi.org/10.1007/978-3-030-80031-4_19

- $\mathcal{M}_1$ is the set of symmetrizable integer matrices: those matrices in $\mathcal{M}_0$ that additionally satisfy the cycle condition (17.9).
- We define $\mathcal{M}_2$ to be the subset of $\mathcal{M}_1$ comprising those symmetrizable integer matrices that are connected.
- Let $\mathcal{M}_2^+$ be the subset of $\mathcal{M}_2$ comprising the positive definite matrices in $\mathcal{M}_2$.

Note that membership of $\mathcal{M}_0$, $\mathcal{M}_1$ or $\mathcal{M}_2$ is completely independent of the values of the diagonal entries: these can be varied freely without affecting whether or not the matrix is in any given $\mathcal{M}_i$. In particular, we can add any multiple of the identity matrix without affecting whether or not a matrix is connected, or satisfies either or both of (17.5) and (17.9). On the other hand, the diagonal entries certainly affect whether or not a matrix is in $\mathcal{M}_2^+$.

We now define a parallel chain

$$\mathcal{Q}_0 \supseteq \mathcal{Q}_1 \supseteq \mathcal{Q}_2 \supseteq \mathcal{Q}_2^+$$

of sets of matrices whose entries come from the larger set $\sqrt{\mathbb{N}_0}$. All the matrices in $\mathcal{Q}_0, \ldots, \mathcal{Q}_2^+$ will be symmetric.

- The set $\mathcal{Q}_0$ comprises symmetric matrices with elements from $\sqrt{\mathbb{N}_0}$ whose diagonal entries are in $\mathbb{Z}$.
- $\mathcal{Q}_1$ is the set of those $B = (b_{ij})$ in $\mathcal{Q}_0$ that satisfy the rational cycle condition (17.11).
- The set $\mathcal{Q}_2$ comprises those elements of $\mathcal{Q}_1$ that are connected.
- $\mathcal{Q}_2^+$ is the set of those elements of $\mathcal{Q}_2$ that are positive definite.

Exercise 19.2 Show that symmetrization map φ from (17.10) maps $\mathcal{M}_i$ to $\mathcal{Q}_i$ ($i = 0$, 1, 2) and maps $\mathcal{M}_2^+$ to $\mathcal{Q}_2^+$. Show that $\varphi : \mathcal{M}_i \to \mathcal{Q}_i$ is surjective for each i (for $i = 1$ use Lemma 17.14), and that $\varphi : \mathcal{M}_2^+ \to \mathcal{Q}_2^+$ is surjective.

Our first result in this chapter (proofs start in the next subsection) is that any element of $\mathcal{Q}_2^+$ has trace at least $2n - 1$.

Proposition 19.3 *Let $B = (b_{ij})$ be a connected symmetric $n \times n$ matrix with entries in $\sqrt{\mathbb{N}_0}$. Suppose also that B is positive definite, and satisfies the rational cycle condition (17.11). Then $\mathrm{trace}(B) \geq 2n - 1$.*

Using the symmetrization map φ that sends each $\mathcal{M}_i$ to the corresponding $\mathcal{Q}_i$, we shall deduce our main theorem of this chapter.

Theorem 19.4 *Let A be an $n \times n$ positive definite connected symmetrizable integer matrix. Then $\mathrm{trace}\, A \geq 2n - 1$. If moreover A is asymmetric, then $\mathrm{trace}\, A \geq 2n$.*

19.1.2 Proof of Proposition 19.3

We start with a lemma.

Lemma 19.5 *Suppose that $B = (b_{ij})$ is a connected symmetric positive definite $n \times n$ matrix with entries in $\sqrt{\mathbb{N}_0}$ that satisfies the rational cycle condition (17.11). Let $\mathbf{e}_1, ..., \mathbf{e}_n$ be a basis for $\mathbb{R}^n$, and let $\langle \cdot, \cdot \rangle$ be the positive definite symmetric bilinear form defined via $\langle \mathbf{e}_i, \mathbf{e}_j \rangle = b_{ij}$. Define $\mathbf{e}_2' = \mathbf{e}_2 - b_{12}\mathbf{e}_1$, and let B' be the matrix of $\langle \cdot, \cdot \rangle$ with respect to the basis $\mathbf{e}_1, \mathbf{e}_2', \mathbf{e}_3, ..., \mathbf{e}_n$. Then:*

- *B' is positive definite;*
- *B' has all entries in $\sqrt{\mathbb{N}_0}$;*
- *B' satisfies the rational cycle condition.*

Before commencing the proof, we define the **core** of a positive integer n, written $\mathrm{core}(n)$ as follows. Write $n = rs^2$ with r square-free. Then $\mathrm{core}(n) = r$.

Proof The form is positive definite, so B' is positive definite. We may assume that $b_{12} \neq 0$, else the result is trivial ($B = B'$).

The only entries of B' that are not simply copied from B are those in the second row and column. The new diagonal entry is

$$\langle \mathbf{e}_2 - b_{12}\mathbf{e}_1, \mathbf{e}_2 - b_{12}\mathbf{e}_1 \rangle = b_{22} - 2b_{12}^2 + b_{12}^2 b_{11} \in \mathbb{Z}. \tag{19.1}$$

For the new off-diagonal entries b_{2i}' ($i \neq 2$), we have

$$b_{2i}' = b_{2i} - b_{12}b_{1i}. \tag{19.2}$$

The rational cycle condition for B gives

$$b_{12}b_{2i}b_{i1} \in \mathbb{Q}. \tag{19.3}$$

We now consider three cases.

Case (i): $b_{1i} = 0$. Then (19.2) gives $b_{2i}' = b_{2i}$. In particular,

$$\mathrm{core}(b_{2i}'^2) = \mathrm{core}(b_{2i}^2). \tag{19.4}$$

Case (ii): $b_{1i} \neq 0$, $b_{2i} \neq 0$. Then (19.3) gives $\mathrm{core}(b_{2i}^2) = \mathrm{core}(b_{12}^2 b_{1i}^2)$, and then (19.2) gives either $b_{2i}' = 0$, or again equation (19.4).

Case (iii): $b_{1i} \neq 0$, $b_{2i} = 0$. Then (19.2) gives

$$b_{2i}'^2 = b_{12}^2 b_{1i}^2. \tag{19.5}$$

In all cases, we have $b_{2i}' \in \sqrt{\mathbb{N}_0}$.

Finally, we must check the rational cycle condition for B'. Let C be any cycle in the graph of B' (vertices $1, ..., n$, weighted edges b_{ij}' between vertices i and j) with all edge weights nonzero. We shall show that there is a closed walk W in the graph of B (edge weights b_{ij}) such that the product of the edge weights of W equals a nonzero rational multiple of the product of the edge weights of C. The rational cycle condition for B' then follows from that for B.

Here then is how to construct W from C. Any edge in C not involving vertex 2 is simply copied to W (weights in B and B' agree for such edges). If C includes an edge $(2, i)$ with either $b_{1i} = 0$ or $(b_{1i} \neq 0$ and $b_{2i} \neq 0)$, then we are in either Case (i) or Case (ii) above, and (19.4) holds. We include the edge $(2, i)$ in W, with weight b_{2i} agreeing with b'_{2i} up to a nonzero rational multiple. If C includes an edge $(2, i)$ and $b_{1i} \neq 0$ and $b_{2i} = 0$ (Case (iii) above), then we include the edges $(2, 1)$ and $(1, i)$ in W, in place of $(2, i)$ in C; by (19.5), the product of the edge weights $b_{21}b_{1i}$ equals b'_{2i}. $\qquad\qquad\Box$

The proof of Proposition 19.3 is now just as for Theorem 15.2. We suppose that B is a minimal counterexample: $n \times n$, connected, positive definite, trace less than $2n - 1$, with n minimal, and with the trace minimal for this n. The only new feature is that B has entries in $\sqrt{\mathbb{N}_0}$.

We perform the basis change given in the proof of Theorem 15.2 to produce a matrix B'. (In particular, we start by permuting the vertices so that $b_{11} = 1$.) By Lemma 19.5, B' is positive definite, has all entries in $\sqrt{\mathbb{N}_0}$, and satisfies the rational cycle condition. Moreover from (19.1) with $b_{11} = 1$ we see that the new diagonal entry b'_{22} is strictly smaller than b_{22}, so B' has smaller trace than B.

By our minimality condition in choosing B, we must have that B' is not connected. As in the proof of Theorem 15.2 we check that B' has exactly two components, and one of the two components would give a smaller counterexample to the Proposition, contradicting our minimality assumptions for B.

19.1.3 Corollaries, Including Theorem 19.4

On applying the map φ, the first part of Theorem 19.4 follows immediately from Proposition 19.3. If A is a connected symmetrizable matrix that is positive definite, i.e., if $A \in \mathcal{M}_2^+$, then $\varphi(A) \in \mathcal{Q}_2^+$ has the same trace as A. Being in $\mathcal{Q}_2^+ \subseteq \mathcal{Q}_1$, $\varphi(A)$ certainly satisfies the rational cycle condition, and the Proposition applies to give $\text{trace}(\varphi(A)) \geq 2n - 1$, and hence $\text{trace}(A) \geq 2n - 1$.

For the second part, we note that Lemma 15.10 applies with unchanged proof to the symmetrization of A. For any minimal-trace example, all off-diagonal entries of its symmetrization are $0, -1$ or 1. This implies that the same holds for A itself, which implies that A is symmetric. Hence, we get an improved lower bound of $2n$ in the asymmetric case.

Exercise 19.6 This exercise shows that the lower bound of $2n$ in the asymmetric case is best-possible. Let A be the adjacency matrix of the weighted digraph

on n vertices. Show that $A + 2I$ is positive definite, symmetrizable, connected, and has trace $2n$. (Hint: to show that $A + 2I$ is positive definite, consider its symmetrization S; show that this is the Gram matrix for the linearly independent list $\sqrt{2}e_1$, $e_1 + e_2, e_2 + e_3, \ldots, e_{n-1} + e_n$ and use Lemma 16.34.)

As in the symmetric case, we get an immediate corollary concerning possible traces of minimal polynomials.

Corollary 19.7 *Let P be a monic irreducible polynomial with integer coefficients, degree n, and with all zeros real and strictly positive. If the trace of P is strictly less than $2n - 1$, then P is not the minimal polynomial of a symmetrizable integer matrix, and nor is P the minimal polynomial of any symmetric matrix with entries in $\sqrt{\mathbb{N}_0}$ that satisfies the rational cycle condition (17.11).*

We can also settle the analogue of the Schur-Siegel-Smyth trace problem in our setting, just as was done for integer symmetric matrices. The mean trace is defined just as for symmetric matrices.

Corollary 19.8 *Let X be the set of mean traces of connected positive definite symmetrizable integer matrices, and let Y be the set of mean traces of connected positive definite real symmetric matrices that have entries in $\sqrt{\mathbb{N}_0}$ and satisfy the rational cycle condition (17.11). Then $X = Y$, and the smallest limit point of X is 2.*

Proof If $A \in \mathcal{M}_2^+$, then $\varphi(A) \in \mathcal{Q}_2^+$ has the same mean trace. Conversely, if $S \in \mathcal{Q}_2^+$, then by Lemma 17.14 there is a symmetrizable matrix $A \in \mathcal{M}_1$ with $\varphi(A) = S$. Since φ preserves the eigenvalues of symmetrizable matrices, and preserves the underlying graph, $A \in \mathcal{M}_2^+$ and has the same mean trace as S. Thus $X = Y$.

The rest is as in the symmetric case (see Corollary 15.5): 2 is a limit point, and bounding the mean trace strictly below 2 gives a bound for n (applying Theorem 19.4); one then bounds the diagonal entries, and finally the off-diagonal ones (using positive definiteness). □

One consequence of all this is that if a symmetrizable positive definite connected $n \times n$ integer matrix has trace $2n - 1$ then it is in fact symmetric. Thus, for example, if $x^2 - 3x + 1$ is presented as the characteristic polynomial of an integer symmetrizable matrix then that matrix must be symmetric.

Exercise 19.9 Suppose that $x^2 - 3x + 1$ is the characteristic polynomial of the matrix $\begin{pmatrix} a & b \\ c & d \end{pmatrix}$, where $a, b, c, d \in \mathbb{Z}$ and b and c have the same sign. Verify that $b = c$.

19.1.4   The Structure of Minimal-Trace Examples

We saw in Chap. 15 (specifically Sect. 15.3) a method to glue together minimal-trace symmetric examples to produce larger minimal-trace symmetric examples, and that working up to equivalence all minimal-trace symmetric examples can be produced this way (starting from the trivial 1×1 case). Indeed we work with strong equivalence rather than equivalence, as changing the signs of all eigenvalues does not preserve positive definiteness.

The asymmetric case is more delicate. Let $C = (c_{ij})$ be an $n \times n$ positive definite connected symmetrizable integer matrix that is *not* symmetric, and has trace $2n$ (necessarily $n \geq 2$).

We cannot assume that there is a diagonal entry equal to 1, and deal first with the special case where every diagonal entry equals 2. Take two vertices in the digraph corresponding to C that are as far apart as possible in terms of the minimal length of a path between them. Unless $n = 2$, deleting a suitable choice of one of these vertices will leave a subgraph that is not only connected but remains asymmetric (if there is only one asymmetric edge it must be an isthmus, by the cycle condition). The matrix corresponding to this subgraph is connected, positive definite, and has minimal trace in this asymmetric case (all diagonal entries equal 2). Hence, working up to equivalence, we can 'grow' all minimal-trace examples in this subcase from smaller ones, starting from the 2×2 cases

$$\begin{pmatrix} 2 & 2 \\ 1 & 2 \end{pmatrix} \quad \text{and} \quad \begin{pmatrix} 2 & 3 \\ 1 & 2 \end{pmatrix},$$

or their transposes. These grow to give just the following two 3×3 examples (up to strong equivalence and transposition):

$$\begin{pmatrix} 2 & 1 & 0 \\ 2 & 2 & 1 \\ 0 & 1 & 2 \end{pmatrix} \quad \text{and} \quad \begin{pmatrix} 2 & 1 & 1 \\ 2 & 2 & 2 \\ 1 & 1 & 2 \end{pmatrix}.$$

Exercise 19.10 Find all the connected positive definite asymmetric 4×4 integer matrices that have all diagonal entries equal to 2, up to strong equivalence and transposition.

Now consider the case where C has some diagonal entry equal to 1. Working up to equivalence, we may assume that $c_{11} = 1$ and $c_{12} > 0$. Let $S = (s_{ij})$ be the symmetrization of C. Performing the basis change of the proof of Proposition 19.3, the matrix S changes to S', where $\text{trace}(S') < \text{trace}(S)$. The argument in the proof of that Proposition shows that S' has at most two components, and there are two possibilities: $\text{trace}(S') = 2n - 2$ or $\text{trace}(S') = 2n - 1$.

If $\text{trace}(S') = 2n - 2$, then it is not connected and must decompose into exactly two components, say A $(r \times r)$ and B $(s \times s)$, where $\text{trace}(A) = 2r - 1$, $\text{trace}(B) = 2s - 1$. Then, A and B are symmetric minimal-trace examples (the last part of Theorem 19.4). We have in this case that $s_{12} = \sqrt{2}$ (since trace$(S') = $ trace$(S) - 2$ and sgn$(s_{12}) = $ sgn(c_{12})). After permuting, we see that S is built from A and B in essentially the same way as in our symmetric construction of Chap. 15, but with

$$s_{ij} = \begin{cases} b_{r+1,r+1} + 2 & i = r+1, \ j = r+1, \\ a_{1,j}\sqrt{2} & i = r+1, \ 1 \le j \le r, \\ a_{i,1}\sqrt{2} & 1 \le i \le r, \ j = r+1. \end{cases} \tag{19.6}$$

Note the factors of $\sqrt{2}$ (the change of basis here is to replace $\mathbf{f}_1$ by $\mathbf{f}_1 + \sqrt{2}\mathbf{e}_1$), and the addition of 2 to the $(r+1, r+1)$ diagonal entry rather than 1. We need to recover C from its symmetrization S. The subgraph corresponding to A must be symmetric, by its trace, and similarly for B (adjusting the special diagonal entry does not break the symmetry). Thus, the only asymmetry comes in the $(r+1)$th row and column. Here, we have $c_{i,r+1}c_{r+1,i} = s_{i,r+1}^2 = 2a_{i,r+1}^2$, which is either 0 or 2. If 2, then we need to choose which of $c_{i,r+1}$ and $c_{r+1,i}$ is $2\,\mathrm{sgn}(s_{i,r+1})$ and which is $\mathrm{sgn}(s_{i,r+1})$. Since A and B are symmetric, we must either always put the factor of 2 in the row, or always in the column, to satisfy the cycle condition (17.9):

vertical paths are symmetric; horizontal edges have sign symmetry

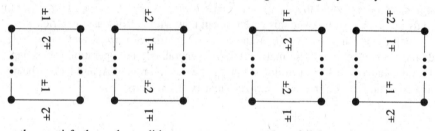

these satisfy the cycle condition   these fail the cycle condition

Hence, the values for c_{ij} are as for s_{ij} in (19.6) except for

$$c_{ij} = \begin{cases} 2a_{1,j} & i = r+1, \ 1 \le j \le r, \\ a_{i,1} & 1 \le i \le r, \ j = r+1, \end{cases}$$

or the transpose of this.

If trace$(S') = 2n - 1$, then $s_{12} = 1$. Note that *a priori* S' might be connected in this case. But if so, then being minimal trace all off-diagonal entries would have modulus at most 1, so would equal 0, 1 or -1, and reversing the change of basis would produce the same conclusion for S (as in the proof of Lemma 15.10), implying that C was symmetric, which it is not. So we still must have S' falling into two components, say A' ($r \times r$) and B' ($s \times s$), with the first vertex of S' corresponding to the first vertex of A'. Let A and B be the subgraphs of C corresponding to A', B', respectively. We permute rows/vertex labels so that the first r rows of C correspond to A (and with $a_{11} = 1$) and the final s rows correspond to B. Our challenge is to complete the first r

entries in row $r + 1$ and column $r + 1$. There are two subcases: $\text{trace}(A') = 2r - 1$, $\text{trace}(B') = 2s$, or $\text{trace}(A') = 2r$, $\text{trace}(B') = 2s - 1$. We shall use the operation $*$ of Sect. 15.3, but anticipating an imminent generalisation we shall call this operation $*_1$.

If $\text{trace}(A') = 2r - 1$, then A must be symmetric, and $C = A *_1 B$. (Here B must be asymmetric, else C would be symmetric.)

If $\text{trace}(A') = 2r$, then it is a minimal-trace asymmetric example with a 1 on the diagonal. We have

$$s_{r+1,i} = s_{1,i} \ (1 \le i \le r), \quad s_{i,r+1} = s_{i,1} \ (1 \le i \le r). \tag{19.7}$$

For $1 \le i \le r$ we have $c_{r+1,i} c_{i,r+1} = s_{1,i} s_{i,1} = s_{1,i}^2$, which is known, and the signs of the $c_{i,j}$ are all known, but in the asymmetric case this formula does not tell us how the factors of $s_{1,i}^2$ are to be shared between $c_{r+1,i}$ and $c_{i,r+1}$. The cycle condition for C for the triangle $1, i, r + 1$ gives

$$c_{1,i} c_{i,r+1} c_{r+1,1} = c_{i,1} c_{r+1,i} c_{1,r+1},$$

which with $c_{1,r+1} = c_{r+1,1} = 1$ gives $c_{1,i} c_{i,r+1} = c_{i,1} c_{r+1,i}$. Together with (19.7) and sign symmetry, we find that $c_{r+1,i} = c_{1,i} = a_{1,i}$ and $c_{i,r+1} = c_{i,1} = a_{i,1}$ $(1 \le i \le r)$, so that $C = A *_1 B$ holds in this case too, but now with A being asymmetric.

For an $r \times r$ matrix $A = (a_{ij})$ and an $s \times s$ matrix $B = (b_{ij})$, and any $k \ge 1$, define the $(r + s) \times (r + s)$ matrix $A *_k B$ (generalising the operation $*$ of Chap. 15; note that $*$ is just $*_1$) as follows. Let $A_{1,.}$ be the first row of A, and let $A_{.,1}$ be its first column. Let $O_{p,q}$ be the $p \times q$ zero matrix. Then define

$$A *_k B = \left(\begin{array}{c|cc} A & A_{.,1} & O_{r,s-1} \\ \hline kA_{1,.} & b_k & \\ & & B \\ O_{s-1,r} & & \end{array} \right), \quad \text{where } b_k = b_{11} + ka_{11}.$$

The matrix A fills the top left corner of $A *_k B$, and the matrix B fills the bottom right corner except that its $(1, 1)$ entry is replaced by b_k as shown.

The above construction of minimal-trace examples can be summarised using the $*_k$ operator. An asymmetric minimal-trace positive definite connected symmetrizable integer matrix C can be grown from smaller ones (unless it is 1×1, or 2×2 with each diagonal entry equal to 2) in one of four ways:

- if all diagonal entries of C equal 2, then it can be grown from an induced submatrix having the same properties;
- $C = A *_2 B$, where A and B are minimal-trace symmetric examples and $a_{11} = 1$;
- $C = A *_1 B$, where A is a minimal-trace symmetric example (with $a_{11} = 1$) and B is a minimal-trace asymmetric example;
- $C = B *_1 A$, where A is a minimal-trace symmetric example and B is a minimal-trace asymmetric example (with $b_{11} = 1$).

For example,

$$(1) *_2 (1) = \begin{pmatrix} 1 & 1 \\ 2 & 3 \end{pmatrix}$$

and

$$(1) *_2 \begin{pmatrix} 2 & 1 \\ 1 & 1 \end{pmatrix} = \begin{pmatrix} 1 & 1 & 0 \\ 2 & 4 & 1 \\ 0 & 1 & 1 \end{pmatrix}$$

and

$$(1) *_1 \begin{pmatrix} 1 & 1 \\ 2 & 3 \end{pmatrix} = \begin{pmatrix} 1 & 1 & 0 \\ 1 & 2 & 1 \\ 0 & 2 & 3 \end{pmatrix}$$

and

$$\begin{pmatrix} 1 & 1 & 0 \\ 1 & 2 & 1 \\ 0 & 2 & 3 \end{pmatrix} *_1 (1) = \begin{pmatrix} 1 & 1 & 0 & 1 \\ 1 & 2 & 1 & 1 \\ 0 & 2 & 3 & 0 \\ 1 & 1 & 0 & 2 \end{pmatrix}$$

are asymmetric symmetrizable connected positive definite integer matrices that have minimal trace for their size.

Exercise 19.11 Find all asymmetric positive definite connected 3×3 symmetrizable integer matrices that have trace 6 but do not have every diagonal entry equal to 2.

19.2 Notes

Much of this chapter is adapted from [MS20b].

19.3 Glossary

Much of the terminology and notation was introduced in Chap 17, and we refer the reader to the Glossary for that chapter in addition to the short glossary here.

$\mathcal{M}_0, \mathcal{M}_1, \mathcal{M}_2, \mathcal{M}_2^+$ Fix some n. The set $\mathcal{M}_0$ is the set of all $n \times n$ sign-symmetric integer matrices, i.e., the set of $n \times n$ integer matrices for which the symmetrization map is defined. The subset $\mathcal{M}_1 \subseteq \mathcal{M}_0$ is the set of symmetrizable $n \times n$ matrices. Thus, in addition to sign symmetry, these matrices satisfy the cycle condition. The subset $\mathcal{M}_2 \subseteq \mathcal{M}_1$ is the set of $n \times n$ connected symmetrizable integer matrices, and the subset $\mathcal{M}_2^+ \subseteq \mathcal{M}_2$ comprises the positive definite $n \times n$ connected symmetrizable integer matrices.

$\mathcal{Q}_0, \mathcal{Q}_1, \mathcal{Q}_2, \mathcal{Q}_2^+$ These are the images under the symmetrization map of $\mathcal{M}_0, \mathcal{M}_1$, $\mathcal{M}_2, \mathcal{M}_2^+$ respectively.

core.   The core of a positive integer n is what is left once all possible square factors have been stripped out. If $n = rs^2$ with r square-free, then the core of n is r.

positive definite.   A symmetrizable matrix is positive definite if all its eigenvalues are strictly positive.

Chapter 20
Salem Numbers from Graphs and Interlacing Quotients

.

20.1 Introduction

The notion of the Mahler measure of a matrix grew out of the study of Salem numbers that appeared as Mahler measures of graphs (see [MS99b, MS05a]). In the first part of this chapter, we look at some special cases of this construction of Salem numbers. (The notes mention some further work in this area.) We show that certain limit points of these Salem numbers are Pisot numbers, again giving just a very special case of a more general result.

The second part of the chapter turns to an interlacing construction for Salem numbers, which historically grew from the graph construction but is considerably more general. Some special cases that are treated in this chapter are needed as a tool for use in the next chapter. Again there is much more known about such interlacing constructions, and some of the more general theory is pointed to in notes.

20.2 Salem Graphs

How should we define a Salem graph? It should surely be something that provides us with an attached Salem number, but it turns out that it will be convenient to relax even this apparently essential condition slightly. The minimal polynomial of a Salem number is reciprocal, and we have seen that we can attach a reciprocal polynomial to a graph G on n vertices, namely $R_G(z) = z^n \chi_G(z + 1/z)$, where $\chi_G(x)$ is the characteristic polynomial of the graph (the characteristic polynomial of its adjacency matrix). A first stab at a definition of a Salem graph might be to require $R_G(z)$ to be the minimal polynomial of a Salem number. But this is a difficult definition to work with, as there is the question of irreducibility of $R_G(z)$.

Recall that a Salem number is a real algebraic integer τ satisfying:

© Springer Nature Switzerland AG 2021
J. McKee and C. Smyth, *Around the Unit Circle*, Universitext,
https://doi.org/10.1007/978-3-030-80031-4_20

  (i) $\tau > 1$;

  (ii) τ and $1/\tau$ are (Galois) conjugates;

  (iii) all Galois conjugates of τ other than τ itself lie in the unit disc $\{z : |z| \leq 1\}$;

  (iv) τ has a Galois conjugate on the unit circle.   (20.1)

We shall see how we can readily construct monic reciprocal integer polynomials from graphs having the first three of these properties, but that the fourth is not automatic.

Exercise 20.1 Let τ be a real algebraic integer satisfying the first three properties of a Salem number, (i)–(iii), as in (20.1). Show that if τ does not satisfy (iv), then τ is a quadratic Pisot number. (It is immediate that τ is a Pisot number. The point is to show that it is quadratic.)

With the above in mind, we are now ready to make our definition. A graph G is called a **Salem graph** if:

  (i) G is connected; and

  (ii) $M(G)$ is either a Salem number or a reciprocal (hence quadratic) Pisot number.

In other words, a connected graph G is a Salem graph if its Mahler measure τ satisfies properties (i)–(iii) of the properties of a Salem number in (20.1). Following [MS05a], we say that a Salem graph is **trivial** if its Mahler measure is a quadratic Pisot number. Of course we have $M(G) = M(R_G(z))$ by definition.

Exercise 20.2 For a polynomial $P(z) \in \mathbb{Z}[z]$, show that $M(P(z^m)) = M(P(z))$ for all $m \geq 1$. (This is a very special case of Proposition 2.9.)

If G is a bipartite graph, show that $R_G(z)$ is a polynomial in z^2, and hence that $M(G) = M(R_G(\sqrt{z}))$.

Lemma 20.3 *Let G be a connected graph and let λ be its largest eigenvalue. (By Theorem B.6, λ is a simple eigenvalue (multiplicity 1).) Then, G is a Salem graph if and only if one of the following two cases holds:*

    (i) G is bipartite, and has only one eigenvalue greater than 2;

    or (ii) G is nonbipartite, and has only one eigenvalue outside the interval $[-2, 2]$ (which

    must be λ, and must be greater than 2 by Theorem B.6).

Proof Suppose first that G is bipartite. Then $\chi_G(x)$ contains only odd powers of x or only even powers of x, and $R_G(z)$ contains only even powers of z. Then

$$M(G) = M(R_G(z)) = M(R_G(\sqrt{z})).$$

If we define the real algebraic integer $\tau > 1$ by

$$\sqrt{\tau} + 1/\sqrt{\tau} = \lambda,$$

then τ is a zero of $R_G(\sqrt{z})$, and G is a Salem graph if and only if $R_G(\sqrt{z})$ has no zeros outside the unit disc other than τ (for then $\tau = M(G)$ and τ satisfies our first three Salem properties (i)–(iii); otherwise $M(G) > \tau$ includes contributions from at least two distinct real zeros of $R_G(\sqrt{z})$ greater than 1). In terms of the eigenvalues of G, this condition is equivalent to G having no eigenvalue greater than 2 other than λ.

The nonbipartite case is even simpler. We define the real algebraic integer $\tau > 1$ by

$$\tau + 1/\tau = \lambda.$$

Then τ is a zero of $R_G(z)$, and G is a Salem graph if and only if $R_G(z)$ has no zeros outside the unit disc other than τ (we cannot have $R_G(z) = R_G(z^m)$ for any $m > 1$, as $m = 2$ would imply G bipartite, and $m > 2$ would imply nonreal zeros outside the unit disc), which is equivalent to λ being the only eigenvalue outside the interval $[-2, 2]$. □

Lemma 20.4 *Let G be a Salem graph, and let λ be the unique eigenvalue of G satisfying $\lambda > 2$. Define $\tau > 1$ to be the larger root of $\sqrt{\tau} + 1/\sqrt{\tau} = \lambda$ in the bipartite case, and the larger root of $\tau + 1/\tau = \lambda$ in the nonbipartite case, so that $M(G) = \tau$ in either case. Then either $\lambda^2 \in \mathbb{Z}$, or τ is a Salem number.*

Proof In the nonbipartite case, λ corresponds to a pair of reciprocal zeros τ, $1/\tau$ of the reciprocal polynomial $R_G(z)$, with say $\tau > 1$ (and $0 < 1/\tau < 1$). Other eigenvalues are in the interval $[-2, 2]$, and these correspond to zeros of $R_G(z)$ that have modulus 1. Hence, τ is a Salem number unless either $\tau \in \mathbb{Z}$ or $1/\tau$ is its only Galois conjugate. Now τ cannot be an integer, else $1/\tau$ would not be an algebraic integer ($\tau > 1$). Hence if τ were not a Salem number then $\lambda = \tau + 1/\tau$ would be a Galois-invariant algebraic integer, so would be in $\mathbb{Z}$, and so certainly we would have $\lambda^2 \in \mathbb{Z}$.

In the bipartite case, the pair of eigenvalues $\pm\lambda$ correspond to two reciprocal pairs of zeros of $R_G(z)$, say θ, $1/\theta$ and $-\theta$, $-1/\theta$, where $\theta > 1$. Other zeros of $R_G(z)$ have modulus 1. Here $\theta + 1/\theta = \lambda$. Put $\tau = \theta^2 > 1$. Then $\sqrt{\tau} + 1/\sqrt{\tau} = \lambda$, and the conjugates of τ are the squares of conjugates of θ, for which the possibilities are τ, $1/\tau$, and perhaps some conjugates of modulus 1. Thus, τ is a Salem number unless $1/\tau$ is its only Galois conjugate (as in the bipartite case we cannot have $\tau \in \mathbb{Z}$). We now have $\lambda^2 = (\sqrt{\tau} + 1/\sqrt{\tau})^2 = \tau + 2 + 1/\tau$, from which we see that $\lambda^2 \in \mathbb{Z}$ (it is a Galois-invariant algebraic integer). □

In view of the above lemma, we see that a Salem graph is trivial if and only if its largest eigenvalue λ satisfies $\lambda^2 \in \mathbb{Z}$. If G is a Salem graph, then we write $\tau(G)$ for the Salem number (or quadratic Pisot number) associated with G by Lemma 20.4. In the nonbipartite (respectively, bipartite) case, $R_G(z)$ (respectively, $R_G(\sqrt{z})$) is either the minimal polynomial of $\tau(G)$ or is a product of the minimal polynomial of $\tau(G)$ and a cyclotomic polynomial.

Exercise 20.5 Show that if λ is the largest eigenvalue of a connected graph G, $\lambda^2 \in \mathbb{Z}$ and $\lambda \notin \mathbb{Z}$, then G is bipartite.

20.3 Examples of Salem Graphs

20.3.1 Nonbipartite Examples

The complete graph on n vertices, K_n, has characteristic polynomial

$$(x + 1)^{n-1}(x - n + 1),$$

so it is a nonbipartite Salem graph for all $n \geq 4$ (for $n \leq 3$, K_n is cyclotomic; for $n \leq 2$ it is bipartite). Indeed K_n is a trivial Salem graph, since its largest eigenvalue $\lambda = n - 1$ is in $\mathbb{Z}$.

The minimal noncyclotomic charged signed graphs of Chap. 13 (specifically Sect. 13.1) include some nonbipartite Salem graphs. The smallest (in terms of numbers of vertices, not Mahler measures) are $4j$ and $4B$:

The corresponding Salem numbers are $1.5061\ldots$ and $2.0810\ldots$, having minimal polynomials $z^6 - z^5 - z^3 - z + 1$ and $z^4 - z^3 - 2z^2 - z + 1$.

Exercise 20.6 Look through the complete list of minimal noncyclotomic charged signed graphs in Sect. 13.1, determine which are nonbipartite Salem graphs, and compute their Mahler measures.

20.3.2 Bipartite Examples

Suppose that a connected graph G has the following three properties:

- G is bipartite;
- G is not cyclotomic;
- there is some vertex v in G such that deleting v (and incident edges) leaves a cyclotomic subgraph (not necessarily connected).

Then, by interlacing, G has at most one eigenvalue greater than 2, and hence (since not cyclotomic), exactly one (and, being bipartite, G also has exactly one eigenvalue less than -2). Hence G is a Salem graph.

For a very simple example, take $a_1, \ldots, a_r$ to be positive integers (for some $r \geq 3$), and let $T(a_1, \ldots, a_r)$ be the starlike tree with r branches, lengths $a_1, \ldots, a_r$. What this means is that G has a 'central' vertex v, the deletion of v leaves r components, with the ith component a path on a_i vertices (length $a_i - 1$), and with v joined to one of the endvertices of each path. Deleting the vertex v leaves a graph whose components are all cyclotomic, so either G is cyclotomic or it is a Salem graph. The graph $T(1, 2, 3, 5)$ is drawn below.

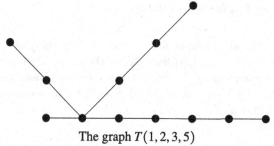

The graph $T(1,2,3,5)$

The characteristic polynomial of $T(1, 2, 3, 5)$ is

$$x^2(x-1)(x+1)(x^8 - 10x^6 + 32x^4 - 37x^2 + 11),$$

and the reciprocal polynomial is

$$(z^2+1)^2(z^2-z+1)(z^2+z+1)(z^{16} - 2z^{14} - 3z^{10} - z^8 - 3z^6 - 2z^2 + 1).$$

The largest real zero of $R_{T(1,2,3,5)}(\sqrt{z})$ is a Salem number, degree 8, minimal polynomial $z^8 - 2z^7 - 3z^5 - z^4 - 3z^3 - 2z + 1$. The other polynomial factors of the polynomial $R_{T(1,2,3,5)}(\sqrt{z})$ are cyclotomic. We have an example of a Salem graph, noting that its characteristic polynomial is not in this case irreducible.

From the structure of cyclotomic graphs, we know that for $r \geq 3$ and $1 \leq a_1 \leq \cdots \leq a_r$, the tree $T(a_1, \ldots, a_r)$ is not cyclotomic unless one of the following holds:

- $r = 3$ and $a_1 = a_2 = 1$;
- $r = 3$, $a_1 = 1$, $a_2 = 2$, and $2 \leq a_3 \leq 5$;
- $r = 3$, $a_1 = 1$, and $a_2 = a_3 = 3$;
- $r = 3$ and $a_1 = a_2 = a_3 = 2$;
- $r = 4$ and $a_1 = a_2 = a_3 = a_4 = 1$.

Exercise 20.7 Verify the above claim, that for $r \geq 3$ the above list describes precisely which $T(a_1, \ldots, a_r)$ are cyclotomic.

Exercise 20.8 Show that if any two of $a_1, \ldots, a_r$ are equal, then the characteristic polynomial of $T(a_1, \ldots, a_r)$ is not irreducible.

Exercise 20.9 Show that $T(1, 2, 5, 5)$ is a trivial Salem graph.

Exercise 20.10 Find suitable $a_1, \ldots, a_r$ such that $R_{T(a_1,\ldots,a_r)}(\sqrt{z})$ is the minimal polynomial of Lehmer's number.

Exercise 20.11 For $3 \leq r \leq 5$ and $1 \leq a_1 \leq \cdots \leq a_r \leq 6$, compute all Mahler measures $\tau(T(a_1, \ldots, a_r))$. Verify that the only cyclotomic graphs amongst these are those predicted above. Which are trivial Salem graphs? Which have 'small' Mahler measures (less than 1.3 is the usual definition of small when referring to Mahler measures)?

20.3.3 Finding Cyclotomic Factors

One issue, both practical and theoretical, arising from the study of Salem graphs is that $R_G(z)$ (or $R_G(\sqrt{z})$ in the bipartite case) might have cyclotomic factors. How can we compute these efficiently? This is a question which was considered earlier, in Sect. 5.2. We see that Algorithm 1 finds all the cyclotomic factors of $R_G(z)$, which can then be factored out.

20.4 Attaching Pendant Paths

20.4.1 A General Construction

For this subsection we briefly turn to a more general setting, considering arbitrary weighted digraphs (digraphs whose directed edges and vertices are given integral weights), rather than simply graphs. We refer to Appendix B.2 for relevant definitions. In the next subsection we shall return to the main focus of Salem graphs, and show that certain limit points of graph Salem numbers are Pisot numbers.

Let G_0 be a weighted digraph with adjacency matrix A_0. Choose any vertex v, corresponding to some row of A_0, which we may suppose is the first. Attaching a pendant path of length m to G_0 at v means the following. We form a new weighted digraph G_m which has all the weighted vertices and edges of G_0, along with m new vertices $v_1, \ldots, v_m$ (all weighted 0) and $2m$ new edges:

- edges of weight 1 from v_i to v_{i+1} ($1 \le i \le m - 1$);
- edges of weight 1 from v_{i+1} to v_i ($1 \le i \le m - 1$);
- edges of weight 1 from v_1 to v and from v to v_1.

For example, using the drawing conventions from Appendix B.2,

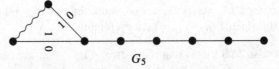

For $m \ge 0$, let A_m be the adjacency matrix of G_m. Reordering the vertices appropriately, we have

$$
A_m = \begin{pmatrix}
0 & 1 & & & & & \\
1 & 0 & 1 & & & & \\
 & 1 & 0 & 1 & & & \\
 & & \ddots & \ddots & \ddots & & \\
 & & & 1 & 0 & 1 & \\
 & & & & 1 & & \\
\hline
 & & & & & & A_0
\end{pmatrix},
$$

where the top left block is $m \times m$, and any entries not shown are 0. Let $X_m(x)$ be the characteristic polynomial of A_m.

Expanding along the first row, we compute

$$X_m(x) = x X_{m-1}(x) - X_{m-2}(x) \quad (m \geq 2).$$

Substituting $x = z + 1/z$ and clearing denominators gives

$$R_m(z) = (z^2 + 1) R_{m-1}(z) - z^2 R_{m-2}(z) \quad (m \geq 2).$$

Inductively, we compute

$$R_m(z) = \frac{z^{2(t+1)} - 1}{z^2 - 1} R_{m-t}(z) - \frac{z^{2t} - 1}{z^2 - 1} z^2 R_{m-t-1}(z) \quad (m \geq 2, \, 1 \leq t \leq m - 1).$$

Putting $t = m - 1$ gives

$$R_m(z) = \frac{z^{2m} - 1}{z^2 - 1} R_1(z) - \frac{z^{2(m-1)} - 1}{z^2 - 1} z^2 R_0(z) \quad (m \geq 2),$$

$$= \frac{1}{z^2 - 1} \left(z^{2m} P(z) - P^*(z) \right),$$

where
$$P(z) = R_1(z) - R_0(z), \qquad P^*(z) = -z^2 R_0(z) + R_1(z).$$

Exercise 20.12 For a nonzero polynomial $Q(z) \in \mathbb{Z}[z]$, degree d, define (as in Sect. A.1 of Appendix A)

$$Q^*(z) = z^d Q(1/z).$$

Thus Q^* is produced from Q by reversing the order of the coefficients. (If Q has zero constant term, then the degree of Q^* is strictly less than d.) In the computation of $R_m(z)$ given above, verify that P^* is produced from P by this process, and that P^* has degree less than that of P.

Show that if Q is irreducible (in $\mathbb{Z}[x]$), then either Q^* is irreducible, also of degree d, or $Q = \pm z$ and $Q^* = \pm 1$ (a unit).

If $Q(z) = \prod_i Q_i(z)^{a_i}$ is the factorisation of Q into powers of irreducible polynomials, show that $Q^*(z) = \prod_i (Q_i^*(z))^{a_i}$.

We now consider the limit as $m \to \infty$ of the Mahler measure of G_m, namely $M(R_m)$, where we have

$$(z^2 - 1) R_m(z) = z^{2m} P(z) - P^*(z). \tag{20.2}$$

Put $Q(y, z) = y^2 P(z) - P^*(z)$. Then

$$M(G_m) = M(R_m(z)) = M((z^2 - 1) R_m(z)) = M(z^{2m} P(z) - P^*(z)) = M(Q(z^m, z)).$$

In the special case where each G_m is a Salem graph, we shall see by an easy argument in the next subsection that the sequence of Mahler measures $(M(G_m))$ converges to a limit (and that the limit is a Pisot number). In this more general setting, we appeal to a theorem of Boyd ([Boy81b], and see [Law83] for a more general result). Boyd showed that $\lim_{m \to \infty} M(Q(z^m, z))$ exists, and equals the 2-variable Mahler measure

$$\int_0^1 \int_0^1 \log \left| Q(e^{2\pi it}, e^{2\pi iu}) \right| \, dt \, du \,.$$

20.4.2   An Application to Salem Graphs

We now return to the case of graphs, with G_0 a graph, and G_m produced by attaching a pendant path of length m to some vertex of G. If G_m is a Salem graph for all sufficiently large m, then the analysis of the previous subsection simplifies considerably. We still have (20.2). When m is large enough, $R_m(z)$ (or $R_m(\sqrt{z})$ in the bipartite case) has a unique zero τ_m outside the unit disc. From this it follows that $P(z)$ has a unique zero outside the unit disc, say θ (apply Rouché to the circle $|z| = 1 + \varepsilon$ for small enough $\varepsilon > 0$ and large enough m). Moreover by Theorem B.6, $\tau_m < \tau_{m+1}$ for all m (sufficiently large that G_m is a Salem graph and τ_m is defined).

Now θ is either a Salem number or a Pisot number. If θ were a Salem number, then it would be a zero of P^*, and hence of R_m for all m. This would imply that $\tau_m = \theta$ for all sufficiently large m, contradicting $\tau_m < \tau_{m+1}$. Hence θ is a Pisot number, $P = P_0$ and P is irreducible. The only contribution to the Mahler measure of G_m is (for large enough m) from the unique zero τ_m of R_m outside the unit circle, and this tends to θ as $m \to \infty$. Hence, in this Salem graph case, one has $\tau_m = M(G_m) \to \theta$ as $m \to \infty$, and θ is a Pisot number.

For example, take any $T(a_1, \ldots, a_r, b)$ with $1 \le a_1 \le \cdots \le a_r$ and let $b \to \infty$. Unless $r = 2$ and $a_1 = a_2 = 1$, the starlike tree is a Salem graph for all large enough b, and the limit of the sequence of Salem numbers is a Pisot number.

20.5   Interlacing Quotients

20.5.1   Rational Interlacing Quotients

We shall say that a rational function $Q(x)/P(x) \in \mathbb{Z}(x)$ is a *rational interlacing quotient*, or simply an *RIQ*, if it is of the shape

$$\frac{\gamma \prod\limits_{i=1}^{n-1} (x - \beta_i)}{\prod\limits_{i=1}^{n} (x - \alpha_i)} \tag{20.3}$$

for some $n \geq 1$, some $\gamma > 0$, $\gamma \in \mathbb{Z}$, and with the α_i and β_i satisfying the strict interlacing condition

$$\alpha_1 < \beta_1 < \alpha_2 < \beta_2 < \cdots < \beta_{n-1} < \alpha_n . \tag{20.4}$$

Thus we have $\alpha_i < \beta_i$ for all relevant i, and $\beta_i < \alpha_{i+1}$.

One natural way to produce RIQs is to start with a monic $P(x) \in \mathbb{Z}[x]$ that has only real zeros, and put $Q(x) = P'(x)$. When reduced to lowest terms, $Q(x)/P(x)$ is an RIQ. For example, $(2x - 5)/(x^2 - 5x + 6)$ is an RIQ. (This construction, and a later one, explains the alphabetical ordering of P and Q, since we pick P first.)

One could generalise this to allow $Q(x)/P(x) \in \mathbb{R}(x)$, but our only interest will be in the case where both $P(x)$ and $Q(x)$ are in $\mathbb{Z}[x]$, and P is monic. Hence, the poles of an RIQ will be totally real algebraic *integers*, and the zeros will be totally real algebraic *numbers* (there is no requirement that Q is monic, only that P is monic). If all the zeros and poles of a rational interlacing quotient $Q(x)/P(x)$ lie in the interval $[-2, 2]$, then we say that $Q(x)/P(x)$ is a **cyclotomic RIQ**. In this case the *poles* will correspond to zeros of unity: $z^{\deg(P)} P(z + 1/z)$ will have only roots of unity as its zeros. The zeros of a cyclotomic RIQ correspond to algebraic numbers all of whose conjugates lie on the unit circle, but since the numerator need not be monic, these algebraic numbers on the unit circle need not be roots of unity. Neither P nor Q need be irreducible.

We record some simple properties and an alternative characterisation.

Lemma 20.13 *A rational interlacing quotient $Q(x)/P(x) \in \mathbb{Z}(x)$ given by (20.3) satisfies*

(i) $Q(x)/P(x) \to 0$ as $x \to \pm\infty$;
(ii) *for any $c > 0$, the equation $Q(x)/P(x) = c$ has exactly n real roots $\gamma_1, \ldots, \gamma_n$, where $\alpha_i < \gamma_i < \beta_i$ for $i = 1, \ldots, n - 1$, and $\alpha_n < \gamma_n$;*
(iii) *away from its poles, the derivative of an RIQ is strictly negative;*
(iv) *one can write*

$$\frac{Q(x)}{P(x)} = \sum_{i=1}^{n} \frac{\gamma_i}{x - \alpha_i}$$

where all the γ_i are strictly positive.

Moreover, any rational function $Q(x)/P(x) \in \mathbb{Z}(x)$ with P monic and of the shape given by (iv) is an RIQ.

Proof Part (i) is clear, as the degree of $P(x)$ is greater than the degree of $Q(x)$.

If $Q(x)/P(x) = c$, then $cP(x) - Q(x) = 0$, and this equation has degree exactly n, so has at most n real roots. On each interval (α_i, β_i) (for $1 \leq i \leq n - 1$), the rational function $Q(x)/P(x)$ ranges from $+\infty$ to 0, so by the intermediate value theorem $Q(x)/P(x) = c$ has at least one solution in that interval. Similarly for the interval (α_n, ∞). By counting, each such interval must account for precisely one solution.

Similarly $Q(x)/P(x) = c$ has precisely n solutions whenever $c < 0$, with one in the interval $(-\infty, \alpha_1)$, and one in each of the intervals (β_i, α_{i+1}) $(1 \leq i \leq n - 1)$. And $Q(x)/P(x) = 0$ has precisely $n - 1$ solutions, namely the β_i. In particular, for any c, the equation $Q(x)/P(x) = c$ never has repeated roots. It follows that $Q(x)/P(x)$ can never have zero derivative. Since it is easy to see that $Q(x)/P(x)$ is decreasing near its poles, it must have strictly negative derivative everywhere that it is defined.

Writing $Q(x)/P(x)$ as a sum of partial fractions gives the form (iv), where by (iii) the γ_i must be strictly positive. Alternatively, one can see directly from the interlacing of the α_i and β_i that each γ_i is positive.

Any function of the shape (iv) is decreasing where defined, so has a zero between each two consecutive poles, and hence is an RIQ with

$$P(x) = \prod_{i=1}^{n}(x - \alpha_i), \quad Q(x) = \sum_{i=1}^{n} \gamma_i \prod_{j \neq i}(x - \alpha_j),$$

provided that $P(x), Q(x) \in \mathbb{Z}[x]$ (with all the $\gamma_i > 0$, one sees that γ in (20.3) is also positive). □

Exercise 20.14 Suppose that a rational function $\varphi(x) \in \mathbb{Z}(x)$ with monic denominator satisfies the following three further properties:

- any poles of $\varphi(x)$ are simple;
- $\varphi(x) \to 0$ as $x \to \pm\infty$;
- $\varphi(x)$ is strictly decreasing where defined.

Show that $\varphi(x)$ is an RIQ. Give examples that are not RIQs satisfying each possible pair of the above three properties but not all three.

By interlacing, real symmetric matrices provide us with a rich source of RIQs. If A is a real symmetric matrix, and B is the submatrix obtained by deleting the ith row and the ith column of A (for some choice of i), then interlacing tells us that $\chi_B(x)/\chi_A(x)$, expressed in lowest terms, is an RIQ. The examples produced in this way are special in that the numerator is also monic. The next lemma shows how from a collection of such RIQs one can produce others, with nonmonic numerators.

Lemma 20.15 If $\phi_1, \ldots, \phi_t$ are RIQs, and $\delta_1 > 0, \ldots, \delta_t > 0$ are integers, then $\delta_1\phi_1 + \cdots + \delta_t\phi_t$ is an RIQ. In particular, the sum of two RIQs is an RIQ.

Proof Using the partial fraction characterisation of Lemma 20.13, this is immediate. □

20.5.2 Circular Interlacing Quotients

We say that $G(z)/F(z) \in \mathbb{Z}(z)$ is a *circular interlacing quotient*, or a *CIQ*, if:

- $F(z)$ is monic, and $G(z)$ has positive leading coefficient;
- $F(1) = 0$;
- $F(z)$ and $G(z)$ have the same degree, say n;
- all zeros of F and G are simple, and lie on the unit circle;
- if the zeros of F have arguments $0 = \theta_1 < \cdots < \theta_n < 2\pi$ and the zeros of G have arguments $0 \leq \phi_1 < \cdots < \phi_n < 2\pi$, then

$$0 = \theta_1 < \phi_1 < \theta_2 < \phi_2 < \cdots < \theta_n < \phi_n < 2\pi .$$

In particular, we see that $F(z)$ and $G(z)$ have no common zeros.

Exercise 20.16 Show that if $G(z)/F(z)$ is a CIQ, then so is $G(z^m)/F(z^m)$ for any $m \geq 1$.

Conversely, suppose that $G(z)/F(z)$ is a CIQ, and for some m we have $G(z) = Q(z^m)$, $F(z) = P(z^m)$. Show that $Q(z)/P(z)$ is a CIQ.

Trivially $(z + 1)/(z - 1)$ is a CIQ, and hence by the previous exercise so is $(z^m + 1)/(z^m - 1)$ for any m.

Note that we have the asymmetric requirements that $F(1) = 0$ and that $F(z)$ is monic. One could relax these conditions, but in our applications we shall always need them, so we include them as part of the definition of a CIQ. In particular, the poles of a CIQ are roots of unity, but the zeros might not be (although they are algebraic numbers all of whose conjugates lie on the unit circle).

Exercise 20.17 Let $G(z)/F(z)$ be a CIQ. Any zeros of G or F other than ± 1 come in complex conjugate pairs. Deduce that $F(-1)G(-1) = 0$. More precisely, if the common degree of F and G is even, then $F(-1) = 0$, so that $(z^2 - 1) \mid F(z)$; if the common degree of F and G is odd, then $G(-1) = 0$.

20.5.3 From CIQs to Cyclotomic RIQs

Suppose that G/F is a CIQ, with the arguments of the zeros of F being $0 = \theta_1 < \cdots < \theta_n < 2\pi$ and the arguments of the zeros of G being $0 < \phi_1 < \cdots < \phi_n < 2\pi$. We have

$$0 = \theta_1 < \phi_1 < \theta_2 < \cdots < \theta_n < \phi_n < 2\pi .$$

Since zeros other that ± 1 come in complex conjugate pairs, there is symmetry in the above:

if $n = 2d$, then $\theta_{d+1} = \pi$, and for $2 \leq j \leq d$ there holds $\theta_j = 2\pi - \theta_{n-j+2}$;

if $n = 2d + 1$, then $\theta_j = 2\pi - \theta_{n-j+2}$ for $2 \le j \le d$.

Since the function $\cos(x)$ is decreasing on the interval $[0, \pi]$, if we define α_j and β_j by

$$\alpha_j := 2\cos\theta_j, \quad \beta_j := 2\cos\phi_j,$$

then we have

$$n = 2d: \quad -2 = \alpha_{d+1} < \beta_d < \alpha_d < \cdots < \beta_1 < \alpha_1 = 2$$
$$n = 2d + 1: \quad -2 = \beta_{d+1} < \alpha_{d+1} < \beta_d < \alpha_d < \cdots < \beta_1 < \alpha_1 = 2$$

Hence, if γ is the leading coefficient of $G(z)$,

$$\frac{Q(x)}{P(x)} = \frac{\gamma \prod_{i=1}^{d}(x - \beta_i)}{\prod_{i=1}^{d+1}(x - \alpha_i)} \tag{20.5}$$

is an RIQ, with the special extra properties that all zeros and poles lie in the interval $[-2, 2]$, with a pole at $x = 2$. (To see that $P(x)$ and $Q(x)$ have integer coefficients, note that they can be obtained by computing the resultants of $F(z)$ and $G(z)$ with $z^2 - xz + 1$ and then removing repeated factors). Note that when $n = 2d + 1$ is odd, we ignore $\beta_{d+1} = -2$ in order to get an RIQ (the resulting g still has integer coefficients; we have simply removed the factor $x + 2$, and what remains has integer coefficients).

A picture may help to reinforce the distinction between the cases $n = 2d$ and $n = 2d + 1$.

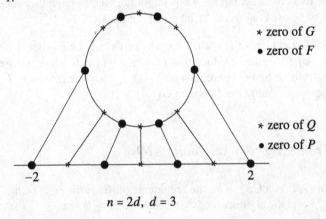

* zero of G

• zero of F

* zero of Q

• zero of P

$n = 2d,\ d = 3$

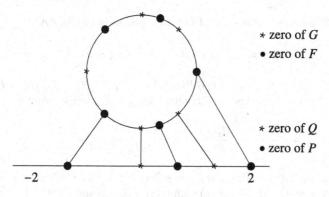

$$n = 2d + 1, \ d = 2$$

Next we shall show that

$$\frac{Q(x)}{P(x)} = \frac{zG(z)}{(z^2 - 1)F(z)},\qquad(20.6)$$

where $x = z + 1/z$. Consider first the case $n = 2d$. Then $(z^2 - 1) \mid F(z)$, and pulling out that factor (corresponding to the zeros α_1 and α_{d+1}) gives

$$
\begin{aligned}
\frac{zG(z)}{(z^2 - 1)F(z)} &= \frac{z}{(z^2 - 1)^2} \frac{\gamma \prod_{j=1}^{d}(z^2 - 2\cos\varphi_j z + 1)}{\prod_{j=2}^{d}(z^2 - 2\cos\theta_j z + 1)} \\
&= \frac{z^2}{(z^2 - 1)^2} \frac{\gamma \prod_{j=1}^{d}(x - \beta_j)}{\prod_{j=2}^{d}(x - \alpha_j)} \\
&= \frac{1}{(x^2 - 4)} \cdot \frac{Q(x)}{P(x)/(x^2 - 4)} \\
&= \frac{Q(x)}{P(x)}.
\end{aligned}
$$

When $n = 2d + 1$ is odd, we pull out a factor of $(z + 1)$ from $G(z)$ (corresponding to β_{d+1}) and a factor of $(z - 1)$ out of $F(z)$ (corresponding to α_1) to give

$$
\begin{aligned}
\frac{zG(z)}{(z^2 - 1)F(z)} &= \frac{z}{(z - 1)^2} \frac{\gamma \prod_{j=1}^{d}(z^2 - 2\cos\varphi_j z + 1)}{\prod_{j=2}^{d+1}(z^2 - 2\cos\theta_j z + 1)} \\
&= \frac{z}{(z - 1)^2} \frac{\gamma \prod_{j=1}^{d}(x - \beta_j)}{\prod_{j=2}^{d+1}(x - \alpha_j)} \\
&= \frac{1}{(x - 2)} \cdot \frac{Q(x)}{P(x)/(x - 2)} \\
&= \frac{Q(x)}{P(x)}.
\end{aligned}
$$

Conversely, suppose that $Q(x)/P(x)$ is an RIQ as in (20.5), with

$$-2 \leq \alpha_{d+1} < \beta_d < \alpha_d < \cdots < \beta_1 < \alpha_1 = 2.$$

We have $Q(x)/P(x) = Q(z + 1/z)/P(z + 1/z) = z^{d+1}Q(z + 1/z)/z^{d+1}P(z + 1/z)$, and noting that $z - 1$ divides the denominator ($\alpha_1 = 2$), we can write

$$\frac{Q(x)}{P(x)} = \frac{zG_0(z)}{(z-1)F_0(z)}$$

for some $F_0(z), G_0(z) \in \mathbb{Z}[z]$. If $\alpha_{d+1} = -2$, then $z + 1$ divides $F_0(z)$, and we have $Q(x)/P(x) = zG(z)/(z^2 - 1)F(z)$ with $G(z) = G_0(z)$ and $F(z) = F_0(z)/(z+1)$. If $\alpha_{d+1} > -2$, then we have $Q(x)/P(x) = zG(z)/(z^2 - 1)F(z)$, this time with $G(z) = (z+1)G_0(z)$ and $F(z) = F_0(z)$. In either case, G/F is a CIQ.

We can use this correspondence to show that the set of CIQs is closed under addition.

Lemma 20.18 *Suppose that G_1/F_1 and G_2/F_2 are CIQs. Then so is their sum* $(F_2G_1 + F_1G_2)/F_1F_2$.

This is initially surprising, and very useful: it is perhaps not transparent even that $G_2F_1 + F_1G_2$ has all zeros on the unit circle.

Proof Let Q_1/P_1 and Q_2/P_2 be the special RIQs corresponding to G_1/F_1 and G_2/F_2, so that $Q_i(x)/P_i(x) = zG_i(z)/(z^2 - 1)F_i(z)$. By Lemma 20.15, $Q_1/P_1 + Q_2/P_2 = Q/F$ is a an RIQ, and indeed is a cyclotomic RIQ since the poles lie amongst those of Q_1 and Q_2. And $Q(2) = 0$. Let G/F be the corresponding CIQ. Then

$$zG/(z^2 - 1)F = Q/P = Q_1/P_1 + Q_2/P_2 = zG_1/(z^2 - 1)F_1 + zG_2/(z^2 - 1)F_2,$$

so $G_1/F_1 + G_2/F_2 \, (= G/F)$ is a CIQ. $\square$

Exercise 20.19 Show that for any $m \geq 1$, $\big((z^{m+1} - 1)/(z - 1)\big)/(z^m - 1)$ is a CIQ. Deduce that

$$\frac{2x^4 + 4x^3 + 5z^2 + 4z + 2}{z^4 + z^3 - z - 1}$$

is a CIQ.

Lemma 20.20 *Let P_r denote the path on r vertices ($r - 1$ edges; $r \geq 1$). Then*

$$R_{P_{n-1}}(\sqrt{z}) = (z^n - 1)/(z - 1) = z^{n-1} + z^{n-2} + \cdots + 1.$$

Proof The eigenvalues of P_{n-1} are all simple, and are given by $2\cos(j\pi/n)$, where $1 \leq j \leq n - 1$ [BH12, Sect. 1.4.4]. Hence, the zeros of $R_{P_{n-1}}$ are $e^{\pm j\pi i/n}$ ($1 \leq j \leq n - 1$). Pairing the factors $z - e^{\pi i j/n}$ and $z - e^{\pi i (n+j)/n}$ ($1 \leq j \leq n - 1$), we see that

$$R_{P_{n-1}}(z) = \prod_{j=1}^{n-1}(z^2 - e^{2\pi i j/n}),$$

whence the advertised formula for $R_{P_{n-1}}(\sqrt{z})$. □

Exercise 20.21 As in the previous lemma, let P_r be the path on r vertices.
(i) Show that for $m, n \geq 2$,

$$\frac{Q(x)}{P(x)} = \frac{\chi_{P_{m-1}}(x)\chi_{P_{n-1}}(x)}{\chi_{P_{m+n-1}}(x)},$$

reduced to lowest terms, is a cyclotomic RIQ that has no poles at ± 2. (Consider deleting a suitable vertex from P_{m+n-1}.)
(ii) Deduce that

$$\frac{\chi_{P_{m+n-1}}(x)}{(x^2 - 4)\chi_{P_{m-1}}(x)\chi_{P_{n-1}}(x)},$$

reduced to lowest terms, is a cyclotomic RIQ with a pole at 2.
(iii) Show that the corresponding CIQ is

$$\frac{z^{2m+2n} - 1}{(z^{2m} - 1)(z^{2n} - 1)}.$$

(iv) Deduce that

$$\frac{z^{m+n} - 1}{(z^m - 1)(z^n - 1)},$$

reduced to lowest terms, is a CIQ.

20.5.4 Salem Numbers from Interlacing Quotients

Lemma 20.22 *Let $G(z)/F(z)$ be a CIQ with both G and F having integer coefficients. (As always, F must be monic, and the top coefficient of G must be positive.) Then*

$$(z^2 - 1)F(z) - zG(z)$$

is the minimal polynomial of a Salem number (or perhaps a reciprocal Pisot number), possibly multiplied by a cyclotomic polynomial.

Proof Let $Q(x)/P(x)$ be the RIQ corresponding to the circular interlacing quotient $G(z)/F(z)$. Let n be the degree of P. By Lemma 20.13(ii), there are n solutions to $Q(x)/P(x) = 1$. All but one of these lie in $[-2, 2)$ (remember that P has a zero at 2), and one lies in the interval $(2, \infty)$. Via $x = z + 1/z$ and (20.6), these correspond to roots of the equation $zG(z)/(z^2-1)F(z) = 1$, with all but two of the roots on the unit circle, and the other two roots being τ and $1/\tau$ for some real $\tau > 1$. If τ has no conjugates other than $1/\tau$ then it is a Pisot number; otherwise it is a Salem number. $\qquad$ $\square$

Exercise 20.23 Show that

$$\frac{z^8 + z^7 - z^5 - z^4 - z^3 + z + 1}{(z-1)(z+1)(z^2+z+1)(z^4+z^3+z^2+z+1)}$$

is a CIQ. Write this as $G(z)/F(z)$ in lowest terms, and verify that the polynomial $(z^2-1)F(z) - zG(z)$ is the minimal polynomial of Lehmer's number $\tau_1 = 1.17628\ldots$.

Show that

$$\frac{z^8 - z^7 + z^6 - z^5 + z^4 - z^3 + z^2 - z + 1}{z^8 - 1}$$

is a CIQ. Write this as $G(z)/F(z)$ in lowest terms, and verify that the polynomial $(z^2-1)F(z) - zG(z)$ is the minimal polynomial of τ_1^2.

20.6   Notes

In the original paper [MS05a], the reciprocal polynomial of a bipartite graph was actually defined to be what we are here calling $R_G(\sqrt{z})$. Rather than have separate definitions for the bipartite and nonbipartite cases, we have chosen here to keep a uniform definition of $R_G(z)$ and instead comment on the extra structure of this polynomial when G is bipartite.

The bipartite examples of Sect. 20.3 have the property that deleting a single vertex leaves a cyclotomic graph. Gumbrell and McKee [GM14] gave a complete classification of all such (connected) Salem graphs, both bipartite and nonbipartite. The smallest number of vertices that must be deleted from a Salem graph to produce a cyclotomic graph is one measure of its combinatorial distance from being cyclotomic. The example K_n ($n > 3$) shows that this number of vertices can be arbitrarily large.

Salem numbers were constructed from starlike trees in [MS99b]. The more general notion of a Salem graph was introduced in [MS05a]. It was shown there that every limit point of the set of graph Salem numbers (those attached to Salem graphs) is a Pisot number, proving a special case of a conjecture of Boyd [Boy77] that this should be true for the set of all Salem numbers. For all graphs that have Mahler measure below the golden ratio, see the papers of Cooley, McKee, and Smyth [MS05a, CMS14].

The interlacing construction appeared first in [MS05b], where it was used to show that there are Salem numbers of every trace. For much more on applying interlacing to Salem and Pisot numbers, see [MS12b, MS13].

20.7 Glossary

$\tau(G)$ If G is a Salem graph, then $\tau(G) = M(G)$, and this is either a Salem number or (if the Salem graph is trivial) a reciprocal quadratic Pisot number.

attaching a pendant path.   See the entry for pendant path.

CIQ.   A circular interlacing quotient.

circular interlacing quotient.   A rational function $G(z)/F(z) \in \mathbb{Z}(z)$ is a circular interlacing quotient (CIQ) if G and F have the same degree, F is monic, all their zeros are simple, $F(1) = 0$, and the zeros of F and G interlace on the unit circle. This interlacing property means that as one travels round the unit circle one encounters zeros of F and G alternately.

cyclotomic RIQ.   This is a rational interlacing quotient for which all zeros and poles lie in the interval $[-2, 2]$.

pendant path.   Let G_0 be a weighted digraph, and let v_0 be a vertex of G_0. Construct the weighted digraph G by taking a path P on vertices $v_1, \ldots, v_r$ (so that there is a weight-1 edge from v_i to v_j if and only if $|j - i| = 1$) and joining this to G_0 by adding weight-1 edges from v_0 to v_1 and from v_1 to v_0. Then the path on vertices $v_0, \ldots, v_r$ is a pendant path in G, and this process of moving from G_0 to G is called attaching a pendant path to G_0.

RIQ.   A rational interlacing quotient.

rational interlacing quotient.   A rational interlacing quotient (RIQ) is a rational function $g(x)/f(x) \in \mathbb{Z}[x]$ where the degree of f is one more than the degree of g, the zeros of f and g are simple and strictly interlace, the top coefficient of g is positive, and f is monic (see (20.3) and (20.4)).

Salem graph.   A connected graph that has single eigenvalue $\lambda > 2$, and is either bipartite (in which case it has the eigenvalue $-\lambda$ and all other eigenvalues lie in $[-2, 2]$) or is nonbipartite and has no eigenvalues below -2 (so that all eigenvalues other than λ lie in $[-2, 2]$).

trivial Salem graph.   A Salem graph G is called trivial if its unique eigenvalue greater than 2, λ, satisfies $\lambda^2 \in \mathbb{Z}$. Then, the attached number $\tau(G)$ is a reciprocal quadratic Pisot number rather than a Salem number.

Chapter 21
Minimal Polynomials of Integer Symmetric Matrices

21.1 Introduction

In this penultimate chapter, we bring together several of the tools developed in previous chapters to address the question: which polynomials are minimal polynomials of integer symmetric matrices? There are some easy necessary conditions. If $P(x)$ is the minimal polynomial of some integer symmetric matrix A, then,

$$P(x) \in \mathbb{Z}[x] \,;$$
$$P(x) \text{ is monic} \,;$$
$$P(x) \text{ is totally real} \,;$$
$$P(x) \text{ is separable (has no multiple zeros in } \mathbb{R}) \,. \tag{21.1}$$

The first two of these conditions arise simply from the fact that $P(x)$ must be a factor of the characteristic polynomial of A. These would be satisfied by the minimal polynomial of any square integer matrix, whether or not symmetric. The third condition follows from the fact that all the eigenvalues of an integer symmetric matrix are real. The final point (that $P(x)$ is separable) is a consequence of the diagonalisability of an integer symmetric matrix (over $\mathbb{Q}$).

Estes and Guralnick [EG93, Corollary C] showed that if the degree of $P(x)$ is at most 4, then the conditions (21.1) are in fact sufficient. Take, for example, the polynomial $x^2 - 3$. Now this is not the characteristic polynomial of any integer symmetric matrix (Exercise 21.1(i)), but it satisfies (21.1), and has degree only 2, so by the Estes–Guralnick result (or see Exercise 21.1(ii)) it must be the minimal polynomial of some larger matrix. One quickly discovers that

$$A = \begin{pmatrix} 1 & 1 & 1 & 0 \\ 1 & -1 & 0 & 1 \\ 1 & 0 & -1 & -1 \\ 0 & 1 & -1 & 1 \end{pmatrix}$$

© Springer Nature Switzerland AG 2021
J. McKee and C. Smyth, *Around the Unit Circle*, Universitext,
https://doi.org/10.1007/978-3-030-80031-4_21

satisfies $A^2 = 3I$, and hence A has minimal polynomial $x^2 - 3$.

Exercise 21.1 (i) Show that $x^2 - d$ is the characteristic polynomial of an integer symmetric matrix if and only if d is the sum of two squares of integers. (This was part of Exercise 19.1.)

(ii) Show that $x^2 - d$ is the minimal polynomial of an integer symmetric matrix for any $d \geq 0$. (Hint: use induction. Note that if $M^2 = dI$, then

$$\begin{pmatrix} M & I \\ I & -M \end{pmatrix}^2 = (d + 1)I \,,$$

where I is used for the identity matrix of whatever size is needed.)

Estes and Guralnick wondered whether the conditions (21.1) were in fact sufficient to imply that $P(x)$ is the minimal polynomial of some integer symmetric matrix [EG93, p. 84]. Although we now know that this conjecture is false, we state it as a conjecture here for ease of reference.

Conjecture 21.2 Suppose that $P(x) \in \mathbb{Z}[x]$ is a monic separable polynomial that has all zeros real. Then there is some integer symmetric matrix A for which $P(x)$ is the minimal polynomial.

Finding counterexamples to Conjecture 21.2 initially seems daunting: a candidate $P(x)$ of degree d is potentially the minimal polynomial of $rd \times rd$ matrices for arbitrarily large r, so one cannot organise a naive finite search. Nevertheless, we shall see that there are certain necessary conditions for $P(x)$ to be the minimal polynomial of an integer symmetric matrix that are not so obvious as the above, but can be applied to eliminate the possibility of arbitrarily large matrices using a finite amount of work:

- the discriminant cannot be too small (Sect. 21.2);
- if the span is small, then a finite search is in principle possible (Sect. 21.3);
- the trace cannot be too small (Sect. 21.4);
- $P(x)$ must be interlaced (Sect. 21.5), i.e., there exists some totally real monic polynomial $Q(x) \in \mathbb{Z}[x]$ whose zeros interlace those of P (see Sect. 21.2).

Using a combination of the span and trace constraints, we shall find counterexamples to the conjecture for every degree greater than 5. The case of degree 5 remains stubbornly open.

Problem 21.3 (*open problem*) Is there a degree-5 polynomial $P(x)$ satisfying (21.1) that is not the minimal polynomial of an integer symmetric matrix?

21.2 Small Discriminant

Dobrowolski [Dob08] found a lower bound for the discriminant of the minimal polynomial of an integer symmetric matrix in terms of the degree of the polynomial, independent of the size of the matrix, in the case where the minimal polynomial is irreducible. Following Yatsyna [Yat16, Yat19], we give a slightly more general bound that applies to a wider class of polynomials which includes all irreducible minimal polynomials of integer symmetric matrices.

Suppose that $P(x) \in \mathbb{Z}[x]$ is a monic polynomial of degree n, with all zeros real. Say the zeros are $\alpha_1 \leq \cdots \leq \alpha_n$. We say that $P(x)$ is *interlaced* if there exists a *monic* polynomial $Q(x) \in \mathbb{Z}[x]$ of degree $n - 1$ with zeros $\beta_1 \leq \cdots \leq \beta_{n-1}$ all real such that

$$\alpha_1 \leq \beta_1 \leq \alpha_2 \leq \beta_2 \leq \cdots \leq \beta_{n-1} \leq \alpha_n .$$

For example, if $P(x)$ is the characteristic polynomial of an integer symmetric matrix A, and $Q(x)$ is the characteristic polynomial of the matrix formed by deleting the last row and column of A, then $P(x)$ is interlaced (with $Q(x)$ serving to illustrate this, by interlacing, Theorem B.1). Note the requirement that $Q(x)$ is monic. If $P(x)$ is totally real, then the zeros of $P'(x)$ interlace with those of $P(x)$, but unless $P(x)$ is linear $P'(x)$ is not monic.

Exercise 21.4 Show that if $P(x)$ is the minimal polynomial of an integer symmetric matrix and $P(x)$ is irreducible, then $P(x)$ is interlaced. (If A is the matrix, then A has characteristic polynomial $C = P^r$, for some r. Let C_1 be the characteristic polynomial of the matrix produced by deleting the first row and column of A. Show that P^{r-1} divides C_1, and put $Q = C_1/P^{r-1}$. Show that Q has degree one less than the degree of P, and that its zeros interlace those of P.)

Lemma 21.5 *Let $P(x) \in \mathbb{Z}[x]$ be a monic separable polynomial of degree n with all zeros real. Say $\alpha_1, ..., \alpha_n$ are the zeros of P (distinct, since P is separable). For $i = 1, ..., n$, define*

$$P_i(x) = P(x)/(x - \alpha_i) \in \mathbb{R}[x]. \tag{21.2}$$

If $Q(x) \in \mathbb{R}[x]$ is any monic polynomial of degree $n - 1$ whose zeros are all real and interlace with those of P, then there exist nonnegative real numbers $\lambda_1, ..., \lambda_n$ such that

$$\sum_{i=1}^{n} \lambda_i = 1 ,$$

$$\text{and} \quad Q(x) = \sum_{i=1}^{n} \lambda_i P_i(x) .$$

Proof Observe that for each i we have $P_i(\alpha_i) = P'(\alpha_i)$. The zeros of P' interlace with those of P, and both P' and Q have positive leading coefficient, hence

$$P'(\alpha_i)Q(\alpha_i) \geq 0$$

for each i, with equality only if $Q(\alpha_i) = 0$. (We never have $P'(\alpha_i) = 0$, since P is separable.) We can therefore define $\lambda_i \geq 0$ for $i = 1, ..., n$ by

$$\lambda_i := Q(\alpha_i)/P'(\alpha_i) = Q(\alpha_i)/P_i(\alpha_i).$$

Put $F(x) := Q(x) - \sum_{i=1}^{n} \lambda_i P_i(x)$. Then, $F(x)$ has degree at most $n - 1$ and vanishes at the n distinct points $\alpha_1, ..., \alpha_n$. Hence $F(x) = 0$, and $Q(x) = \sum_{i=1}^{n} \lambda_i P_i(x)$. Since $Q(x)$ and all the $P_i(x)$ are monic, we have also that $\sum_{i=1}^{n} \lambda_i = 1$. □

Exercise 21.6 Show that the necessary condition for interlacing given in Lemma 21.5 is also sufficient: if $\lambda_i \geq 0$ for $i = 1, ..., n$, $\sum \lambda_i = 1$ and $Q = \sum \lambda_i P_i \in \mathbb{Z}[x]$, then P is interlaced.

Theorem 21.7 *Let $P(x) \in \mathbb{Z}[x]$ be an irreducible monic separable polynomial of degree n that is interlaced. Then Δ_P, the discriminant of P, satisfies $|\Delta_P| \geq n^n$.*

Proof Let $\alpha_1, ..., \alpha_n$ be the zeros of P (real and distinct). Note from (A.3) that $\Delta_P = \prod_{i=1}^{n} P'(\alpha_i)$.

Since P is interlaced, there is some $Q(x) \in \mathbb{Z}[x]$ a monic polynomial of degree $n - 1$ whose zeros are real and interlace those of P. By Lemma 21.5, there are nonnegative real numbers $\lambda_1, ..., \lambda_n$ such that

$$\sum_{i=1}^{n} \lambda_i = 1, \quad Q = \sum_{i=1}^{n} \lambda_i P_i,$$

where $P_1, ..., P_n$ are defined by (21.2) as in Lemma 21.5, and in particular we note again that $P_i(\alpha_i) = P'(\alpha_i)$ for each i. Then, since

$$Q(\alpha_i) = \sum_j \lambda_j P_j(\alpha_i) = \lambda_i P_i(\alpha_i),$$

we have from (A.2) that

$$|\mathrm{res}_x(P, Q)| = \left| \prod_{i=1}^{n} Q(\alpha_i) \right|$$

$$= \left| \prod_{i=1}^{n} \lambda_i P'(\alpha_i) \right|$$

$$= \left(\prod_{i=1}^{n} \lambda_i \right) \cdot \left| \prod_{i=1}^{n} P'(\alpha_i) \right|$$

$$\leq \left(\frac{1}{n} \sum_{i=1}^{n} \lambda_i \right)^n |\Delta_P|$$

$$= \frac{1}{n^n}|\Delta_P|.$$

Since P is irreducible, $|\operatorname{res}_x(P, Q)| \geq 1$, and hence $|\Delta_P| \geq n^n$. □

It is a challenge to find irreducible, monic, separable, totally real polynomials of degree n and discriminant smaller than n^n, but they do exist. Simon [Sim99] gave an explicit estimate for the size of the discriminant of the nth cyclotomic polynomial, recalled in the following exercise.

Exercise 21.8 (i) ([Dob08]) Show that $\Delta^2_{\chi_A} \mid \Delta_{R_A}$. (Hint: show that

$$(\alpha_i - \alpha_j)(\alpha_i^{-1} - \alpha_j^{-1})(\alpha_i - \alpha_j^{-1})(\alpha_i^{-1} - \alpha_j)$$

is a perfect square in $\mathbb{Z}[\alpha_i + \alpha_i^{-1}, \alpha_j + \alpha_j^{-1}]$.)

(ii) For a reciprocal polynomial $P(x) \in \mathbb{Z}[x]$, zeros $\alpha_1, \alpha_1^{-1}, \ldots, \alpha_n, \alpha_n^{-1}$, define $\widetilde{P}(x) := \prod_{i=1}^n (x - \alpha_i - \alpha_i^{-1})$. For example, $\chi_A(x) = \widetilde{R}_A(x)$. Simon [Sim99, Proposition 2.1] shows that if n is the product of all primes up to X, then

$$|\Delta_{\Phi_n}|^{1/\varphi(n)} \sim e^{2\gamma} \varphi(n) \log\log\varphi(n) / \log\varphi(n)$$

as $X \to \infty$ (here Φ_n is the nth cyclotomic polynomial (degree $\varphi(n)$), and γ is Euler's constant). Use part (i) to deduce that there exist n such that the totally real polynomial $\widetilde{\Phi}_n$ breaks the discriminant bound of Theorem 21.7, and hence is not the minimal polynomial of an integer symmetric matrix.

(iii) Find the smallest n such that $\widetilde{\Phi}_n$ breaks the discriminant bound of Theorem 21.7. What is the degree of this $\widetilde{\Phi}_n$?

Thus, we get some counterexamples to Conjecture 21.2, indeed infinitely many of them, but the degrees are all unpleasantly large. For the exact formula

$$\Delta_{\widetilde{\Phi}_n} = \left(cn^{\varphi(n)} / \prod_{p|n} p^{\varphi(n)/(p-1)} \right)^{1/2},$$

where the product is over the primes p dividing n, and

$$c = \begin{cases} 1/p & n = p^\ell, \ 2p^\ell \ (p > 2), \\ 1/4 & n = 2^\ell, \\ 1 & \text{otherwise}, \end{cases}$$

see [Rob64], which uses [Leh30].

Problem 21.9 (*open problem*) What is the smallest n for which there exists an irreducible monic separable totally real polynomial of degree n with discriminant smaller than n^n?

21.3 Small Span

Recall from Chap. 16 that the span of a totally real polynomial is the difference between the largest and smallest zeros: it is the length of the shortest real interval in which all the zeros lie. Recall also that a polynomial is said to have *small* span if its span is strictly less than 4. In this section, we show that if a monic separable irreducible totally real polynomial $P(x) \in \mathbb{Z}[x]$ has small span, then there is an algorithm to determine whether or not it is the *minimal* polynomial of an integer symmetric matrix. This algorithm is only practical for fairly small degrees, but it proves useful in providing low-degree counterexamples to Conjecture 21.2.

Replacing $P(x)$ by $\pm P(\pm x + t)$, we can suppose that small-span P has all its zeros in the interval $[-2, 2.5)$ (Exercise 16.4). There are two distinct flavours: perhaps $P(x)$ is cyclotomic, having all its zeros in the interval $[-2, 2]$, or perhaps not.

The span of a matrix is the difference between its largest and smallest eigenvalues. Although defined in terms of the characteristic polynomial, this span naturally relates to the minimal polynomial too.

Lemma 21.10 *Let $P(x)$ be the minimal polynomial of an integer symmetric matrix A. Then the span of A equals the span of P.*

Proof Every zero of the characteristic polynomial is a zero of the minimal polynomial, and vice versa.   □

The usefulness of this lemma is that, given P, we do not need to know the size of A: the span is determined by the minimal polynomial.

21.3.1 The Cyclotomic Case

Suppose that an *irreducible* polynomial $P(x)$ has all its zeros in the interval $[-2, 2]$. Let A be an integer symmetric matrix that has minimal polynomial P. If A is not connected, then each component of A has minimal polynomial P (here using that P is irreducible). Hence, there is a connected integer symmetric matrix that has minimal polynomial P.

We can try to find all possibilities for A by growing. There is an infinite supply of connected cyclotomic A, but for each small span (*strictly* less than 4) there are finitely many possibilities. This is not a totally trivial observation, so we shall prove it.

Lemma 21.11 *Take any real number $s < 4$. Then, there are only finitely many connected cyclotomic integer symmetric matrices that have span at most s.*

Proof There are only finitely many submatrices of the sporadic maximal connected cyclotomic examples, so we are left to deal with the infinite families of charged signed graphs (Theorem 6.31). As the number of vertices of a connected cyclotomic charged signed graph grows to infinity, inspection of the infinite families shows that the length of the longest uncharged path must grow to infinity. Any uncharged path is equivalent to an unsigned uncharged path, and the largest eigenvalue of such a path tends to 2 as the number of the vertices in the path grows to infinity, and the smallest eigenvalue tends to -2 ([BH12, Sect. 1.4.4]). Hence, by interlacing, as the number of vertices of a connected cyclotomic charged signed graph grows to infinity, the span tends to 4. Hence, only finitely many connected cyclotomic integer symmetric matrices have span at most s (we worked up to equivalence, but each equivalence class is finite). $\qquad\qquad\qquad\qquad\qquad\qquad\qquad\qquad\qquad\qquad\qquad\qquad\qquad$ $\square$

We have then a process of finding all connected integer symmetric matrices that have minimal polynomial P (and have shown that there are only finitely many!). We grow connected cyclotomic integer symmetric matrices until the span exceeds the span of P. We shall see some examples below: even for degree 6 the computations are substantial.

21.3.2 The Noncyclotomic Case

This case is computationally much more straightforward. Suppose that $P(x)$ is our target polynomial, degree d, span less than 4, but not span-equivalent to a polynomial that has all its zeros in the interval $[-2, 2]$. After Theorem 16.39, we need only (re)grow small-span integer symmetric matrices up to 12-vertex examples, or rather the largest multiple of d that is at most 12. Either we find a matrix whose minimal polynomial is $P(x)$, or the theorem tells us that none exists. Note that the examples shown in Sect. 16.4 are the *maximal* ones, so it is not enough simply to look at those.

21.3.3 Some Counterexamples to Conjecture 21.2

Consider the three polynomials

$$P_1(x) = x^7 - x^6 - 7x^5 + 5x^4 + 15x^3 - 5x^2 - 10x - 1,$$
$$P_2(x) = x^7 - 8x^5 + 19x^3 - 12x - 1,$$
$$P_3(x) = x^7 - 2x^6 - 6x^5 + 11x^4 + 11x^3 - 17x^2 - 6x + 7.$$

These are all totally real, irreducible, and have spans (truncated, not rounded) 3.96002, 3.97044, 3.97129, respectively. None is cyclotomic.

We need to grow small-span integer symmetric matrices up to 7×7 (and can restrict the interval of eigenvalues to the one defined by the zeros of each P_i in turn, which greatly reduces the computational effort). None are found that have minimal polynomial any of the P_i. By the classification Theorem 16.39 these P_i cannot be minimal polynomials of 14×14 or larger integer symmetric matrices either. They provide counterexamples to Conjecture 21.2.

Next consider the three cyclotomic examples

$$Q_1(x) = x^6 - x^5 - 6x^4 + 6x^3 + 8x^2 - 8x + 1,$$
$$Q_2(x) = x^6 - 7x^4 + 14x^2 - 7,$$
$$Q_3(x) = x^6 - 6x^4 + 9x^2 - 3. \tag{21.3}$$

These all have all their zeros in the interval $I = [-1.970, 1.970]$. Running our echelon growing code (Exercise 6.22), restricting to all eigenvalues being in this interval, we find that there are no connected integer symmetric matrices having 19 or more rows that have all their eigenvalues in this interval. (After Sect. 16.3 we can restrict to adjacency matrices of charged signed graphs.) Looking at the output for 6×6, 12×12 and 18×18 matrices, none of the Q_i appears as a minimal polynomial. They provide three more counterexamples to Conjecture 21.2.

Here is a little more detail of the search. Up to equivalence, there are only two connected charged signed graphs on 6 vertices that have all their eigenvalues in the interval I. Their minimal polynomials (in these cases also their characteristic polynomials) are shown below the digraphs:

$$(x^2 - 2)(x^4 - 4x^2 + 1) \qquad\qquad x^6 + x^5 - 5x^4 - 4x^3 + 6x^2 + 3x - 1$$

The next possibility for one if the Q_i to appear as a minimal polynomial is when the charged signed graph has 12 vertices. Up to equivalence there are again just two examples. Their characteristic polynomials are show under the digraphs, and their minimal polynomials are of course the squarefree parts of these:

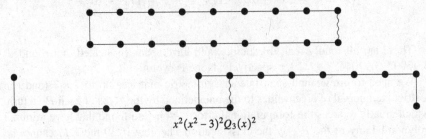

$$(x^2 - 2)^2(x^4 - 4x^2 + 1)^2 \qquad\qquad x^2(x^2 - 3)^2 Q_3(x)$$

We see that $Q_3(x)$ appears as a *factor* of a minimal polynomial, but none of the Q_i is equal to a minimal polynomial.

Next we reach degree 18. Again there are two examples up to equivalence, but this time they are cospectral. We see $Q_3(x)$ making another appearance as a factor of a minimal polynomial, but no Q_i is a minimal polynomial:

$$x^2(x^2 - 3)^2 Q_3(x)^2$$

Although the procedure for testing a whether or not a polynomial is the minimal polynomial of an integer symmetric matrix in this cyclotomic case is finite, we see that even for degree 6 it is rather painful, requiring growing up to 19 vertices in the above example.

21.4 Small Trace

We saw in Chap. 15, Corollary 15.4, that if a totally positive irreducible monic polynomial $P(x) \in \mathbb{Z}[x]$ has degree d and trace strictly less than $2d - 1$, then P cannot be the minimal polynomial of an integer symmetric matrix: it gives a counterexample to Conjecture 21.2.

For example (and we shall need these later), consider the polynomials

$$
\begin{aligned}
P_{17}(x) = x^{17} &- 32x^{16} + 464x^{15} - 4035x^{14} + 23476x^{13} - 96534x^{12} + 288970x^{11} \\
&- 639202x^{10} + 1049959x^9 - 1276648x^8 + 1136674x^7 - 727064x^6 \\
&+ 324455x^5 - 96702x^4 + 18002x^3 - 1871x^2 + 87x - 1, \\
P_{18}(x) = x^{18} &- 34x^{17} + 527x^{16} - 4933x^{15} + 31141x^{14} - 140233x^{13} + 464719x^{12} \\
&- 1152689x^{11} + 2156049x^{10} - 3041595x^9 + 3215905x^8 - 2515820x^7 \\
&+ 1426785x^6 - 569065x^5 + 152586x^4 - 25648x^3 + 2401x^2 - 99x + 1.
\end{aligned}
$$

One may check that these are both irreducible, all zeros are real and positive, and the traces are too small for them to be minimal polynomials of integer symmetric matrices (Corollary 15.4). These provide counterexamples to Conjecture 21.2.

From where were those polynomials magically produced? There is a technique described in [McK11] for computing totally positive algebraic integers of small trace, and this was applied to produce these examples.

Exercise 21.12 Study the method of [McK11], and implement it.

21.5 Polynomials that are Not Interlaced

We have noted that if $P(x)$ is the minimal polynomial of an integer symmetric matrix, then it is interlaced: if P has degree d and zeros $\alpha_1 < \cdots < \alpha_d$ (P is separable), then there is a monic $Q \in \mathbb{Z}[x]$, degree $d - 1$, zeros $\beta_1 < \cdots < \beta_{d-1}$ such that

$$
\alpha_1 \le \beta_1 \le \alpha_2 \le \cdots \le \alpha_{d-1} \le \beta_{d-1} \le \alpha_d.
$$

(If P is irreducible then all inequalities are strict.) If we can find a monic totally real polynomial in $\mathbb{Z}[x]$ that is *not* interlaced, then we have a counterexample to Conjecture 21.2.

Exercise 21.13 (*computational exercise*)

Write a program to test if a polynomial is interlaced. For example, one could use Robinson's algorithm ([Rob64, MS04]) to home in on finitely many possibilities for a polynomial that might achieve interlacing.

Yatsyna noted that the degree-6 polynomial Q_1 of (21.3) is not interlaced. Indeed he proved rather more ([Yat19, Theorem 1.2], [Yat16]): if $d > 20$, $d \neq 30$, d is not a prime, nor twice a prime, and d is squarefree, then the minimal polynomial of $e^{2\pi i/d} + e^{-2\pi i/d}$ is not interlaced. The polynomial Q_1 is the case $d = 21$.

Exercise 21.14 What are the smallest degrees of Yatsyna's counterexamples to Conjecture 21.2? Use your code to verify that the polynomials are not interlaced.

21.6   Counterexamples for all Degrees Greater than 5

We have seen counterexamples to the Estes–Guralnick Conjecture (Conjecture 21.2) for degrees 6 and 7. We now show that there are counterexamples for all larger degrees, leaving only the degree 5 case open. The main argument will only apply once the degree is large enough, so we start by cataloguing some small-degree examples, that have either small span or small trace. Rhin's table [Rhi19] of small span polynomials is a useful resource, combined with the work of Chap. 16.

21.6.1   Degrees 8 to 16

The following irreducible totally real polynomials have small span, are not cyclotomic, and no power of them appears as the characteristic polynomial of an integer symmetric matrix. For degrees 13, 14, 15 (since these are not cyclotomic examples) this is immediate from Theorem 16.39. For degree 12, we merely need look at the degree-12 examples in Sect. 16.4. For degrees 8 to 11, we need to (re)grow small-span matrices up to the relevant size, checking that the given polynomial never appears as a minimal polynomial. In each case this can be done using echelon growing, allowing zeros in the interval spanned by the zeros of the target polynomial. To save space, the polynomial $\sum_{i=0}^{n} a_i x^i$ is represented by the vector $[a_n, a_{n-1}, \ldots, a_0]$.

degree	polynomial
8	[1, -4, -2, 21, -6, -33, 12, 13, -1]
9	[1, -2, -7, 14, 15, -30, -10, 19, 2, -1]
10	[1, -1, -10, 9, 35, -28, -49, 35, 21, -15, 1]
11	[1, -2, -10, 19, 37, -65, -61, 95, 42, -53, -9, 7]
12	[1, -2, -11, 20, 48, -74, -105, 123, 118, -88, -61, 21, 11]
13	[1, -4, -5, 33, 2, -103, 24, 150, -42, -101, 22, 26, -3, -1]
14	[1, -3, -10, 33, 38, -141, -68, 293, 56, -301, -15, 135, -1, -17, -1]
15	[1, -7, 7, 50, -103, -127, 393, 129, -692, -35, 597, -1, -217, -10, 15, 1]
16	[1, -1, -16, 13, 106, -66, -373, 164, 742, -203, -818, 109, 451, -10, -93, -6, 1]

All of these are counterexamples to Conjecture 21.2.

Exercise 21.15 Look through the polynomials in Rhin's table [Rhi19], and use the methods of Chap. 16 to find further counterexamples to Conjecture 21.2.

21.6.2 Degree 20

For degree 20, we shall find a counterexample using the interlacing construction of Chap. 20. By Lemma 20.18, the sum of the two CIQs of Exercises 20.21 and 20.23 is a CIQ, and by Lemma 20.22 the polynomial

$$P_{a,b}(z) = (z^2 - 1)(z + 1)(z^3 - 1)(z^5 - 1)(z^a - 1)(z^b - 1)$$
$$- z\left((z^a - 1)(z^b - 1)(z - 1)(z^8 + z^7 - z^5 - z^4 - z^3 + z + 1)\right.$$
$$\left. + (z^{a+b} - 1)(z + 1)(z^3 - 1)(z^5 - 1)\right)$$

is the product of a cyclotomic polynomial and a Salem polynomial (the minimal polynomial of a Salem number; actually this might be a reciprocal quadratic Pisot number). Now $P_{a,b}(z)$ has trace 1 and is divisible by $(z - 1)^3$. If $P_{a,b}(z)/(z - 1)^3$ happens to be irreducible, then it is the minimal polynomial of a Salem number θ of trace -2. The corresponding totally positive algebraic integer $\theta + 1/\theta + 2$, if of degree d, would have trace $2d - 2$, and so would provide a counterexample to Conjecture 21.2 (by Corollary 15.4).

Taking $a = 13$ and $b = 19$, the above construction gives a Salem number of degree 40, trace -2, and hence a counterexample to the conjecture of degree 20, trace $38 < 2 \times 20 - 1$:

$$x^{20} - 38x^{19} + 665x^{18} - 7109x^{17} + 51925x^{16} - 274605x^{15} + 1087174x^{14}$$
$$- 3286104x^{13} + 7665466x^{12} - 13859066x^{11} + 19399916x^{10}$$
$$- 20897816x^{9} + 17129522x^{8} - 10501109x^{7} + 4696415x^{6}$$
$$- 1478958x^{5} + 311554x^{4} - 40653x^{3} + 2912x^{2} - 94x + 1 .$$

Exercise 21.16 (i) Verify that when $(a, b) = (7, 19)$ or $(11, 17)$, the polynomial $R_{a,b}(z)$ is reducible, and one does not produce counterexamples from these two pairs.

(ii) Find some other pairs that *do* provide counterexamples.

21.6.3 Degree 19 and All Degrees Greater than 20

Exercise 21.17 (i) Take any positive integers $p_1, \ldots, p_5, n$. Using Exercise 20.21(ii) and Lemma 20.18 show that

$$\frac{z^{p_1+p_2} - 1}{(z^{p_1} - 1)(z^{p_2} - 1)} + \frac{z^{p_3+p_4} - 1}{(z^{p_3} - 1)(z^{p_4} - 1)} + \frac{z^{p_5+n} - 1}{(z^{p_5} - 1)(z^{n} - 1)} ,$$

when reduced to lowest terms, is a CIQ. (The idiosyncratic choice of parameter names will make sense later.)

(ii) Show that permuting the p_i in (i) does not change the CIQ (hint: partial fractions).

Take $p_1 = 2$ and any primes $p_2, \ldots, p_5$, and $n > 1$ with

$$\gcd(2 p_2 p_3 p_4 p_5, n) = 1, \tag{21.4}$$

and consider the CIQ (Exercise 21.17)

$$\frac{Q(z)}{P(z)} = \frac{(z^{2+p_2} - 1)}{(z^2 - 1)(z^{p_2} - 1)} + \frac{(z^{p_3+p_4} - 1)}{(z^{p_3} - 1)(z^{p_4} - 1)} + \frac{(z^{p_5+n} - 1)}{(z^{p_5} - 1)(z^{n} - 1)} ,$$

where Q/P is in lowest terms. The conditions on the p_i and n imply that

$$P(z) = (z^{n} - 1) \left(\prod_{i=1}^{5} (z^{p_i} - 1) \right) / (z - 1)^5 ,$$

$$Q(z) = \big((z^{p_1+p_2} - 1)(z^{p_3} - 1)(z^{p_4} - 1)(z^{p_5} - 1)(z^{n} - 1)$$
$$+ (z^{p_3+p_4} - 1)(z^{p_1} - 1)(z^{p_2} - 1)(z^{p_5} - 1)(z^{n} - 1)$$
$$+ (z^{p_5+n} - 1)(z^{p_1} - 1)(z^{p_2} - 1)(z^{p_3} - 1)(z^{p_4} - 1) \big) / (z - 1)^5 .$$

Applying Lemma 20.22, we have that

$$R(z) = (z^2 - 1)P(z) - zQ(z) = F(z)G(z),$$

where $F(z)$ is cyclotomic (or 1), and $G(z)$ is the minimal polynomial of a Salem number (or reciprocal quadratic Pisot number, or perhaps $G(z) = 1$). Now

$$
\begin{aligned}
R(z) &= (z^2 - 1)P(z) - zQ(z) \\
&= (z^{n-3+\sum_{i=1}^{5} p_i} + 5z^{n-4+\sum_{i=1}^{5} p_i} + \cdots) - (3z^{n-5+\sum_{i=1}^{5} p_i} + \cdots) \\
&= z^{n-3+\sum_{i=1}^{5} p_i} + 2z^{n-4+\sum_{i=1}^{5} p_i} + \cdots .
\end{aligned}
$$

Suppose we are lucky enough to find that $R(z)$ is irreducible: then it is the minimal polynomial of a Salem number, degree $\sum_{i=1}^{5} p_i + n - 3$ (even, since $p_1 = 2$ and n must be odd) and trace -2.

If $\widetilde{R}(z)$ (irreducible in the case that R is) were the minimal polynomial of an integer symmetric matrix, then it would be the minimal polynomial of any component of that matrix (here using irreducibility of $\widetilde{R}$), so we would have $\widetilde{R}$ being the minimal polynomial of a connected $d \times d$ integer symmetric matrix A. All eigenvalues of A would be at least -2, indeed strictly greater than -2 by irreducibility of $\widetilde{R}(z)$. Adding $2I$ to A would give a positive definite connected matrix. Put $n = 2m + 1$. The degree d of the characteristic polynomial of $A + 2I$ would be

$$d = m - 1 + \frac{1}{2} \sum_{i=1}^{5} p_i .$$

For any eigenvalue λ of $A + 2I$ we would have $\lambda = 2 + \theta + 1/\theta$ for some zero θ of R, and taking traces gives $\mathrm{trace}(\lambda) = 2d - 2$. We would therefore have a counterexample to Conjecture 21.2, by Corollary 15.4.

Which choices (p_2, p_3, p_4, p_5, n) give irreducible $R(z)$? Let us fix p_2, p_3, p_4, p_5, and ask which n give irreducible $R(z)$. Put $y = z^n$, and define

$$P_2(y, z) = (y - 1)\left(\prod_{i=1}^{5}(z^{p_i} - 1)\right)/(z - 1)^4 ,$$

$$
\begin{aligned}
Q_2(y, z) = &\big((z^{p_1+p_2} - 1)(z^{p_3} - 1)(z^{p_4} - 1)(z^{p_5} - 1)(y - 1) \\
&+ (z^{p_3+p_4} - 1)(z^{p_1} - 1)(z^{p_2} - 1)(z^{p_5} - 1)(y - 1) \\
&+ (yz^{p_5} - 1)(z^{p_1} - 1)(z^{p_2} - 1)(z^{p_3} - 1)(z^{p_4} - 1)\big)/(z - 1)^4 ,
\end{aligned}
$$

so that $P(z) = P_2(z^n, z)/(z - 1)$ and $Q(z) = Q_2(z^n, z)/(z - 1)$, and

$$(z - 1)R(z) = (z^2 - 1)P_2(z^n, z) - zQ_2(z^n, z).$$

If $R(z)$ is reducible, then it has a cyclotomic factor, and then both z and z^n are roots of unity, so that (y, z) would be a *cyclotomic point* (i.e., one for which both coordinates are roots of unity) on the curve

$$C_0 : \quad (z^2 - 1)P_2(y, z) - zQ_2(y, z) = 0.$$

We now apply the technique of Sect. 5.3 for finding cyclotomic points on curves, noting that if z is a root of unity which is also a zero of R, then $y = z^n$ is also a root of unity (and n is odd, from (21.4)). Thus, by Exercise 5.12, precisely one of the points $(-y, -z)$, (y^2, z^2), $(-y^2, -z^2)$ is also on C_0. Thus any cyclotomic points lie not only on C_0 but also on one of the three related curves:

$$C_1 : \quad (z^2 - 1)P_2(-y, -z) + zQ_2(-y, -z) = 0,$$
$$C_2 : \quad (z^4 - 1)P_2(y^2, z^2) - z^2 Q_2(y^2, z^2) = 0,$$
$$C_3 : \quad (z^4 - 1)P_2(-y^2, -z^2) + z^2 Q_2(-y^2, -z^2) = 0.$$

For example, take $(p_1, \ldots, p_5) = (2, 3, 5, 7, 19)$. Eliminating y between C_0 and C_1 (take resultants with y as the variable) one gets a polynomial in z that has no cyclotomic factors. The same holds for C_0 and C_3. On the other hand, eliminating y between C_0 and C_2 gives a whole clutch of cyclotomic factors:

$$\Phi_2, \ \Phi_3, \ \Phi_5, \ \Phi_7, \ \Phi_{11}, \ \text{and } \Phi_{19}. \tag{21.5}$$

One can ask for which n do any of these appear as factors of R. To ask if Φ_m is a factor R is a question that depends only on $n \pmod{m}$, given that the p_i are fixed. For if $n_1 \equiv n_2 \pmod{m}$, then $z^{n_1} - 1 \equiv z^{n_2} - 1 \pmod{z^m - 1}$, and hence $z^{n_1} - 1 \equiv z^{n_2} - 1 \pmod{\Phi_m}$. Apart from Φ_{11}, the only troublesome Φ_m in the list (21.5) are with $m = p_i$; for these m one finds that we require $n \not\equiv 0 \pmod{p_i}$, which is something we were excluding anyway. The case Φ_{11} is more interesting: one finds that Φ_{11} is a factor of R (given our choice of p_i) if and only if $n \equiv 10 \pmod{11}$.

Exercise 21.18 Verify the above computation.

Having determined which values of n need to be excluded to prevent cyclotomic factors appearing in R, one can convert these to excluded values of d, using

$$n = 2d + 3 - p_1 - p_2 - p_3 - p_4 - p_5.$$

In our particular example, $p_1 + \cdots + p_5 = 36$, so that $2d = n + 33$. The restriction on $n \pmod 2$ is automatic for any d, but the other modular restrictions give $d \not\equiv 0 \pmod 3$, $d \not\equiv 4 \pmod 5$, $d \not\equiv 6 \pmod 7$, $d \not\equiv 5 \pmod{11}$ and $d \not\equiv 7 \pmod{19}$.

This example provides one of the lines of the following table, for the moment ignoring the final column. The reader may check that the other 13 examples of p_1, ..., p_5 lead to the stated modular restrictions on d.

$p_1, \ldots, p_5$					Excluded n (and $p_i \nmid n$)	Excluded d modulo:							Cumulative classes covered
						3	5	7	11	13	17	19	
2	3	5	7	11		2	0	2	7				2015520
2	3	5	7	13		0	1	3		7			3480648
2	3	5	7	17		2	3	5			7		4071624
2	3	5	7	19	10 mod 11	0	4	6	5			7	4497954
2	3	5	11	13		2	3		10	9			4575108
2	3	5	11	17	12 mod 13	1	0		1	4	9		4780638
2	3	5	11	19		2	1		2			9	4802832
2	3	5	13	17		2	1			12	10		4804881
2	3	5	17	19		2	4				13	12	4805171
2	3	7	11	13		0	5	6	0	10			4832519
2	3	7	11	19		0		2	3			10	4847940
2	3	7	13	17		0		2		0	11		4849604
2	3	7	13	19		1		3		1		11	4849833
2	3	7	17	19		0		5			14	13	4849845

For each choice of $p_1, \ldots, p_5$ in the above table, the restrictions on d rule out certain residue classes modulo $3 \times 5 \times 7 \times 11 \times 13 \times 17 \times 19 = 4849845$. The final column ('Cumulative classes covered') indicates how many residue classes modulo 4849845 have been covered by at least one of the choices up to that point. Thus, there are 2015520 values of d (mod 4849845) for which d is not excluded by the choice of p_i in the first row; in total there are 3480648 classes that are not excluded by at least one of the choices in the first two rows and so on. We see that including all 14 rows covers all possible residue classes modulo 4849845.

Taking any number d, there will therefore be at least one of the lines for which none of the excluded moduli for d come into force. Using that line, one may construct a counterexample to the Estes–Guralnick Conjecture using the p_i in the given row, with $n = 2d + 3 - \sum p_i$, provided this is greater than 1 (if $n = 1$ then the coprimeness conditions check out, but the formula for the trace of R is wrong).

For example, let us try $d = 315149$. This is excluded in the first line of the table ($d \equiv 2$ (mod 3)), but the second line works. Taking $(p_1, p_2, p_3, p_4, p_5) = (2, 3, 5, 7, 13)$ gives $n = 628291$, which is not divisible by any of 2, 3, 5, 7, 13.

If $d > 23$, then $2d > 46$, and for any of the lines of the table we have $2d + 3 - \sum p_i > 46 + 3 - 48 = 1$. This establishes the following Lemma.

Lemma 21.19 *For any degree $d > 23$ there is a monic irreducible polynomial $P(x) \in \mathbb{Z}[x]$ that has all its zeros real but such that $P(x)$ is not the minimal polynomial of any integer symmetric matrix.*

For $d = 19, 21, 22, 23$, we may use the following p_i:

d	p_1	p_2	p_3	p_4	p_5	n
19	2	3	5	7	11	13
21	2	3	5	7	11	17
22	2	3	5	7	11	19
23	2	3	5	7	13	19

21.6.4 All Together Now

We have seen that there are counterexamples for degrees 6–23. Together with Lemma 21.19 we have completed the proof of the main result of this chapter.

Theorem 21.20 *There are counterexamples to the conjecture of Estes and Guralnick for every degree greater than or equal to 6.*

Proof For degrees 6 and 7, see Sect. 21.3.3.
  For degrees 8–16, see Sect. 21.6.1.
  For degrees 17 and 18, see Sect. 21.4.
  For degree 19, see Sect. 21.6.3.
  For degree 20, see Sect. 21.6.2.
  For degrees 21–23, see Sect. 21.6.3.
  For degrees greater than 23, see Lemma 21.19.   □

21.7 Notes

The reciprocal polynomials used to establish Lemma 21.19 are in fact the minimal polynomials of Salem numbers of trace -2. These, along with some low-degree examples, were used in [MY14] to show that there are Salem numbers of degree $2d$ and trace -2 for every $d \geq 12$. In that paper, the only quintuples $(p_1, \ldots, p_5)$ that were permitted were those that led simply to the exclusions $p_i \nmid n$. By allowing other quintuples to appear, the number of quintuples on our covering set has been reduced from 15 in [MY14] to 14 here.

  The technique to exclude cyclotomic points to produce infinitely many examples of Salem numbers of trace -2 appeared first in [MS04].

21.8 Glossary

$\widetilde{P}$ If $P(z)$ is a monic integer reciprocal polynomial, not divisible by $z - 1$ or $z + 1$, then its zeros appear in reciprocal pairs: $\alpha_1, 1/\alpha_1, \alpha_2, 1/\alpha_2, \ldots, \alpha_d, 1/\alpha_d$. The degree of P here is $2d$. The polynomial $\widetilde{P}$ is the degree-d monic integer polynomial whose zeros are $\alpha_1 + 1/\alpha_1, \ldots, \alpha_d + 1/\alpha_d$. One may compute $\widetilde{P}$ by noting that its square is the resultant of $P(z)$ and $z^2 - xz + 1$ (with z as the variable).

$\Phi_n(z)$ The nth cyclotomic polynomial. Its zeros are the primitive nth roots of unity in $\mathbb{C}$. The degree of $\Phi_n(z)$ is $\varphi(n)$ (Euler's totient function). For each n, the polynomial $\Phi_n(z)$ is irreducible over $\mathbb{Q}$.

cyclotomic point. A point (x, y) on a curve for which both x and y are roots of unity.

Estes–Guralnick Conjecture. The (false) conjecture (Conjecture 21.2) that if a monic integer polynomial is totally real and separable then it is the minimal polynomial of an integer symmetric matrix. Estes and Guralnick proved this to be true for polynomials of degree up to 4. There are known counterexamples for polynomials of every degree strictly greater than 5. The degree-5 case remains open.

interlaced. Let $P(x)$ be a monic integer polynomial that has all zeros real; suppose its zeros are $\alpha_1 \leq \cdots \leq \alpha_n$. The polynomial P is said to be *interlaced* if there is a monic integer polynomial Q that has real zeros $\beta_1 \leq \cdots \leq \beta_{d-1}$ that interlace with those of P:

$$\alpha_1 \leq \beta_1 \leq \alpha_2 \leq \beta_2 \leq \cdots \leq \alpha_{d-1} \leq \beta_{d-1} \leq \alpha_d .$$

Note the requirement that Q is monic. There exist monic integer polynomials P with all zeros real that are *not* interlaced. Indeed Yatsyna showed that there are infinitely many such: all these polynomials are necessarily counterexamples to the Estes–Guralnick Conjecture.

Chapter 22
Breaking Symmetry

We have seen how powerful interlacing is as a tool for understanding small Mahler measures of integer symmetric matrices (Chaps. 6, 7, 9, 13) and symmetrizable integer matrices (Chaps. 17, 18, 19). The power of interlacing in this context stems from the fact that we can 'grow' matrices from smaller ones that have smaller Mahler measure (or at least not larger). The results are both impressive yet disappointing: we can prove the analogue of Lehmer's Conjecture, we can settle the trace problem, but we can produce only a tiny number of small Mahler measures (8 out of 236 known below 1.25 (Table D.1), 16 out of more than 8000 known below 1.3 that have degree at most 180 [Mos11]).

Recently [CM21] some explorations by Coyston and McKee have been made that break the symmetry (and break symmetrizability). Interlacing is lost, but by restricting to special classes of digraphs explicit formulas can be exploited, allowing a restricted form of growing. Nearly all the known small Mahler measures have been found (no doubt with more to come). At the time of writing, 235 of the 236 known Mahler measures below 1.25 have been found via digraphs, and 8336 below 1.3 with degree at most 180, including one that had not been found before.

The digraphs used lie in families whose reciprocal polynomials are of the shape studied in Chap. 2, namely polynomials in z, z^{m_1}, ... z^{m_r} for some r and some parameters $m_1, \ldots, m_r$. Allowing the m_i to grow in a variety of ways, restricting to one-dimensional families, 57 of the 61 small limit points in Table D.2 are found.

The approach adopted in [CM21] was to take a simple family of cyclotomic charged signed graphs and then change one or two entries in the adjacency matrix to break the symmetry, and indeed to break sign symmetry too so that the resulting matrix would not be symmetrizable. The new matrix is nearly cyclotomic in a combinatorial sense of nearness, and might be a fruitful object of study for finding small Mahler measures. We content ourselves here with illustrating the approach by an example. Letting

© Springer Nature Switzerland AG 2021
J. McKee and C. Smyth, *Around the Unit Circle*, Universitext,
https://doi.org/10.1007/978-3-030-80031-4_22

indicate an edge that has been subdivided by the addition of t vertices, the digraph

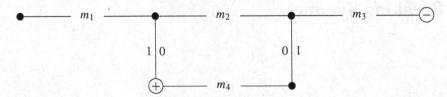

is obtained from the union of two cyclotomic paths (subgraphs of the cylindrical tessellations of Chap. 6), with two entries of the adjacency matrix then changed from 0 to 1 in an asymmetric manner. Here there are four growing paths. Adapting a deletion-contraction theorem of Rowlinson [Row87] to the digraph setting, one can show that the reciprocal polynomial of this digraph, after scaling by factors that have Mahler measure 1, is of the shape $Q(z, z^{m_1}, \ldots, z^{m_4})$ for some integer polynomial Q. If the m_i grow to infinity in a generic manner, then one gets a sequence of Mahler measures that converge to the 5-variable Mahler measure of $Q(z, z_1, \ldots, z_4)$. Instead, one can restrict the m_i to satisfy any number of linear constraints and thereby produce different limits of measures. The approach in [CM21] was to constrain the vectors $(m_1, \ldots, m_4)$ to lie on 'good' lines along which the Mahler measures were relatively small. In this manner (and using a small selection of nearly cyclotomic families of digraphs) many thousands of Mahler measures below 1.3 were found for polynomials of degree up to 180, including one that had not been seen before, and from the limits of such sequences 57 of the 61 limit points in Table D.2 were produced.

Problem 22.1 *(research problem)* Find all the 61 limit points in Table D.2 by following this approach (the ones missed by [CM21] are numbers 21, 34, 39 and 58 in that table), and find all the 236 measures in Table D.1 (the one missed by [CM21] is the third smallest of degree 48, number 64 in Table D.1).

Problem 22.2 *(research problem)* For some manageable class of digraphs, carry out part or all of the programme successfully accomplished for integer symmetric matrices by considering the following questions. What are the cyclotomic examples? What are the minimal noncyclotomic examples? What is the spectrum of smallest measures?

Appendix A
Algebraic Background

A.1   Self-Reciprocal Polynomials

For a *real* polynomial $P(z)$ of degree d, its **reciprocal polynomial** is the polynomial $z^d P(1/z)$, denoted $P^*(z)$. Also, $P(z)$ is called **self-reciprocal**, or simply **reciprocal**, if $P^*(z) = c_P P(z)$ identically, for some constant c_P. A polynomial is called **nonreciprocal** if it is not self-reciprocal.

More generally, a real polynomial $F(z_1, \ldots, z_k)$ in k variables is called **self-reciprocal** if the quotient $F(z_1, \ldots, z_k)/F(z_1^{-1}, \ldots, z_k^{-1})$ is a monomial in $z_1, \ldots, z_k$. The following easy exercises describe self-reciprocal polynomials in more detail.

Exercise A.1 Show that if $P(z) = a \prod_{i=1}^{d}(z - \alpha_i)$, then $P^*(z) = a \prod_{i=1}^{d}(1 - \alpha_i z)$.

Exercise A.2 Show that if $P(z)$ has degree d, and $z^r \parallel P(z)$, then $P^*(z)$ has degree $d - r$.

Exercise A.3 Show that, for $P(z)$ self-reciprocal, the constant c_P equals 1 unless one has $(z - 1)^e \parallel P(z)$ with e odd, in which case $c_P = -1$.

If $c_P = 1$ then the coefficients of $P(z)$ form a *palindromic sequence*: they read the same backwards as forwards.

Exercise A.4 Show that, for $P(z)$ self-reciprocal, its zeros are a multiset union of the sets $\{1\}$, $\{-1\}$, $\{\omega, -\omega\}$ for ω a zero of modulus 1 with $\omega^2 \neq 1$, and sets $\{\alpha, \overline{\alpha}, 1/\alpha, 1/\overline{\alpha}\}$ for α a zero of modulus not equal to 1 or 0.

Exercise A.5 Suppose that $P(z) \in \mathbb{Q}[z]$ is self-reciprocal, monic and irreducible over $\mathbb{Q}$. Show that either $P(z) = z - 1$, $P(z) = z + 1$ or $P(z)$ is of even degree with palindromic coefficient sequence.

Exercise A.6 Show that for any given real polynomial $\prod_{i=1}^{d}(z - \alpha_i)$ the polynomial
$$\prod_{\substack{1 \le i \le d: \\ |\alpha_i| = 1}} (z - \alpha_i)$$
is real and self-reciprocal.

© Springer Nature Switzerland AG 2021
J. McKee and C. Smyth, *Around the Unit Circle*, Universitext,
https://doi.org/10.1007/978-3-030-80031-4

Before our next result, we need the following.

Exercise A.7 Suppose that we are given $\varepsilon > 0$ and a polynomial $f(z) + t R(t, z) \in$ $\mathbb{C}[z, t]$ such that, for fixed t real with $0 < t < \varepsilon$, the polynomial has exactly h zeros in $|z| < 1$. Using Rouché's Theorem (Theorem C.1), show that $f(z)$ has at most h zeros in $|z| < 1$.

The main result of this section is as follows.

Theorem A.8 *Suppose that $P(z)$ is real and self-reciprocal. Then $P(z)$ and $P'(z)$ have the same number of zeros outside the unit circle.*

Proof Take $P(z)$ of degree d to be self-reciprocal. From differentiating $z^d P(1/z) = c_P P(z)$ and using $c_P^2 = 1$, we readily obtain

$$d\, P(z) = z P'(z) + c_P (P')^*(z). \tag{A.1}$$

Let n be the number of zeros of $P(z)$ in $|z| > 1$, and n' be the number of zeros of $(P')^*(z)$ in $|z| < 1$. We need to prove that $n = n'$. For then both $P(z)$ and $P'(z)$ have n zeros in $|z| > 1$.

We first show that $n \le n'$. Indeed, on $|z| = 1$ we have

$$|z P'(z)| = |P'(z)| = |(P')^*(z)| = |c_P (P')^*(z)|.$$

By Rouché's Theorem (Theorem C.1)

$$c_P (P')^*(z) + (1 - t) z P'(z)$$

has n' zeros in $|z| < 1$. But, using (A.1), this simplifies to $d P(z) - t z P'(z)$ which, by Exercise A.7 with $\varepsilon = 1$, has at most n zeros in $|z| < 1$. So $n \le n'$.

To complete the proof, let us show that $n' \le n$. We choose ε so that for $0 < t < \varepsilon$ the polynomial $f(z) := P((1 - t)z)$ has the same number n of zeros in $|z| < 1$ as $P(z)$ (see Exercise A.9 below). Then, because $P(z)$ is self-reciprocal, we have

$$c_P f^*(z) = c_P z^d P((1 - t)/z) = (1 - t)^d P(z/(1 - t)).$$

Then, by Rouché's Theorem, we see that the polynomial

$$\begin{aligned} H(t, z) :&= f(z) - (1 - t)^d c_P f^*(z) \\ &= P((1 - t)z) - (1 - t)^{2d} P(z/(1 - t)) \\ &= t \frac{\partial H}{\partial t}(0, z) + t^2 R(t, z) \end{aligned}$$

say, has n zeros in $|z| < 1$. Hence so does

$$\frac{1}{t} H(t, z) := \frac{\partial H}{\partial t}(0, z) + t R(t, z)$$
$$= (-2z P'(z) + 2d\, P(z)) + t R(t, z)$$
$$= 2c_P (P')^*(z) + t R(t, z),$$

using (A.1). Then, Exercise A.7 shows that $n' \le n$. Hence $n = n'$. □

Exercise A.9 Show that ε in the above proof can be chosen to be $1 - \max_{|\alpha| < 1} |\alpha|$, where the maximum is taken over all zeros α of $P(z)$ that lie in $|z| < 1$.

Exercise A.10 Suppose that $P(z)$ is self-reciprocal of even degree $d = 2d'$. Show that $P(z)$ can be written in the form

$$P(z) = z^{d'} Q(z + z^{-1})$$

for some polynomial $Q(x)$. Show too that $Q(x)^2 = \text{res}_z(P(z), z^2 - xz + 1)$. (See Sect. A.2 immediately below for the definition of $\text{res}_z(P, Q)$.)

A.2 Resultant Essentials

Given two real polynomials,

$$P(z) = p_0 z^d + p_1 z^{d-1} + \cdots + p_d = p_0 \prod_{j=1}^{d} (z - \alpha_j)$$

$$Q(z) = q_0 z^e + q_1 z^{e-1} + \cdots + q_e = q_0 \prod_{j=1}^{e} (z - \beta_k)$$

say, of degrees d, e, respectively, we take a polynomial $A(z) = a_0 z^{e-1} + a_1 z^{e-2} + \cdots + a_{e-1}$ of degree at most $e - 1$ and a polynomial $B(z) = b_0 z^{d-1} + b_1 z^{d-2} + \cdots + b_{d-1}$ of degree at most $d - 1$, and define the linear map $R_{P,Q} : \mathbb{R}^{d+e} \to \mathbb{R}^{d+e}$ which takes $a_0, \ldots, a_{e-1}, b_0, \ldots, b_{d-1}$ to the coefficients of $z^{d+e-1}, z^{d+e-2}, \ldots, z, 1$ of $AP + BQ$. Then

$$\text{res}_z(P, Q) := \det(R_{P,Q}) = \begin{vmatrix} p_0 & 0 & \cdots & 0 & q_0 & 0 & \cdots & 0 \\ p_1 & p_0 & \cdots & 0 & q_1 & q_0 & \cdots & 0 \\ p_2 & p_1 & \ddots & 0 & q_2 & q_1 & \ddots & 0 \\ \vdots & \vdots & \ddots & p_0 & \vdots & \vdots & \ddots & q_0 \\ p_d & p_{d-1} & \cdots & \vdots & q_e & q_{e-1} & \cdots & \vdots \\ 0 & p_d & \ddots & \vdots & 0 & q_e & \ddots & \vdots \\ \vdots & \vdots & \ddots & p_{d-1} & \vdots & \vdots & \ddots & q_{e-1} \\ 0 & 0 & \cdots & p_d & 0 & 0 & \cdots & q_e \end{vmatrix}$$

is called the *(Sylvester) resultant* of P and Q. If $\mathrm{res}_z(P, Q) = 0$ then we can choose nonzero polynomials A and B in z such that $AP + BQ = 0$. Since $\deg A < \deg Q$ it follows that P and Q must have a nonconstant polynomial factor. On the other hand, if $\mathrm{res}_z(P, Q) \neq 0$ then by Cramer's Rule (Exercise A.11) there are polynomials A and B such that $AP + BQ = \mathrm{res}_z(P, Q)$. Further, if P and Q have integer coefficients, then so do A and B.

We shall need the following standard property of $\mathrm{res}_z(P, Q)$:

$$\mathrm{res}_z(P, Q) = p_0^e q_0^d \prod_{\substack{1 \le j \le d \\ 1 \le k \le e}} (\alpha_j - \beta_k) = p_0^e \prod_{j=1}^{d} Q(\alpha_j) = (-1)^{de} q_0^d \prod_{k=1}^{e} P(\beta_k).$$

(A.2)

In particular, note that if P and Q both have integer coefficients then $\mathrm{res}_z(P, Q)$ is an integer, and furthermore if P and Q have no common zero then $\mathrm{res}_z(P, Q) \neq 0$.

The *discriminant* Δ_P *of* P is defined as $\mathrm{res}_z(P, P')$. Thus from (A.2) we have that

$$\Delta(P) = p_0^{d-1} \prod_{j=1}^{d} P'(\alpha_j).$$

(A.3)

Exercise A.11 Prove *Cramer's Rule*: Let A be an $n \times n$ matrix with entries in a field F, and $\mathbf{c} \in F^n$. Then if $\det A \neq 0$ the equation $A\mathbf{x} = \mathbf{c}$ has a unique solution $\mathbf{x} \in F^n$, where $(\det A)\mathbf{x} = (D_1, D_2, \ldots, D_n)^\mathsf{T}$, where D_j is the determinant of A but with its jth column replaced by $\mathbf{c}$.

Exercise A.12 Given polynomials P, Q, A, B and C with $\deg C < \deg P + \deg Q$, and an identity $AP + BQ = C$, show how A and B can be modified so that $\deg A < \deg Q$ and $\deg B < \deg P$ while $AP + BQ = C$ still holds.

Exercise A.13 Let $P(x) = x^2 + 2$ and $Q(x) = x^2 + 2x + 2$. Show that $\mathrm{res}_x(P, Q) = 8$ but that there are polynomials $A(x)$, $B(x)$ with integer coefficients such that $AP + BQ = 4$.

Exercise A.14 For given coprime $P, Q \in \mathbb{Z}[x]$, an alternative definition of their resultant (Bézout resultant) is the smallest positive integer r such that $AP + BQ = r$ for some $A, B \in \mathbb{Z}[x]$. Show that this r divides the Sylvester resultant. (Clearly r can be computed using Cramer's Rule.)

A.3   Valuation Essentials

A *valuation* of a field K is a mapping of K to $\mathbb{R}_{\ge 0}$ that takes $x \in K$ to its valuation $|x| \in \mathbb{R}_{\ge 0}$, and which has the following properties:

- $|x| = 0$ for $x \in K$ if and only if $x = 0$;
- $|xx'| = |x| \cdot |x'|$ for $x, x' \in K$;

- $|x + x'| \leq |x| + |x'|$ for $x, x' \in K$.

If a valuation also satisfies

$$|x + x'| \leq \max(|x|, |x'|) \text{ for } x, x' \in K, \qquad (A.4)$$

then it is called a ***nonarchimedean*** valuation. Otherwise, it is called an ***archimedean*** valuation. A nonarchimedean valuation has the nice property that

$$\text{if } |x| < |x'| \text{ then } |x + x'| = |x'|. \qquad (A.5)$$

The valuations of the field $\mathbb{Q}$ are the following:

- the standard absolute value $|x|$, sometimes written $|x|_\infty$ (an archimedean valuation);
- for every prime p the nonarchimedean valuation $|x|_p$ which for $x = \pm p^k r/s \in \mathbb{Q}$ with $\gcd(rs, p) = 1$, has $|x|_p = p^{-k}$.

Every nonzero $a \in \mathbb{Q}$ satisfies the following product rule:

$$|a| \cdot \prod_{\text{all primes } p} |a|_p = 1. \qquad (A.6)$$

This special case of Proposition A.15 is easily seen by factorising $x = \pm \prod_p p^{k_p}$ and using the valuation definitions.

One can use each valuation of $\mathbb{Q}$ to define Cauchy sequences in $\mathbb{Q}$, with respect to that valuation. Equivalence classes of Cauchy sequences (those whose difference tends to 0 with respect to that valuation) then define the ***completion*** ($\mathbb{Q}_\infty = \mathbb{R}$ or $\mathbb{Q}_p$) of $\mathbb{Q}$ with respect to that valuation. These are the smallest fields containing $\mathbb{Q}$ for which all Cauchy sequences tend to a limit in the field.

For a number field K, regarded as a finite extension $K = \mathbb{Q}(\alpha)$, the valuations of K are divided into infinitely many finite sets, each finite set 'lying above' a valuation of $\mathbb{Q}$. To find the valuations lying above $|\ |_p$, factorise the minimal polynomial of α over $\mathbb{Q}$ into irreducible factors (assumed monic) over $\mathbb{Q}_p$ (where p, for this paragraph only, might be ∞). Each such factor, $f_\mathfrak{p}(z)$ say, defines a finite extension $\mathbb{Q}_\mathfrak{p} = \mathbb{Q}_p[z]/(f_\mathfrak{p}(z))$ of $\mathbb{Q}_p$ that contains (an isomorphic copy) of K. For $\beta \in \mathbb{Q}_\mathfrak{p}$, take its minimal polynomial, $f_{\beta,\mathfrak{p}}(z)$ say, over $\mathbb{Q}_p$, and define $|\beta|_\mathfrak{p} := |f_{\beta,\mathfrak{p}}(0)|_p^{1/\deg f_{\beta,\mathfrak{p}}}$. This defines a valuation $|\ |_\mathfrak{p}$ on $\mathbb{Q}_\mathfrak{p}$, and hence on K. For $\alpha \in K$, the minimal polynomial $f_\alpha(z)$ over $\mathbb{Q}$ factorises over $\mathbb{Q}_p$ as $f_\alpha(z) = \prod_\mathfrak{p} f_{\alpha,\mathfrak{p}}(z)$. Then from $f_\alpha(0) = \prod_\mathfrak{p} f_{\alpha,\mathfrak{p}}(0)$, the product rule (A.6) for $f_\alpha(0)$ and the definition of $|\alpha|_\mathfrak{p}$ we have that

$$1 = \prod_p |f_\alpha(0)|_p = \prod_p \prod_{\mathfrak{p} \text{ above } p} |f_{\alpha,\mathfrak{p}}(0)|_p = \prod_p \prod_{\mathfrak{p} \text{ above } p} |\alpha|_\mathfrak{p}^{\deg f_{\alpha,\mathfrak{p}}}. \qquad (A.7)$$

This is one form of the Product Rule for $\alpha \in K$. Again we stress that here, and here only, $p = \infty$ is included. Generally, we use p only for finite primes.

An alternative, equivalent approach to valuations of algebraic numbers α is the following. For p prime let $\overline{\mathbb{Q}}_p$ be an algebraic closure of $\mathbb{Q}_p$. Extend the valuation $|\ |_p$ to the whole of $\overline{\mathbb{Q}}_p$. Let $\alpha = \alpha_1, \alpha_2, \ldots, \alpha_d$ be the conjugates of α, and embed them into $\overline{\mathbb{Q}}_p$. Then, using the Newton polygon (see below), we obtain d valuations $|\alpha_i|_p$ for each prime p, and d valuations $|\alpha_i|$ by embedding the α_i into $\mathbb{C}$. Then from the fact that $\prod_i \alpha_i \in \mathbb{Q}$ we obtain from (A.6) the following.

Proposition A.15 (The Product Rule) *For every nonzero algebraic number α, we have*

$$\prod_i \left(|\alpha_i| \prod_p |\alpha_i|_p \right) = 1, \qquad\qquad (A.8)$$

the products being taken over all primes p and all conjugates α_i of α.

Of course the valuations $|\alpha_i|$ and $|\alpha_j|$ are the same if $\alpha_j = \overline{\alpha_i}$. Similarly, the valuations $|\alpha_i|_p$ and $|\alpha_j|_p$ are the same if α_i and α_j are conjugate over $\mathbb{Q}_p$. This shows the equivalence of (A.7) and (A.8).

Proposition A.16 *Suppose that α is an algebraic number with conjugates $\alpha = \alpha_1, \alpha_2, \ldots, \alpha_d$. If for $j = 1, \ldots, d$ we have $|\alpha_j| = 1$ and $|\alpha_j|_p = 1$ for all primes p then α is a root of unity.*

Proof For every prime p all coefficients a_i of the minimal polynomial of α are sums of products of the α_j, so have $|a_i|_p \leq 1$, and therefore are all integers. Hence, the result follows from Kronecker's First Theorem (Theorem 1.3). $\qquad\square$

With the help of the Product Rule, we have the following corollary.

Corollary A.17 *Every algebraic number α, not a root of unity, has a conjugate, α_j say, for which $|\alpha_j|_p > 1$ for some prime p, or for $p = \infty$.*

The Newton polygon enables us to read off the p-adic valuations of the zeros of a polynomial with coefficients in $\mathbb{Q}_p$.

Proposition A.18 (The Newton polygon) *Let p be a prime. Suppose that we are given a sequence of r integer powers of p, $v_1 > v_2 > \cdots > v_r > 0$, a sequence of integers $0 = k_0 < k_1 < \cdots < k_r = d$ and a polynomial $P(z) \in \mathbb{Q}_p[z]$ given by*

$$P(z) = z^d + a_1 z^{d-1} + \cdots + a_d = \prod_{j=1}^d (z - \alpha_j),$$

where for $j = 1, \ldots, d$ we have $|\alpha_j|_p = v_\ell$ if $k_{\ell-1} < j \leq k_\ell$. Then

$$|a_{k_\ell}|_p = \prod_{j \leq k_\ell} |\alpha_j|_p$$

and if $k_{\ell-1} < i \le k_\ell$ *then*

$$|a_i|_p \le |a_{k_{\ell-1}}|_p v_\ell^{i-k_{\ell-1}}.$$

The proof just uses the properties (A.4) and (A.5) of the p-adic valuation. The valuations $|\alpha_j|_p$ are then readily deduced from the $|a_i|_p$, as follows. Define the *Newton polygon of P* to be the convex hull of the points $(i, \log |a_i|_p)$ $(i = 0, \ldots, d)$. The extreme points are the points $(k_\ell, \log |a_{k_\ell}|_p)$ $(\ell = 0, \ldots, r)$, and there are $k_\ell - k_{\ell-1}$ zeros α_j of P having p-adic valuation $(|a_{k_\ell}|_p/|a_{k_{\ell-1}}|_p)^{1/(k_\ell - k_{\ell-1})}$. This expression equals v_ℓ.

Exercise A.19 Show that if α is an algebraic number that is not an algebraic integer then $|\alpha'|_p > 1$ for some conjugate α' of α and some prime p. (Use the Newton polygon of the minimal polynomial of α.)

A.4   Galois Theory Essentials

In this section, we record basic results from Galois theory that we need (see, for example, [DF04, Chap. 14]). Let K be a normal extension of $\mathbb{Q}$, which means an extension $K = \mathbb{Q}(\beta_1, \ldots, \beta_n)$ of $\mathbb{Q}$ generated by some algebraic number β_1 and its (Galois) conjugates $\beta_1, \ldots, \beta_n$. The field automorphisms of K that fix $\mathbb{Q}$ pointwise form a group $\mathrm{Gal}(K/\mathbb{Q})$ under functional composition. For any algebraic number $\alpha \in K$, its conjugates α_i also lie in K. Then $\mathrm{Gal}(K/\mathbb{Q})$ acts transitively on the set of α_i: that is, for any given indices i, j there is an automorphism of $\mathrm{Gal}(K/\mathbb{Q})$ that maps α_i to α_j. Because of this fact, the only elements of K that every automorphism of $\mathrm{Gal}(K/\mathbb{Q})$ fixes pointwise are those with only one conjugate, namely the elements of $\mathbb{Q}$.

Now every automorphism in $\mathrm{Gal}(K/\mathbb{Q})$ is determined by its effect on $\beta_1, \ldots, \beta_n$, and so it is isomorphic to a subgroup of $\mathfrak{S}_n$, the full symmetric group of permutations of n elements.

A *symmetric polynomial* of $z_1, \ldots, z_n$ is a polynomial with rational coefficients that is fixed pointwise by all elements of $\mathfrak{S}_n$ acting on $z_1, \ldots, z_n$. Now a symmetric polynomial of $\beta_1, \ldots, \beta_n$ will certainly be fixed pointwise by all automorphisms in $\mathrm{Gal}(K/\mathbb{Q})$, and so must belong to $\mathbb{Q}$. Thus we have the following.

Proposition A.20 *If two monic integer polynomials P and Q have no common zero then their resultant is at least 1 in modulus.*

For, by (A.2), their resultant is a symmetric function of the zeros of P, and so is a nonzero rational number. However, it is also an algebraic integer, so is a nonzero element of $\mathbb{Z}$.

If L is a finite extension of $\mathbb{Q}$, then its *normal closure* is the smallest normal extension of $\mathbb{Q}$ containing L.

A.5   Algebraic Numbers and Algebraic Integers

Proposition A.21 (The companion matrix of an algebraic number) *Let α be an algebraic number having minimal polynomial $z^d + a_1 z^{d-1} + \cdots + a_d$. Then, α is an eigenvalue of the $d \times d$ matrix*

$$C := \begin{pmatrix} 0 & 1 & 0 & 0 & \cdots & 0 \\ 0 & 0 & 1 & 0 & \cdots & 0 \\ 0 & 0 & 0 & 1 & \cdots & 0 \\ \vdots & \vdots & \vdots & \vdots & \cdots & \vdots \\ 0 & 0 & 0 & 0 & \cdots & 1 \\ -a_d & -a_{d-1} & -a_{d-2} & -a_{d-3} & \cdots & -a_1 \end{pmatrix}.$$

*The matrix C is the **companion matrix** of α.*

This follows immediately from the easily seen fact that $(1, \alpha, \ldots, \alpha^{d-1})^{\mathsf{T}}$ is an eigenvector of C with eigenvalue α.

Proposition A.22 *The algebraic integers form a ring.*

Proof Suppose that α is an algebraic integer of degree a and β is an algebraic integer of degree b. Let C_α be the companion matrix of α and C_β be the companion matrix of β; see Proposition A.21. Let I_a be the $a \times a$ identity matrix and I_b be the $b \times b$ identity matrix. Then, the tensor product matrix $C_\alpha \otimes C_\beta$ (see [DF04, Sect. 11.2]) has $\alpha\beta$ as an eigenvalue, and the matrix $C_\alpha \otimes I_b + I_a \otimes C_\beta$ has $\alpha + \beta$ as an eigenvalue. Both are, of course, integer matrices, so both $\alpha\beta$ and $\alpha + \beta$ are zeros of monic polynomials with integer coefficients. $\square$

Lemma A.23 (Dobrowolski's Lemma [Dob79]) *Let p be a prime number, and α be an algebraic integer with conjugates $\alpha_1 = \alpha, \alpha_2, \ldots, \alpha_d$. Then, $\prod_{i,j=1}^{d}(\alpha_i^p - \alpha_j)$ is an integer multiple of p^d.*

Proof Let $f(z) \in \mathbb{Z}(z)$ be the minimal polynomial of α. Then (by the Multinomial Theorem and Fermat's Little Theorem) $(f(z))^p = f(z^p) - pf_p(z)$, where $f_p(z)$ is some polynomial having integer coefficients. Since $f(\alpha_i) = 0$ we have $\prod_{i=1}^{d} f(\alpha_i^p) = p^d \prod_{i=1}^{d} f_p(\alpha_i)$. Note that $f_p(\alpha_i)$ is an algebraic integer. Thus, this latter product is also an algebraic integer, and, because it is a symmetric function of the α_i, it is a rational integer. $\square$

Corollary A.24 *If α in Lemma A.23 is neither 0 nor a root of unity then*

$$\left| \prod_{i,j=1}^{d} (\alpha_i^p - \alpha_j) \right| \geq p^d.$$

Exercise A.25 More generally, for p and α as in Lemma A.23, and integers $\ell > m \geq 0$ show that $\prod_{i,j=1}^{d}(\alpha_i^{p^\ell} - \alpha_j^{p^m})$ is an integer multiple of $p^{(m+1)d}$.

Complementing Dobrowolski's Lemma is a related result for algebraic numbers.

Lemma A.26 *Let α_1 and α_2 be nonzero conjugate algebraic numbers, and $r > s$ be positive integers, with $\alpha_1^r = \alpha_2^s$. Then α_1 is a root of unity.*

Proof Suppose first that α_1 is a non-integer. Then from Exercise A.19 there is some prime p such that α_1 has a conjugate, α_3 say, with $|\alpha_3|_p > 1$. Let α_4 be a conjugate of α_1 with $|\alpha_4|_p$ maximal. Thus, the number $\alpha_1^r - \alpha_2^s$ has a conjugate, $\alpha_4^r - \alpha_5^s$ say, for some conjugate α_5 of α_1. But then $|\alpha_4^r|_p > |\alpha_5^s|_p$, so that $\alpha_4^r - \alpha_5^s$, and hence $\alpha_1^r - \alpha_2^s$, is nonzero; a contradiction.

Knowing now that α_1 is an algebraic integer, we apply the above argument with the usual absolute value on $\mathbb{C}$, under the assumption that α_1 has a conjugate α_4 of maximal modulus $|\alpha_4| > 1$. This again gives a contradiction. Hence all conjugates of α_1 have modulus 1, and so, by Kronecker's First Theorem (Theorem 1.3), α_1 is a root of unity. □

For any algebraic number α, we define the **mean trace** $\overline{\mathrm{tr}}(\alpha)$ of α by

$$\overline{\mathrm{tr}}(\alpha) := (\mathrm{trace}\,(\alpha))/[\mathbb{Q}(\alpha) : \mathbb{Q}] \tag{A.9}$$

This is the mean of the conjugates of α. We need the following basic property of the mean trace.

Lemma A.27 *For any algebraic numbers α, γ, we have*

$$\overline{\mathrm{tr}}(\alpha + \gamma) = \overline{\mathrm{tr}}(\alpha) + \overline{\mathrm{tr}}(\gamma). \tag{A.10}$$

Proof Let F be the normal closure of $\mathbb{Q}(\alpha, \gamma)$. Then

$$\overline{\mathrm{tr}}(\alpha) = \frac{1}{[F : \mathbb{Q}]} \sum_{\sigma \in \mathrm{Gal}(F/\mathbb{Q})} \sigma(\alpha),$$

from which, using the corresponding formula for γ and for $\alpha + \gamma$, the result follows. □

A.5.1 The Gorškov Polynomials

We now define the **Gorškov polynomials** [Gor59] $G_n(x)$ by $G_0(x) = x - 1$ and, for $n \geq 1$,

$$G_n(x) = x^{2^{n-1}} G_{n-1}\left(x + \frac{1}{x} - 2\right). \tag{A.11}$$

Also let $\beta^{(0)} = 1$, and define $\beta^{(n)} > 1$ for $n \geq 1$ by

$$\beta^{(n)} + \frac{1}{\beta^{(n)}} = 2 + \beta^{(n-1)}. \tag{A.12}$$

Thus, $\beta^{(n)}$ is a zero of $G_n(z)$. Also, all conjugates of $\beta^{(n)}$ are real and positive.

Proposition A.28 (Irreducibility of the Gorškov polynomials) *For $n \geq 0$ the polynomial $G_n(x)$ is irreducible of degree 2^n.*

Thus, we immediately see that $\beta^{(n)}$ has degree 2^n.

Proof The proof is essentially due to Wirsing—see [Mon94, Th. 4, p. 187]. We can clearly assume that $n \geq 2$. Assume now that for some $n \geq 2$ the polynomial G_n is reducible. Take n to be the smallest integer with this property. So $G_{n-1}(x)$ is irreducible, and $\beta^{(n-1)}$ has degree 2^{n-1} over $\mathbb{Q}$. From (A.12) and the reducibility of G_n we have that $\deg \beta^{(n)} = 2^{n-1}$. Thus if $\beta^{(n)}$ were conjugate to $1/\beta^{(n)}$ we would have (again from (A.12)) that $\deg \beta^{(n-1)} = 2^{n-2}$, a contradiction. So $G_n(x)$ divides into irreducible factors $G_n(x) = r(x)r(0)r^*(x)$, where $r^*(x) := x^{2^{n-1}} r(1/x)$ is the reciprocal polynomial of $r(x)$. The term $r(0) = \pm 1$ is included to make $r(x)r(0)r^*(x)$ monic. In fact, as r has even degree and no negative zeros, we must have $r(0) = 1$. Hence

$$G_n(-1) = r(-1)r^*(-1) = r(-1)^2(-1)^{2^{n-1}} \equiv 0 \text{ or } 1 \pmod{3}.$$

On the other hand, from (A.11),

$$G_n(-1) = (-1)^{2^{n-1}} G_{n-1}(-4)$$
$$\equiv G_{n-1}(-1) \equiv G_{n-2}(-1) \equiv \cdots \equiv G_1(-1) \equiv 2 \pmod{3}.$$

Thus such an n cannot exist. ∎

To study the algebraic integers $\beta^{(n)}$ in more detail, it is helpful to consider $\gamma^{(n)} := \sqrt{\beta^{(n)}}$, as in the following exercise.

Exercise A.29 Show that $\gamma^{(0)} = 1$ and for $n \geq 1$ that $\gamma^{(n)} - 1/\gamma^{(n)} = \gamma^{(n-1)}$. Prove that $\gamma^{(n)}$ has degree 2^n (the same as that of $\beta^{(n)}$) and that when the conjugates of $\gamma^{(n)}$ are arranged in ascending order of magnitude they alternate in sign, with $-1/\gamma^{(n)}$ being the first, and $\gamma^{(n)}$ the last.

Thus, on defining

$$H(x) = x - \frac{1}{x}, \tag{A.13}$$

we have $H^n(\gamma^{(n)}) = 1$, where H^n denotes the n-fold iterate of H.

Let μ_n denote the atomic measure on $\mathbb{R}$ that has weight 2^{-n} at each conjugate of $\gamma^{(n)}$. Then, the weak limit $\mu = \lim_{n \to \infty} \mu_n$ defined by $\mu(A) = \lim_{n \to \infty} \mu_n(A)$

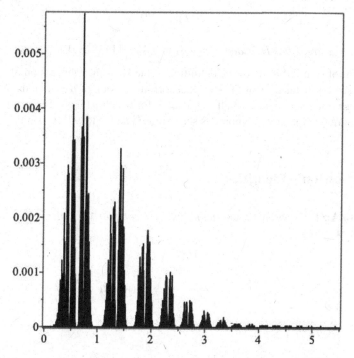

Fig. A.1 The density function of the moduli of the 32768 conjugates of $\sqrt{\beta^{(15)}} = 5.4297\cdots$. These values have been plotted as a bar graph of the number of such values in the intervals formed by dividing $[0, 5.5]$ into 10000 subintervals of equal length

for any Borel subset $A \subset \mathbb{R}$ exists and is H-invariant in the sense that when $H|_A$ is one-to-one then

$$\mu(H(A)) = 2\mu(A). \tag{A.14}$$

Because of the 'alternating signs' phenomenon in Exercise A.29, the signs of the conjugates of $\gamma^{(n)}$ are known, so we can concentrate on their moduli. The graph in Fig. A.1 shows the atomic measure defined by the moduli of the conjugates of $\gamma^{(15)} = \sqrt{\beta^{(15)}}$. Modulo signs, it approximates the limit measure μ, whose highly irregular distribution is explained by the following results of Davie and Smyth.

Theorem A.30 ([DS89])

(a) For any conjugate γ' of $\gamma^{(n)}$ and sufficiently small neighbourhood I of γ' of length $|I|$ there is a constant $c_1 = c_1(\gamma', n) > 0$ independent of I such that

$$\mu(I) < e^{-c_1/|I|^2}. \tag{A.15}$$

(b) For any root a of $H^n(x) = x$ and sufficiently small neighbourhood I of a there is a constant $c_2 = c_2(a, n) > 0$ independent of I such that

$$\mu(I) > |I|^{c_2}. \qquad (A.16)$$

(c) The measure μ has Hausdorff dimension $0.800611138269168784\cdots$.

The proof is in [DS89]. For the definition of the Hausdorff dimension of a subset E of $\mathbb{R}$, see for instance [Fal03]. The Hausdorff dimension of μ is then the infimum of the Hausdorff dimensions of all sets $E \subset \mathbb{R}$ for which $\mu(E) = 1$.

For more on Gorškov polynomials see [Mon94] and [FRS97, Appendix].

A.6   Newton's Identities

Theorem A.31 (Newton's Identities) *Take any complex variables $\alpha_1, \ldots, \alpha_d$, and define for $k \geq 0$*

$$s_k := \alpha_1^k + \cdots + \alpha_d^k,$$

$$t_k := (-1)^k \sum_{1 \leq i_1 < \cdots < i_k \leq d} \alpha_{i_1} \cdots \alpha_{i_k}, \qquad (\text{with } t_k = 0 \text{ if } k > d).$$

Then, for each $k \geq 1$, we have

$$s_k + \sum_{r=1}^{k-1} t_r s_{k-r} + k t_k = 0. \qquad (A.17)$$

Proof We have formally that

$$
\begin{aligned}
\sum_{k=0}^{\infty} \frac{s_k}{z^k} &= \sum_{i=1}^{d} \sum_{k=0}^{\infty} \left(\frac{\alpha_i}{z} \right)^k \\
&= z \sum_{i=1}^{d} \frac{1}{z - \alpha_i} \\
&= \frac{z P'(z)}{P(z)},
\end{aligned}
$$

where

$$P(z) := \prod_{i=1}^{d} (z - \alpha_i) = z^d + t_1 z^{d-1} + t_2 z^{d-2} + \cdots + t_d.$$

Then comparing the coefficients of z^{d-k} of $P(z) \sum_{k=0}^{\infty} (s_k/z^k)$ and $z P'(z)$, and using $s_0 = d$, gives (A.17). $\qquad \square$

In practice, we apply this to conjugate sets of algebraic integers $\alpha_1, \ldots, \alpha_d$. Then, the identities (A.17) show that these coefficients t_k of $P(z)$ are determined by the s_k

$(0 \le k \le d)$. In fact, we can give both the s_k in terms of the t_k, and the t_k in terms of the s_k, explicitly.

Proposition A.32 *For $k = 1, 2 \ldots$, we have*

$$s_k = (-1)^k \begin{vmatrix} t_1 & 1 & 0 & 0 & \cdots & 0 & 0 \\ 2t_2 & t_1 & 1 & 0 & \cdots & 0 & 0 \\ 3t_3 & t_2 & t_1 & 1 & 0 & \cdots & 0 \\ \vdots & \vdots & \ddots & \ddots & \ddots & \ddots & \vdots \\ (k-2)t_{k-2} & t_{k-3} & \ddots & \ddots & t_1 & 1 & 0 \\ (k-1)t_{k-1} & t_{k-2} & \ddots & \ddots & t_2 & t_1 & 1 \\ kt_k & t_{k-1} & \cdots & \cdots & t_3 & t_2 & t_1 \end{vmatrix} \tag{A.18}$$

and

$$t_k = \frac{(-1)^k}{k!} \begin{vmatrix} s_1 & 1 & 0 & 0 & \cdots & 0 & 0 \\ s_2 & s_1 & 2 & 0 & \cdots & 0 & 0 \\ s_3 & s_2 & s_1 & 3 & 0 & \cdots & 0 \\ \vdots & \vdots & \ddots & \ddots & \ddots & \ddots & \vdots \\ s_{k-2} & s_{k-3} & \ddots & \ddots & s_1 & k-2 & 0 \\ s_{k-1} & s_{k-2} & \ddots & \ddots & s_2 & s_1 & k-1 \\ s_k & s_{k-1} & \cdots & \cdots & s_3 & s_2 & s_1 \end{vmatrix}. \tag{A.19}$$

Proof We write (A.17) for $r = 1, \ldots, k$ as

$$\begin{pmatrix} 1 & 0 & 0 & 0 & \cdots & 0 & 0 \\ t_1 & 1 & 0 & 0 & \cdots & 0 & 0 \\ t_2 & t_1 & 1 & 0 & \cdots & 0 & 0 \\ t_3 & t_2 & t_1 & 1 & 0 & \cdots & 0 \\ \vdots & \vdots & \ddots & \ddots & \ddots & \ddots & \vdots \\ t_{k-2} & t_{k-3} & \ddots & t_2 & t_1 & 1 & 0 \\ t_{k-1} & t_{k-2} & \cdots & \cdots & t_2 & t_1 & 1 \end{pmatrix} \begin{pmatrix} s_1 \\ s_2 \\ s_3 \\ s_4 \\ \vdots \\ s_{k-1} \\ s_k \end{pmatrix} = \begin{pmatrix} -t_1 \\ -2t_2 \\ -3t_3 \\ -4t_4 \\ \vdots \\ -(k-1)t_{k-1} \\ -kt_k \end{pmatrix}.$$

Then, we can evaluate s_k as a determinant, by Cramer's Rule (Exercise A.11). This determinant can then be written in the form (A.18) by moving its rightmost column to be the leftmost one.

Alternatively, we write (A.17) for $r = 1, \ldots, k$ as

$$
\begin{pmatrix}
1 & 0 & 0 & 0 & \cdots & 0 \\
s_1 & 2 & 0 & 0 & \cdots & 0 \\
s_2 & s_1 & 3 & 0 & \cdots & 0 \\
\vdots & & \ddots & \ddots & \ddots & \vdots \\
s_{k-2} & s_{k-3} & \ddots & s_1 & k-1 & 0 \\
s_{k-1} & s_{k-2} & \cdots & s_2 & s_1 & k
\end{pmatrix}
\begin{pmatrix}
t_1 \\ t_2 \\ t_3 \\ t_4 \\ \vdots \\ t_{k-1} \\ t_k
\end{pmatrix}
=
\begin{pmatrix}
-s_1 \\ -s_2 \\ -s_3 \\ -s_4 \\ \vdots \\ -s_{k-1} \\ -s_k
\end{pmatrix}. \tag{A.20}
$$

Then t_k can be evaluated as a determinant, using Cramer's Rule in a similar way, to give (A.19). □

Remark Let us now drop the requirement that $t_k = 0$ for $k > d$, and so let $\{t_k\}_{k \in \mathbb{N}}$ be any sequence of integers. Then, (A.18) gives an integer sequence $\{s_k\}_{k \in \mathbb{N}}$. In particular, we see immediately from (A.20) that if $t_k = 1$ for all k then $s_k = -1$ for all k!

In the other direction, which integer sequences $\{s_k\}_{k \in \mathbb{N}}$ give rise to an integer sequence $\{t_k\}_{k \in \mathbb{N}}$ using (A.19)? Well, assuming that $s_1, s_2, \ldots, s_{k-1}$ have been chosen so that $t_1, t_2, \ldots, t_{k-1}$ are integers, we then see from (A.17) that t_k will be an integer provided that we choose

$$
s_k \equiv -\sum_{r=1}^{k-1} t_r s_{k-r} \pmod{k}. \tag{A.21}
$$

But, as we see from Theorem A.33 below, we also have

$$
s_k \equiv -\sum_{\ell \mid k, \ell < k} s_\ell \, \mu\left(\frac{k}{\ell}\right) \pmod{k},
$$

which specifies the equivalence class of $s_k \pmod{k}$ solely in terms the s_ℓ for $\ell < k$.

A.7 Jänichen's Generalisation of Fermat's Little Theorem

Fermat's Little Theorem $a^p \equiv a \pmod{p}$ for integers a and primes p is of course very well known. Less well known is Gauss's generalisation $\sum_{\ell \mid n} a^\ell \mu(n/\ell) \equiv 0 \pmod{n}$ for $a, n \in \mathbb{Z}, n > 0$. Both are contained in the following result of Jänichen.

Theorem A.33 (Jänichen [Jän21]) *Let $P(z) = \prod_{i=1}^d (z - \alpha_i) \in \mathbb{Z}[z]$, and*

$$
s_k := \sum_{i=1}^d \alpha_i^k \quad (k = 1, 2, \ldots),
$$

as before. Then for all $n \in \mathbb{N}$ we have (with μ the Möbius function)

$$\sum_{\ell|n} s_\ell \, \mu\left(\frac{n}{\ell}\right) \equiv 0 \pmod{n}.$$

For the proof, we need the following simple lemma.

Lemma A.34 *Any integer power series*

$$1 + a_1 z + a_2 z^2 + \cdots + a_n z^n + \cdots \tag{A.22}$$

can be written formally as a product $\prod_{k=1}^{\infty} \left(1 - z^k\right)^{b_k}$, *where the exponents* b_k *are also all integers.*

For, inductively, if

$$\prod_{k=1}^{n-1} \left(1 - z^k\right)^{b_k} \equiv 1 + a_1 z + a_2 z^2 + \cdots + a_{n-1} z^{n-1} + a_n' z^n \pmod{z^{n+1}}$$

say, then a_n' is an integer and

$$\left(1 - z^n\right)^{a_n' - a_n} \prod_{k=1}^{n-1} \left(1 - z^k\right)^{b_k} \equiv 1 + a_1 z + a_2 z^2 + \cdots + a_{n-1} z^{n-1} + a_n z^n \pmod{z^{n+1}}.$$

Proof (of Theorem A.33) We use the lemma to write the reciprocal polynomial of $P(z)$ as

$$\prod_{i=1}^{d} (1 - \alpha_i z) = \prod_{k=1}^{\infty} \left(1 - z^k\right)^{b_k},$$

where the exponents b_k are all integers. Taking logs, we have

$$\sum_{i=1}^{d} \log(1 - \alpha_i z) = \sum_{k=1}^{\infty} b_k \log(1 - z^k).$$

Then, comparing coefficients of z^n in the Maclaurin expansions of both sides gives

$$-\frac{s_n}{n} = \sum_{k|n} -\frac{b_k}{n/k},$$

or $s_n = \sum_{k|n} k b_k$. Then, we apply Möbius inversion ([Apo76, Theorem 2.9]) to obtain $\sum_{\ell|n} s_\ell \, \mu\left(\frac{n}{\ell}\right) = n b_n$. $\qquad\square$

An alternative, combinatorial proof of Jänichen's Theorem is given in [Smy86].

Given a monic integer polynomial $P(z) = \prod_{i=1}^{d}(z - \alpha_i)$, let $P_\ell(z)$ denote the polynomial $\prod_{i=1}^{d}(z - \alpha_i^\ell)$. Because its coefficients are symmetric functions of the α_i, we have $P_\ell(z) \in \mathbb{Z}[z]$.

Corollary A.35 *For all $\ell, n \in \mathbb{N}$ we have*

$$\sum_{\ell \mid n} P_\ell(z)\mu\left(\frac{n}{\ell}\right) \equiv 0 \pmod{n}.$$

Proof The coefficient of z^{d-1} of $\sum_{\ell \mid n} P_\ell(z)\mu(n/\ell)$ is $-\sum_{\ell \mid n} s_\ell\, \mu(n/\ell) \equiv 0$ (mod n). Similarly, the coefficient of z^{d-j} of $\sum_{\ell \mid n} P_\ell(z)\mu(n/\ell)$ is also a multiple of n. This is seen by replacing P by the polynomial of degree $\binom{d}{j}$ whose zeros are the products $\alpha_{i_1}\alpha_{i_2}\cdots\alpha_{i_j}$ for all possible j-element subsets $\{i_1, i_2, \ldots, i_j\}$ of $\{1, 2, \ldots, d\}$, and then considering its coefficient of $z^{\binom{d}{j}-1}$. $\qquad\square$

In particular, $P_p(z) \equiv P(z) \pmod{p}$ for any prime p (Schönemann [Sch39]).

Remark Note that when the series $1 + a_1 z + a_2 z^2 + \cdots + a_n z^n + \cdots$ in (A.22) is actually the nth cyclotomic polynomial $\Phi_n(z)$, then from (5.2) we see that

$$\Phi_n(z) = \prod_{k=1}^{\infty} \left(1 - z^k\right)^{b_k},$$

where $b_k = \mu(n/k)$ if $k \mid n$, and 0 otherwise. More generally, any product representation $\prod_{k=1}^{\infty} \left(1 - z^k\right)^{b_k}$ of the series (A.22) will have only finitely many of the b_k nonzero if and only if this series is the Maclaurin expansion of a quotient of (products of) cyclotomic polynomials.

Exercise A.36 Suppose that formally

$$\prod_{i=1}^{d}(1 - \alpha_i z) = \prod_{k=1}^{\infty} \left(1 - z^k\right)^{b_k}, \tag{A.23}$$

where $\prod_{i=1}^{d}(1 - \alpha_i z) \in \mathbb{Z}[z]$. Show that then

$$\prod_{i=1}^{d}(1 - \alpha_i^2 z) = \prod_{\substack{k \geq 1 \\ \text{odd}}} \left(1 - z^k\right)^{b_k + 2b_{2k}} \prod_{\substack{k \geq 2 \\ \text{even}}} \left(1 - z^k\right)^{2b_{2k}};$$

$$\prod_{i=1}^{d}(1 + \alpha_i z) = \prod_{\substack{k \geq 1 \\ \text{odd}}} \left(1 - z^k\right)^{-b_k} \prod_{2^1 \| k} \left(1 - z^k\right)^{b_{k/2} + b_k} \prod_{4 \mid k} \left(1 - z^k\right)^{b_k};$$

$$\prod_{i=1}^{d}(1 + \alpha_i^2 z) = \prod_{\substack{k \geq 1 \\ \text{odd}}} \left(1 - z^k\right)^{-b_k - 2b_{2k}} \prod_{2^1 \| k} \left(1 - z^k\right)^{b_{k/2} + 2b_k + 2b_{2k}} \prod_{4 \mid k} \left(1 - z^k\right)^{2b_{2k}}.$$

Exercise A.37 Show that if formally

$$\prod_{k=1}^{\infty} \left(1 - z^k\right)^{b_k} = \prod_{k=1}^{\infty} \left(1 + z^k\right)^{c_k}$$

then

$$c_n = -(b_n + b_{n/2} + b_{n/2^2} + \cdots + b_{n/2^\ell}), \quad \text{where } 2^\ell \parallel n$$

and, in the other direction,

$$b_n = \begin{cases} -c_n & n \text{ odd}; \\ -c_{n/2} - c_n & n \text{ even}. \end{cases}$$

Find all the c_k when only one b_n is nonzero; find all the b_k when only one c_n is nonzero.

Thus any series $1 + \sum_{n=1}^{\infty} a_n z^n$ can also be written in the form $\prod_{k=1}^{\infty} \left(1 + z^k\right)^{c_k}$, where, as with the exponents b_k, the exponents c_k are integers when the coefficients a_n are integers.

Appendix B
Combinatorial Background

In this appendix we state, mostly without proof, the combinatorial background results that we shall require. We also establish certain nomenclature and notation that we shall use in the main text.

B.1 Interlacing

Let A be an $n \times n$ complex matrix. An **eigenvector** of A is a nonzero vector $\mathbf{v} \in \mathbb{C}^n$ such that for some $\lambda \in \mathbb{C}$ there holds $A\mathbf{v} = \lambda\mathbf{v}$. The number λ is called an **eigenvalue** of A. The eigenvalues are the zeros of the **characteristic polynomial** of A, namely the monic polynomial $\det(xI - A)$, and this polynomial is usually denoted χ_A.

Here I is the identity matrix. The **minimal polynomial** of A is the monic polynomial $P(x) \in \mathbb{C}[x]$ of least degree such that $P(A)$ is the zero matrix. If A is a **real symmetric matrix** (i.e., the entries are in $\mathbb{R}$ and A equals its transpose A^{T}), then the zeros of $P(x)$ are real and distinct, comprising all the distinct zeros of χ_A.

Our first key result is the Interlacing Theorem. Essentially going back to Cauchy [Cau29] (for a short proof, see [Fis05]), we shall need the interlacing theorem only for real symmetric matrices, and we shall not be using the full strength of the theorem even in that restricted setting. We make explicit in a couple of corollaries the chief applications of the theorem for our purposes, after stating the theorem in its full generality. A complex matrix A is said to be **hermitian** if $A^{\mathsf{T}} = \overline{A}$. A familiar consequence of this property is that all eigenvalues of A are real, for if $A\mathbf{v} = \lambda\mathbf{v}$ with $\mathbf{v} \neq \mathbf{0}$, then $\overline{\mathbf{v}}^{\mathsf{T}} A\mathbf{v} = \lambda\overline{\mathbf{v}}^{\mathsf{T}}\mathbf{v}$, and on taking the transpose of the complex conjugate of both sides and using that A is hermitian one deduces that $\lambda = \overline{\lambda}$.

Theorem B.1 (Interlacing Theorem) *Let A be a hermitian matrix, with eigenvalues $\lambda_1 \leq \lambda_2 \leq \cdots \leq \lambda_n$. Pick any row i, and let B be the matrix formed by deleting row i and column i from A. Then, the eigenvalues of B interlace with those of A: if B has eigenvalues $\mu_1 \leq \cdots \leq \mu_{n-1}$, then*

© Springer Nature Switzerland AG 2021
J. McKee and C. Smyth, *Around the Unit Circle*, Universitext,
https://doi.org/10.1007/978-3-030-80031-4

$$\lambda_1 \leq \mu_1 \leq \lambda_2 \leq \mu_2 \leq \cdots \leq \mu_{n-1} \leq \lambda_n.$$

We shall usually make use of this result to make one of the two deductions given here as corollaries, both of which are immediate from the theorem.

Corollary B.2 *Let A be a real symmetric matrix. Pick any row i, and let B be the matrix formed by deleting row i and column i from A. Then, the largest eigenvalue of A is at least as large as the largest eigenvalue of B.*

Corollary B.3 *Let A be a real symmetric matrix. Pick any row i, and let B be the matrix formed by deleting row i and column i from A. If all the eigenvalues of A lie in some real interval I, then all the eigenvalues of B lie in I.*

B.2 Graph Theory

A *graph* $G = G(V, E)$ consists of a finite set of *vertices*, V, along with a finite set of *edges*, E, where each edge is an unordered pair of distinct vertices. Thus, the elements of E are of the shape (v, w), where v and w are distinct elements of V, and the pair is unordered, meaning that $(v, w) = (w, v)$. One frequently conveys this information by drawing the graph using a solid disc ● to represent a vertex and drawing a line between two discs if the corresponding pair of vertices forms an edge. One speaks of the two vertices that form an edge as being *adjacent*. If G has n vertices $v_1, \ldots, v_n$, then the *adjacency matrix*, A_G, is the $n \times n$ matrix (a_{ij}), where $a_{ij} = 1$ if v_i is adjacent to v_j, and $a_{ij} = 0$ if v_i is not adjacent to v_j. Of course if we change the ordering of the vertices then the rows and columns of A_G get permuted: the adjacency matrix of a graph is determined only up to conjugation by a *permutation matrix* (a matrix with a single nonzero entry in each row and in each column, and with each nonzero entry being $+1$).

A *signed graph* $G = G(V, E)$ consists of a finite set of vertices, V, along with a finite set of *signed edges*, E, where each element of E is of the form (v, w, ε), where (v, w) is a pair of distinct elements of V, $\varepsilon \in \{-1, 1\}$, and we regard (v, w, ε) and (w, v, ε) as being equal; moreover each unordered pair (v, w) appears in at most one signed edge (v, w, ε). We refer to ε as the *sign* of the (signed) edge (v, w, ε). Where it does not confuse, we may speak simply of the (signed) edge (v, w). The adjacency matrix is defined as before, but now with entries ± 1 indicating the signs of the corresponding edges. When drawing pictures (some examples will soon follow), we use a wavy line to indicate an edge that has sign -1.

A graph is also a signed graph.

A *charged signed graph* is a signed graph in which the vertices may have *charges* 0, 1 or -1. If a vertex has charge $\varepsilon = \pm 1$, then the corresponding diagonal entry in the adjacency matrix is ε and we speak of a *charged vertex*. If a vertex has charge 0 it is called *neutral*, and the diagonal entry in the adjacency matrix is 0. It is sometimes convenient to think of a charged vertex as corresponding to a directed loop on that vertex, weighted by its charge. In particular, we regard a charged vertex as being adjacent to itself; we may think of the charged vertices as corresponding to some extra edges in E of the shape (v, v, ε), where v is the charged vertex and ε is its charge.

A signed graph is also a charged signed graph.

A **weighted digraph** $G = G(V, A)$ consists of a finite set of **weighted vertices**, V (each element of V has a weight attached; for us this will always be an integer, positive, negative or zero), along with a finite set of **weighted arcs**, A, where each element of A is of the form (v, w, δ), where (v, w) is an *ordered* pair of vertices, and δ is a weight attached to that arc (for us this will always be an integer). Each ordered pair of vertices (v, w) has precisely one corresponding element $(v, w, \delta) \in A$ (perhaps $\delta = 0$). It is sometimes convenient, instructive, and otherwise helpful to represent the information carried by an arbitrary $n \times n$ integer matrix $A = (a_{ij})$ using a weighted digraph G_A. From the matrix A, we produce G_A as follows. There is a vertex corresponding to each row. The diagonal entry a_{ii} gives the weight attached to the ith vertex. An off-diagonal entry a_{ij} corresponds to an arc (i, j, a_{ij}). We may simply refer to G_A as the graph corresponding to A, although strictly it is a weighted digraph.

We may draw G_A as follows, making the picture less cluttered by recording the information corresponding to arcs from i to j and from j to i on a single edge between those vertices. For distinct i and j, if at least one of a_{ij} and a_{ji} is nonzero, then we draw a line (an edge) between the vertices corresponding to rows i and j. We put the values of a_{ij} and a_{ji} as labels on either side of this edge, with the convention that the label a_{ij} appears on the left of the edge as we move from i to j (we drive on the left). We label the vertex corresponding to row i with the diagonal value a_{ii}:

$$\overset{a_{ij}}{\underset{a_{ji}}{(a_{ii})\text{——}(a_{jj})}}.$$

(The orientation of the edge labels is like markings on the road: if we are driving on the left side of the road then they appear the right way up.) If we are given G_A and try to reconstruct A then we may get the rows and columns shuffled, but we shall regard all such shuffled matrices as equivalent anyway.

The matrix A is referred to as the **adjacency matrix** of G_A, with the caveat that the information in G_A does not give us the order of the rows/columns of the matrix: as before the adjacency matrix is determined only up to conjugation by a permutation matrix.

For example, consider the matrix

$$A = \begin{pmatrix} 1 & 2 & 3 \\ -1 & 0 & 1 \\ 0 & 1 & -2 \end{pmatrix}.$$

The graph G_A is pictured below, following the above conventions on labels:

$$(B.1)$$

The **underlying graph** of G_A is the graph whose vertices are those of G_A, with (i, j) an edge $(i \neq j)$ whenever $a_{ij} \neq 0$. There will be occasions when we wish to carry information about the diagonal entries of A into the underlying graph, but in such cases we shall make it explicit that we are doing so; unless we make such an explicit exception, an underlying graph is simply a graph, with zeros on the diagonal of its adjacency matrix.

Many of our matrices will be symmetric, and often the entries will all be either 0, 1 or -1. In these circumstances, we have some simplifications to our pictures to avoid clutter: an unlabelled vertex ● indicates 0 as the corresponding diagonal entry in the matrix; a vertex drawn as ⊕ indicates $+1$ as the corresponding diagonal entry; a vertex drawn as ⊖ indicates -1 as the corresponding diagonal entry; if $a_{ij} = a_{ji} = 1$ (for some distinct i and j), then we draw an unlabelled edge between the corresponding vertices. The above example (B.1) can now be drawn using fewer labels:

If both edge labels are 0, then the edge is simply not drawn at all.

If the matrix A is symmetric, and all entries are from the set $\{-1, 0, 1\}$, then G_A is a charged signed graph as defined above, where the word 'charged' refers to the three types of vertex, and the word 'signed' refers to the two possible signs for each edge. For charged signed graphs, we avoid all edge labels by using an unwavy edge to indicate that the corresponding off-diagonal entry is $+1$ (a **positive edge**), and a wavy edge for -1 (a **negative edge**). For example, the charged signed graph

represents the matrix

$$\begin{pmatrix} 1 & 1 & 1 \\ 1 & 0 & -1 \\ 1 & -1 & -1 \end{pmatrix}.$$

If G_A is a charged signed graph for which all vertices are neutral, then it is a signed graph, as defined above. If G_A is a signed graph for which all edges are positive, then it is a graph, as defined above.

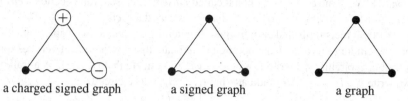

a charged signed graph   a signed graph   a graph

Of course any graph is an example of a signed graph, and any signed graph is an example of a charged signed graph.

If $G = G(V, E)$ and $H = H(W, F)$ are graphs, then we say that H is a **subgraph** of G if both $W \subseteq V$ and $F \subseteq E$ hold. This notion extends naturally to signed graphs and charged signed graphs. Mostly, we shall be interested in *induced* subgraphs: $H = H(W, F)$ is an **induced subgraph** of $G = G(V, E)$ if both (i) H is a subgraph of G, and (ii) whenever $v, w \in W$ and $(v, w) \in E$, we have $(v, w) \in F$. Thus, an induced subgraph is one that is obtained by removing a number of vertices (possibly none) along with any edges that involve those vertices. Again this notion extends to charged signed graphs.

Exercise B.4 Draw charged signed graphs to represent the following matrices:

$$\begin{pmatrix} 0 & 1 & 1 & 1 \\ 1 & 0 & 0 & 0 \\ 1 & 0 & 0 & 0 \\ 1 & 0 & 0 & 0 \end{pmatrix}, \quad \begin{pmatrix} 0 & 1 & 0 & 0 \\ 1 & -1 & 1 & 0 \\ 0 & 1 & 0 & 1 \\ 0 & 0 & 1 & 1 \end{pmatrix}, \quad \begin{pmatrix} 1 & -1 & -1 & 1 \\ -1 & -1 & 1 & 1 \\ -1 & 1 & -1 & 1 \\ 1 & 1 & 1 & 1 \end{pmatrix}.$$

A **walk** in a charged signed graph is a sequence of vertices $i_1, \ldots, i_r$ with i_j adjacent to i_{j+1} for $1 \leq j < r$. Note that since a charged vertex is adjacent to itself, neighbouring vertices in a walk can actually be the same vertex v repeated, provided that v is charged. A charged signed graph is said to be **connected** if for any pair of distinct vertices i and j there is a walk 'from i to j', i.e., there is a walk $i_1, \ldots, i_r$ with $i_1 = i$ and $i_r = j$. If G_A is connected, then in matrix language A is **indecomposable**, but we may simply speak of A as being connected. If G_A is not connected then it falls into at least two **components**, where a component is a maximal induced subgraph in which any two distinct vertices are joined by a walk. If A is not connected, then by ordering the vertices so that those in each component are grouped together, the matrix A is transformed to one in block diagonal form

$$P^{-1}AP = \begin{pmatrix} A_1 & O & \cdots & O \\ O & A_2 & \cdots & O \\ \vdots & \vdots & \ddots & \vdots \\ O & O & \cdots & A_r \end{pmatrix}, \tag{B.2}$$

where there are $r \geq 2$ components, and P is a permutation matrix that achieves the desired reordering of vertices. If the vertices of a walk $i_1, \ldots, i_r$ are distinct, then the walk is called a ***path***; if the edges are distinct it is called a ***trail***. If the last vertex of a walk is the same as the first, then it called ***closed***. A closed trail is called a ***circuit***. A circuit $i_1, \ldots, i_r, i_1$ with $i_1, \ldots, i_r$ distinct is called a ***cycle***.

We shall meet trails and circuits quite frequently, especially in Chap. 9, and it will be convenient to have a visual notation to indicate that a sequence of vertices forms a trail or a circuit. We write $i_1 \bullet i_2 \bullet \cdots \bullet i_r$ to indicate that $i_1, \ldots, i_r$ is a trail, and we write $i_1 : i_2 : \cdots : i_r$ to indicate that $i_1, \ldots, i_r, i_1$ is a circuit.

Example B.5 Let G be the charged signed graph shown, with labels $a, \ldots, f$ being

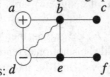

used to identify the vertices: d
Then

$$a \bullet b \bullet d \bullet d \bullet e \bullet f$$

is a trail in G (note that we regard the charge on vertex d as an edge that we can use at most once in a trail; in the notation set out above, $(d, d, -1)$ is an edge), and

$$a : a : b : d$$

is a circuit in G (the edges are $(a, a, 1)$, $(a, b, 1)$, $(b, d, -1)$ and $(a, d, 1)$, with the final edge being understood from the notation for a circuit). Note that $a : a : b : d$ is not a cycle, as a is repeated, but $a : b : d$ is a cycle.

B.3   Perron–Frobenius Theory

When all the entries of a matrix are real and nonnegative, then the powerful tool of Perron–Frobenius theory can be exceedingly useful. See, for example, [BH12, §2.2]. Our interest will almost entirely be restricted to applications to graph eigenvalues, and since the statement of the theorem becomes somewhat simpler in this case, we restrict to what we actually need (but in the latter part of the theorem we do allow arbitrary nonnegative integer square matrices).

Theorem B.6 (Perron–Frobenius) *Let $A = (a_{ij})$ be the adjacency matrix of a connected graph, with at least one edge, and let λ be the largest eigenvalue of A. Then*

- $\lambda > 0$;
- *there is an eigenvector $\mathbf{v}$ for λ that has all entries real and strictly positive;*
- *if $\mathbf{w}$ is a vector with all entries real and strictly positive such that $A\mathbf{w} = \mu\mathbf{w}$, for some $\mu \in \mathbb{R}$, then $\mu = \lambda$ and $\mathbf{w}$ is a scalar multiple of $\mathbf{v}$;*

- the largest eigenvalue of any proper induced subgraph (connected or not) is strictly smaller than λ.

Now let $A = (a_{ij})$ be a square matrix that has nonnegative integer entries, and for which the underlying graph is connected. Let λ be the largest modulus of any eigenvalue of A. Then

- λ is in fact an eigenvalue of A;
- if $\mathbf{x} = (x_i)$ is a nonzero vector with all components at least 0 such that $(A\mathbf{x})_i \geq \lambda x_i$ for all i, then $\mathbf{x}$ is an eigenvector for A, with eigenvalue λ, and $x_i > 0$ for all i.

Appendix C
Tools from the Theory of Functions

We recall without proof some results that we need from the theory of complex functions.

Theorem C.1 (Rouché's Theorem [Tit39, 3.42]) *Let C be a closed contour. Suppose that f and g are analytic inside and on C, and that $|f(z)| > |g(z)|$ on C. Then, $f(z)$ and $f(z) - g(z)$ have the same number of zeros inside C.*

Theorem C.2 (Parseval's Identity [Apo74, Exercise 16.12]) *Suppose that $f(z)$ is analytic in the disc $|z - a| < R$, with Taylor expansion $f(z) = \sum_{n=0}^{\infty} a_n (z - a)^n$. Then for $0 \leq r < R$ there holds*

$$\int_0^1 |f(a + re^{2\pi it})|^2 \, dt = \frac{1}{2\pi} \int_0^{2\pi} |f(a + re^{it})|^2 \, dt = \sum_{n=0}^{\infty} |a_n|^2 r^{2n} .$$

We shall generally use this with $a = 0$ and $r = 1$, giving $\int_0^1 |f(e^{2\pi it})|^2 \, dt = \sum_{n=0}^{\infty} |a_n|^2$.

Theorem C.3 (The Maximum Modulus Principle [Tit39, 5.1]) *Let C be a simple closed contour. Suppose that $f(z)$ is analytic inside and on C. If $|f(z)| \leq M$ for z on C, then the same inequality holds for z in the interior of C. Moreover, unless $f(z)$ is constant we have $f(z) < M$ for z in the interior of C.*

On one occasion, we will need a version of the Maximum Modulus Principle for subharmonic rather than analytic functions. A *real-valued* continuous function of a complex variable, $f(z)$, is called **subharmonic** on an open subset of the complex plane if for every closed disc in that region the function value at the centre is bounded above by the average value on the boundary of the disc: $f(z) \leq \int_0^1 f(z + re^{2\pi it}) \, dt$, where z is at the centre of the disc and r is the radius. Extending $f(z)$ to its boundary by continuity (allowing $-\infty$ as a value), the maximum of $f(z)$ is attained on the boundary. See the relevant chapter of [Kra92] for details.

© Springer Nature Switzerland AG 2021
J. McKee and C. Smyth, *Around the Unit Circle*, Universitext,
https://doi.org/10.1007/978-3-030-80031-4

Appendix D
Tables

The other appendices have no prerequisites within this book, but this appendix concerns tables of Mahler measures, and requires the definition of the Mahler measure of a polynomial with integer coefficients (Chap. 1), the definition of the Mahler measure of a Laurent polynomial in more than one variable (Chap. 2), and the notion of the shortness of a polynomial (Sect. 1.7).

D.1 Small Mahler Measures

Let $\mathcal{L}$ be the set of all Mahler measures of polynomials with integer coefficients. The following Table D.1 gives all known elements of $\mathcal{L}$ in $(1, 1.25)$. The shortest known polynomial with the given measure is shown. Of course there is also an irreducible monic irreducible polynomial P_n^{irr} with that measure. The polynomials P_n^{irr} can be computed easily from the short polynomials P_n in the table, by dividing out their cyclotomic factors, using Algorithm 1 of Chap. 5. The degree of P_n^{irr} is given in the table. The column 'out' shows the number of zeros of P_n^{irr} lying outside the unit circle. The cutoff 1.25 has been chosen because there is a probable limit point of such Mahler measures at $1.2554\cdots$—see Table D.2.

The polynomials P_n have been chosen to be as short as could be found, but may not all be as short as possible. All have one irreducible noncyclotomic factor, whose degree is listed. As we see in Chap. 1, these are very often too long to list conveniently. Most of the polynomials listed have cyclotomic factors. (Those that have no such factors can be identified from the fact that the degree listed is that of the short polynomial.) All measures are rounded to eight decimal places. They were found by Derrick Lehmer, David Boyd and Michael Mossinghoff.

© Springer Nature Switzerland AG 2021
J. McKee and C. Smyth, *Around the Unit Circle*, Universitext,
https://doi.org/10.1007/978-3-030-80031-4

Table D.1 Known Mahler measures in (1, 1.25)

n	Mahler measure	Degree	Out	Short polynomial $P_n(z)$	Shortness
1	1.17628082	10	1	$z^{12} - z^7 - z^6 - z^5 + 1$	5
2	1.18836815	18	1	$z^{19} - z^{15} - z^{13} + z^6 + z^4 - 1$	6
3	1.20002652	14	1	$z^{20} + z^{19} - z^{10} + z + 1$	5
4	1.20139619	18	2	$z^{22} - z^{17} - z^{15} + z^7 + z^5 - 1$	6
5	1.20261674	14	1	$z^{14} - z^{12} + z^7 - z^2 + 1$	5
6	1.20501985	22	2	$z^{27} + z^{25} - z^{21} + z^6 - z^2 - 1$	6
7	1.20795003	28	2	$z^{36} - z^{32} + z^{21} - z^{15} + z^4 - 1$	6
8	1.21282418	20	2	$z^{30} - z^{22} - z^{21} + z^9 + z^8 - 1$	6
9	1.21499570	20	2	$z^{21} - z^{16} - z^{15} + z^6 + z^5 - 1$	6
10	1.21639166	10	1	$z^{10} - z^6 + z^5 - z^4 + 1$	5
11	1.21839636	20	2	$z^{35} + z^{28} + z^{27} - z^8 - z^7 - 1$	6
12	1.21885515	24	2	$z^{29} - z^{22} - z^{19} + z^{10} + z^7 - 1$	6
13	1.21905751	24	2	$z^{39} - z^{37} + z^{21} + z^{18} - z^2 + 1$	6
14	1.21944688	18	3	$z^{25} - z^{21} - z^{15} + z^{10} + z^4 - 1$	6
15	1.21972086	18	1	$z^{19} - z^{17} - z^{11} + z^8 + z^2 - 1$	6
16	1.22028744	34	3	$z^{43} - z^{33} - z^{29} + z^{14} + z^{10} - 1$	6
17	1.22344738	38	3	$z^{47} - z^{41} + z^{28} - z^{19} + z^6 - 1$	6
18	1.22377745	26	3	$z^{42} + z^{33} - z^{26} - z^{16} + z^9 + 1$	6
19	1.22427891	16	2	$z^{18} + z^{10} + z^9 + z^8 + 1$	5
20	1.22550342	18	2	$z^{22} + z^{18} + z^{11} + z^4 + 1$	5
21	1.22561985	30	2	$z^{32} - z^{28} + z^{19} - z^{13} + z^4 - 1$	6
22	1.22581053	30	3	$z^{41} - z^{31} - z^{27} + z^{14} + z^{10} - 1$	6
23	1.22609289	26	2	$z^{29} - z^{23} + z^{19} - z^{10} + z^6 - 1$	6
24	1.22649330	36	2	$z^{37} - z^{35} - z^{20} + z^{17} + z^2 - 1$	6
25	1.22699376	20	2	$z^{24} + z^{14} - z^{13} - z^{11} + z^{10} + 1$	6
26	1.22778556	12	2	$z^{18} - z^{11} + z^9 - z^7 + 1$	5
27	1.22814077	30	3	$z^{41} - z^{31} - z^{28} - z^{13} - z^{10} + 1$	6
28	1.22948281	36	4	$z^{53} - z^{40} - z^{35} + z^{18} + z^{13} - 1$	6
29	1.22956646	22	3	$z^{30} - z^{25} - z^{23} - z^7 - z^5 + 1$	6
30	1.22999904	34	3	$z^{40} + z^{29} - z^{28} - z^{12} + z^{11} + 1$	6
31	1.23026327	38	3	$z^{42} - z^{34} + z^{27} + z^{15} - z^8 + 1$	6
32	1.23029547	42	4	$z^{53} - z^{39} + z^{37} - z^{16} + z^{14} - 1$	6
33	1.23039143	10	1	$z^{10} + z^7 + z^5 + z^3 + 1$	5
34	1.23074301	46	3	$z^{55} - z^{53} - z^{29} + z^{26} + z^2 - 1$	6
35	1.23134277	18	3	$z^{22} + z^{17} - z^{14} - z^8 + z^5 + 1$	6
36	1.23220295	48	4	$z^{58} - z^{50} - z^{35} + z^{23} + z^8 - 1$	6
37	1.23261355	20	1	$z^{25} + z^{24} + z^{16} - z^9 - z - 1$	6
38	1.23262878	28	2	$z^{32} - z^{29} - z^{27} + z^{22} + z^{10} - z^5 - z^3 + 1$	8
39	1.23367200	38	3	$z^{48} - z^{44} - z^{27} - z^{21} - z^4 + 1$	6

(continued)

Table D.1 (continued)

n	Mahler measure	Degree	Out	Short polynomial $P_n(z)$	Shortness
40	1.23434837	52	4	$z^{61} - z^{55} + z^{35} + z^{26} - z^6 + 1$	6
41	1.23444383	24	2	$z^{35} - z^{33} + z^{19} - z^{16} + z^2 - 1$	6
42	1.23450034	26	3	$z^{28} + z^{27} - z^{16} + z^{14} - z^{12} + z + 1$	7
43	1.23525671	16	2	$z^{21} - z^{16} - z^{14} + z^7 + z^5 - 1$	6
44	1.23549604	46	3	$z^{53} - z^{51} - z^{28} - z^{25} - z^2 + 1$	6
45	1.23566458	22	1	$z^{35} - z^{33} + z^{30} + z^5 - z^2 + 1$	6
46	1.23576110	42	4	$z^{52} - z^{41} - z^{32} + z^{20} + z^{11} - 1$	6
47	1.23608337	32	4	$z^{44} + z^{37} + z^{36} - z^{22} + z^8 + z^7 + 1$	7
48	1.23619847	32	3	$z^{39} - z^{30} - z^{25} - z^{14} - z^9 + 1$	6
49	1.23622792	32	4	$z^{52} - z^{40} + z^{35} - z^{17} + z^{12} - 1$	6
50	1.23624956	40	3	$z^{43} - z^{37} - z^{26} - z^{17} - z^6 + 1$	6
51	1.23631793	16	1	$z^{16} + z^{15} - z^8 + z + 1$	5
52	1.23656692	54	5	$z^{65} - z^{51} - z^{43} + z^{22} + z^{14} - 1$	6
53	1.23657922	34	3	$z^{39} - z^{29} - z^{27} + z^{12} + z^{10} - 1$	6
54	1.23667481	44	4	$z^{51} - z^{37} - z^{36} + z^{15} + z^{14} - 1$	6
55	1.23680831	28	2	$z^{35} + z^{31} - z^{24} - z^{11} + z^4 + \cdot 1$	6
56	1.23750482	26	1	$z^{34} - z^{32} + z^{29} + z^5 - z^2 + 1$	6
57	1.23763483	46	5	$z^{63} - z^{48} - z^{41} - z^{22} - z^{15} + 1$	6
58	1.23768413	58	5	$z^{69} - z^{59} - z^{42} - z^{27} - z^{10} + 1$	6
59	1.23804010	48	4	$z^{53} - z^{43} - z^{34} + z^{19} + z^{10} - 1$	6
60	1.23843163	56	4	$z^{64} - z^{60} + z^{35} - z^{29} + z^4 - 1$	6
61	1.23870898	56	4	$z^{59} - z^{53} - z^{34} + z^{25} + z^6 - 1$	6
62	1.23950577	42	4	$z^{50} + z^{37} - z^{34} + z^{16} - z^{13} - 1$	6
63	1.23974797	54	6	$z^{74} - z^{55} - z^{50} + z^{24} + z^{19} - 1$	6
64	1.23986133	48	4	$z^{85} + z^{83} + z^{65} + z^{54} - z^{49} + z^{36} - z^{31} - z^{20} - z^2 - 1$	10
65	1.24006186	60	4	$z^{69} - z^{67} - z^{36} + z^{33} + z^2 - 1$	6
66	1.24025418	26	3	$z^{40} - z^{39} - z^{25} - z^{15} - z + 1$	6
67	1.24037907	50	3	$z^{51} - z^{49} - z^{27} + z^{24} + z^2 - 1$	6
68	1.24069964	28	2	$z^{34} - z^{31} - z^{27} + z^{20} + z^{14} - z^7 - z^3 + 1$	8
69	1.24072642	12	1	$z^{13} - z^9 - z^7 + z^6 + z^4 - 1$	6
70	1.24077063	18	2	$z^{26} + z^{19} - z^{18} + z^8 - z^7 - 1$	6
71	1.24098340	54	5	$z^{62} - z^{46} + z^{43} + z^{19} - z^{16} + 1$	6
72	1.24137253	68	6	$z^{80} - z^{68} + z^{49} - z^{31} + z^{12} - 1$	6
73	1.24142203	70	5	$z^{80} - z^{76} - z^{43} - z^{37} - z^4 + 1$	6
74	1.24154108	58	5	$z^{67} - z^{57} - z^{41} + z^{26} + z^{10} - 1$	6
75	1.24178857	72	6	$z^{83} - z^{73} + z^{49} + z^{34} - z^{10} + 1$	6
76	1.24190216	58	6	$z^{73} + z^{53} - z^{51} + z^{22} - z^{20} - 1$	6
77	1.24197438	50	4	$z^{54} - z^{46} + z^{33} - z^{21} + z^8 - 1$	6

(continued)

Table D.1 (continued)

n	Mahler measure	Degree	Out	Short polynomial $P_n(z)$	Shortness
78	1.24221705	58	5	$z^{70} - z^{62} - z^{41} - z^{29} - z^8 + 1$	6
79	1.24236214	52	5	$z^{61} - z^{44} - z^{43} - z^{18} - z^{17} + 1$	6
80	1.24261029	58	5	$z^{64} - z^{52} + z^{41} + z^{23} - z^{12} + 1$	6
81	1.24277574	76	5	$z^{85} - z^{83} - z^{44} - z^{41} - z^2 + 1$	6
82	1.24287866	50	5	$z^{62} + z^{49} - z^{38} - z^{24} + z^{13} + 1$	6
83	1.24294012	32	2	$z^{33} - z^{31} - z^{18} + z^{15} + z^2 - 1$	6
84	1.24297921	70	7	$z^{85} - z^{63} - z^{59} + z^{26} + z^{22} - 1$	6
85	1.24302777	60	6	$z^{73} - z^{56} - z^{47} + z^{26} + z^{17} - 1$	6
86	1.24312870	30	3	$z^{41} - z^{35} - z^{25} - z^{16} - z^6 + 1$	6
87	1.24321004	72	6	$z^{86} - z^{78} + z^{49} - z^{37} + z^8 - 1$	6
88	1.24347762	16	2	$z^{18} - z^{13} + z^9 - z^5 + 1$	5
89	1.24348626	60	4	$z^{67} - z^{65} + z^{35} - z^{32} + z^2 - 1$	6
90	1.24356479	46	4	$z^{51} - z^{41} + z^{33} - z^{18} + z^{10} - 1$	6
91	1.24368275	46	5	$z^{61} - z^{47} - z^{39} + z^{22} + z^{14} - 1$	6
92	1.24387880	28	2	$z^{29} - z^{23} - z^{20} + z^9 + z^6 - 1$	6
93	1.24393593	56	6	$z^{73} - z^{55} - z^{50} + z^{23} + z^{18} - 1$	6
94	1.24427148	60	6	$z^{74} - z^{58} - z^{49} + z^{25} + z^{16} - 1$	6
95	1.24427364	70	5	$z^{73} - z^{67} - z^{41} + z^{32} + z^6 - 1$	6
96	1.24441450	40	4	$z^{41} - z^{39} - z^{22} + z^{19} + z^2 - 1$	6
97	1.24459886	62	6	$z^{72} - z^{52} + z^{51} - z^{21} + z^{20} - 1$	6
98	1.24461706	18	3	$z^{20} - z^{12} + z^{11} + z^9 - z^8 + 1$	6
99	1.24472917	76	6	$z^{81} - z^{71} - z^{48} + z^{33} + z^{10} - 1$	6
100	1.24480245	22	2	$z^{22} + z^{19} - z^{11} + z^3 + 1$	5
101	1.24490168	28	2	$z^{32} + z^{31} + z^{16} + z + 1$	5
102	1.24491415	40	4	$z^{49} - z^{38} - z^{31} + z^{18} + z^{11} - 1$	6
103	1.24495359	86	7	$z^{97} - z^{87} - z^{56} - z^{41} - z^{10} + 1$	6
104	1.24496773	82	8	$z^{97} - z^{75} + z^{65} + z^{32} - z^{22} + 1$	6
105	1.24504823	72	8	$z^{97} + z^{77} - z^{59} - z^{38} + z^{20} + 1$	6
106	1.24518114	22	3	$z^{24} - z^{13} - z^{12} - z^{11} + 1$	5
107	1.24537084	82	5	$z^{83} - z^{81} - z^{43} + z^{40} + z^2 - 1$	6
108	1.24537290	42	3	$z^{49} - z^{47} + z^{26} - z^{23} + z^2 - 1$	6
109	1.24537474	80	6	$z^{89} - z^{83} + z^{49} + z^{40} - z^6 + 1$	6
110	1.24537655	78	8	$z^{95} - z^{69} + z^{67} + z^{28} - z^{26} + 1$	6
111	1.24539780	66	7	$z^{83} - z^{61} - z^{57} + z^{26} + z^{22} - 1$	6
112	1.24542392	42	3	$z^{44} - z^{40} + z^{25} + z^{19} - z^4 + 1$	6
113	1.24555411	68	6	$z^{75} - z^{61} - z^{48} + z^{27} + z^{14} - 1$	6
114	1.24560232	22	2	$z^{26} + z^{24} + z^{13} + z^2 + 1$	5
115	1.24561355	28	3	$z^{37} - z^{27} + z^{26} - z^{11} + z^{10} - 1$	6
116	1.24567897	54	5	$z^{61} - z^{47} - z^{41} + z^{20} + z^{14} - 1$	6
117	1.24583146	34	3	$z^{38} - z^{30} + z^{25} + z^{13} - z^8 + 1$	6

(continued)

Table D.1 (continued)

n	Mahler measure	Degree	Out	Short polynomial $P_n(z)$	Shortness
118	1.24592220	88	8	$z^{102} - z^{86} - z^{63} + z^{39} + z^{16} - 1$	6
119	1.24594065	78	8	$z^{95} + z^{73} - z^{61} - z^{34} + z^{22} + 1$	6
120	1.24605524	74	7	$z^{84} - z^{64} + z^{57} + z^{27} - z^{20} + 1$	6
121	1.24610814	20	2	$z^{26} - z^{23} - z^{15} + z^{11} + z^3 - 1$	6
122	1.24628966	58	4	$z^{60} - z^{56} + z^{33} - z^{27} + z^4 - 1$	6
123	1.24629516	60	6	$z^{71} - z^{52} - z^{49} + z^{22} + z^{19} - 1$	6
124	1.24633583	72	7	$z^{83} - z^{61} + z^{58} - z^{25} + z^{22} - 1$	6
125	1.24639320	26	2	$z^{28} - z^{25} - z^{24} + z^{20} + z^8 - z^4 - z^3 + 1$	8
126	1.24640313	90	9	$z^{107} - z^{81} - z^{73} + z^{34} + z^{26} - 1$	6
127	1.24648199	70	7	$z^{85} - z^{67} - z^{56} - z^{29} - z^{18} + 1$	6
128	1.24658557	86	7	$z^{92} - z^{80} + z^{55} + z^{37} - z^{12} + 1$	6
129	1.24662722	90	6	$z^{99} - z^{97} + z^{51} + z^{48} - z^2 + 1$	6
130	1.24664687	78	7	$z^{89} - z^{75} - z^{55} + z^{34} + z^{14} - 1$	6
131	1.24675708	68	7	$z^{83} + z^{64} - z^{53} + z^{30} - z^{19} - 1$	6
132	1.24682890	78	8	$z^{94} + z^{71} - z^{62} + z^{32} - z^{23} - 1$	6
133	1.24687827	20	2	$z^{21} - z^{17} - z^{14} + z^7 + z^4 - 1$	6
134	1.24691881	64	4	$z^{65} - z^{63} - z^{34} + z^{31} + z^2 - 1$	6
135	1.24704952	36	5	$z^{43} + z^{32} - z^{29} + z^{14} - z^{11} - 1$	6
136	1.24713354	94	9	$z^{109} - z^{87} - z^{71} + z^{38} + z^{22} - 1$	6
137	1.24719987	82	9	$z^{105} + z^{78} - z^{71} + z^{34} - z^{27} - 1$	6
138	1.24721290	70	5	$z^{76} - z^{72} + z^{41} + z^{35} - z^4 + 1$	6
139	1.24734756	70	7	$z^{82} + z^{59} - z^{58} - z^{24} + z^{23} + 1$	6
140	1.24736510	100	8	$z^{111} - z^{101} + z^{63} + z^{48} - z^{10} + 1$	6
141	1.24739576	98	9	$z^{113} - z^{95} - z^{70} - z^{43} - z^{18} + 1$	6
142	1.24740604	58	6	$z^{72} - z^{57} - z^{44} + z^{28} + z^{15} - 1$	6
143	1.24749078	68	5	$z^{71} - z^{65} + z^{40} - z^{31} + z^6 - 1$	6
144	1.24752151	106	7	$z^{115} - z^{113} - z^{59} + z^{56} + z^2 - 1$	6
145	1.24757476	30	3	$z^{37} - z^{29} - z^{23} + z^{14} + z^8 - 1$	6
146	1.24757582	78	7	$z^{86} - z^{70} + z^{55} + z^{31} - z^{16} + 1$	6
147	1.24760870	26	2	$z^{28} - z^{22} + z^{19} - z^9 + z^6 - 1$	6
148	1.24763294	84	8	$z^{95} - z^{73} - z^{64} + z^{31} + z^{22} - 1$	6
149	1.24765460	70	6	$z^{79} - z^{69} + z^{47} - z^{32} + z^{10} - 1$	6
150	1.24765479	52	4	$z^{55} - z^{49} - z^{32} + z^{23} + z^6 - 1$	6
151	1.24768588	90	10	$z^{116} - z^{85} - z^{80} + z^{36} + z^{31} - 1$	6
152	1.24769691	70	6	$z^{76} - z^{64} + z^{47} - z^{29} + z^{12} - 1$	6
153	1.24769875	74	8	$z^{94} - z^{70} + z^{65} - z^{29} + z^{24} - 1$	6
154	1.24771237	50	5	$z^{59} - z^{43} - z^{41} + z^{18} + z^{16} - 1$	6
155	1.24771845	78	9	$z^{107} - z^{85} - z^{65} - z^{42} - z^{22} + 1$	6
156	1.24781622	56	6	$z^{71} - z^{53} + z^{49} - z^{22} + z^{18} - 1$	6
157	1.24782157	96	10	$z^{119} - z^{93} + z^{79} + z^{40} - z^{26} + 1$	6

(continued)

Table D.1 (continued)

n	Mahler measure	Degree	Out	Short polynomial $P_n(z)$	Shortness
158	1.24794442	74	5	$z^{81} - z^{79} + z^{42} - z^{39} + z^2 - 1$	6
159	1.24794798	86	9	$z^{106} + z^{83} - z^{66} - z^{40} + z^{23} + 1$	6
160	1.24795303	54	6	$z^{71} + z^{55} - z^{45} - z^{26} + z^{16} + 1$	6
161	1.24796570	84	6	$z^{92} - z^{88} + z^{49} - z^{43} + z^4 - 1$	6
162	1.24799688	96	8	$z^{103} - z^{89} - z^{62} + z^{41} + z^{14} - 1$	6
163	1.24801361	16	2	$z^{17} - z^{13} - z^{12} + z^5 + z^4 - 1$	6
164	1.24808813	80	8	$z^{96} - z^{76} + z^{63} - z^{33} + z^{20} - 1$	6
165	1.24828489	94	7	$z^{103} - z^{97} - z^{56} - z^{47} - z^6 + 1$	6
166	1.24830443	80	8	$z^{93} - z^{67} - z^{66} + z^{27} + z^{26} - 1$	6
167	1.24832114	112	10	$z^{127} - z^{109} + z^{77} - z^{50} + z^{18} - 1$	6
168	1.24837841	98	11	$z^{127} - z^{95} - z^{85} - z^{42} - z^{32} + 1$	6
169	1.24840541	86	9	$z^{105} - z^{81} - z^{67} + z^{38} + z^{24} - 1$	6
170	1.24846635	112	10	$z^{130} - z^{114} - z^{77} + z^{53} + z^{16} - 1$	6
171	1.24854822	108	10	$z^{124} - z^{104} + z^{77} - z^{47} + z^{20} - 1$	6
172	1.24855707	94	7	$z^{98} - z^{90} + z^{55} + z^{43} - z^8 + 1$	6
173	1.24855960	98	7	$z^{108} - z^{104} - z^{57} - z^{51} - z^4 + 1$	6
174	1.24860739	78	6	$z^{82} - z^{74} + z^{47} - z^{35} + z^8 - 1$	6
175	1.24861117	22	3	$z^{26} + z^{19} + z^{13} + z^7 + 1$	5
176	1.24866635	96	6	$z^{97} - z^{95} - z^{50} + z^{47} + z^2 - 1$	6
177	1.24869886	116	8	$z^{119} - z^{113} - z^{64} + z^{55} + z^6 - 1$	6
178	1.24872877	104	8	$z^{114} - z^{106} - z^{63} + z^{51} + z^8 - 1$	6
179	1.24878388	86	9	$z^{105} - z^{79} - z^{72} - z^{33} - z^{26} + 1$	6
180	1.24881875	26	3	$z^{27} - z^{23} - z^{17} + z^{10} + z^4 - 1$	6
181	1.24883439	104	11	$z^{129} - z^{102} - z^{79} - z^{50} - z^{27} + 1$	6
182	1.24884233	94	9	$z^{106} - z^{82} + z^{71} + z^{35} - z^{24} + 1$	6
183	1.24888703	106	11	$z^{127} - z^{93} - z^{89} + z^{38} + z^{34} - 1$	6
184	1.24893573	66	6	$z^{73} - z^{59} + z^{47} + z^{26} - z^{14} + 1$	6
185	1.24895229	94	10	$z^{115} + z^{83} - z^{81} - z^{34} + z^{32} + 1$	6
186	1.24896795	118	9	$z^{130} - z^{122} - z^{71} - z^{59} - z^8 + 1$	6
187	1.24903094	34	3	$z^{45} - z^{38} - z^{34} + z^{11} + z^7 - 1$	6
188	1.24903130	122	8	$z^{124} - z^{120} + z^{65} - z^{59} + z^4 - 1$	6
189	1.24903162	88	8	$z^{97} - z^{79} - z^{62} + z^{35} + z^{18} - 1$	6
190	1.24904317	52	6	$z^{70} - z^{53} - z^{46} + z^{24} + z^{17} - 1$	6
191	1.24908706	86	9	$z^{104} + z^{79} - z^{68} - z^{36} + z^{25} + 1$	6
192	1.24909048	96	10	$z^{115} - z^{86} - z^{77} + z^{38} + z^{29} - 1$	6
193	1.24909625	106	9	$z^{114} - z^{98} + z^{69} + z^{45} - z^{16} + 1$	6
194	1.24909954	68	7	$z^{81} + z^{60} - z^{55} + z^{26} - z^{21} - 1$	6
195	1.24913695	90	9	$z^{104} - z^{76} + z^{73} + z^{31} - z^{28} + 1$	6
196	1.24919996	108	7	$z^{113} - z^{111} + z^{58} - z^{55} + z^2 - 1$	6
197	1.24920997	40	4	$z^{49} - z^{39} - z^{32} + z^{17} + z^{10} - 1$	6
198	1.24924343	104	8	$z^{109} - z^{99} - z^{62} + z^{47} + z^{10} - 1$	6

(continued)

Table D.1 (continued)

n	Mahler measure	Degree	Out	Short polynomial $P_n(z)$	Shortness
199	1.24926496	106	11	$z^{126} + z^{93} - z^{86} - z^{40} + z^{33} + 1$	6
200	1.24926942	80	7	$z^{87} - z^{73} + z^{54} - z^{33} + z^{14} - 1$	6
201	1.24928809	110	9	$z^{117} - z^{103} - z^{69} + z^{48} + z^{14} - 1$	6
202	1.24929633	90	9	$z^{107} - z^{85} - z^{70} - z^{37} - z^{22} + 1$	6
203	1.24929845	136	10	$z^{141} - z^{131} - z^{78} + z^{63} + z^{10} - 1$	6
204	1.24931954	54	5	$z^{66} - z^{58} + z^{39} + z^{27} - z^{8} + 1$	6
205	1.24932226	114	11	$z^{131} - z^{105} - z^{85} + z^{46} + z^{26} - 1$	6
206	1.24932674	58	5	$z^{63} - z^{53} - z^{39} + z^{24} + z^{10} - 1$	6
207	1.24934093	118	12	$z^{139} - z^{105} + z^{95} - z^{44} + z^{34} - 1$	6
208	1.24935215	28	3	$z^{36} - z^{32} + z^{29} + z^{18} + z^{7} - z^{4} + 1$	7
209	1.24941177	130	9	$z^{140} - z^{136} - z^{73} - z^{67} - z^{4} + 1$	6
210	1.24942089	68	8	$z^{93} + z^{72} - z^{59} - z^{34} + z^{21} + 1$	6
211	1.24942643	94	11	$z^{127} - z^{98} - z^{81} - z^{46} - z^{29} + 1$	6
212	1.24944489	66	7	$z^{83} - z^{65} - z^{55} - z^{28} - z^{18} + 1$	6
213	1.24945683	78	8	$z^{92} - z^{67} - z^{64} + z^{28} + z^{25} - 1$	6
214	1.24946871	118	11	$z^{135} - z^{113} - z^{84} - z^{51} - z^{22} + 1$	6
215	1.24949933	26	3	$z^{27} - z^{25} - z^{15} + z^{12} + z^{2} - 1$	6
216	1.24950246	70	9	$z^{86} + z^{67} - z^{54} - z^{32} + z^{19} + 1$	6
217	1.24952344	108	12	$z^{137} + z^{100} - z^{95} - z^{42} + z^{37} + 1$	6
218	1.24956071	88	7	$z^{93} - z^{83} + z^{54} - z^{39} + z^{10} - 1$	6
219	1.24960982	120	8	$z^{129} - z^{127} - z^{66} + z^{63} + z^{2} - 1$	6
220	1.24965403	70	7	$z^{81} - z^{59} - z^{57} + z^{24} + z^{22} - 1$	6
221	1.24966071	102	10	$z^{116} - z^{88} + z^{79} - z^{37} + z^{28} - 1$	6
222	1.24968344	124	10	$z^{136} - z^{124} + z^{77} - z^{59} + z^{12} - 1$	6
223	1.24968830	32	4	$z^{40} - z^{33} + z^{32} - z^{8} + z^{7} - 1$	6
224	1.24974750	86	10	$z^{117} + z^{93} - z^{71} - z^{46} + z^{24} + 1$	6
225	1.24981127	122	13	$z^{150} + z^{117} - z^{94} - z^{56} + z^{33} + 1$	6
226	1.24983423	48	5	$z^{59} - z^{46} - z^{37} - z^{22} - z^{13} + 1$	6
227	1.24983510	114	12	$z^{137} + z^{103} - z^{91} - z^{46} + z^{34} + 1$	6
228	1.24989662	54	7	$z^{64} + z^{47} - z^{44} - z^{20} + z^{17} + 1$	6
229	1.24990794	88	9	$z^{103} - z^{74} - z^{73} - z^{30} - z^{29} + 1$	6
230	1.24991903	114	12	$z^{137} - z^{99} + z^{97} - z^{40} + z^{38} - 1$	6
231	1.24992460	136	12	$z^{152} - z^{132} - z^{91} + z^{61} + z^{20} - 1$	6
232	1.24992702	88	10	$z^{116} - z^{91} - z^{72} + z^{44} + z^{25} - 1$	6
233	1.24993433	136	9	$z^{145} - z^{143} - z^{74} - z^{71} - z^{2} + 1$	6
234	1.24994466	132	12	$z^{149} - z^{127} + z^{91} + z^{58} - z^{22} + 1$	6
235	1.24997052	116	10	$z^{125} - z^{107} - z^{76} + z^{49} + z^{18} - 1$	6
236	1.24998636	82	9	$z^{103} - z^{77} - z^{69} + z^{34} + z^{26} - 1$	6

Table D.2 Known Mahler measures in (1, 1.37) of irreducible dimension-2 polynomials

n	Mahler measure	Polynomial $Q_n(x, y)$	Length
1	1.255434	$yx^4 - x^3 - y - y^{-1} - x^{-3} + y^{-1}x^{-4}$	6
2	1.285735	$x + y + 1 + y^{-1} + x^{-1}$	5
3	1.309098	$yx^2 + x^2 + yx - 1 + y^{-1}x^{-1} + x^{-2} + y^{-1}x^{-2}$	7
4	1.315693	$yx^6 + x^5 + y + y^{-1} + x^{-5} + y^{-1}x^{-6}$	6
5	1.324718	$y^4x^5 + x + y + y^{-1} + x^{-1} + y^{-4}x^{-5}$	6
6	1.325372	$x^3 + y^2x^2 + y^3 - y^{-3} - y^{-2}x^{-2} - x^{-3}$	6
7	1.332051	$x^5 - yx^4 + y + y^{-1} - y^{-1}x^{-4} + x^{-5}$	6
8	1.332396	$x^2 + x + y + y^{-1} + x^{-1} + x^{-2}$	6
9	1.338137	$x^3 + yx + y^3 - y^{-3} - y^{-1}x^{-1} - x^{-3}$	6
10	1.340000	$yx^8 - x^7 - y - y^{-1} - x^{-7} + y^{-1}x^{-8}$	6
11	1.340507	$yx^6 + x + y + y^{-1} + x^{-1} + y^{-1}x^{-6}$	6
12	1.348652	$y^4x^5 + x^3 + y^3 + y^{-3} + x^{-3} + y^{-4}x^{-5}$	6
13	1.349716	$x^7 + y^2x^5 + y^7 + y^{-7} + y^{-2}x^{-5} + x^{-7}$	6
14	1.350015	$x^7 + yx^6 + y + y^{-1} + y^{-1}x^{-6} + x^{-7}$	6
15	1.350317	$x^5 - x^3 + y + y^{-1} - x^{-3} + x^{-5}$	6
16	1.351146	$yx^8 - x^5 - y - y^{-1} - x^{-5} + y^{-1}x^{-8}$	6
17	1.352468	$x^{10} + y^9x^9 + y^{10} + y^{-10} + y^{-9}x^{-9} + x^{-10}$	6
18	1.353698	$x^3 + yx + y + y^{-1} + y^{-1}x^{-1} + x^{-3}$	6
19	1.356748	$yx^8 - x^3 - y - y^{-1} - x^{-3} + y^{-1}x^{-8}$	6
20	1.356786	$x^5 + y^4x^4 + y^5 - y^{-5} - y^{-4}x^{-4} - x^{-5}$	6
21	1.358130	$yx^3 + x^3 + yx^2 - x - 1 - x^{-1} + y^{-1}x^{-2} +$ $x^{-3} + y^{-1}x^{-3}$	9
22	1.358546	$yx^8 - x - y - y^{-1} - x^{-1} + y^{-1}x^{-8}$	6
23	1.359208	$y^9x^9 + x^8 - y^8 - y^{-8} + x^{-8} + y^{-9}x^{-9}$	6
24	1.359376	$y^7x^6 - x^5 - y^5 - y^{-5} - x^{-5} + y^{-7}x^{-6}$	6
25	1.359812	$yx^{12} + x^{11} - y + y^{-1} - x^{-11} - y^{-1}x^{-12}$	6
26	1.359816	$x^7 + x^5 + y + y^{-1} + x^{-5} + x^{-7}$	6
27	1.359914	$y^7x^8 + x + y + y^{-1} + x^{-1} + y^{-7}x^{-8}$	6
28	1.360221	$yx^{10} + x^7 + y + y^{-1} + x^{-7} + y^{-1}x^{-10}$	6
29	1.361956	$y^7x^4 - x - y - y^{-1} - x^{-1} + y^{-7}x^{-4}$	6
30	1.362724	$x^5 + y^3x^3 + y^5 - y^{-5} - y^{-3}x^{-3} - x^{-5}$	6
31	1.363651	$x^4 + x + y + y^{-1} + x^{-1} + x^{-4}$	6
32	1.364200	$x^5 - y^2x^3 + y^5 + y^{-5} - y^{-2}x^{-3} + x^{-5}$	6
33	1.364436	$yx^2 + x + y^3 + y^{-3} + x^{-1} + y^{-1}x^{-2}$	6
34	1.364546	$yx^{14} + x^{13} - y + y^{-1} - x^{-13} - y^{-1}x^{-14}$	6
35	1.364656	$y^{11}x^{11} + x^{10} + y^{10} + y^{-10} + x^{-10} + y^{-11}x^{-11}$	6
36	1.365062	$x^9 - x^5 + y + y^{-1} - x^{-5} + x^{-9}$	6
37	1.365270	$y^9x^8 + x^7 + y^7 + y^{-7} + x^{-7} + y^{-9}x^{-8}$	6
38	1.365469	$x^5 + y^2x^2 + y^5 - y^{-5} - y^{-2}x^{-2} - x^{-5}$	6
39	1.365985	$y^2x^2 + yx^2 + y^2x + yx - y^{-1}x - yx^{-1}$ $+ y^{-1}x^{-1} + y^{-2}x^{-1} + y^{-1}x^{-2} + y^{-2}x^{-2}$	10

(continued)

Table D.2 (continued)

n	Mahler measure	Polynomial $Q_n(x, y)$	Length
40	1.366146	$yx^{10} + x^3 + y + y^{-1} + x^{-3} + y^{-1}x^{-10}$	6
41	1.366299	$y^2x^7 - x^3 - y - y^{-1} - x^{-3} + y^{-2}x^{-7}$	6
42	1.366402	$y^2x^9 - x^5 - y - y^{-1} - x^{-5} + y^{-2}x^{-9}$	6
43	1.366436	$y^7x^8 + x^3 + y^3 + y^{-3} + x^{-3} + y^{-7}x^{-8}$	6
44	1.366571	$x^5 + yx + y^5 - y^{-5} - y^{-1}x^{-1} - x^{-5}$	6
45	1.366808	$yx^{10} + x + y + y^{-1} + x^{-1} + y^{-1}x^{-10}$	6
46	1.366883	$x^5 - yx^2 + y + y^{-1} - y^{-1}x^{-2} + x^{-5}$	6
47	1.366991	$x^7 + y^6x^6 + y^7 - y^{-7} - y^{-6}x^{-6} - x^{-7}$	6
48	1.367511	$x^7 + y^4x^5 + y^7 + y^{-7} + y^{-4}x^{-5} + x^{-7}$	6
49	1.367799	$yx^{16} + x^{15} - y + y^{-1} - x^{-15} - y^{-1}x^{-16}$	6
50	1.367855	$x^{11} + y^7x^4 + y^{11} + y^{-11} + y^{-7}x^{-4} + x^{-11}$	6
51	1.368132	$x^9 - y^2x^5 + y^9 + y^{-9} - y^{-2}x^{-5} + x^{-9}$	6
52	1.368196	$x^{13} + yx^{12} - y + y^{-1} - y^{-1}x^{-12} - x^{-13}$	6
53	1.368214	$y^9x^9 + x^2 + y^2 + y^{-2} + x^{-2} + y^{-9}x^{-9}$	6
54	1.368343	$yx^3 + x^3 + yx^2 + x^2 - 1 + x^{-2} + y^{-1}x^{-2} + x^{-3} + y^{-1}x^{-3}$	9
55	1.368397	$yx^8 - x^3 - y^3 - y^{-3} - x^{-3} + y^{-1}x^{-8}$	6
56	1.368747	$yx^{12} - x^7 - y - y^{-1} - x^{-7} + y^{-1}x^{-12}$	6
57	1.368922	$y^{11}x^{10} - x^9 - y^9 - y^{-9} - x^{-9} + y^{-11}x^{-10}$	6
58	1.368949	$yx^{14} + x^{11} - y + y^{-1} - x^{-11} - y^{-1}x^{-14}$	6
59	1.368979	$x^9 - y^7x^4 + y^9 + y^{-9} - y^{-7}x^{-4} + x^{-9}$	6
60	1.369489	$y^{11}x^6 - x - y - y^{-1} - x^{-1} + y^{-11}x^{-6}$	6
61	1.369782	$x^5 + x^4 + y + y^{-1} + x^{-4} + x^{-5}$	6

D.2 Known Small Mahler Measures of Two-Variable Polynomials

In speaking of Mahler measures of two-variable polynomials we generally implicitly mean something more subtle, in light of the following exercise.

Exercise D.1 If M is the Mahler measure of an irreducible 1-variable polynomial with integer coefficients, show that M is also the Mahler measure of an irreducible 2-variable integer polynomial.

If $P(x, y) = \sum_{i,j} c_{ij}x^i y^j$ is a Laurent polynomial, then $\mathcal{C}(P)$ is defined to be the convex hull in $\mathbb{R}^2$ of those points (i, j) for which $c_{ij} \neq 0$. We wish to restrict to those P for which $\mathcal{C}(P)$ does not lie on any line — it is a two-dimensional polytope. We then say that P has dimension 2. (For more on dimension, see Sect. 2.2.) Restricting to irreducible polynomials that have dimension 2, the smallest known Mahler measure is $1.2554\cdots$, the first entry in Table D.2.

Table D.2 gives the 61 smallest known Mahler measures of irreducible 2-dimensional polynomials with integer coefficients. We expect that these are all limit points of the set $\mathcal{L}$, but this is not known in all cases. The cutoff value 1.37 was chosen in view of the 3-dimensional measure $M(x + 1/x + y + 1/y + z + 1/z) = M(1 + x + y) = 1.38135\cdots$, as noted in (2.9). The measures in this table are rounded to six decimal places. They were found by Boyd and Mossinghoff [BM05] and El Otmani, Maul, Rhin, and Sac-Épée [EORSE19, EOMRSE21].

References

[ABP06]   Aguirre, J., Bilbao, M., Peral, J.C.: The trace of totally positive algebraic integers.
           Math. Comp. **75**(253), 385–393 (2006)
[AD99]   Amoroso, F., David, S.: Le problème de Lehmer en dimension supérieure. J. Reine
           Angew. Math. **513**, 145–179 (1999)
[AD00]   Amoroso, F., Dvornicich, R.: A lower bound for the height in abelian extensions. J.
           Number Theory **80**(2), 260–272 (2000)
[AM16]   Amoroso, F., Masser, D.: Lower bounds for the height in Galois extensions. Bull.
           Lond. Math. Soc. **48**(6), 1008–1012 (2016)
[Amo08]   Amoroso, F.: On the Mahler measure in several variables. Bull. Lond. Math. Soc.
           40(4), 619–630 (2008)
[Amo18]   Amoroso, F.: Mahler measure on Galois extensions. Int. J. Number Theory **14**(6),
           1605–1617 (2018)
[AP08]   Aguirre, J., Peral, J.C.: The trace problem for totally positive algebraic integers.
           In: Number Theory and Polynomials. London Mathematical Society Lecture Note
           series, pp. 1–19. Cambridge University Press, Cambridge (2008). With an appendix
           by Jean-Pierre Serre
[Apo74]   Apostol, T..M.: Mathematical Analysis, 2nd edn. Addison-Wesley (1974)
[Apo76]   Apostol, T.M.: Introduction to Analytic Number Theory. Undergraduate Texts in
           Mathematics, Springer, New York-Heidelberg (1976)
[AV19]   Akhtari, S., Vaaler, J.D.: Lower bounds for Mahler measure that depend on the
           number of monomials. Int. J. Number Theory **15**(7), 1425–1436 (2019)
[Axl19]   Axler, C.: On the sum of the first n prime numbers. J. Théor. Nombres Bordeaux
           31(2), 293–311 (2019)
[Bae87]   Baernstein, A.: II. Dubinin's symmetrization theorem. In: Complex Analysis, I (Col-
           lege Park, Md., 1985–86). Lecture Notes in Mathematics, vol. 1275, pp. 23–30.
           Springer, Berlin (1987)
[Baz77]   Bazylewicz, A.: On the product of the conjugates outside the unit circle of an algebraic
           integer. Acta Arith. **30**(1), 43–61 (1976/77)
[Baz88]   Bazylewicz, A.: An extension of a result of C. J. Smyth to polynomials in several
           variables. Acta Arith. **50**(2), 211–214 (1988). Corrigendum: Acta Arith. **60**(4), 417
           (1992)
[BDM07]   Borwein, P., Dobrowolski, E., Mossinghoff, M.J.: Lehmer's problem for polynomials
           with odd coefficients. Ann. Math. **166**(2), 347–366 (2007)
[BE96]   Borwein, P., Erdélyi, T.: The integer Chebyshev problem. Math. Comp. **65**(214),
           661–681 (1996)

© Springer Nature Switzerland AG 2021
J. McKee and C. Smyth, *Around the Unit Circle*, Universitext,
https://doi.org/10.1007/978-3-030-80031-4

[BH05]   Bhargava, M., Hanke, J.: Universal quadratic forms and the 290-theorem. preprint (2005)

[BH12]   Brouwer, A.E., Haemers, W.H.: Spectra of Graphs. Springer (2012)

[BHM04]   Borwein, P., Hare, K.G., Mossinghoff, M.J.: The Mahler measure of polynomials with odd coefficients. Bull. London Math. Soc. 36(3), 332–338 (2004)

[BM71]   Blanksby, P.E., Montgomery, H.L.: Algebraic integers near the unit circle. Acta Arith. 18, 355–369 (1971)

[BM96]   Boyd, D.W., Daniel Mauldin, R.: The order type of the set of Pisot numbers. Topol. Appl. 69(2), 115–120 (1996)

[BM05]   Boyd, D.W., Mossinghoff, M.J.: Small limit points of Mahler's measure. Exp. Math. 14(4), 403–414 (2005)

[BMV07]   Borwein, P., Mossinghoff, M.J., Vaaler, J.D.: Generalizations of Gonçalves' inequality. Proc. Amer. Math. Soc. 135(1), 253–261 (2007)

[Boy77]   Boyd, D.W.: Small Salem numbers. Duke Math. J. 44, 315–328 (1977)

[Boy78]   Boyd, D.W.: Variations on a theme of Kronecker. Canad. Math. Bull. 21(2), 129–133 (1978)

[Boy79]   Boyd, D.W.: On the successive derived sets of the Pisot numbers. Proc. Amer. Math. Soc. 73(2), 154–156 (1979)

[Boy81a]   Boyd, D.W.: Kronecker's theorem and Lehmer's problem for polynomials in several variables. J. Number Theory 13(1), 116–121 (1981)

[Boy81b]   Boyd, D.W.: Speculations concerning the range of Mahler's measure. Canad. Math. Bull. 24(4), 453–469 (1981)

[Boy98]   Boyd, D.W.: Mahler's measure and special values of l-functions. Exp. Math. 7(1), 37–82 (1998)

[BPP03]   Borwein, P.B., Pinner, C.G., Pritsker, I.E.: Monic integer Chebyshev problem. Math. Comp. 72(244), 1901–1916 (2003)

[Bre51]   Breusch, R.: On the distribution of the roots of a polynomial with integral coefficients. Proc. Amer. Math. Soc. 2, 939–941 (1951)

[BRV02]   Boyd, D.W., Rodriguez-Villegas, F.: Mahler's measure and the dilogarithm. I. Canad. J. Math. 54(3), 468–492 (2002)

[BRV05]   Boyd, D.W., Rodriguez-Villegas, F.: Mahler's measure and the dilogarithm (II) (2005). arXiv:math/0308041v2

[BS66]   Borevich, A.I., Shafarevich, I.R.: Number Theory. Translated from the Russian by Newcomb Greenleaf. Pure and Applied Mathematics, vol. 20. Academic Press, New York-London (1966)

[BS02]   Beukers, F., Smyth, C.J.: Cyclotomic points on curves. In: Number theory for the millennium. I (Urbana, IL, 2000), pp. 67–85. A K Peters, Natick, MA (2002)

[BZ97]   Beukers, F., Zagier, D.: Lower bounds of heights of points on hypersurfaces. Acta Arith. 79(2), 103–111 (1997)

[BZ01]   Bombieri, E., Zannier, U.: A note on heights in certain infinite extensions of $\mathbb{Q}$. Atti Accad. Naz. Lincei Cl. Sci. Fis. Mat. Natur. Rend. Lincei (9) Mat. Appl. 12, 5–14 (2002), 2001

[BZ16]   Bertin, M.J., Zudilin, W.: On the Mahler measure of a family of genus 2 curves. Math. Z. 283(3–4), 1185–1193 (2016)

[BZ20]   Brunault, F., Zudilin, W.: Many Variations of Mahler Measures. A Lasting Symphony. Australian Mathematical Society Lecture Series, vol. 28. Cambridge University Press, Cambridge (2020)

[Cas65]   Cassels, J.W.S.: An Introduction to Diophantine Approximation. Cambridge Tracts in Mathematics and Mathematical Physics, vol. 45. Cambridge University Press, Cambridge (1965)

[Cas69]   Cassels, J.W.S.: On a conjecture of R. M. Robinson about sums of roots of unity. J. Reine Angew. Math. 238, 112–131 (1969)

[Cau29]   Cauchy, A.L.: Sur l'équation à l'aide de laquelle on détermine les inégalités séculaires des mouvements des planètes. In: Oeuvres Complètes. volume 9 of IIième série, pp. 174–195. Gauthier-Villars, Paris (1829)

[CF10]   Cassels, J.W.S., Fröhlich, A. (eds.): In: Algebraic Number Theory. London Mathe-
           matical Society, London (2010). Papers from the conference held at the University
           of Sussex, Brighton, September 1–17, 1965, Including a list of errata
[CFS10]   Capparelli, S., Del Fra, A., Sciò, C.: On the span of polynomials with integer coef-
           ficients. Math. Comp. **79**, 967–981 (2010)
[Che52]   Chebyshev, P.: Mémoire sur les nombres premiers. J. Math. Pures. Appl. **1**(17),
           366–390 (1852)
[CM21]   Coyston, J., McKee, J.: Small Mahler measures from digraphs. to appear in Exp.
           Math. (2021)
[CMS11]   Calegari, F., Morrison, S., Snyder, N.: Cyclotomic integers, fusion categories, and
           subfactors. Comm. Math. Phys. **303**(3), 845–896 (2011)
[CMS14]   Cooley, J., McKee, J., Smyth, C.J.: Non-bipartite graphs of small Mahler measure.
           J. Comb. Number Theory **5**(2), 53–64 (2014)
[CS83]   Cantor, D.C., Straus, E.G.: On a conjecture of D. H. Lehmer. Acta Arith. **42**(1),
           97–100 (1982/83). Correction: Acta Arith. **42**(3), 327 (1982/83)
[DD04]   Dixon, J.D., Dubickas, A.: The values of Mahler measures. Mathematika **51**(1–2),
           131–148 (2004)
[Den97]   Deninger, C.: Deligne periods of mixed motives, K-theory and the entropy of certain
           Z^n-actions. J. Amer. Math. Soc. **10**(2), 259–281 (1997)
[DF04]   Dummit, D..S.., Foote, R..M.: Abstract Algebra, 3rd edn. Wiley (2004)
[Dim19]   Dimitrov, V.: A proof of the Schinzel-Zassenhaus conjecture on polynomials (2019).
           arXiv:1912.12545
[Dir37]   Dirichlet, P.G.L.: Beweis des Satzes, dass jede unbegrenzte arithmetische Progres-
           sion, deren erstes Glied und Differenz ganze Zahlen ohne gemeinschaftlichen Factor
           sind, unendlich viele Primzahlen enthält. Abhandlungen der Königlichen Preußis-
           chen Akademie der Wissenschaften zu Berlin **48**, 45–71 (1837)
[DLS83]   Dobrowolski, E., Lawton, W., Schinzel, A.: On a problem of Lehmer. In: Studies in
           pure mathematics, pp. 135–144. Birkhäuser, Basel (1983)
[DM05]   Dubickas, A., Mossinghoff, M.J.: Auxiliary polynomials for some problems regard-
           ing Mahler's measure. Acta Arith. **119**(1), 65–79 (2005)
[Dob78]   Dobrowolski, E.: On the maximal modulus of conjugates of an algebraic integer.
           Bull. Acad. Polon. Sci. Sér. Sci. Math. Astronom. Phys. **26**(4), 291–292 (1978)
[Dob79]   Dobrowolski, E.: On a question of Lehmer and the number of irreducible factors of
           a polynomial. Acta Arith. **34**(4), 391–401 (1979)
[Dob91]   Dobrowolski, E.: Mahler's measure of a polynomial in function of the number of its
           coefficients. Canad. Math. Bull. **34**(2), 186–195 (1991)
[Dob06]   Dobrowolski, E.: Mahler's measure of a polynomial in terms of the number of its
           monomials. Acta Arith. **123**(3), 201–231 (2006)
[Dob08]   Dobrowolski, E.: A note on integer symmetric matrices and Mahler's measure.
           Canad. Math. Bull. **51**(1), 57–59 (2008)
[Doc01a]   Doche, C.: On the spectrum of the Zhang-Zagier height. Math. Comp. **70**(233),
           419–430 (2001)
[Doc01b]   Doche, C.: Zhang-Zagier heights of perturbed polynomials. J. Théor. Nombres Bor-
           deaux **13**(1), 103–110 (2001). 21st Journées Arithmétiques (Rome, 2001)
[Dre98]   Dresden, G.P.: Orbits of algebraic numbers with low heights. Math. Comp. **67**(222),
           815–820 (1998)
[DS89]   Davie, A.M., Smyth, C.J.: On a limiting fractal measure defined by conjugate alge-
           braic integers. In: Groupe de Travail en Théorie Analytique et Élémentaire des Nom-
           bres, 1987–1988, vol. 89, Publ. Math. Orsay, pp. 93–103. Univ. Paris XI, Orsay
           (1989)
[DS01a]   Dubickas, A., Smyth, C.J.: On the metric Mahler measure. J. Number Theory **86**(2),
           368–387 (2001)
[DS01b]   Dubickas, A., Smyth, C.J.: The Lehmer constants of an annulus. J. Théor. Nombres
           Bordeaux **13**(2), 413–420 (2001)

[DS17]   Dobrowolski, E., Smyth, C.J.: Mahler measures of polynomials that are sums of a bounded number of monomials. Int. J. Number Theory **13**(6), 1603–1610 (2017)

[Dub84]   Dubinin, V.N.: Change of harmonic measure in symmetrization. Mat. Sb. (N.S.) **124**(166)(2), 272–279 (1984)

[Dub93]   Dubickas, A.: On a conjecture of A. Schinzel and H. Zassenhaus. Acta Arith. **63**(1), 15–20 (1993)

[Dub97]   Dubickas, A.: The maximal conjugate of a non-reciprocal algebraic integer. Liet. Mat. Rink. **37**(2), 168–174 (1997)

[Dub00]   Dubickas, A.: On the measure of a nonreciprocal algebraic number. Ramanujan J. **4**(3), 291–298 (2000)

[Dub14]   Dubinin, V.N.: Condenser capacities and symmetrization in geometric function theory. Springer, Basel (2014). Translated from the Russian by Nikolai G. Kruzhilin

[Dus18]   Dusart, P.: Explicit estimates of some functions over primes. Ramanujan J. **45**(1), 227–251 (2018)

[EG93]   Estes, D.R., Guralnick, R.M.: Minimal polynomials of integral symmetric matrices. Linear Algebra Appl. **192**, 83–99 (1993)

[EOMRSE13] El Otmani, S., Maul, A., Rhin, G., Sac-Épée, J.-M.: Integer linear programming applied to determining monic hyperbolic irreducible polynomials with integer coefficients and span less than 4. J. de Théorie des Nombres de Bordeaux **25**(1), 71–78 (2013)

[EOMRSE21] El Otmani, S., Maul, A., Rhin, G., Sac-Épée, J.-M.: Finding new limit points of Mahler measure by methods of missing data restoration. BAU J. Sci. Technol. **2**(2) Article 10 (2021)

[EORSE19]   El Otmani, S., Rhin, G., Sac-Épée, J.-M.: Finding new limit points of Mahler's measure by genetic algorithms. Exp. Math. **28**(2), 129–131 (2019)

[Fal03]   Falconer, K.: Fractal Geometry, 2nd edn. Wiley, Hoboken, NJ (2003). Mathematical foundations and applications

[Fek23]   Fekete, M.: Über die Verteilung der Wurzeln bei gewissen algebraischen Gleichungen mit ganzzahligen Koeffizienten. Math. Z. **17**(1), 228–249 (1923)

[Fek30a]   Fekete, M.: Über den transfiniten Durchmesser ebener Punktmengen i. Math. Z. **32**(1), 108–114 (1930)

[Fek30b]   Fekete, M.: Über den transfiniten Durchmesser ebener Punktmengen ii. Math. Z. **32**(1), 215–221 (1930)

[Fis05]   Fisk, S.: A very short proof of Cauchy's interlace theorem. Amer. Math. Monthly **112**, 118 (2005)

[Fla95]   Flammang, V.: Sur le diamètre transfini entier d'un intervalle à extrémités rationnelles. Ann. Inst. Fourier (Grenoble) **45**(3), 779–793 (1995)

[Fla96]   Flammang, V.: Two new points in the spectrum of the absolute Mahler measure of totally positive algebraic integers. Math. Comp. **65**(213), 307–311 (1996)

[Fla15]   Flammang, V.: The Mahler measure and its areal analog for totally positive algebraic integers. J. Number Theory **151**, 211–222 (2015)

[Fla16]   Flammang, V.: Une nouvelle minoration pour la trace absolue des entiers alg'ebriques totalement positifs. hal-01346165 (2016)

[Fla19]   Flammang, V.: The absolute trace of totally positive algebraic integers. Int. J. Number Theory **15**(1), 173–181 (2019)

[FR15]   Flammang, V., Rhin, G.: On the absolute Mahler measure of polynomials having all zeros in a sector. III. Math. Comp. **84**(296), 2927–2938 (2015)

[FRS97]   Flammang, V., Rhin, G., Smyth, C.J.: The integer transfinite diameter of intervals and totally real algebraic integers. J. Théor. Nombres Bordeaux **9**(1), 137–168 (1997)

[FRW11]   Flammang, V., Rhin, G., Wu, Q.: The totally real algebraic integers with diameter less than 4. Moscow J. Comb. Number Theory **1**, 21–32 (2011)

[FS55]   Fekete, M., Szegö, G.: On algebraic equations with integral coefficients whose roots belong to a given point set. Math. Z. **63**, 158–172 (1955)

[GG83]   Glashoff, K., Gustafson, S.-A.: Linear Optimization and Approximation. Applied Mathematical Sciences, vol. 45. Springer, New York (1983). An introduction to the theoretical analysis and numerical treatment of semi-infinite programs

[GL98]   Goberna, M.A., López, M.A.: Linear Semi-Infinite Optimization. Wiley Series in Mathematical Methods in Practice, vol. 2. Wiley, Chichester (1998)

[GL17]   Goberna, M.A., López, M.A.: Recent contributions to linear semi-infinite optimization. 4OR **15**(3), 221–264 (2017)

[GL18]   Goberna, M.A., López, M.A.: Recent contributions to linear semi-infinite optimization: an update. Ann. Oper. Res. **271**(1), 237–278 (2018)

[GM14]   Gumbrell, L., McKee, J.: A classification of all 1-Salem graphs. LMS J. Comput. Math. **17**(1), 582–594 (2014)

[Gol69]   Goluzin, G.M.: Geometric Theory of Functions of a Complex Variable. Translations of Mathematical Monographs, vol. 26. American Mathematical Society, Providence, R.I. (1969)

[Gor59]   Gorškov, D.S.: On the distance from zero on the interval [0, 1] of polynomials with integral coefficients. In: Proceedings of the Third All Union Mathematical Congress (Moscow, 1956), vol. 4, pp. 5–7. Akad. Nauk SSSR, Moscow (1959)

[Gre12a] Greaves, G.: Cyclotomic matrices over quadratic integer rings. Ph.D. thesis, Royal Holloway, University of London (2012)

[Gre12b] Greaves, G.: Cyclotomic matrices over real quadratic integer rings. Linear Algebra Appl. **437**(9), 2252–2261 (2012)

[Gre12c] Greaves, G.: Cyclotomic matrices over the Eisenstein and Gaussian integers. J. Algebra **372**, 560–583 (2012)

[Gre15]   Greaves, G.: Small-span Hermitian matrices over quadratic integer rings. Math. Comp. **84**(291), 409–424 (2015)

[GT13]   Greaves, G., Taylor, G.: Lehmer's conjecture for Hermitian matrices over the Eisenstein and Gaussian integers. Electron. J. Combin. **20**(1), 42 (2013)

[Hay66]   Hayman, W.K.: Transfinite Diameter and Its Applications, vol. 45, MatScience Report. Institute of Mathematical Sciences, Madras, second printing (1966). Notes by K.R. Unni

[HS93]   Höhn, G., Skoruppa, N.-P.: Un résultat de Schinzel. J. Théor. Nombres Bordeaux **5**(1), 185 (1993)

[HS97]   Habsieger, L., Salvy, B.: On integer Chebyshev polynomials. Math. Comp. **66**(218), 763–770 (1997)

[HS06]   Hare, K.G., Smyth, C.J.: The monic integer transfinite diameter. Math. Comp. **75**(256), 1997–2019 (2006)

[Jän21]   Jänichen, W.: Über die Verallgemeinerung einer Gauß'schen Formel aus der Theorie der höheren Kongruenzen. Sitzungsber. Berlin. Math. Ges. **20**, 21–39 (1921)

[Jen99]   Jensen, J.L.W.V.: Sur un nouvel et important théorème de la théorie des fonctions. Acta Math. **22**(1), 359–364 (1899)

[Jon35]   Jones, B.W.: A table of Eisenstein-reduced positive ternary quadratic forms of determinant ≤ 200. Bull. Natl. Res. Counc. **97** (1935)

[Jon68]   Jones, A.J.: Sums of three roots of unity. Proc. Camb. Philos. Soc. **64**, 673–682 (1968)

[Kna92]   Knapp, A.W.: Elliptic curves. Mathematical Notes, vol. 40. Princeton University Press, Princeton, NJ (1992)

[Kou16]   Kouachi, S.: The Cauchy interlace theorem for symmetrizable matrices (2016). arXiv:1603.04151v1 [math.DS]

[Kra92]   Krantz, S.G.: Function Theory of Several Complex Variables. AMS Chelsea (1992)

[Kro57]   Kronecker, L.: Zwei Sätze über Gleichungen mit ganzzahligen Coefficienten. J. Reine Angew. Math. **53**, 173–175 (1857)

[Kro81]   Kronecker, L.: Zur Theorie der Elimination einer Variablen aus zwei algebraischen Gleichungen. Monatsberichte d. k. preuss. Akademie d. Wiss. zu Berlin, pp. 535–600 (1881)

[Lag70]   Lagrange, P.G.L.: Démonstration d'un théorème d'arithmétique. Mém. Acad. Roy.
            Sci. Berlin **123** (1770). Also Oeuvres. vol. 3 1869, pp. 189–201

[Lal03]   Lalín, M.N.: Some examples of Mahler measures as multiple polylogarithms. J.
            Number Theory **103**(1), 85–108 (2003)

[Lal06a]   Lalín, M.N.: Mahler measure of some n-variable polynomial families. J. Number
            Theory **116**(1), 102–139 (2006)

[Lal06b]   Lalín, M.N.: On certain combination of colored multizeta values. J. Ramanujan Math.
            Soc. **21**(1), 115–127 (2006)

[Lan86]   Langevin, M.: Minorations de la maison et de la mesure de Mahler de certains entiers
            algébriques. C. R. Acad. Sci. Paris Sér. I Math. **303**(12), 523–526 (1986)

[Law77]   Lawton, W.: Asymptotic properties of roots of polynomials—preliminary report.
            In: Proceedings of the Seventh National Mathematics Conference (Dept. Math.,
            Azarabadegan Univ., Tabriz, 1976), pp. 212–218. Azarabadegan Univ., Tabriz (1977)

[Law83]   Lawton, W.M.: A problem of Boyd concerning geometric means of polynomials. J.
            Number Theory **16**(3), 356–362 (1983)

[Leh30]   Lehmer, E.: A numerical function applied to cyclotomy. Bull. Amer. Math. Soc. **36**,
            291–298 (1930)

[Leh33]   Lehmer, D.H.: Factorization of certain cyclotomic functions. Ann. of Math. (2) **34**(3),
            461–479 (1933)

[Lou83]   Louboutin, R.: Sur la mesure de Mahler d'un nombre algébrique. C. R. Acad. Sci.
            Paris Sér. I Math. **296**(16), 707–708 (1983)

[Lox72]   Loxton, J.H.: On the maximum modulus of cyclotomic integers. Acta Arith. **22**,
            69–85 (1972)

[Lox75]   Loxton, J.H.: On two problems of R. W. Robinson about sums of roots of unity. Acta
            Arith. **26**, 159–174 (1974/75)

[LS07]   López, M., Still, G.: Semi-infinite programming. Eur. J. Oper. Res. **180**(2), 491–518
            (2007)

[Mah60]   Mahler, K.: An application of Jensen's formula to polynomials. Mathematika **7**,
            98–100 (1960)

[Mah62]   Mahler, K.: On some inequalities for polynomials in several variables. J. London
            Math. Soc. **37**, 341–344 (1962)

[Mah64]   Mahler, K.: An inequality for the discriminant of a polynomial. Michigan Math. J.
            11, 257–262 (1964)

[Mai00]   Maillot, V.: Géométrie d'Arakelov des variétés toriques et fibrés en droites inté-
            grables. *Mém. Soc. Math. Fr. (N.S.)*, 80:vi+129 (2000)

[Mas16]   Masser, David: Auxiliary polynomials in number theory. Cambridge Tracts in Math-
            ematics, vol. 207. Cambridge University Press, Cambridge (2016)

[McK08]   McKee, J.: Conjugate algebraic numbers on conics: a survey. In: McKee, J., Smyth,
            C. (eds.) Number Theory and Polynomials. London Mathematical Society Lecture
            Note Series, vol. 352, pp. 211–240. London Mathematical Society, Cambridge (2008)

[McK10]   McKee, J.: Small-span characteristic polynomials of integer symmetric matrices. In:
            Hanrot, G., Morain, F., Thomé, E. (eds.) Algorithmic Number Theory: 9th Inter-
            national Syposium. ANTS-IX. Lecture Notes in Computer Science, vol. 6197, pp.
            270–284. Springer, Nancy, France (2010)

[McK11]   McKee, J.: Computing totally positive algebraic integers of small trace. Math. Comp.
            80, 1041–1052 (2011)

[ME05]   Murty, M.R., Esmonde, J.: Problems in Algebraic Number Theory. Graduate Texts
            in Mathematics, vol. 190, 2nd edn. Springer, New York (2005)

[Mér99]   Méray, Ch.: Sur un déterminant dont celui de vandermonde n'est qu'un cas partic-
            ulier. Revue de Mathématiques Spéciales **9**, 217–219 (1899)

[Mig78]   Mignotte, M.: Entiers algébriques dont les conjugués sont proches du cercle unité. In:
            Séminaire Delange-Pisot-Poitou, 19e année: 1977/78, Théorie des nombres, Fasc. 2,
            pp. Exp. No. 39, 6. Secrétariat Math., Paris (1978)

[Mit20]   Mitchell, L.: A trace bound for integer-diagonal positive semidefinite matrices. Spec. Matrices **8**, 14–16 (2020)

[Mon94]   Montgomery, H.L.: Ten lectures on the interface between analytic number theory and harmonic analysis. CBMS Regional Conference Series in Mathematics, vol. 84. Published for the Conference Board of the Mathematical Sciences, Washington, DC; by the American Mathematical Society, Providence, RI (1994)

[Mos11]   Mossinghoff, M.: (2011). http://wayback.cecm.sfu.ca/~mjm/Lehmer/lists/index.html

[MOS20]   McKee, J., Oh, B.-K., Smyth, C.: The Cassels heights of cyclotomic integers (2020). arXiv:2007.00270

[MRW08]   Mossinghoff, M., Rhin, G., Wu, Q.: Minimal Mahler measures. Exp. Math. **17**(4), 451–458 (2008)

[MS99a]   Maz'ya, V., Shaposhnikova, T.O.: Jacques Hadamard: A Universal Mathematician. American Mathematical Society (1999)

[MS99b]   McKee, J., Smyth, C.: Salem numbers and Pisot numbers from stars. In: Győry, K., Iwaniec, H., Urbanowicz, J. (eds.) Number Theory in Progress, vol. 1, pp. 309–320. de Gruyter (1999)

[MS04]   McKee, J., Smyth, C.: Salem numbers of trace −2 and traces of totally positive algebraic integers. In: Algorithmic Number Theory. Lecture Notes in Computer Science, vol. 3076, pp. 327–337. Springer, Berlin (2004)

[MS05a]   McKee, J., Smyth, C.: Salem numbers, Pisot numbers, Mahler measure, and graphs. Exp. Math. **14**(2), 211–229 (2005)

[MS05b]   McKee, J., Smyth, C.: There are Salem numbers of every trace. Bul. Lond. Math. Soc. **37**, 25–36 (2005)

[MS07]   McKee, J., Smyth, C.: Integer symmetric matrices having all their eigenvalues in the interval [−2, 2]. J. Algebra **317**, 260–290 (2007)

[MS12a]   McKee, J., Smyth, C.: Integer symmetric matrices of small spectral radius and small Mahler measure. Int. Math. Res. Not. **2012**(1), 102–136 (2012)

[MS12b]   McKee, J., Smyth, C.: Salem numbers and Pisot numbers via interlacing. Can. J. Math. **64**(2), 345–367 (2012)

[MS13]   McKee, J., Smyth, C.: Single polynomials that correspond to pairs of cyclotomic polynomials with interlacing zeros. Cent. Eur. J. Math. **11**(5), 882–899 (2013)

[MS20a]   McKee, J., Smyth, C.: Symmetrizable integer matrices having all their eigenvalues in the interval [−2, 2]. Algebr. Comb. **3**(3), 775–789 (2020)

[MS20b]   McKee, J., Smyth, C.: Symmetrizable matrices, quotients and the trace problem. Linear Algebra Appl. **600**, 60–81 (2020)

[MY14]   McKee, J., Yatsyna, P.: A trace bound for positive definite connected integer symmetric matrices. Linear Algebra Appl. **444**, 227–230 (2014)

[Neu99]   Neukirch, J.: Algebraic Number Theory. Springer (1999)

[New72]   Newman, M.: Integral Matrices. Pure and Applied Mathematics, vol. 45. Academic Press, New York-London (1972)

[Nip91]   Nipp, G.L.: Quaternary Quadratic Forms. Springer, New York (1991). Computer generated tables. https://doi.org/10.1007/978-1-4612-3180-6

[Not78]   Notari, C.: Sur le produit des conjugués à l'extérieur du cercle unité d'un nombre algébrique. C. R. Acad. Sci. Paris Sér. A-B **286**(7), A313–A315 (1978)

[O'M58]   O'Meara, O.T.: The integral representations of quadratic forms over local fields. Amer. J. Math. **80**, 843–878 (1958)

[O'M63]   O'Meara, O.T.: Introduction to Quadratic Forms. Die Grundlehren der mathematischen Wissenschaften, Bd. 117. Academic Press, Inc., Publishers, New York; Springer, Berlin-Göttingen-Heidelberg (1963)

[Pól28]   Pólya, G.: Über gewisse notwendige Determinantenkriterien für die Fortsetzbarkeit einer Potenzreihe. Math. Ann. **99**(1), 687–706 (1928)

[Pot18]   Pottmeyer, L.: Small totally p-adic algebraic numbers. Int. J. Number Theory **14**(10), 2687–2697 (2018)

[Pri05a]   Pritsker, I.E.: The Gelfond-Schnirelman method in prime number theory. Canad. J. Math. **57**(5), 1080–1101 (2005)

[Pri05b]   Pritsker, I.E.: Small polynomials with integer coefficients. J. Anal. Math. **96**, 151–190 (2005)

[Pri21]   Pritsker, I.E.: House of algebraic integers symmetric about the unit circle (2021). arXiv:2101.06710

[Ran95]   Ransford, T.: Potential Theory in the Complex Plane. London Mathematical Society Student Texts, vol. 28. Cambridge University Press, Cambridge (1995)

[Ray87]   Ray, G.A.: Relations between Mahler's measure and values of L-series. Canad. J. Math. **39**(3), 694–732 (1987)

[Rei74]   Reidemeister, K.: Knotentheorie. Reprint of 1932 original. Springer (1974)

[Rhi19]   Rhin, G.: Complete list of totally real algebraic integers with diameter <4 and degree from 2 to 15 which are not of cosine type (2019). http://www.iecl.univ-lorraine.fr/~Georges.Rhin/. Accessed: 2019-06-18

[Rob62]   Robinson, R.M.: Intervals containing infinitely many sets of conjugate algebraic integers. In: Mathematical Analysis and Related Topics: Essays in Honor of George Pólya, pp. 305–315. Stanford (1962)

[Rob64]   Robinson, R.M.: Algebraic equations with span less than 4. Math. Comp. **18**, 547–559 (1964)

[Rob65]   Robinson, R.M.: Some conjectures about cyclotomic integers. Math. Comp. **19**, 210–217 (1965)

[Row87]   Rowlinson, P.: A deletion-contraction algorithm for the chracteristic polynomial of a multigraph. Proc. Royal Soc. Edinburgh **105A**, 153–160 (1987)

[RS95]   Rhin, G., Smyth, C.: On the absolute Mahler measure of polynomials having all zeros in a sector. Math. Comp. **64**(209), 295–304 (1995)

[RS97]   Rhin, G., Smyth, C.J.: On the Mahler measure of the composition of two polynomials. Acta Arith. **79**(3), 239–247 (1997)

[RS02]   Rahman, Q.I., Schmeisser, G.: Analytic Theory of Polynomials. London Mathematical Society Monographs. New Series, vol. 26. The Clarendon Press, Oxford University Press, Oxford (2002)

[Ruz99]   Ruzsa, I.Z.: On Mahler's measure for polynomials in several variables. In: Number Theory in Progress, vol. 1 (Zakopane-Kościelisko, 1997), pp. 431–444. de Gruyter, Berlin (1999)

[RV99]   Rodriguez-Villegas, F.: Modular Mahler measures. I. In: Topics in Number Theory (University Park, PA, 1997). Mathematics Applications, vol. 467, pp. 17–48. Kluwer Acadamic Publication, Dordrecht (1999)

[RVTV04]   Rodriguez-Villegas, F., Toledano, R., Vaaler, J.D.: Estimates for Mahler's measure of a linear form. Proc. Edinb. Math. Soc. (2), **47**(2), 473–494 (2004)

[RW08]   Rhin, G., Wu, Q.: Integer transfinite diameter and computation of polynomials. In: Number theory and polynomials. London Mathematical Society Lecture Note Series, vol. 352, pp. 277–285. Cambridge Univ. Press, Cambridge (2008)

[RW13]   Robinson, F., Wurtz, M.: On the magnitudes of some small cyclotomic integers. Acta Arith. **160**(4), 317–332 (2013)

[RZ14]   Rogers, M., Zudilin, W.: On the Mahler measure of $1 + X + 1/X + Y + 1/Y$. Int. Math. Res. Not. IMRN **2014**(9), 2305–2326 (2014)

[Sal44]   Salem, R.: A remarkable class of algebraic integers. Proof of a conjecture of Vijayaraghavan. Duke Math. J. **11**, 103–108 (1944)

[Sal63]   Salem, R.: Algebraic Numbers and Fourier Analysis. D. C. Heath and Co., Boston, Mass (1963)

[Sam06]   Samuels, C.L.: Lower bounds on the projective heights of algebraic points. Acta Arith. **125**(1), 41–50 (2006)

[Sam11]   Samuels, C.L.: The infimum in the metric Mahler measure. Canad. Math. Bull. **54**(4), 739–747 (2011)

[Sam20]   Samart, D.: A functional identity for Mahler measures of non-tempered polynomials (2020). arXiv:2006.09922v1

[Sch39]   Schönemann, T.: Theorie der symmetrischen Functionen der Wurzeln einer Gleichung. Allgemeine Sätze über Congruenzen nebst einigen Anwendungen derselben. (Schlußder Abhandlung). J. Reine Angew. Math. **19**, 289–308 (1839)

[Sch18]   Schur, I.: Über die Verteilung der Wurzeln bei gewissen algebraischen Gleichungen mit ganzzahligen Koeffizienten. Math. Z. **1**(4), 377–402 (1918)

[Sch66]   Schinzel, A.: On sums of roots of unity. Solution of two problems of R. M. Robinson. Acta Arith. **11**, 419–432 (1966)

[Sch73]   Schinzel, A.: On the product of the conjugates outside the unit circle of an algebraic number. Acta Arith. **24**, 385–399 (1973)

[Sch82]   Schinzel, A.: Selected topics on polynomials. University of Michigan Press, Ann Arbor, Mich (1982)

[Sch00]   Schinzel, A.: Polynomials with special regard to reducibility. Encyclopedia of Mathematics and its Applications, vol. 77. Cambridge University Press, Cambridge (2000). With an appendix by Umberto Zannier

[Sel49]   Selberg, A.: An elementary proof of Dirichlet's theorem about primes in an arithmetic progression. Ann. Math. **2**(50), 297–304 (1949)

[Ser19]   Serre, J.-P.: Distribution asymptotique des valeurs propres des endomorphismes de Frobenius [d'après Abel, Chebyshev, Robinson, . . .]. Astérisque, Séminaire Bourbaki. **2017/2018**, **414**(1146), 379–426 (2019)

[Sie44]   Siegel, C.L.: Algebraic integers whose conjugates lie in the unit circle. Duke Math. J. **11**, 597–602 (1944)

[Sie45]   Siegel, C.L.: The trace of totally positive and real algebraic integers. Ann. Math. **2**(46), 302–312 (1945)

[Sim99]   Simon, D.: Construction de polynômes de petits discriminants. C. R. Acad. Sci. Paris **329**, 465–468 (1999)

[Smi70]   Smith, J.H.: Some properties of the spectrum of a graph. In: Combinatorial Structures and their Applications, Proc. Conf. Calgary 1969, pp. 403–406. ed. R. Guy *et al.* (1970)

[Smy71]   Smyth, C.J.: On the product of the conjugates outside the unit circle of an algebraic integer. Bull. Lond. Math. Soc. **3**, 169–175 (1971)

[Smy72]   Smyth, C.J.: Topics in the theory of numbers. PhD thesis, University of Cambridge (1972)

[Smy80]   Smyth, C.J.: On the measure of totally real algebraic integers. J. Austral. Math. Soc. Ser. A **30**(2), 137–149 (1980/81)

[Smy81a]   Smyth, C.J.: A Kronecker-type theorem for complex polynomials in several variables. Canad. Math. Bull. **24**(4), 447–452 (1981). Addenda and Errata, ibid. **25**(4), 504 (1982)

[Smy81b]   Smyth, C.J.: On measures of polynomials in several variables. Bull. Austral. Math. Soc. **23**(1), 49–63 (1981). Corrigendum with G. Myerson, ibid. **26**(1), 317–319 (1982)

[Smy81c]   Smyth, C.J.: On the measure of totally real algebraic integers. II. Math. Comp. **37**(155), 205–208 (1981)

[Smy84]   Smyth, C.J.: The mean values of totally real algebraic integers. Math. Comp. **42**(166), 663–681 (1984)

[Smy86]   Smyth, C.J.: Additive and multiplicative relations connecting conjugate algebraic numbers. J. Number Theory **23**(2), 243–254 (1986)

[Smy87]   Smyth, C.J.: Solution of Elementary Problem e3100. Amer. Math. Monthly **94**(6), 552 (1987)

[Smy99]   Smyth, C.J.: An inequality for polynomials. In: Number Theory (Ottawa, ON, 1996). CRM Proc. Lecture Notes, vol. 19, pp. 315–321. Amer. Math. Soc, Providence, RI (1999)

[Smy02]   Smyth, C.J.: An explicit formula for the Mahler measure of a family of 3-variable polynomials. J. Théor. Nombres Bordeaux **14**(2), 683–700 (2002)

[Smy08]   Smyth, C.J.: The Mahler measure of algebraic numbers: a survey. In: Number Theory and Polynomials. London Mathematical Society Lecture Note Series, vol. 352, pp. 322–349. Cambridge University Press, Cambridge (2008)

[Smy18]   Smyth, C.J.: Closed sets of Mahler measures. Proc. Amer. Math. Soc. **146**(6), 2359–2372 (2018)

[Spe08]   Speyer, D.: The sign of the Gauss sum (2008). https://sbseminar.wordpress.com/2008/10/11/the-sign-of-the-gauss-sum/

[Sta20]   Stacy, E.: Totally p-adic numbers of degree 3. In: Galbraith, S.D. (ed.) ANTS XIV. Proceedings of the Fourteenth Algorithmic Number Theory Symposium, University of Auckland, 2020, number 4 in Open Book Series, pp. 387–402. Mathematical Sciences Publishers, Berkeley (2020)

[Ste78a]   Stewart, C.L.: On a theorem of Kronecker and a related question of Lehmer. In: Séminaire de Théorie des Nombres 1977–1978, pp. Exp. No. 7, 11. CNRS, Talence (1978)

[Ste78b]   Stewart, C.L.: Algebraic integers whose conjugates lie near the unit circle. Bull. Soc. Math. France **106**(2), 169–176 (1978)

[Ste04]   Sternberg, S.: Lie Algebras (2004). http://people.math.harvard.edu/~shlomo/docs/lie_algebras.pdf

[SZ65]   Schinzel, A., Zassenhaus, H.: A refinement of two theorems of Kronecker. Michigan Math. J. **12**, 81–85 (1965)

[SZ09]   Stan, F., Zaharescu, A.: Siegel's trace problem and character values of finite groups. J. Reine Angew. Math. **637**, 217–234 (2009)

[Sze15]   Szegö, G.: Ein Grenzwertsatz über die Toeplitzschen Determinanten einer reellen positiven Funktion. Math. Ann. **76**(4), 490–503 (1915)

[Tay10]   Taylor, G.: Cyclotomic Matrices and Graphs. PhD thesis, University of Edinburgh (2010)

[Tay11]   Taylor, G.: Cyclotomic matrices and graphs over the ring of integers of some imaginary quadratic fields. J. Algebra **331**, 523–545 (2011)

[Tay12]   Taylor, G.: Lehmer's conjecture for matrices over the ring of integers of some imaginary quadratic fields. J. Number Theory **132**, 590–607 (2012)

[Thu12]   Thue, A.: Über eine Eigenschaft, die keine transzendente Größe haben kann. Christiania Vidensk. selsk. Skrifter. **2**(20), 15 pp. (1912)

[Tit39]   Titchmarsh, E.C.: The Theory of Functions, 2nd edn. Oxford University Press, Oxford (1939)

[Tsu75]   Tsuji, M.: Potential Theory in Modern Function Theory. Chelsea Publishing Co., New York (1975). Reprinting of the 1959 original

[Van03]   Vandervelde, S.: A formula for the Mahler measure of $axy + bx + cy + d$. J. Number Theory **100**(1), 184–202 (2003)

[vI17]   van Ittersum, J.-W.M.: A group-invariant version of Lehmer's conjecture on heights. J. Number Theory **171**, 145–154 (2017)

[Vou96]   Voutier, P.: An effective lower bound for the height of algebraic numbers. Acta Arith. **74**(1), 81–95 (1996)

[Wal00]   Waldschmidt, M.: Diophantine approximation on linear algebraic groups. Grundlehren der Mathematischen Wissenschaften [Fundamental Principles of Mathematical Sciences], vol. 326. Springer, Berlin (2000). Transcendence properties of the exponential function in several variables

[Wu03]   Qiang, W.: On the linear independence measure of logarithms of rational numbers. Math. Comp. **72**(242), 901–911 (2003)

[WWW21]   Wang, C., Wu, J., Wu, Q.: The totally positive algebraic integers with small trace. Math. Comp. **90**(331), 2317–2332 (2021)

[Yat16]   Yatsyna, P.: Integer symmetric matrices: counterexamples to Estes-Guralnick's conjecture. PhD thesis, Royal Holloway, University of London (2016)

[Yat19]   Yatsyna, P.: A lower bound for the rank of a universal quadratic form with integer coefficients in a totally real number field. Comment. Math. Helv. **94**, 221–239 (2019)

[Zag90]   Zagier, D.: The Bloch-Wigner-Ramakrishnan polylogarithm function. Math. Ann. **286**(1–3), 613–624 (1990)

[Zag93]   Zagier, D.: Algebraic numbers close to both 0 and 1. Math. Comp. **61**(203), 485–491 (1993)

[Zha92]   Zhang, S.: Positive line bundles on arithmetic surfaces. Ann. of Math. (2) **136**(3), 569–587 (1992)

Index

© Springer Nature Switzerland AG 2021

J. McKee and C. Smyth, *Around the Unit Circle*, Universitext,

https://doi.org/10.1007/978-3-030-80031-4

Printed in the United States
by Baker & Taylor Publisher Services